Theory of categories

Theory of categories

Nicolae Popescu
National Institute for Scientific and Technical Creation
Bucharest, Romania

Liliana Popescu
University of Bucharest, Romania

Editura Academiei
București, România

Sijthoff & Noordhoff
International Publishers

1979

ISBN-13: 978-94-009-9552-9 e-ISBN-13: 978-94-009-9550-5
DOI: 10.1007/978-94-009-9550-5

Sijthoff & Noordhoff International Publishers. Alphen aan den Rijn. The Netherlands

Edited by Professor Dr. Leo F. Boron and Professor Dr. Charles O. Christenson, both of the University of Idaho, Moscow, U.S.A.

Contents

Chapter 2

COMPLETION OF CATEGORIES

Chapter 3

ALGEBRAIC CATEGORIES

Chapter 4

ABELIAN CATEGORIES

Preface

Although it is a relatively young branch of mathematics, category theory has already achieved important results that are dispersed in a great number of papers and gathered in some monographs. For this reason, to write a new monograph on the theory of categories is easy, due to the abundance of material, but also difficult, due to the great quantity of ideas and results.

In this work we try to give an exposition of some of the ideas and results of the theory of categories. We use current terminology and build as simple a framework as possible, but nevertheless sufficient to enable the reader of this book to understand most of the research papers devoted to this theory.

In order to read this book effectively, the reader is assumed to possess some knowledge of set theory, as well as some elementary facts from algebra and general topology. However, the reader should have mathematical maturity.

The first chapter deals with the general study of categories. Here the basic notions are introduced and most of the fundamental results are proved.

The second chapter discusses the problem of the completion of categories. Briefly said, this means the embedding of a given category into a category that possesses some additional properties, especially those connected with the existence of limits and colimits.

The third chapter covers the algebraic categories, i.e. *categories of universal algebras and their general study, as well as the study of the functors between them.*

The formal results from the theory of abelian categories are presented in the fourth chapter. We remark that the theory of abelian categories is the most developed part of the theory of categories and perhaps has the most numerous applications. We believe that the reader interested in a deeper study of abelian categories, as well as in their applications, might profitably consult N. Popescu [3]. *Due to the existence of that monograph, the authors considered it possible to reduce the space devoted to abelian categories in the present book.*

In the field of category theory there are some very good books. We mention Mitchell [2], *Bucur-Deleanu* [1], *MacLane* [3], *Pareigis* [1], *Schubert* [1], *etc. The results contained in these books (taken together) cover most of the results presented by us (as well as other results, which we omit). However due to the selection and organization of material, we consider our work to be useful for those who wish to know some aspects and results of the theory of categories.*

We mention that the notions of category and functor were originally defined and studied by Eilenberg and MacLane [2]. *Remarkable results and ideas were also contributed by Buchsbaum, Freyd, Gabriel, Grothendieck, Hilton, Isbell, Kan, Kuroš, Lambek, Lawvere, Mitchell, Schubert, Ulmer, etc.*

The source of the present work lies in a seminar on the theory of categories, conducted by N. Popescu at the Institute of Mathematics of the Academy of the Socialist Republic of Romania during the years 1964—1966. Of course, the results have since been thoroughly completed and revised.

The authors are profoundly indebted and thank all those who have assisted them in the preparation of the present book: the members of the above-mentioned seminar, especially Mr. Dorin Popescu, who essentially contributed to the second chapter; the cited (and the non-cited) authors, who, by their papers, made possible the development of the theory of categories and, as a result, the writing of this book; Miss Magdalena Vatamaniuc and Mr. Tiberiu Spircu, who made many critical remarks on its contents and on the English style of the book; Mr. Paul Cernovodeanu, who manifested much patience in typing the manuscript; all those who, by direct or indirect manifestation, deed, word or thought, have given their contribution. The authors also wish to thank Professor Dr. Leo F. Boron and Professor Dr. Charles O. Christenson, both of the University of Idaho, Moscow, U.S.A. for editing the book.

N. Popescu

Liliana Popescu

Note to the reader

All chapters are closely related and thus the reader should study the book chapter by chapter.

Each section begins with a brief summary of its contents and some references. The purpose of the references is to indicate some works where the main ideas and results of the section occur and also to help the reader interested in other aspects of the problems with which the section deals.

All theorems, propositions, lemmas, corollaries and notes are numbered consecutively in a single series in each section. The end (or absence) of a proof is indicated by a large asterisk, *.

References to the results stated in previous sections are given as follows: if the expression "according to Proposition 1.6" occurs in Ch. 3 this means that we refer to Proposition 1.6 stated in Section 3.1. Other references are obvious.

Generally, even terms in current use have been defined. In all cases, the definition can be traced through the subject index. It should be noted that we define (and indicate in the index) a great number of notions and their duals, as an aid to the reader.

For some of the more usual symbols and notation, the following list (Some Terminology, Notation and Conventions) should be consulted.

Some Terminology, Notation and Conventions

The letters **N**, **Z**, $\mathbf{Z}_n$, **Q**, **R**, stand as usual for the set (group or ring) of nonnegative integers, integers, integers modulo n, rational numbers and real numbers, respectively. The following is a list of some important symbols and notations:

$\mathcal{S}et$: the category of sets (Section 1.1)

$\mathcal{O}b\mathcal{C}$: the class of objects of the category $\mathcal{C}$ (Section 1.1)

$\mathcal{M}or(\mathcal{C})$: the class of all morphism of the category $\mathcal{C}$ (Section 1.1)

$\mathcal{T}op$: the category of topological spaces (Section 1.1)

$\mathcal{G}r$: the category of groups (Section 1.2)

$\mathcal{A}b$: the category of abelian groups (Section 1.2)

H^X (resp. H_X): $\mathcal{C} \to \mathcal{S}et$ the functor $H^X(Y) = \mathcal{C}(X, Y)$ (resp. the cofunctor $H_X(Y) = \mathcal{C}(Y, X)$) (Section 1.3)

Id$\mathcal{C}$: the identity functor of a category (Section 1.3)

$[\mathcal{I}, \mathcal{C}]$: the category of functors from the small category $\mathcal{I}$ into $\mathcal{C}$ (Section 1.3)

$\varprojlim F$: the limit of the functor F (Section 1.6)

$\varinjlim F$: the colimit of the functor F (Section 1.6)

$\prod_i X_i$: the product of a family of objects (Section 1.7)

$\coprod_i X_i$: the coproduct of a family of objects (Section 1.7)

$\{\mathcal{C}, \mathcal{S}et\}$: the category of all proper functors (Section 2.1)

$\mathcal{T}h$: the category of algebraic theories (Section 3.1)

$\mathcal{A}^a$: the category of $\mathcal{A}$-algebras (Section 3.2)

$\subseteq$: inclusion (with the possibility of equality)

$\subset$: strict (proper) inclusion

Chapter 1

Categories and functors

1.1. The notion of a category. Examples. Duality

In this section the notion of a category is defined and some elementary examples are given. For a discussion of some aspects of the definition of a category and also for a variety of other examples, the reader can refer to: Bucur-Deleanu [1], Gabriel [1], Grothendieck—Verdier [1], Kuroš-Livšic—Šulgeifer [1], Lawvere [3], Gabriel—Zisman [1], Eilenberg—MacLane [2], Mitchell [2], Pareigis [1], Sonner [1], Schubert [1], Bourbaki [1], [3].

A *category*, which we shall denote by $\mathscr{C}$, consists of:

a) A class, $\mathcal{O}b\mathscr{C}$, whose elements are called *objects* in $\mathscr{C}$.

b) For each ordered pair (X, Y) of objects from $\mathscr{C}$, a set $\mathscr{C}(X, Y)$ and if $(X, Y) \neq (X', Y')$ then $\mathscr{C}(X, Y)$ and $\mathscr{C}(X', Y')$ are disjoint.

c) For each ordered triple (X, Y, Z) of objects from $\mathscr{C}$ a map $m(X, Y, Z): \mathscr{C}(X, Y) \prod \mathscr{C}(Y, Z) \to \mathscr{C}(X, Z)$, subject to the further conditions:

d) If $f \in \mathscr{C}(X, Y)$, $g \in \mathscr{C}(Y, Z)$ and $h \in \mathscr{C}(Z, T)$ then

$$m(X, Y, T)(f, m(Y, Z, T)(g, h)) = m(X, Z, T)(m(X, Y, Z)(f, g), h)$$

(associativity), and

e) For each $X \in \mathcal{O}b\mathscr{C}$, there is an element $1_X \in \mathscr{C}(X, X)$, called the *identity morphism*, or *identity* of X, such that if $f \in \mathscr{C}(Y, X)$ and $g \in \mathscr{C}(X, Z)$ then $m(Y, X, X)(1_X, f) = f$ and $m(X, X, Z)(g, 1_X) = g$.

The set $\mathscr{C}(X, Y)$ is called the set of *morphisms* of X into Y. We shall frequently write $f: X \to Y$ or $X \xrightarrow{f} Y$ instead of $f \in \mathscr{C}(X, Y)$. If $f: X \to Y$, then X is the *domain* and Y is the *codomain* of f. We shall also call f a *morphism* from X to Y.

If $f: X \to Y$ and $g: Y \to Z$ we denote $m(X, Y, Z)(f, g)$ by $g \circ f$ or gf. m is called the *composition* map of morphisms.

In this simpler notation d) can be restated: If $f: X \to Y$, $g: Y \to Z$ and $h: Z \to T$ then $h(gf) = (hg)f$. Part e) may be restated: for each $X \in \mathcal{O}b\mathscr{C}$ there exists $1_X: X \to X$ such that for all $f: Y \to X$ and all $g: X \to Z$, $1_X f = f$ and $g 1_X = g$.

In general, objects will be denoted by capital Latin letters and morphisms by small Latin letters. That X is an object of $\mathscr{C}$ will be expressed by $X \in \mathscr{C}$; also, $f \in \mathscr{C}$ means that f is a morphism between two objects of $\mathscr{C}$, i.e. there exist $X, Y \in \mathscr{C}$ such that $f \in \mathscr{C}(X, Y)$; by axiom b), the objects X and Y are uniquely determined by f. Sometimes in speaking of a morphism f we shall denote by $d(f)$ the domain and by $c(f)$ the codomain of f, *i.e.* $f: d(f) \to c(f)$. The class of all morphisms in $\mathscr{C}$ is denoted by $\mathscr{M}or(\mathscr{C})$.

We note that $\mathscr{C}(X, Y)$ may be empty if $X \neq Y$, whereas from a) we conclude that $\mathscr{C}(X, X)$ is always nonempty.

PROPOSITION 1.1. *Let $\mathscr{C}$ be a category. Then for any object X of $\mathscr{C}$, the identity morphism 1_X is unique.*

Proof. Suppose that $1'_X$ is another identity; then we have

$$1'_X = 1'_X 1_X = 1_X. \quad \divideontimes$$

A category $\mathscr{C}$ is called *small* if $\mathscr{O}b\mathscr{C}$ is a set.

Examples of categories

Example 1.2. The category $\mathscr{S}et$. $\mathscr{O}b\mathscr{S}et$ is the class of all sets. If X and Y are sets, we define $\mathscr{S}et(X, Y)$ to be the set of all maps defined on X and taking values in Y. The composition of morphisms is the usual composition of maps. If X is a set, then 1_X is the identity map, i.e., $1_X(x) = x$ for every $x \in X$.

Generally, we denote the *empty set* by $\emptyset$. Note that for each set X, $\mathscr{S}et(\emptyset, X)$ contains precisely one element, called the *empty map*, whereas $\mathscr{S}et(X, \emptyset)$ is empty if $X \neq \emptyset$.

Example 1.3. The category $\mathscr{T}op$. A topological space is a pair (X, t), where X is a set and t is a family of subsets of X such that the following axioms hold:

i) If $(X_i)_{i \in \mathscr{I}}$ is a family of elements of t, then $\bigcup X_i \in t$.

ii) If X_1, $X_2 \in t$, then $X_1 \cap X_2 \in t$.

iii) X and $\emptyset$ are elements of t.

The elements of t are called *open sets* of X. A *continuous map* from a topological space (X, t) into a topological space (X', t') is a map $f: X \to X'$ such that $f^{-1}(Y) \in t$ for each element Y of t'. Sometimes we shall write simply X instead of (X, t).

The objects of the category $\mathscr{T}op$ are topological spaces and the morphisms are the continuous maps between them. The composition of morphisms is the usual composition of maps and the identity morphisms are the identity maps.

Example 1.4. The category $\mathscr{R}el$. An *equivalence relation* on a set X is a subset R of the direct (cartesian) product $X \prod X$ such that the following axioms are satisfied:

i) For each $x \in X$, $(x, x) \in R$.

ii) If $(x, y) \in R$ then $(y, x) \in R$.

iii) If (x, y) and $(y, z) \in R$, then $(x, z) \in R$.

The objects of the category $\mathscr{R}el$ are the pairs (X, R), where X is a set and R is an equivalence relation on X; a morphism $f: (X, R) \to (X', R')$ of $\mathscr{R}el$ is a map $f: X \to X'$ such that if $(x, y) \in R$, then $(f(x), f(y)) \in R'$. The composition of morphisms is the usual composition of maps and the identity morphisms are the identity maps.

Example 1.5. The category $\mathcal{M}on$. A category with a single object is called a *monoid*. Thus, a monoid may be defined as a set M together with a map

$$m: M \coprod M \to M,$$

called the *law of composition* (denote $m(x, y)$ by xy). m is associative and has a neutral element e. A morphism of monoids is a map $f: M \to M'$ such that $f(xy) = f(x)f(y)$ for each pair $x, y \in M$, and $f(e) = e'$ (i.e. f preserves compositions and the neutral element). The objects of the category $\mathcal{M}on$ are monoids and the morphisms are morphisms of monoids. The composition of morphisms is the usual composition of maps and the identity morphisms are the identity maps.

Example 1.6. The category $\mathcal{P}re$. A *preordered set* is a set X together with a relation "$\leqslant$" which is reflexive ($x \leqslant x$ for each $x \in X$) and transitive ($x \leqslant y$ and $y \leqslant z \Rightarrow x \leqslant z$ for all $x, y, z \in X$). By a *nondecreasing map* defined on a preordered set X and taking values in a preordered set Y we mean a map $f: X \to Y$ such that $x \leqslant y$ implies $f(x) \leqslant f(y)$ for each pair $x, y \in X$. The objects of the category $\mathcal{P}re$ are the preordered sets and the morphisms are the nondecreasing maps. The composition of morphisms is the usual composition of maps.

Example 1.7. The category $\mathcal{O}rd$. A preordered set X is called *ordered*, if the relation "$\leqslant$" is also antisymmetric, i.e. $x \leqslant y$ and $y \leqslant x$ imply $x = y$ for all $x, y \in X$. The objects of the category $\mathcal{O}rd$ are the ordered sets, the morphisms are the nondecreasing maps, and the composition of morphisms is the usual composition of maps.

Example 1.8. The category $\mathcal{S}$. For each natural number n, we denote $(0, 1, 2, \ldots, n)$ by S_n; assume that S_n is ordered by the natural ordering. The objects of the category $\mathcal{S}$ are S_n, $n = 0, 1, \ldots$, the morphisms are the nondecreasing maps and the composition is the usual composition of maps. $\mathcal{S}$ is called the *simplicial* category.

Example 1.9. The category $\mathcal{S}^+$. The objects of $\mathcal{S}^+$ are S_n, $n = 0, 1, \ldots$ (as defined in the preceding example). $S^+(S_n, S_m)$ is the set of all nondecreasing maps $f: S_n \to S_m$ such that $f(0) = 0$.

Example 1.10. The *point category* is denoted by $\{0\}$ and is defined as a category with a single object denoted by 0 and a single morphism, 1_0.

Example 1.11. The *category arrow* is denoted by $\{0 \to 1\}$ and is defined as a category with two objects 0,1 and three morphisms: the identity morphisms, and the morphism $0 \to 1$.

Example 1.12. The category *double arrow* is denoted by $\{0 \rightrightarrows 1\}$ and has two objects, 0, 1, and four morphisms: the identity morphisms and two other morphisms both having domain 0 and codomain 1.

Example 1.13. *Duality*. Given a category $\mathscr{C}$, a new category $\mathscr{C}^0$, called the *dual category* of $\mathscr{C}$, is defined as follows:

i) The objects of the category $\mathscr{C}^0$ coincide with the objects of the category $\mathscr{C}$, that is $\mathcal{O}b\,\mathscr{C}^0 = \mathcal{O}b\,\mathscr{C}$. If X is an object of $\mathscr{C}$, then the same object regarded as being in the category $\mathscr{C}^0$ is denoted by X^0 and is called the *dual* of X.

ii) The set of morphisms $\mathscr{C}^0(X^0, Y^0)$ is identical with $\mathscr{C}(Y, X)$. If $f \in \mathscr{C}(X, Y)$ then the same morphism when considered as an element in $\mathscr{C}^0(Y^0, X^0)$, will be denoted by f^0 (f^0 is called the *dual* of f).

iii) The composition map

$$m^0(X^0, Y^0, Z^0): \mathscr{C}^0(X^0, Y^0) \prod \mathscr{C}^0(Y^0, Z^0) \to \mathscr{C}^0(X^0, Z^0)$$

is defined as follows: if $f^0 \in \mathscr{C}^0(X^0, Y^0)$ and $g^0 \in \mathscr{C}^0(Y^0, Z^0)$, then $m^0(X^0, Y^0, Z^0)$ $(f^0, g^0) = g^0 f^0 = (m(Z, Y, X)(g, f))^0 = (fg)^0$. Clearly, $(\mathscr{C}^0)^0 \equiv \mathscr{C}$.

Associating with each category $\mathscr{C}$ its dual category $\mathscr{C}^0$ enables us to dualize each notion or statement concerning a category $\mathscr{C}$ into a corresponding notion or statement concerning the dual category $\mathscr{C}^0$. From a practical point of view, this is the procedure of "reversing the arrows". Thus, we get the following *duality principle*.

"Let P be a notion or statement about categories; then there is a dual notion or statement P^0 which is also a notion or statement about categories. P^0 is called the *dual* of P."

Clearly, when we apply this duality principle, we must realize that we dualize not only the conclusions of the statements about categories, but also the hypotheses. When a new notion is introduced, we tacitly assume that the corresponding dual notion is also defined.

Example 1.14. Let $\mathscr{C}$ be a category. A *subcategory* of $\mathscr{C}$ is a category $\mathscr{C}'$ which satisfies the following conditions:

i) $\mathcal{O}b\,\mathscr{C}' \subseteq \mathcal{O}b\,\mathscr{C}$.

ii) For every $X, Y \in \mathcal{O}b\,\mathscr{C}'$, $\mathscr{C}'(X, Y) \subseteq \mathscr{C}(X, Y)$.

iii) The composition of morphisms in $\mathscr{C}'$ is the restriction of the composition of morphisms in $\mathscr{C}$.

iv) The identity morphisms in $\mathscr{C}'$ are also the identity morphisms in $\mathscr{C}'$ For each $X \in \mathcal{O}b\,\mathscr{C}'$, $1_X \in \mathscr{C}(X, X)$ is also in $\mathscr{C}'(X, X)$.

A subcategory $\mathscr{C}'$ of the category $\mathscr{C}$ is said to be *full* if for each pair (X, Y) of objects of $\mathscr{C}'$ we have that

$$\mathscr{C}'(X, Y) = \mathscr{C}(X, Y).$$

The category $\mathcal{O}rd$ is a full subcategory of $\mathscr{P}re$, while $\mathscr{S}^+$ is a subcategory (but not a full subcategory) of $\mathscr{S}$.

Example 1.15. If $\{\mathscr{C}_i\}_{i \in \mathscr{I}}$ is a set of categories, we define the *product category* $\prod_{i \in \mathscr{I}} \mathscr{C}_i$ as follows.

An object of $\prod_{i \in \mathscr{I}} \mathscr{C}_i$ is a family of the form $\{X_i\}_{i \in \mathscr{I}}$ where for each $i \in \mathscr{I}$, X_i is an object of $\mathscr{C}_i$ and a morphism is

$$(\prod_{i \in \mathscr{I}} \mathscr{C}_i)(\{X_i\}_i, \{Y_i\}_i) = \prod_{i \in \mathscr{I}} \mathscr{C}_i(X_i, Y_i).$$

(The second member of this equality is the cartesian product of sets (see Bourbaki [1]).)

The composition of morphisms is defined argumentwise.

Example 1.16. Let $\mathscr{C}$ be a category and let X be an object of $\mathscr{C}$. Denote by $\mathscr{C}/X$ the following category, which is called the *category of objects over* X:

An object of $\mathscr{C}/X$ is a couple (Y,f), where $Y \in \mathcal{O}b\,\mathscr{C}$ and $f \in \mathscr{C}(Y, X)$. A morphism $g:(Y,f) \to (Y',f')$ in $\mathscr{C}/X$ is a morphism $g\colon Y \to Y'$ in $\mathscr{C}$ such that $f'g = f$. The composition of morphisms in $\mathscr{C}/X$ is the restriction of the composition of morphisms in $\mathscr{C}$.

Also, we can define the category $X/\mathscr{C}$ (*the category of objects under* X): an object of $X/\mathscr{C}$ is a pair (f, Y), where $Y \in \mathcal{O}b\,\mathscr{C}$ and $f \in \mathscr{C}(X, Y)$. A morphism and the composition of morphisms are defined in the obvious way.

A category $\mathscr{C}$ is called *discrete* if $\mathscr{C}$ has only identity morphisms (i.e. for each object X, $\mathscr{C}(X, X) = \{1_X\}$ and $\mathscr{C}(X, Y) = \emptyset$ if $X \neq Y$). Clearly, every discrete category can be regarded as a class. Conversely, every class may be interpreted as a discrete category. In particular, any set will be considered a discrete category and a discrete category is small if and only if it is a set.

A category whose morphisms form a finite set is called a *finite category*.

A diagram of the form

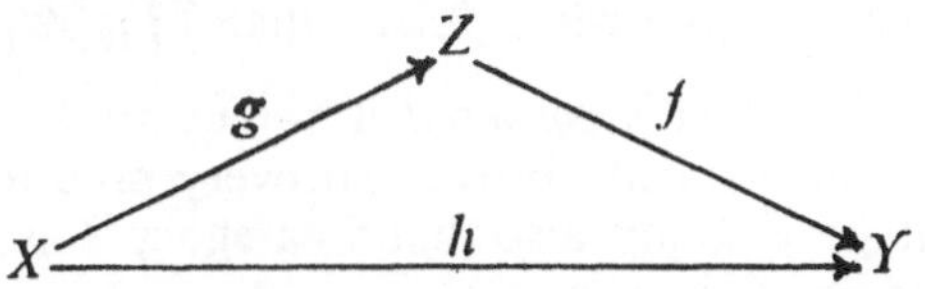

in the category $\mathscr{C}$ is called *commutative* if $fg = h$; in this case, we also say that h *factors through* Z, or that g *and* f factor h. A diagram of the form

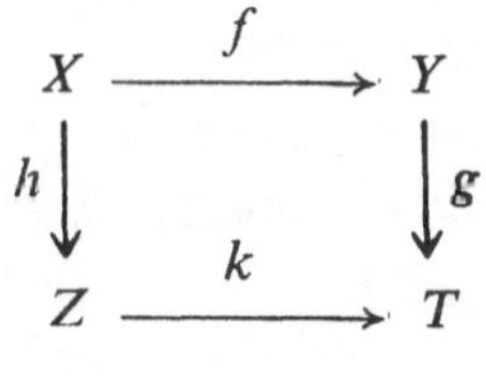

is *commutative* if $kh = gf$. Later we shall give a general definition of diagram and commutativity, but the above types and some combinations of them will be all we need at present.

The above diagrams will be called *usual diagrams*.

Note 1.17. If in the definition of a category we remove the requirement that $\mathscr{C}(X, Y)$ be a set for all X and Y, i.e., if we permit $\mathscr{C}(X, Y)$ to be a class (rather han a set) for some (or all) X and Y, then we obtain the notion of an *illegitimate category*.

Although the topic of this book is concerned only with categories (sometimes called *legitimate categories*), the illegitimate categories often arise in a natural manner. We do not subscribe to the notion that, apparently, there exists only a formal difference between categories and illegitimate categories. Our opinion is that the difference is very profound but the justification of this opinion is beyond the scope of this book.

Exercises

1.1. Let $\mathscr{C}$ be a category. Define the category $\mathscr{M}orf(\mathscr{C})$ as follows. An object of $\mathscr{M}orf(\mathscr{C})$ is a triple (X, f, Y), where $f: X \to Y$ is a morphism in $\mathscr{C}$. The set $\mathscr{M}orf(\mathscr{C})((X, f, Y), (X', f', Y'))$ is defined as the set of couples (u, v) where $u: X \to X'$ and $v: Y \to Y'$ are morphisms in $\mathscr{C}$ such that $vf = f'u$. The composition of morphisms is defined canonically. Prove that $\mathscr{M}orf(\mathscr{C})$ is actually a category (called the *category of morphisms* in $\mathscr{C}$). Calculate the categories $\mathscr{M}orf(\{0 \to 1\})$ and $\mathscr{M}orf(\{0 \rightrightarrows 1\})$.

1.2. Let $\{\mathscr{C}_i\}_{i \in \mathscr{I}}$ be a class of categories. Then we construct a new category $\coprod_{i\in\mathscr{I}}\mathscr{C}_i$ (the *coproduct of categories*) as follows. The objects of $\coprod_{i\in\mathscr{I}}\mathscr{C}_i$ are disjoint unions of the classes $\mathscr{O}b\mathscr{C}_i$. If $X, Y \in \mathscr{O}b(\coprod_{i\in\mathscr{I}}\mathscr{C}_i)$, then $(\coprod_{i\in\mathscr{I}}\mathscr{C}_i)(X, Y) = \mathscr{C}_j(X, Y)$, if $X, Y \in \mathscr{O}b\mathscr{C}_j$ and empty otherwise. Prove that $\coprod_{i\in\mathscr{I}}\mathscr{C}_i$ is indeed a category.

1.3. A category $\mathscr{C}$ is called *preordered* if for every $X, Y \in \mathscr{C}$, the set $\mathscr{C}(X, Y)$ contains no more than one element. Prove that every preordered set may be regarded as a preordered category. Show also that a category $\mathscr{C}$ is preordered if and only if every usual diagram in $\mathscr{C}$ is commutative.

1.4. In the simplicial category $\mathscr{S}$, consider the following morphisms: $\partial_n^i: \mathscr{S}_{n-1} \to \mathscr{S}_n$ is the increasing injection which does not take the value $i \in S_n$. $\sigma_n^i: S_{n+1} \to S_n$ is the nondecreasing surjection which takes twice the value $i \in S_n$. Prove that these morphisms satisfy the following relations.

$$\partial_{n+1}^j \partial_n^i = \partial_{n+1}^i \partial_n^{j-1}, \quad i < j,$$

$$\sigma_n^j \sigma_{n+1}^i = \sigma_n^i \sigma_{n+1}^{j+1}, \quad i \leqslant j,$$

$$\sigma_{n-1}^j \partial_n^i = \begin{cases} \partial_{n-1}^i \sigma_{n-2}^{j-1}, & i < j, \\ 1_{S_{n-1}}, & i = j \text{ or } i = j+1, \\ \partial_{n-1}^{i-1} \sigma_{n-2}^j, & i > j+1. \end{cases}$$

Prove that every morphism $f: S_m \to S_n$ of $\mathscr{S}$ can be written in one and only one way as

$$f = \partial_n^{i_s} \partial_{n-1}^{i_{s-1}} \dots \partial_{n-t+1}^{i_1} \sigma_{m-t}^{j_t} \dots \sigma_{m-2}^{j_2} \sigma_{m-1}^{j_1}$$

with $n \geqslant i_s > \dots > i_1 \geqslant 0$, $0 \leqslant j_t < \dots < j_1 \leqslant m$ and $m - t + s = n$.

1.5. Show that if $f: X \to Y$ is a morphism of $\mathscr{S}et$ and $X \neq \emptyset$, then there is a map $g: Y \to X$ such that $f = fgf$ and that the same result is valid in $\mathscr{S}$.

1.2. Special morphisms in a category

This section deals with some important morphisms in a category. These morphisms are generalizations of the usual notions of injective, surjective and bijective maps. Some aspects of these notions are also discussed in Bucur—Deleanu [1], Grothendieck [1], Kuroš-Livšic-Šulgeifer [1], MacLane [3], Mitchell [2], Pareigis [1], Schubert [1], Buchsbaum [1], Kelly [2], Baron [1].

Let $f: X \to Y$ be a morphism in a category $\mathscr{C}$. f is called a *monomorphism* if $fu = fv$ implies $u = v$ for all pairs of morphisms u, v (of course, u and v must have the same domain, and must have codomain X), i.e. f is left cancellable. Dually, f is called an *epimorphism* if f^0, the dual of f, is a monomorphism, i.e. if $uf = vf$ implies that $u = v$ for each pair of morphisms u, v (of course, u and v must have the domain Y, and they must have the same codomain). A morphism f is called a *bimorphism* if it is both a monomorphism and an epimorphism.

Example 2.1. A morphism $f: X \to Y$ in $\mathscr{S}et$ is a monomorphism if and only if f is an *injection*, i.e. for any x, $y \in X$, $x \neq y$ implies that $f(x) \neq f(y)$. Indeed, suppose that f is a monomorphism and that there are $x, y \in X$, $x \neq y$ such that $f(x) = f(y)$. Define A to be a singleton with element a and define two maps u, $v: A \to X$ by $u(a) = x$ and $v(a) = y$. It is clear that $fu = fv$, but that $u \neq v$, which is a contradiction. The converse is clear.

Analogously, a morphism $f: X \to Y$ in $\mathscr{S}et$ is an epimorphism if and only if it is a *surjection*, i.e. for any $y \in Y$ there is an $x \in X$ such that $f(x) = y$. Indeed, let us assume that f is an epimorphism that is not a surjection. Then there is an element $\bar{y} \in Y$ such that $\bar{y} \neq f(x)$ for any $x \in X$. Let T be a set with two elements t, t'; define two maps u, v: $Y \to T$ such that $u(y) = t$ for every $y \in Y$ and $v(y) = t$, if $y \neq \bar{y}$ and $v(\bar{y}) = t'$. It is clear that $uf = vf$ and $u \neq v$, which is a contradiction. The converse is obvious.

It is clear that a monomorphism f in a category $\mathscr{C}$ will be a monomorphism in any subcategory of $\mathscr{C}$. However, it is easy to see that a morphism may be a monomorphism in a subcategory of $\mathscr{C}$ without being a monomorphism in $\mathscr{C}$. The same remarks are valid for epimorphisms.

PROPOSITION 2.2. *Let $X \xrightarrow{f} Y \xrightarrow{g} Z$ be morphisms in a category $\mathscr{C}$. If f and g are monomorphisms then gf is also a monomorphism. Moreover, if gf is a monomorphism then f is a monomorphism. Dually, if f and g are epimorphisms then gf is also an epimorphism. Moreover, if gf is an epimorphism then g is an epimorphism.*

Proof. Let f and g be monomorphisms and let $u, v \in \mathscr{C}(T, X)$ be such that $(gf)u = (gf)v$. Thus, $g(fu) = g(fv)$, so that $fu = fv$ since g is a monomorphism, and $u = v$ because f is also a monomorphism.

Assume that gf is a monomorphism and $u, v \in \mathscr{C}(T, X)$ are such that $fu = fv$. Then $g(fu) = g(fv) = (gf)u = (gf)v$ and $u = v$ since gf is a monomorphism, so that f is a monomorphism. ✲

A morphism $f \in \mathscr{C}(X, Y)$ is called a *retraction* or is said to be *right invertible* if there is a morphism $f' : Y \to X$ such that $ff' = 1_Y$, in which case f' is called a *right inverse* of f. Dually, f is called a *coretraction* or is said to be *left invertible* if f^0, the dual of f, is a retraction, i.e. if there is a morphism $f'' : Y \to X$ such that $f''f = 1_X$, in which case f'' is called a *left inverse* of f. Sometimes a coretraction is also called a *section*. Finally, f is called an *isomorphism* or an *invertible morphism* if it is both a retraction and a coretraction.

Example 2.3. In the category $\mathscr{S}et$, any monomorphism (i.e. an injection) whose domain is nonempty, is a coretraction, and any epimorphism (i.e. a surjection) is a retraction. Indeed, if $f: X \to Y$ is an injection and $X \neq \emptyset$, we define the map $g: Y \to X$ such that $g(f(x)) = x$ for each $x \in X$ and $g(y) = x_0$, if $y \neq f(x)$ for any $x \in X$, where x_0 is a fixed element of X. It is clear that $gf = 1_X$. Note that g is not uniquely defined when f is not a surjection and X has at least two elements.

Furthermore, if $f: X \to Y$ is a surjection, then we define the map $h: Y \to X$ by $h(y) = x$, where x is a suitable element of X such that $f(x) = y$. Clearly $fh = 1_Y$. Note also that h is not uniquely defined if f is not an injection.

It is easy to see (compare with Proposition 2.7) that a morphism in $\mathscr{S}et$ is an isomorphism if and only if it is a *bijection* (i.e. both an injection and a surjection). Thus a bimorphism in $\mathscr{S}et$ is an isomorphism.

As a direct consequence of Proposition 2.2, we obtain the following result.

PROPOSITION 2.4. *Every retraction is an epimorphism. Every coretraction is a monomorphism. Every isomorphism is a bimorphism.* ✳

PROPOSITION 2.5. *Let* $X \xrightarrow{f} Y \xrightarrow{g} Z$ *be morphisms of* $\mathscr{C}$. *If f and g are retractions then gf is a retraction. If gf is a retraction then g is a retraction. Dually, if f and g are coretractions then gf is a coretraction. If gf is a coretraction then f is a coretraction.* ✳

COROLLARY 2.6. *The composition of any two isomorphisms is also an isomorphism.* ✳

PROPOSITION 2.7. *A right invertible monomorphism is an isomorphism. Dually, a left invertible epimorphism is an isomorphism.*

Proof. Let $f: X \to Y$ be a right invertible monomorphism and let f' be a right inverse of f, i.e. $ff' = 1_Y$. Then

$$f(f'f) = (ff')f = 1_Y f = f = f 1_X,$$

and hence $f'f = 1_X$, or equivalently f is also left invertible, as claimed. ✳

PROPOSITION 2.8. *If $f: X \to Y$ is an isomorphism, then every left inverse of f is also a right inverse of f.*

Proof. Let f' be a right inverse and f'' a left inverse of f. Then

$$f'' = f''1_Y = f''(ff') = (f''f)f' = 1_X f' = f'. ✳$$

Thus, an isomorphism f has a unique right inverse that is also the only left inverse. This is denoted by f^{-1} and called the *inverse* of f. Moreover, f^{-1} is also an isomorphism and $(f^{-1})^{-1} = f$.

PROPOSITION 2.9. *If fg is a monomorphism and g is right invertible, then f is also a monomorphism.*

Proof. Assume that g' is a right inverse of g. Then $(fg)g' = f(gg') = f$ so that the result follows from Propositions 2.4 and 2.2. ✳

COROLLARY 2.10. *The following are equivalent for any category $\mathscr{C}$:*

a) *Every morphism in $\mathscr{C}$ is right invertible.*

b) *Every morphism in $\mathscr{C}$ is left invertible.*

c) *Every morphism in $\mathscr{C}$ is invertible.*

The proof follows from the above results. ✳

A category as in Corollary 2.10 is called a *groupoid*. A groupoid with a single object is called a *group*. Therefore a group is a monoid in which every morphism has an inverse. Hence, a group can be defined as a nonempty set G together with an associative law of composition, having a neutral element e and for each x an element x^{-1}, called the *inverse* of x, such that $xx^{-1} = x^{-1}x = e$. Thus we can define the *category of groups*, denoted by $\mathscr{Gr}$, whose objects are groups and whose morphisms are morphisms of monoids. It is clear that $\mathscr{Gr}$ is a full subcategory of $\mathscr{Mon}$.

A group A is called *abelian* (or *commutative*) if for any $x, y \in A$, $xy = yx$. The full subcategory of $\mathscr{Gr}$ whose objects are abelian groups will be denoted by $\mathscr{Ab}$. Usually the law of composition in an abelian group will be denoted by "+", i.e. we shall write $x + y$ instead of xy; the inverse of an element x is called the *opposite* of x and is denoted by $-x$. Also the neutral element of an abelian group is denoted by "0" (and called the *zero element*). Clearly, these are purely conventions.

Let $\mathscr{C}$ be a category. A morphism whose codomain is the same as its domain is called an *endomorphism*. For each $X \in \mathscr{C}$, the set $\mathscr{C}(X, X)$ of endomorphisms of X is a monoid and it is denoted by $\mathrm{End}(X)$ or by $\mathrm{End}_{\mathscr{C}}(X)$ if there is danger of confusion. An endomorphism that is also an isomorphism is called an *automorphism*. The set of automorphisms of X is a group and it is denoted by $\mathrm{Aut}(X)$ or by $\mathrm{Aut}_{\mathscr{C}}(X)$.

We say that *X is isomorphic to Y* if there is an isomorphism in the set $\mathscr{C}(X, Y)$ or, equivalently, if there is an isomorphism from X into Y. It should be kept in mind, however, that there may be many isomorphisms from X into Y. However, the above terminology will usually be used with reference to a specific isomorphism $f: X \to Y$. The notation $f: X \xrightarrow{\sim} Y$ will also be used to express the fact that f is an isomorphism; the notation $X \simeq Y$ will mean that X is isomorphic to Y.

It is easy to see that the relation "X is isomorphic to Y" is an equivalence relation on the class $\mathscr{Ob}\,\mathscr{C}$. The equivalence class of an object X relative to this equivalence relation will be denoted by $[X]$ and called the *type* of X.

Proposition 2.11. *Let $f: X \to Y$ be an isomorphism in a category $\mathscr{C}$. For each object Z of $\mathscr{C}$, the map*

$$f_Z: \mathscr{C}(X, Z) \to \mathscr{C}(Y, Z),$$

defined by $f_Z(g) = gf^{-1}$, is a bijection.

The proof is easy and is left for the reader. $\divideontimes$

In the theory of categories, morphisms play the important role, the object being used primarily for the definition of morphisms. Hence two isomorphic objects play essentially the same role. Thus, any result valid for an object X of a category $\mathscr{C}$ is automatically valid for any object X' isomorphic to X.

Exercises

2.1. Prove that in the categories $\mathscr{Top}$, $\mathscr{Rel}$, $\mathscr{Pre}$, $\mathscr{Ord}$, $\mathscr{S}$ and $\mathscr{S}^+$, a morphism is a monomorphism if and only if it is an injection, and an epimorphism if and only if it is a surjection.

2.2. Let $f_i: X_i \to X_{i+1}$, $i = 1, \ldots, n-1$ be monomorphisms (epimorphisms) in the category $\mathscr{C}$. Show that $f_{n-1} \ldots f_1$ is a monomorphism (an epimorphism)

2.3. Prove that in the categories $\mathscr{S}$ and $\mathscr{S}^+$ any monomorphism (epimorphism) is a coretraction (retraction).

2.4. Prove that in each of the categories $\mathscr{Top}$, $\mathscr{Rel}$, $\mathscr{Pre}$ and $\mathscr{Ord}$ there exist bimorphisms which are not isomorphisms. Furthermore, show that in these categories, there exist monomorphisms (epimorphisms) that are not coretractions (retractions). (Hint: On the same set X, two topologies can be defined (resp. equivalence relations, preorder relations) which are comparable but not identical.)

2.5. If the composition gf is left invertible and g and f are epimorphisms show that f and g are also isomorphisms.

2.6. Prove that a morphism $\{f_i\}_{i \in \mathscr{I}}$ in the product category $\prod_{i \in \mathscr{I}} \mathscr{C}_i$ is a monomorphism (epimorphism, retraction, coretraction, isomorphism) if and only if, for each i, f_i is a monomorphism (epimorphism, retraction, coretraction, isomorphism) in the category $\mathscr{C}_i$.

2.7. A topological space (X, t) is called a *Hausdorff space* if, for any $x, x' \in X$, $x \neq x'$, there exist two elements $U, U' \in t$, such that $x \in U$, $x' \in U'$ and $U \cap U' = \emptyset$. Denote by $\mathscr{Hd}$ the category whose objects are Hausdorff spaces and whose morphisms are the continuous maps between them. Prove that a morphism in $\mathscr{Hd}$ can be an epimorphism without being a surjection. In fact, in order that a continuous map $f: X \to Y$ be an epimorphism in $\mathscr{Hd}$, it is sufficient that $f(X)$ be a *dense* subset of Y (i.e. for each nonempty open set U of Y, $U \cap f(X) \neq \emptyset$).

2.8. (Cube lemma). Assume that the four vertical sides and the bottom of the cube

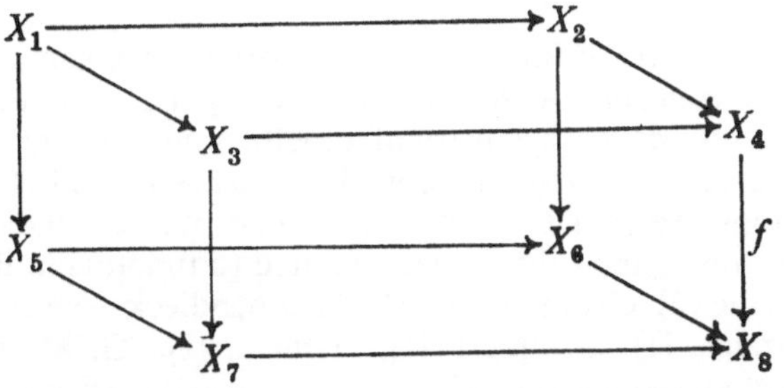

constructed with objects and morphisms of a category, are commutative, and let f be a monomorphism. Show that the top side is also commutative.

2.9. A *ring* is a nonempty set R having two laws of composition, one denoted by "+" (called *addition*), and the other denoted by juxtaposition (called *multiplication*). Relative to addition, R is a group whose neutral element is denoted by 0 (called the *zero element*); also, relative to multiplication R is a monoid, whose neutral element is denoted by 1 (called the *identity element*). Finally, assume that for every $x, y, z \in R$,

$$(x + y)z = xz + yz,$$
$$z(x + y) = zx + zy.$$

Prove that relative to addition R is a commutative group and that $1 \neq 0$ (assuming that R has at least two elements).

Denote the category of rings by $\mathscr{R}g$; the objects of $\mathscr{R}g$ are rings and the morphisms of $\mathscr{R}g$ are ring morphisms, where by a *ring morphism* we mean a map $f: R \to R'$, from the ring R to the ring R', such that for all $x, y \in R$,

$$f(x + y) = f(x) + f(y),$$
$$f(xy) = f(x) f(y),$$

and

$$f(1) = 1'$$

(1 and 1′ are the identity elements of R and R', respectively).

Show that if **Z** is the ring of integers and **Q** is the ring of rational numbers, then the natural inclusion $u: \mathbf{Z} \to \mathbf{Q}$ is a bimorphism but not an isomorphism in $\mathscr{R}g$.

2.10. A category in which every bimorphism is an isomorphism is called a *balanced* category. Prove that the categories $\mathscr{S}et$, $\mathscr{S}$, $\mathscr{S}^+$, $\mathscr{G}r$ and $\mathscr{A}b$ are balanced. (Hint: See Example 16.14.)

2.11. A morphism $f: X \to X$ in a category $\mathscr{C}$ is called *idempotent* if $f^2 = f$; an idempotent f is said to *split* if there exist morphisms $X \xrightarrow{g} Y \xrightarrow{h} X$ such that $f = hg$ and $gh = 1_Y$. Show that all idempotents in $\mathscr{S}et$ split. Show that the same result is valid in $\mathscr{A}b$, $\mathscr{G}r$, $\mathscr{R}g$ and $\mathscr{T}op$.

1.3. Functors

In this section the notions of functor and functorial morphism are presented. These are very important concepts in the theory of categories. Recall that what we call a functor is frequently called a covariant functor and what we call a cofunctor is frequently called a contravariant functor. The reason for this break with tradition is that these notions are in a sense dual and also our terminology is shorter. The reader can find examples and other aspects related to functors in Bucur—Deleanu [1], Eilenberg—MacLane [2], Grothendieck [1], Grothendieck—Verdier [1], MacLane [3], Mitchell [2], Pareigis [1], Schubert [1], Yoneda [1], G. M. Kelly [1].

Let $\mathscr{C}$ and $\mathscr{C}'$ be categories. A *(covariant) functor* F from $\mathscr{C}$ to $\mathscr{C}'$ consists of the following:

a) a map $X \rightsquigarrow F(X)$ which associates with each object X of $\mathscr{C}$ an object $F(X)$ of $\mathscr{C}'$.

b) for each pair (X, Y) of objects of $\mathscr{C}$, a map

$$F(X, Y): \mathscr{C}(X, Y) \to \mathscr{C}'(F(X), F(Y))$$

such that if we write $F(f)$ instead of $F(X, Y)(f)$, then

$$F(1_X) = 1_{F(X)},$$

and

$$F(gh) = F(g)\,F(h).$$

A *cofunctor* (or *contravariant functor*) F from $\mathscr{C}$ to $\mathscr{C}'$ consists of the following:

a′) a map $X \rightsquigarrow F(X)$ which associates with each object X of $\mathscr{C}$ an object $F(X)$ of $\mathscr{C}'$.

b′) for each pair (X, Y) of objects of $\mathscr{C}$, a map

$$F(X, Y): \mathscr{C}(X, Y) \to \mathscr{C}'(F(Y), F(X))$$

such that (if we denote $F(X, Y)(f)$ by $F(f)$)

$$F(1_X) = 1_{F(X)},$$

and

$$F(gh) = F(h)\,F(g).$$

Example 3.1. Let $S: \mathscr{T}op \to \mathscr{S}et$ denote the functor obtained as follows: for each topological space (X, t) let $S(X)$ be the underlying set X; for any continuous map f, let $S(f)$ be the corresponding set map. S is a (covariant) functor often called the *forgetful functor*.

Example 3.2. Let $D: \mathscr{S}et \to \mathscr{T}op$ denote the functor that assigns to each set X the *discrete topological space*, that is, the space in which every subset is open. It is clear that a map $f: X \to Y$ in $\mathscr{S}et$ becomes a continuous map between the discrete spaces $D(X)$ and $D(Y)$.

Example 3.3. Let $P:\mathcal{S}et \to \mathcal{S}et$ denote the cofunctor defined as follows: $P(X)$ is the power set of X, for every $X \in \mathcal{S}et$. If $f: X \to Y$ is a map, then $P(f): P(Y) \to P(X)$ is the map defined by $P(f)(Y') = f^*(Y')$, $Y' \in P(Y)$ (where $f^*(Y') = \{x \in X: f(x) \in Y'\}$). It is easy to check that P is a cofunctor.

Example 3.4. Let X be an object of a category $\mathscr{C}$. Let $H^X:\mathscr{C} \to \mathcal{S}et$ denote the functor defined as follows: for any $Y \in \mathscr{C}$, $H^X(Y) = \mathscr{C}(X, Y)$; if $f: Y \to Y'$ is a morphism in $\mathscr{C}$, then $H^X(f): \mathscr{C}(X, Y) \to \mathscr{C}(X, Y')$ is the map $H^X(f)(g) = fg$; H^X is called the *functor associated with* X.

Example 3.5. Let $H_X:\mathscr{C} \to \mathcal{S}et$ denote the cofunctor defined by $H_X(Y) = \mathscr{C}(Y, X)$ for each $Y \in \mathscr{C}$; if $f \in \mathscr{C}(Y, Y')$, then $H_X(f): \mathscr{C}(Y', X) \to \mathscr{C}(Y, X)$ is the map $H_X(f)(g) = gf$. H_X is called the *cofunctor associated with* X.

Example 3.6. For each category $\mathscr{C}$, there is a canonical cofunctor $D:\mathscr{C} \to \mathscr{C}^0$, defined in the obvious manner by $D(X) = X^0$, $D(f) = f^0$; it is clear that D is a cofunctor — D is called the *duality functor*. Obviously there is also a duality functor from $\mathscr{C}^0$ to $\mathscr{C}$; it is often denoted by D^0.

Example 3.7. Let $F:\mathscr{C} \to \mathscr{C}'$ be either a functor or a cofunctor and let $G:\mathscr{C}' \to \mathscr{C}''$ also be a functor or a cofunctor.

We define a new functor $GF:\mathscr{C} \to \mathscr{C}''$ as follows:

$$GF(X) = G(F(X)), \quad X \in \mathscr{C}$$

$$GF(f) = G(F(f)), \quad f \in \mathscr{C}.$$

GF is called the *composition functor* of the given functors F and G; GF is a functor if F and G are both functors or both cofunctors. Otherwise GF is a cofunctor.

Note that if $F:\mathscr{C} \to \mathscr{C}'$ is a cofunctor, then $FD^0:\mathscr{C}^0 \to \mathscr{C}'$ is a functor so that we can reduce the study of cofunctors to the study of functors which are defined on the dual category.

Example 3.8. If $\mathscr{C}'$ is a subcategory of $\mathscr{C}$, then we may define a functor $I:\mathscr{C}' \to \mathscr{C}$ in the obvious way; this functor is called the *inclusion* functor.

Example 3.9. Let M and M' be monoids; then a functor from M to M' is in fact a morphism of monoids.

Note 3.10. The composition of functors is associative, i.e. if $\mathscr{C} \xrightarrow{F} \mathscr{C}' \xrightarrow{G} \mathscr{C}'' \xrightarrow{H} \mathscr{C}'''$ are functors, then $H(GF) = (HG)F$. Hence with no ambiguity we shall denote the composition by HGF.

Note 3.11. For any category $\mathscr{C}$ there is a functor Id $\mathscr{C}$, or $1_{\mathscr{C}}$, called the *identity functor* of $\mathscr{C}$ and defined by $1_{\mathscr{C}}(X) = X$, $1_{\mathscr{C}}(f) = f$, for all $X \in \mathscr{C}$ and for all $f \in \mathscr{C}$.

The functor $F:\mathscr{C} \to \mathscr{C}'$ is called *faithful* if for each pair of objects X, Y of $\mathscr{C}$ the map $F(X, Y)$ is an injection. A faithful functor which takes distinct objects to distinct objects is called an *embedding*. A functor F is called *full* if $F(X, Y)$

is a surjection for every pair (X, Y) of objects of $\mathscr{C}$. Thus a subcategory $\mathscr{C}_1$ of a category $\mathscr{C}$ is full if and only if the inclusion functor $I: \mathscr{C}_1 \to \mathscr{C}$ is full.

We say that a functor $F: \mathscr{C} \to \mathscr{C}'$ is *representative* if for each $X' \in \mathscr{C}'$ there is an object $X \in \mathscr{C}$ such that $F(X) \simeq X'$.

Let $F, G: \mathscr{C} \to \mathscr{C}'$ be functors. We say that u is a *functorial morphism* from the functor F to the functor G, if for each object X in $\mathscr{C}$ we have a morphism $u_X: F(X) \to G(X)$ such that the diagram

$$\begin{array}{ccc} F(X) & \xrightarrow{u_X} & G(X) \\ {\scriptstyle F(f)}\downarrow & & \downarrow{\scriptstyle G(f)} \\ F(Y) & \xrightarrow{u_Y} & G(Y) \end{array}$$

is commutative for each morphism $f: X \to Y$ in $\mathscr{C}$.

The concept of a functorial morphism between cofunctors is defined similarly. A functorial morphism is commonly called a *natural transformation.*

Example 3.12. Let $G: \mathscr{S}et \to \mathscr{T}op$ be the functor which associates with each set X the topological space $(X, \{X, \emptyset\})$, that is, the only open sets are $\emptyset$ and X itself. Let D be the functor defined in Example 3.2. Then a functorial morphism $u: D \to G$ can be determined as follows: for each set X, let $u_X : D(X) \to G(X)$ be the identity map of X, which is clearly a continuous map, i.e. it can be viewed as a morphism in $\mathscr{T}op$.

Example 3.13. Let D and $C: \mathscr{M}orf(\mathscr{C}) \to \mathscr{C}$ be respectively the functor $D(X, f, Y) = X$, $D(u, v) = u$, and $C(X, f, Y) = Y$, $C(u, v) = v$. Then the morphism $f: D(X, f, Y) \to C(X, f, Y)$ determines a functorial morphism.

Example 3.14. Let $F = PP$, where P is the cofunctor defined in Example 3.3; obviously, F is a functor. For each set X let $u_X: X \to F(X) = P(P(X))$ denote the map defined by $u_X(x) = \{\{x\}\}$, $x \in X$ (here $\{x\}$ is the subset of X which contains x only; likewise $\{\{x\}\}$ is the subset of $P(X)$ formed by $\{x\}$ only). In this manner we get a functorial morphism $u: \mathrm{Id}\,\mathscr{S}et \to F = PP$.

Example 3.15. Let $f: X \to X'$ be a morphism in the category $\mathscr{C}$. We associate with this morphism a functorial morphism $H(f): H_X \to H_{X'}$ in the following way. Let Y be an arbitrary object of $\mathscr{C}$ and let $H(f)_Y: \mathscr{C}(Y, X) \to \mathscr{C}(Y, X')$ denote the map defined by $H(f)_Y(g) = fg$, $g \in \mathscr{C}(Y, X)$. To conclude that we have thus defined a functorial morphism from H_X to $H_{X'}$, we have to check the commutativity of the diagram

$$\begin{array}{ccc} H_X(Y) & \xrightarrow{H(f)_Y} & H_{X'}(Y) \\ {\scriptstyle H_X(g)}\uparrow & & \uparrow{\scriptstyle H_{X'}(g)} \\ H_X(Y') & \xrightarrow{H(f)_{Y'}} & H_{X'}(Y') \end{array}$$

for each morphism $g\colon Y \to Y'$ in $\mathscr{C}$. This holds since, for each $t \in H_X(Y') = \mathscr{C}(Y', X)$, we have that $H(f)^Y H_X(g))(t) = H(f)_Y(tg) = ftg = H_{X'}(g)(ft) = (H_{X'}(g)\,H(f)_{Y'})(t)$.

Example 3.16. Let F, G and $H\colon \mathscr{C} \to \mathscr{C}'$ be functors and let $u\colon F \to G$, $v\colon G \to H$ be functorial morphisms. Let $vu\colon F \to H$ denote the functorial morphism defined as follows; for each $X \in \mathscr{C}$, define $(vu)_X$ to be $v_X u_X$. This is called the *composition of functorial morphisms*. It should be noted that the composition of functorial morphisms as defined above is associative.

Example 3.17. Let $F, G\colon \mathscr{C} \to \mathscr{C}'$ and $H\colon \mathscr{C}' \to \mathscr{C}''$ be functors and let $u\colon F \to G$ be a functorial morphism. Then we denote by $Hu\colon HF \to HG$ the functorial morphism defined by $(Hu)_X = H(u_X)$ for each $X \in \mathscr{C}$. Also, if $T\colon \mathscr{C}'' \to \mathscr{C}$ is a functor, then we denote by $uT\colon FT \to GT$ the functorial morphism defined by $(uT)_X = u_{T(X)}$ for each $X \in \mathscr{C}''$. It may be readily checked that $(Hu)T = H(uT)$. Hence this last morphism is frequently denoted by HuT.

For any functor $f\colon \mathscr{C} \to \mathscr{C}'$ the morphisms $1_{F(X)}\colon F(X) \to F(X)$ define a functorial morphism, denoted $1_F\colon F \to F$, and called the *identity functorial morphism* of F. It is clear that if $u\colon G \to F$ and $v\colon F \to H$ are functorial morphisms, then $1_F u = u$ and $v 1_F = v$.

Let $F, G\colon \mathscr{C} \to \mathscr{C}'$ be functors (or cofunctors). We shall let $[F, G]$ denote the class of all functorial morphisms from F to G. This class need not be a set as we show in the following example:

Example 3.18. Let $\mathscr{I}$ be a class which is not a set (for example, the class of all sets). Define a category $\mathscr{C}$ such that $\mathcal{O}b\,\mathscr{C} = \mathscr{I}$; now, if $X, Y \in \mathscr{I}$, put $\mathscr{C}(X, Y) = \emptyset$ if $X \neq Y$, and let $\mathscr{C}(X, X)$ consist of all formal powers of a morphism f_X, i.e. $\mathscr{C}(X, X) = \{1_X = f_X^0, f_X, f_X^2, \ldots\}$. Composition in $\mathscr{C}(X, X)$ is given by the rule: $f_X^n f_X^m = f_X^{n+m}$. Furthermore, for each $X \in \mathscr{I}$ we choose a natural number n_X. The morphisms $\{f_X^{n_X}\}_{X \in \mathscr{I}}$ give a functorial morphism from $\mathrm{Id}\mathscr{C}$ to $\mathrm{Id}\mathscr{C}$. It is easy to see that $[\mathrm{Id}\mathscr{C}, \mathrm{Id}\mathscr{C}]$ is not a set.

However, there is an important case where the class of all functorial morphisms forms a set. This situation is described by the following result due to Yoneda [1] and Grothendieck [2].

THEOREM 3.19. *Let $F\colon \mathscr{C} \to \mathscr{S}et$ be a functor, $X \in \mathscr{C}$ and $x \in F(X)$. For each $Y \in \mathscr{C}$, let $u_Y^x\colon H^X(Y) \to F(Y)$ denote the map defined by $u_Y^x(f) = F(f)(x)$, $f \in H^X(Y)$. The maps $\{u_Y^x\}_{Y \in \mathscr{C}}$ define a functorial morphism $u^x\colon H^X \to F$. Moreover, the assignment*

$$x \rightsquigarrow u^x$$

defines a bijection between $F(X)$ and $[H^X, F]$. Hence, for each $X \in \mathscr{C}$, there exists a set of functorial morphisms from H^X to F.

Proof. We first prove that u^x is a functorial morphism. For this, let $f\colon Y \to Y'$ be a morphism of $\mathscr{C}$ and consider the diagram

$$\begin{array}{ccc} H^X(Y) & \xrightarrow{\;u_Y^x\;} & F(Y) \\ {\scriptstyle H^X(f)}\downarrow & & \downarrow{\scriptstyle F(f)} \\ H^X(Y') & \xrightarrow{\;u_{Y'}^x\;} & F(Y'). \end{array} \qquad (1)$$

For every $g \in H^X(Y) = \mathscr{C}(X, Y)$,

$$(F(f)u_Y^x)(g) = F(f)(F(g)(x)) = F(fg)(x)$$

and

$$(u_{Y'}^x H^X(f))(g) = u_{Y'}^x(fg) = F(fg)(x),$$

i.e. (1) is commutative, or equivalently, u^x is a functorial morphism.

Now let $v: H^X \to F$ be a functorial morphism. Then we claim that $u = u^x$, where $x = v_X(1_X)$. Indeed, let $Y \in \mathscr{C}$ and $g \in H^X(Y)$; then by the commutativity of the diagram

$$\begin{array}{ccc} H^X(X) & \xrightarrow{v_X} & F(X) \\ {\scriptstyle H^X(g)}\downarrow & & \downarrow{\scriptstyle F(g)} \\ H^X(Y) & \xrightarrow{v_Y} & F(Y) \end{array}$$

we get that $(F(g)v_X)(1_X) = F(g)(v_X(1_X)) = u_Y^x(g) = (v_Y H^X(g))(1_X) = v_Y(g)$, so that $u^x = v$.

Finally, the proof follows, since every functorial morphism $v: H^X \to F$ is completely determined by $v_X(1_X)$, i.e. by a suitable element of $F(X)$. ✳

Note 3.20. The above result also holds for cofunctors. Thus, if $G: \mathscr{C} \to \mathscr{S}et$ is a cofunctor, $X \in \mathscr{C}$ and $x \in G(X)$, then there is a functorial morphism $u^x: H_X \to G$ defined as follows: for each $Y \in \mathscr{C}$ and $g \in H_X(Y)$, define $u_Y^x(g)$ to be $G(g)(x)$. Furthermore, the assignment

$$x \rightsquigarrow u^x$$

defines a bijection between the set $G(X)$ and $[H_X, G]$.

3.21. The morphism u^x, defined above, is called the *functorial morphism associated with* x. Observe that for both functors and cofunctors we have denoted the functorial morphism associated with x by the same symbol u^x. This ambiguous notation should not lead to confusion, since by the context it will be clear whether we are speaking of functors or cofunctors.

3.22 Denote by

$$\varphi_X: [H^X, F] \to F(X)$$

the inverse of the bijection defined in Theorem 3.19. It is clear from the proof that $\varphi_X(v) = v_X(1_X)$ where $v \in [H^X, F]$. The map φ_X is called the *natural bijection.*

Corollary 3.23. *Let* $X, Y \in \mathscr{O}b\,\mathscr{C}$, *and* $f \in \mathscr{C}(X, Y)$. *Denote by* $H^0(f): H^Y \to H^X$, *the functorial morphism associated with* $f \in H^X(Y)$.

The assignment $f \rightsquigarrow H^0(f)$ *defines a bijection between* $\mathscr{C}(X, Y) = H^X(Y)$ *and* $[H^Y, H^X]$.

Likewise, denote by $H(f): H_X \to H_Y$ the functorial morphism associated with $f \in H_Y(X)$. The assignment

$$f \rightsquigarrow H(f)$$

is a bijection between $\mathscr{C}(X, Y) = H_Y(X)$ and $[H_X, H_Y]$. ✳

Another case in which the family of all functorial morphisms between two functors is a set is given by the following result.

PROPOSITION 3.23. *Let $\mathscr{C}$ be a small category. If $F, G: \mathscr{C} \to \mathscr{C}'$ are functors (cofunctors) then the class $[F, G]$ is a set.*

Proof. If $u: F \to G$ is a functorial morphism, then $\{u_X\}_{X \in \mathscr{C}}$ is an element of $\prod_{X \in \mathscr{C}} \mathscr{C}'(F(X), G(X))$. This defines a map

$$t: [F, G] \to \prod_{X \in \mathscr{C}} \mathscr{C}'(F(X), G(X)).$$

The proof follows, since it is easy to check that this map is an injection. ✳

For every two categories $\mathscr{C}$ and $\mathscr{C}'$ let $[\mathscr{C}, \mathscr{C}']$ be the class of all functors from $\mathscr{C}$ to $\mathscr{C}'$. Using the law of composition of functorial morphisms, $[\mathscr{C}, \mathscr{C}']$ comes very close to being a category (it is always an illegitimate category). The only requirement that is missing is that $[F, G]$ need not be a set when $F, G: \mathscr{C} \to \mathscr{C}'$. However, by Proposition 3.23, if we assume that $\mathscr{C}$ is small, then we can speak of the category $[\mathscr{C}, \mathscr{C}']$, called the *category of functors* from $\mathscr{C}$ to $\mathscr{C}'$. When speaking of the category of functors $[\mathscr{C}, \mathscr{C}']$ we will always assume that $\mathscr{C}$ is small. Similarly, when $\mathscr{C}$ is a small category we let $[\mathscr{C}^0, \mathscr{C}']$ denote the category of all cofunctors from $\mathscr{C}$ to $\mathscr{C}'$.

A functor $F: \mathscr{C} \to \mathscr{C}'$ is called a *small functor* into $\mathscr{C}'$ when $\mathscr{C}$ is a small category. Let $F, G: \mathscr{C} \to \mathscr{C}'$ be two functors. Consider a functorial morphism $u: F \to G$ and suppose that for each $X \in \mathscr{C}$ the morphism $u_X: F(X) \to G(X)$ is a monomorphism. Then we call u an *argumentwise functorial monomorphism.* If for each $X \in \mathscr{C}$, u_X is an epimorphism or bimorphism, then we get the concepts of *argumentwise functorial epimorphism*, or *argumentwise functorial bimorphism.*

If $u: F \to G$ is a functorial morphism, such that for all functorial morphisms $s, t: H \to F$ the relation $us = ut$ implies $s = t$, then u is called a *functorial monomorphism.* Analogously, we have the notions of *functorial epimorphism* and *functorial bimorphism.* It is easy to see that an argumentwise functorial monomorphism (epimorphism, bimorphism) is a functorial monomorphism (epimorphism, bimorphism). Exercise 3.6 shows that the converse is not valid.

A functorial morphism $u: F \to G$ is called *left* (*right*) *invertible* if there exists a functorial morphism $v: G \to F$ such that $vu = 1_F$ ($uv = 1_G$). This v is called a *left* (*right*) *inverse* of u. If $u: F \to G$ is left invertible, then, for each $X \in \mathscr{C}$, u_X is a left invertible morphism in $\mathscr{C}'$. The converse is not generally valid. (A simple counterexample may be found in the category $[\{0 \to 1\}, \mathscr{S}et]$.)

PROPOSITION 3.24. *For the functorial morphism $u: F \to G$ the following are equivalent:*

a) u is left and right invertible.

b) There is a functorial morphism $v: G \to F$ such that $uv = 1_G$ and $vu = 1_F$.

c) For each $X \in \mathscr{C}$, u_X is invertible.

Proof. a) ⇒ b). Let u' be a left inverse and u'' a right inverse of u. Then

$$u' = u'1_G = u'(uu'') = (u'u)u'' = 1_F u'' = u''$$

so that any left inverse of u is equal to any right inverse of u.

c) ⇒ b). For each $X \in \mathscr{C}$, let $u_X^{-1}: G(X) \to F(X)$ denote the inverse of u_X. Then the morphisms $\{u_X^{-1}\}_{X \in \mathscr{C}}$ define a functorial morphism $u^{-1}: G \to F$, the inverse of u.

The other implications are obvious. ✳

A functorial morphism as in the previous proposition is called a *functorial isomorphism*. We say that the functors $F, G: \mathscr{C} \to \mathscr{C}'$ are isomorphic if there is a functorial isomorphism from F into G. It must be kept in mind, however, that there may be many isomorphisms from F into G. The notation $u: F \xrightarrow{\sim} G$ will often be used to express the fact that u is an isomorphism; also, the notation $F \simeq G$ will mean that F is isomorphic to G.

A *universal element* for a functor $F: \mathscr{C} \to \mathscr{S}et$ is an ordered pair (x, X) consisting of an object X of $\mathscr{C}$ and an element $x \in F(X)$ with the following property: For each object Y of $\mathscr{C}$ and each element $y \in F(Y)$, there is exactly one morphism $f: X \xrightarrow{\sim} Y$ with $F(f)\ (x) = y$.

PROPOSITION 3.25. (Uniqueness of universal elements). *If (x, X) and (x', X') are universal elements for a functor $F: \mathscr{C} \to \mathscr{S}et$, then there is an isomorphism $f: X \xrightarrow{\sim} X'$ such that $F(f)\,(x) = x'$.*

Proof. Since (x, X) is a universal element for F, there is a morphism $f: X \to X'$ with $x' = F(f)\,(x)$. Since (x', X') is universal, there is also a morphism $f': X' \to X$ with $x = F(f')\,(x')$. Therefore

$$f(1_X)\,(x) = x = F(f')\,(x') = F(f')\,F(f)\,(x) = F(f'f)\,(x) = F(1_X)\,(x);$$

but (x, X) universal means that $f'f$ must be equal to 1_X. In an analogous fashion we get that $ff' = 1_Y$. ✳

A universal element (x, X) for a functor $F: \mathscr{C} \to \mathscr{S}et$ will often be written not as a pair but as $x \in F(X)$ or even just as x, when the object X is clear from the context.

PROPOSITION 3.26. *Let $x \in F(X)$ be a universal element for a functor $F: \mathscr{C} \to \mathscr{S}et$. Then the functorial morphism associated with x*

$$u^x : H^X \to F,$$

is an isomorphism.

Proof. It will suffice to check that, for each object Y of $\mathscr{C}$, the map $u_Y^x: H^X(Y) \to F(Y)$ is a bijection. But this follows easily from the definition of a universal element. ✳

COROLLARY 3.27. *The following are equivalent for a functor $F: \mathscr{C} \to \mathscr{S}et$:*

a) *F has a universal element.*

b) *There is an object X of $\mathscr{C}$ and a functorial isomorphism $u: H^X \xrightarrow{\sim} F$.*

For the proof it suffices to use the proposition above and to observe that H^X has a universal element, namely, $1_X \in H^X(X)$. ✳

A functor $F: \mathscr{C} \to \mathscr{S}et$ having a universal element is called a *representable functor*. The corollary above gives us a complete description of representable functors (up to functorial isomorphism).

The notion of a universal element can also be defined for cofunctors. The above results with the necessary modifications hold for cofunctors. A cofunctor $F: \mathscr{C} \to \mathscr{S}et$ having a universal element is called a *representable cofunctor*.

Clearly a functor that is isomorphic to a representable functor is also representable.

An important example of a category is the *category of categories*, denoted by $\mathscr{C}at$. The objects of $\mathscr{C}at$ are small categories and the morphisms are functors.

The following category of functors is of particular interest. Let (X, t) be a topological space. We consider t to be an ordered set relative to inclusion. Denote $[t^0, \mathscr{S}et]$ by $\mathscr{P}(X)$ and call this category the *category of presheaves* over X. An object H of $\mathscr{P}(X)$, i.e. a presheaf over X, associates with each open subset U of X a set $H(U)$; if $U \subseteq V$ then a map $r_{UV}: H(V) \to H(U)$ is given such that $r_{UV} r_{VW} = r_{UW}$ where $U \subseteq V \subseteq W$ and $r_{UU} = 1_{H(U)}$. The elements of $H(U)$ are called the *sections* of H over U. If $x \in H(U)$ and $V \subseteq U$, then $r_{VU}(x)$ is sometimes denoted by x_V and is called the restriction of x to V.

Note 3.28. We can also consider the illegitimate category $\mathscr{CAT}$, whose objects are arbitrary categories and whose morphisms are functors. We remark that all results valid for $\mathscr{C}at$ are also formally valid for $\mathscr{CAT}$, but these considerations are left for the reader.

Exercises

3.1. Prove that a morphism $f: X \to X'$ in a category $\mathscr{C}$ is a monomorphism (epimorphism) if and only if $H^Y(f)$ $(H_Y(f))$ is a monomorphism in $\mathscr{S}et$ for every $Y \in \mathscr{C}$.

3.2. Let $F: \mathscr{C} \to \mathscr{C}'$ be a faithful functor and $f \in \mathscr{C}$. Show that f is a monomorphism (epimorphism) whenever $F(f)$ is a monomorphism (epimorphism) and also that a diagram

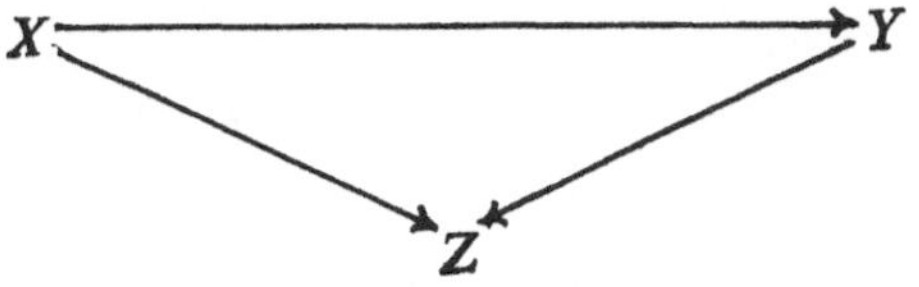

in $\mathscr{C}$ is commutative if and only if the corresponding diagram in $\mathscr{C}'$ is commutative.

3.3. Let $f, g\colon G \to G'$ be two morphisms in $\mathcal{G}r$. (Since any group can be considered as a category, f and g may be considered as functors.) Show that a functorial morphism $u\colon f \to g$ is given if and only if there exists an element $y \in G'$ such that $f(x) = yg(x)y^{-1}$ for all $x \in G$.

3.4. Let $f\colon X \to Y$ be a morphism in $\mathscr{C}$. Show that for each functor $F\colon \mathscr{C} \to \mathscr{S}et$, we get the commutative diagram

$$\begin{array}{ccc} [H^Y, F] & \xrightarrow{\varphi_Y} & F(Y) \\ {\scriptstyle [H^0(f), F]}\uparrow & & \uparrow{\scriptstyle F(f)} \\ [H^X, F] & \xrightarrow{\varphi_X} & F(X) \end{array}$$

where $[H^0(f), F](v) = vH^0(f)$, $v \in [H^X, F]$.

3.5. Let $F\colon \mathscr{C} \to \mathscr{C}'$ be a functor and let $f\colon X \to Y, g\colon Y \to X$ be morphisms in $\mathscr{C}$. Prove that:

a) If F is a faithful, then $F(g)$ is a left (right) inverse of $F(f)$ if and only if g is a left (right) inverse of f.

b) If F is full and faithful, then f has a left (right) inverse if and only if $F(f)$ has a left (right) inverse.

3.6. Let $\mathscr{C}$ be the category whose objects are topological spaces and whose morphisms are continuous maps which are surjections. Let $F, G\colon \{0 \to 1\} \to \mathscr{C}$ be the functors defined by:

a) $F(0)$ is a topological space which consists of more than one point, and $F(0 \to 1)$ is a bimorphism.

b) $G(0) = F(1)$, where $G(1)$ is a topological space that consists of one point and $G(0 \to 1)$ is the unique map $G(0) \to G(1)$.

Let $u\colon F \to G$ denote the functorial morphism with the components $u_0 = F(0 \to 1$ and $u_1 = G(0 \to 1)$. Prove that u is a functorial monomorphism that is not an argumentwise functorial monomorphism.

3.7. Let $f \in \mathscr{C}(X, Y)$. Prove that $H^0(f)\colon H^Y \to H^X$ (resp. $H(f)\colon H_X \to H_Y$) is a functorial isomorphism if and only if f is an isomorphism.

3.8. Let $F, G\colon \mathscr{C} \to \mathscr{S}et$ be functors, (x, X) a universal element for F and (y, Y) a universal element for G. Prove that for each functorial morphism $u\colon F \to G$, there exists a unique morphism $f\colon Y \to X$ such that the diagram

$$\begin{array}{ccc} H^X & \xrightarrow{u^x} & F \\ {\scriptstyle H^0(f)}\downarrow & & \downarrow{\scriptstyle u} \\ H^Y & \xrightarrow{u^y} & G \end{array}$$

is commutative.

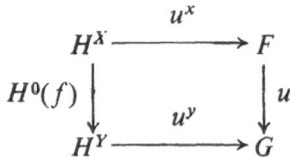

3.9. Show that the monomorphisms in $\mathscr{C}at$ are precisely the faithful functors. What are the epimorphisms?

3.10. Let $F:\mathscr{C}\to\mathscr{C}'$ be a functor. For each small category $\mathscr{A}$, let $[\mathscr{A}, F]:[\mathscr{A},\mathscr{C}]\to\to[\mathscr{A},\mathscr{C}']$ denote the functor defined by $[\mathscr{A}, F](G) = FG$ and $[\mathscr{A}, F](u) = Fu$. Prove that F is faithful if and only if, for each small category $\mathscr{A}$, $[\mathscr{A}, F]$ is faithful, and that if F is full and faithful, then $[\mathscr{A}, F]$ is full and faithful for every small category $\mathscr{A}$.

3.11. Let $H:\mathscr{A}\to\mathscr{A}'$ be a functor between small categories. Let $[H,\mathscr{C}]:[\mathscr{A}',\mathscr{C}]\to\to[\mathscr{A},\mathscr{C}]$ denote the functor defined by $[H,\mathscr{C}](G) = GH$ and $[H,\mathscr{C}](u) = uH$. Prove that if H is representative then $[H,\mathscr{C}]$ is full for every category $\mathscr{C}$.

3.12. Let $G:\mathscr{S}et\to\mathscr{S}et$ be the functor defined as follows: for each set X, $G(X)$ is the power set of X, i.e. $G(X) = P(X)$; if $f: X\to Y$ is a map, then $G(f): G(X)\to G(Y)$ is the direct image map, i.e. $G(f)(X') = f(X')$ for all $X'\in G(X)$. Prove that G is not a representable functor.

3.13. Prove that the cofunctor $P:\mathscr{S}et\to\mathscr{S}et$ defined in Example 3.3 is representable.

3.14. Let $F, G:\mathscr{C}\to\mathscr{C}'$ be functors, where $\mathscr{C}'$ is a preordered category. Show that there exists a necessarily unique functorial morphism $u: F\to G$ if and only if $F(X)\leqslant G(X)$ for every $X\in\mathscr{C}$.

3.15. Let $\{\mathscr{C}_i\}_{i\in\mathscr{I}}$ be a family of categories. For each $j\in\mathscr{I}$, let $P_j:\prod_i\mathscr{C}_i\to\mathscr{C}_j$ denote the functor projection defined by $P_j(\{X_i\}_{i\in\mathscr{I}} = X_j$ and $P_j(\{f_i\}_{i\in\mathscr{I}}) = f_j$. Assume that for each $j\in\mathscr{I}$ a functor $F_j:\mathscr{C}\to\mathscr{C}_j$ is given. Show that there is a unique functor $F:\mathscr{C}\to\prod_{i\in\mathscr{I}}\mathscr{C}_i$ such that $P_jF = F_j$ for all $j\in\mathscr{I}$.

3.16. Let $\{\mathscr{C}_i\}_{1\leqslant i\leqslant n}$ be a family of categories. A functor $F:\prod_i\mathscr{C}_i\to\mathscr{C}$ is called a *multifunctor*. If $n = 2$, then a multifunctor $F:\mathscr{C}\prod\mathscr{C}_2\to\mathscr{C}$ is called a *bifunctor*. For each category $\mathscr{C}$, let $H(?, ?):\mathscr{C}\prod\mathscr{C}\to\mathscr{S}et$ denote the bifunctor defined by $H(?, ?)(X^0,Y) = \mathscr{C}(X, Y)$ for all $(X^0, Y)\in\mathscr{O}b(\mathscr{C}^0\prod\mathscr{C})$ and $H(?,?)(f^0, g) = = H^0_{Y'}(f)(g)$ for each morphism (f^0, g) of $\mathscr{C}^0\prod\mathscr{C}$. Assume that $K:\mathscr{C}\prod\mathscr{C}'\to\mathscr{C}''$ is a bifunctor. For each $X\in\mathscr{C}$, let $K^X:\mathscr{C}'\to\mathscr{C}''$ denote the functor defined by $K^X(X') = K(X, X')$ and $K^X(f) = K(1_X, f)$ for each morphism $f: X'\to Y'$, of $\mathscr{C}'$. In the same fashion, for every $X'\in\mathscr{C}'$ the functor $K^{X'}:\mathscr{C}\to\mathscr{C}''$ is defined. Show that $K^X(X') = K^{X'}(X)$ for each $X\in\mathscr{C}$ and $X'\in\mathscr{C}'$, and $K^{Y'}(f)K^X(g) = K^Y(g)\,K^{X'}(f)$ for all morphisms $f: X\to Y, g: X'\to Y'$ in $\mathscr{C}$ and in $\mathscr{C}'$, respectively. Conversely, let $F^{X'}:\mathscr{C}\to\mathscr{C}''$ and $G^X:\mathscr{C}'\to\mathscr{C}''$ be functors for all $X\in\mathscr{C}$ and $X'\in\mathscr{C}'$. Show that if

$$F^X(X) = G^X(X') \text{ and } F^Y(f)\,G^X(g) = G^Y(g)F^{X'}(f)$$

for all $X, Y \in \mathscr{C}$ and $X', Y' \in \mathscr{C}'$ and all morphisms $f: X \to Y$ and $g: X' \to Y'$, then there is exactly one bifunctor $K: \mathscr{C} \prod \mathscr{C}' \to \mathscr{C}''$ such that $K(X, X') = F^{X'}(X) = G^X(X')$ and $K(f, g) = F^{Y'}(f)\, G^X(g)$. Describe the functorial morphisms between bifunctors.

3.17. Prove that the forgetful functor $S: \mathscr{T}op \to \mathscr{S}et$ defined in Example 3.1 is faithful but not an embedding. Show that, moreover, a functor $F: \mathscr{C} \to \mathscr{C}'$ is an embedding if and only if for each pair of morphisms f and g in $\mathscr{C}$, the condition $F(f) = F(g)$ implies $f = g$.

3.18. Let S be a fixed set and let $? \prod S: \mathscr{S}et \to \mathscr{S}et$ denote the functor defined by $(? \prod S)(X) = X \prod S$ for each $X \in \mathscr{S}et$. If $f: X \to Y$ is a map then $(? \prod S)(f) = f \prod S: X \prod S \to Y \prod S$ is defined by $(f \prod S)(x, s) = (f(x), s)$. For each set X, let

$$u_X: H^S(X) \prod S = (? \prod S)\, H^S(X) \to X$$

denote the map defined by $u_X(f, s) = f(s)$. Show that the maps $\{u_X\}_{X \in \mathscr{S}et}$ define a functorial morphism $u: (? \prod S)H^S \to \mathrm{Id}\, \mathscr{S}et$. Show also that this functorial morphism is always a functorial epimorphism. Moreover, prove that u is a functorial isomorphism if and only if S contains no more than one element.

3.19. Let $F, G: \mathscr{C} \to \mathscr{C}'$ be functors. Show that every functorial morphism $u: F \to G$ defines a function $\bar{u}: \mathscr{M}or\!f(\mathscr{C}) \to \mathscr{M}or\!f(\mathscr{C}')$ which sends each morphism $f: X \to Y$ in $\mathscr{C}$ to the morphism $\bar{u}(f): F(X) \to G(Y)$ in $\mathscr{C}'$ in such a way that $G(g)\, \bar{u}(f) = \bar{u}(g)\, F(f)$ for each composable pair (g, f). Conversely, show that each such function $\bar{u}$ defines a unique functorial morphism $u: F \to G$ such that $u_X = \bar{u}(1_X)$ for all $X \in \mathscr{C}$.

3.20. Prove that any functor $F: \mathscr{S}et \to \mathscr{C}$ takes monomorphisms to monomorphisms and epimorphisms to epimorphisms.

3.21. Let $\mathscr{C}$ and $\mathscr{C}'$ be categories. Show that each functor $H: \mathscr{C} \to [\{0 \to 1\}, \mathscr{C}']$ determines two functors $F, G: \mathscr{C} \to \mathscr{C}'$ and a functorial morphism $u: F \to G$. Show that this assignment $H \rightsquigarrow (F, G, u)$ is a bijection.

3.22. Let $\mathscr{C}$ be a preordered category. Show that for each small category $\mathscr{A}$, the category $[\mathscr{A}, \mathscr{C}]$ is also a preordered category.

3.23. Let $\mathscr{C}$ be a category and let $u, v: \mathrm{Id}\mathscr{C} \to \mathrm{Id}\mathscr{C}$ be functorial morphisms. Prove that $uv = vu$. If $\mathscr{C}$ is one of the categories $\mathscr{S}et$, $\mathscr{T}op$, $\mathscr{M}on$, $\mathscr{G}r$, $\mathscr{A}b$ or $\mathscr{R}g$, prove that the class $[\mathrm{Id}\mathscr{C}, \mathrm{Id}\mathscr{C}]$ is a set. Find this monoid (i.e. $[\mathrm{Id}\mathscr{C}, \mathrm{Id}\mathscr{C}]$) in each of the above cases. (Hint: Observe that in each of these cases we may consider a representable functor $F: \mathscr{C} \to \mathscr{S}et$, and an injection $[\mathrm{Id}\,\mathscr{C}, \mathrm{Id}\,\mathscr{C}] \to [F, F]$.) We remark that the monoid $[\mathrm{Id}\,\mathscr{C}, \mathrm{Id}\,\mathscr{C}]$ is called the *center of* $\mathscr{C}$.

3.24. Let $F: \mathscr{C} \to \mathscr{C}'$ be a functor, $X \in \mathscr{C}$, $X' \in \mathscr{C}'$, and let $u: H^X \to H^{X'}F$ be a functorial morphism. Prove that there is a unique morphism $t: X' \to F(X)$ such that u may be factored by

$$H^X \xrightarrow{\ v\ } H^{F(X)}F \xrightarrow{\ H^0(t)F\ } H^{X'}F$$

where v is canonically defined.

1.4. Equivalence of categories

In this section the concept of equivalence of categories is presented. Two equivalent categories have essentially the same properties. Also the notion of concrete category is defined, and some concrete categories are presented.

References: Eilenberg—MacLane [1], Grothendieck [1], Kuroš—Livšic—Šulgeifer [1], MacLane [3], Mitchell [2], Pareigis [1], Schubert [1], Šulgeifer—Calenko [1].

THEOREM 4.1. *Let $F: \mathscr{C} \to \mathscr{C}'$ be a functor. The following are equivalent:*
1) *F is full, faithful and representative.*
2) *There are a functor $G: \mathscr{C}' \to \mathscr{C}$ and functorial isomorphisms:*

$$u: \mathrm{Id}\mathscr{C}' \to FG, \quad v: GF \to \mathrm{Id}\mathscr{C}.$$

Furthermore, we can choose v such that $Fv = (uF)^{-1}$, and $Gu = (vG)^{-1}$.

Proof. 1) $\Rightarrow$ 2). For each $X' \in \mathscr{C}'$ we choose an object $G(X')$ of $\mathscr{C}$ and an isomorphism $u_{X'}: X' \to FG(X')$. Now if $h: X' \to Y'$ is a morphism in $\mathscr{C}'$, then we can define the morphism

$$u_{Y'} h u_{X'}^{-1}: FG(X') \to FG(Y').$$

Since F is full and faithful there exists a unique morphism $G(h): G(X') \to G(Y')$ such that the diagram

$$\begin{array}{ccc} X' & \xrightarrow{u_{X'}} & FG(X') \\ {\scriptstyle h}\downarrow & & \downarrow{\scriptstyle FG(h)} \\ Y' & \xrightarrow{u_{Y'}} & FG(Y') \end{array} \qquad (2)$$

is commutative.

Using the uniqueness of $G(h)$ and the functorial properties of F it is easy to verify that G is a functor. Also, from (2) we see that the morphisms $\{u_{X'}\}_{X' \in \mathscr{C}'}$ define a functorial morphism:

$$u: \mathrm{Id}\mathscr{C}' \to FG.$$

Furthermore, for each $X \in \mathscr{C}$, $u_{F(X)}: F(X) \to FGF(X)$ is an isomorphism. Thus, since F is full and faithful, there is a unique isomorphism

$$v_X: GF(X) \to X$$

such that $F(v_X) = u_{F(X)}^{-1}$. We claim that the morphisms $\{v_X\}_{X \in \mathscr{C}}$ define a functorial morphism

$$v: GF \to \mathrm{Id}\,\mathscr{C}.$$

If $f: X \to Y$ is a morphism of $\mathscr{C}$, then we may consider the diagram

$$\begin{array}{ccc} GF(X) & \xrightarrow{v_X} & X \\ {\scriptstyle GF(f)}\downarrow & & \downarrow{\scriptstyle f} \\ GF(Y) & \xrightarrow{v_Y} & Y \end{array} \qquad (3)$$

If we apply the functor F to this diagram and use the definition of v_X, we obtain a commutative diagram in $\mathscr{C}'$. Since F is faithful it follows that (3) is commutative, that is, v is a functorial morphism. Finally, it is easy to check the relations $Fv = (uF)^{-1}$, and $Gu = (vG)^{-1}$.

2) $\Rightarrow$ 1). F is faithful; for if $f, g \in \mathscr{C}(X, Y)$ are such that $F(f) = F(g)$, then $fv_X = v_Y GF(f) = v_Y GF(g) = gv_X$, i.e. $f = g$, since v_X is an isomorphism.

F is full; for, if $h \in \mathscr{C}'(F(X), F(Y))$, then by the commutative diagram

$$\begin{array}{ccc} F(X) & \xrightarrow{u_{F(X)}} & FGF(X) \\ {\scriptstyle h}\downarrow & & \downarrow{\scriptstyle FG(h)} \\ F(Y) & \xrightarrow{u_{F(Y)}} & FGF(Y) \end{array}$$

we see that $h = u_{F(Y)}^{-1} FG(h) u_{F(X)} = F(v_Y) FG(h) F(v_X^{-1}) = F(v_Y G(h) v_X^{-1})$.

Now the proof is complete since F is obviously a representative functor. ✳

A functor $F: \mathscr{C} \to \mathscr{C}'$ as in the previous theorem is called an *equivalence*; two categories are called *equivalent* if there is an equivalence between them.

Example 4.2. Consider the preordered set $\mathscr{M}$ as a category. Denote by $\sim$ the relation $x \sim y \Leftrightarrow x \leqslant y$ and $y \leqslant x$. Thus $\sim$ is an equivalence relation on $\mathscr{M}$; let $\overline{\mathscr{M}}$ denote the set $\mathscr{M}/\sim$. It is clear that $\overline{\mathscr{M}}$ is canonically an ordered set, and the canonical map

$$u_{\mathscr{M}}: \mathscr{M} \to \overline{\mathscr{M}}$$

is nondecreasing, i.e. it is a functor. It is easy to see that $u_{\mathscr{M}}$ is in fact an equivalence of categories.

Example 4.3. (Generalization of the previous example). Let $\mathscr{C}$ be a category. In Section 1.2, we noted that the relation "X is isomorphic to X'" is in fact an equivalence relation on the class $\mathscr{Ob}\,\mathscr{C}$. The equivalence classes are called "types". Now we define a new category $\mathscr{C}_1$ whose objects are the types of $\mathscr{C}$. Thus for each type of $\mathscr{C}$ we choose an object X_1 and for every $X \in [X_1]$, we choose a fixed isomorphism $u_X: X_1 \to X$. Let us put $\mathscr{C}_1(X_1, Y_1) = \mathscr{C}(X_1, Y_1)$ and let $F: \mathscr{C} \to \mathscr{C}_1$ denote the functor defined by

$$F(X) = X_1 \text{ if } X \in [X_1].$$

Also, if $f: X \to Y$ is a morphism in $\mathscr{C}$, then define $F(f)$ to be $u_Y^{-1} f u_X$. It is easy to see that F is in fact a functor. By its construction, F is full, faithful and representative, i.e. an equivalence of categories. The category $\mathscr{C}_1$ constructed above, is called a *skeleton* of $\mathscr{C}$. Obviously, any two skeletons of a category $\mathscr{C}$ are in some sense identical (see Exercise 4.5).

An equivalence $F:\mathscr{C} \to \mathscr{C}'$ which produces a bijection between the objects of $\mathscr{C}$ and $\mathscr{C}'$ is called an *isomorphism of categories*. In this case we shall write $\mathscr{C} \simeq \mathscr{C}'$.

A category $\mathscr{C}$ is called *concrete* if it is isomorphic to a subcategory of $\mathscr{S}et$. It is clear that a category $\mathscr{C}$ is concrete if and only if $\mathscr{C}^0$, the dual of $\mathscr{C}$, is concrete.

A family $\{S_i\}_{i\in\mathscr{I}}$ of objects from a category $\mathscr{C}$ is called a *family of separators* if, for each object X of $\mathscr{C}$, there is a subset $\mathscr{I}'$ of $\mathscr{I}$ such that for every pair of distinct morphisms $f, g \in \mathscr{C}(X, Y)$ there is an $i \in \mathscr{I}'$ and an $h \in \mathscr{C}(S_i, X)$ such that $fh \neq gh$. Note that we do not assume that $\mathscr{I}$ is a set, but if this is the case then the family under consideration is called a *set of separators*. An object S of $\mathscr{C}$ is called a *separator* of $\mathscr{C}$ if the set $\{S\}$ is a set of separators. Dually, we have the notions of *set of coseparators* and *coseparator*.

Example 4.4. The additive monoid **N** of the natural numbers is a separator of $\mathscr{M}on$. Indeed, if x is an element of a monoid M, let $m_x: \mathbf{N} \to M$ denote the morphism of monoids defined by

$$m_x(n) = x^n.$$

It is clear that if x, y, $x \neq y$, are elements of M, then $m_x \neq m_y$. Also, if $f: M \to M'$ is a morphism in $\mathscr{M}on$ then $fm_x = m_{f(x)}$ for each $x \in M$. Thus it is easy to check that **N** is a separator of $\mathscr{M}on$.

Example 4.5. The additive group of integers is a separator of $\mathscr{G}r$.

Example 4.6. Every set having more than one element is a coseparator of $\mathscr{S}et$. Indeed, if $f, g: X \to Y$ are two distinct maps, then $f(x) \neq g(x)$ for some element $x \in X$. Now if P is a set having two elements a and b, let $h: Y \to P$ denote the map defined by

$$h(f(x)) = a \quad \text{and} \quad h(y) = b \quad \text{if} \quad y \neq f(x).$$

It is obvious that $hf \neq hg$.

THEOREM 4.7. *Let $\mathscr{C}$ be a category having a set of separators $\{S_i\}_{i\in\mathscr{I}}$. Suppose that for each $X \in \mathscr{C}$ and for each $i \in \mathscr{I}$, $\mathscr{C}(S_i, X) \neq \emptyset$. Then $\mathscr{C}$ is concrete.*

Proof. Let $\{S_i\}_{i\in\mathscr{I}}$ be a set of separators of $\mathscr{C}$. Let $F:\mathscr{C} \to \mathscr{S}et$ denote the functor defined by

$$F(X) = \textstyle\prod_{i\in\mathscr{I}} \mathscr{C}(S_i, X).$$

If $f: X \to Y$ is a morphism in $\mathscr{C}$, then $F(f): \prod_{i\in\mathscr{I}} \mathscr{C}(S_i, X) \to \prod_{i\in\mathscr{I}} \mathscr{C}(S_i, Y)$ is the map defined by

$$F(f)\,(\textstyle\prod_{i\in\mathscr{I}} f_i) = \prod_{i\in\mathscr{I}} (ff_i).$$

The reader can show that F is actually an isomorphism of the categories. ✳

COROLLARY 4.8. *Every category which is equivalent to a small category is concrete.* ✳

COROLLARY 4.9. *Every category $\mathscr{C}$ having a set of coseparators $\{S_i\}_{i\in\mathscr{I}}$ such that $\mathscr{C}(X,S_i) \neq \emptyset$ for each $X \in \mathscr{C}$ and for each $i \in \mathscr{I}$, is concrete.*

Proof. It is clear by Theorem 4.7 that $\mathscr{C}^0$ is concrete, i.e. there exists an embedding:

$$F: \mathscr{C}^0 \to \mathscr{S}et.$$

Furthermore, according to Example 4.6, $(\mathscr{S}et)^0$ is likewise concrete so that there is the embedding

$$G: (\mathscr{S}et)^0 \to \mathscr{S}et.$$

Finally, the composition of functors:

$$\mathscr{C} \xrightarrow{D} \mathscr{C}^0 \xrightarrow{F} \mathscr{S}et \xrightarrow{H} (\mathscr{S}et)^0 \xrightarrow{G} \mathscr{S}et$$

where D and H are dual cofunctors, is an embedding. ✳

The question naturally arises whether every category is concrete. The following example due to Calenko and Šulgeifer [1] shows that the answer to this question is negative.

Example 4.10. Let $\mathscr{C}$ be a category whose class of objects is the class of all ordinal numbers (see Bourbaki [1]) and another symbol ω. Assume that 0, the first ordinal number, is the zero object in this category (i.e. for each $X \in \mathscr{C}$, $\mathscr{C}(0, X)$ and $\mathscr{C}(X, 0)$ are singletons).

If α and β are nonzero ordinal numbers, then $\mathscr{C}(\alpha, \beta)$ contains only the zero morphism when $\alpha \neq \beta$. For any ordinal number α, the set $\mathscr{C}(\alpha, \alpha)$ consists of all triples (α, β, α) where $\beta \leqslant \alpha$. Likewise, for every ordinal number α the set $\mathscr{C}(\alpha, \omega)$ consists of all triples (α, β, ω) where $\beta \leqslant \alpha$ and $\mathscr{C}(\omega, \alpha)$ consists of all triples (ω, β, α) where $\beta \leqslant \alpha$. Finally, $\mathscr{C}(\omega, \omega)$ contains the identity and zero morphisms.

The composition of morphisms is given by the rules:

$$\alpha \xrightarrow{(\alpha,\beta,\omega)} \omega \xrightarrow{(\omega,\beta',\alpha)} \alpha = \alpha \xrightarrow{(\alpha,\ \max(\beta,\beta'),\alpha)} \alpha$$

for every ordinal number α. All other possible compositions are the zero morphisms. It is easy to see that $\mathscr{C}$ is in fact a category.

Now we assert that $\mathscr{C}$ is not concrete. Indeed, assume to the contrary that there is a faithful functor

$$F: \mathscr{C} \to \mathscr{S}et.$$

Let α be an ordinal number whose cardinality is strictly greater than the cardinality of the power set of $F(\omega)$. Thus there exist two distinct ordinals $\beta < \beta_1 \leqslant \alpha$ such that

$$F(\alpha, \beta_1, \omega)(F(\alpha)) = F(\alpha, \beta, \omega)(F(\alpha)).$$

Let $x \in F(\alpha)$; by the above equality, there is an $x' \in F(\alpha)$ such that

$$F(\alpha, \beta, \omega)(x) = F(\alpha, \beta_1, \omega)(x').$$

Hence

$$\begin{aligned} F(\omega, \beta, \alpha)\, F(\alpha, \beta, \omega)(x) &= F(\omega, \beta, \alpha)F(\alpha, \beta_1, \omega)\,(x') \\ &= F(\alpha, \beta_1, \omega)(x') = F(\omega, \beta_1, \alpha)F(\alpha, \beta_1, \omega)(x') \\ &= F(\omega, \beta_1, \alpha)F(\alpha, \beta, \omega)(x) \end{aligned}$$

so that

$$F(\omega, \beta, \alpha)F(\alpha, \beta, \omega) = F(\omega, \beta_1, \alpha)F(\alpha, \beta, \omega),$$

which is a contradiction, since

$$(\omega, \beta, \alpha)(\alpha, \beta, \omega) \neq (\omega, \beta_1, \alpha)(\alpha, \beta, \omega).$$

Note 4.11. There is an example (see Calenko [3]) of a concrete category $\mathscr{C}$, such that for every embedding $F: \mathscr{C} \to \mathscr{S}et$, there is a monomorphism f in $\mathscr{C}$ such that $F(f)$ is not a monomorphism.

4.12. Let $\{X_i\}_{i \in \mathscr{I}}$ be a set of objects of category $\mathscr{C}$. Denote by $\mathscr{C}_1$ the full subcategory of $\mathscr{C}$ whose objects are X_i, $i \in \mathscr{I}$. By Corollary 4.8, there is an embedding $F: \mathscr{C}_1 \to \mathscr{S}et$. We thus see that every category can be regarded "locally" as a concrete category.

Exercises

4.1. Prove that the conditions in Theorem 4.1 are equivalent to:
3) There is a functor $G: \mathscr{C}' \to \mathscr{C}$ and functorial isomorphisms

$$u: FG \to \mathrm{Id}\mathscr{C}', \quad v: \mathrm{Id}\,\mathscr{C} \to GF$$

such that $uF = (Fv)^{-1}$ and $Gu = (vG)^{-1}$.

4.2. Prove that a functor $F: \mathscr{C} \to \mathscr{C}'$ is an equivalence of categories if and only if, for every small category $\mathscr{A}$, the functor $[\mathscr{A}, F]$ is an equivalence.

4.3. Prove: If $H: \mathscr{A} \to \mathscr{A}'$ is an equivalence between small categories then for every category $\mathscr{C}$, the functor $[H, \mathscr{C}]$ is also an equivalence; moreover, if a category $\mathscr{A}$ is equivalent to a small category, then for any two functors $F, G: \mathscr{A} \to \mathscr{C}$, $[F, G]$ is a set.

4.4. Let X be a topological space. Let $\mathscr{F}(X)$ denote the full subcategory of $\mathscr{P}(X)$ whose objects are defined as follows. A presheaf H belongs to $\mathscr{F}(X)$ if it satisfies the following condition:

(f): If U is an open subset of X, and $\{U_i\}_{i \in \mathscr{I}}$ is an open cover of U (i.e. a family of open subsets such that $\cup_{i \in \mathscr{I}} U_i = U$), and if for each $i \in \mathscr{I}$ a section $x_i \in H(U_i)$ is given such that for every pair $i, j \in \mathscr{I}$ we have

$$(x_i)_{U_i \cap U_j} = (x_j)_{U_i \cap U_j},$$

then there is a unique section $x \in H(U)$ such that $x_{U_i} = x_i$ for every i.

The objects of $\mathscr{F}(X)$ are called *sheaves* over X.

Likewise let $\mathscr{E}(x)$ denote the full subcategory of $\mathscr{T}op/X$ defined as follows. An object (Y, p) belongs to $\mathscr{E}(Y)$ if p is a local homeomorphism (i.e. each $y \in Y$ has a neighborhood U such that the restriction of p to U is a homeomorphism between U and $p(U)$, where U has the induced topology).

Let $F: \mathscr{E}(X) \to \mathscr{F}(X)$ denote the functor defined as follows: If $(Y, p) \in \mathscr{E}(X)$, then for every open set V of X, define:

$$F(Y,p)(V) = \{f \mid f \in \mathscr{T}op(V, Y), \text{ such that } pf = 1_V\}.$$

Prove that the functor F is an equivalence of categories.

4.5. Show that any two skeletons of a category are isomorphic.

4.6. Let $\mathscr{A}$ and $\mathscr{B}$ be small categories. Prove that for every category $\mathscr{C}$ we have the following canonical isomorphisms of categories:

$$[\mathscr{A}, [\mathscr{B}, \mathscr{C}]] \simeq [\mathscr{A} \coprod \mathscr{B}, \mathscr{C}] \simeq [\mathscr{B}, [\mathscr{A}, \mathscr{C}]].$$

4.7. Show that any nonempty set is a separator of $\mathscr{S}et$. Likewise, any object of $\mathscr{T}op$, $\mathscr{P}re$, $\mathscr{O}rd$ and $\mathscr{S}$, whose underlying set is nonempty, is a separator. Describe the separators in $\mathscr{S}^+$ and $\mathscr{C}at$.

4.8. Prove that a category $\mathscr{C}$ is concrete if and only if $\mathscr{C}^0$, the dual of $\mathscr{C}$, is concrete.

1.5. Equivalence relations on a category

The concept of an equivalence relation on a category generalizes the notion of an equivalence relation on a monoid. In this section we also present the concept of a free category which generalizes the notion of a free monoid.

References: Bănică—Popescu [1], Chevalley [1], Gabriel—Zisman [1], Grothendieck [1], Liapin [1], MacLane [3], Mitchell [2], Schubert [1].

We say that R is an *equivalence relation* on a category $\mathscr{C}$ if for every pair of objects (X, Y) of $\mathscr{C}$ an equivalence relation $R(X, Y)$ is given on the set $\mathscr{C}(X, Y)$ such that:

If $f, g \in \mathscr{C}(X, Y)$ and $f \sim g(R(X, Y))$ then for each morphism $h: Y \to Z$ ($k: T \to X$) we have

$$hf \sim hg\ (R(X, Z))\ (fk \sim gk\ (R(T, Y))).$$

If R is an equivalence relation on a category $\mathscr{C}$, then we define a new category, denoted by $\mathscr{C}/R$, and called the *factor category* of $\mathscr{C}$ with respect to R. The objects of $\mathscr{C}/R$ are the objects of $\mathscr{C}$; if X is an object of $\mathscr{C}$, then the same object considered as an object of $\mathscr{C}/R$ will be denoted by $\overline{X}$. For any pair $\overline{X}$, $\overline{Y}$ of objects of $\mathscr{C}/R$, we define

$$(\mathscr{C}/R)(\overline{X}, \overline{Y}) = \mathscr{C}(X, Y)/R(X, Y).$$

If $f \in \mathscr{C}(X, Y)$ let $\overline{f}$ denote the class of f modulo $R(X, Y)$. Thus, if $\overline{f}: \overline{X} \to \overline{Y}$ and $\overline{g}: \overline{Y} \to \overline{Z}$ are morphisms of $\mathscr{C}/R$, then $\overline{g}\overline{f} = \overline{gf}$; also $1_{\overline{X}} = \overline{1_X}$ (for every object X).

Finally, let

$$P: \mathscr{C} \to \mathscr{C}/R$$

denote the functor defined by $P(X) = \bar{X}$, $P(f) = \bar{f}$. It is clear that P is a representative and a full functor. We call P the *canonical functor*.

Example 5.1. Let $f, g \in \mathscr{T}\!op(X, Y)$; we say that f is *homotopic* with g if there is a continuous map $h: I \coprod X \to Y$ with $h(0, x) = f(x)$ and $h(1, x) = g(x)$ for all $x \in X$, where I is the interval $[0, 1]$ of the real numbers. The open sets of $I \coprod X$ are arbitrary unions of sets of the form $U \coprod V$, where $U \subseteq I$ and $V \subseteq X$ are open sets. The reader is invited to consult Schubert [2], or any other book on Algebraic Topology, to see that the homotopy of maps defines an equivalence relation on $\mathscr{T}\!op$. The equivalence classes are called *homotopy classes* of continuous maps. The factor category of $\mathscr{T}\!op$ with respect to "homotopy" is denoted by $\mathscr{H}tp$ and called the *homotopy category* of topological spaces.

Example 5.2. Let G be a group; a subcategory of G (i.e. a subset G' of G that is closed under multiplication and contains the identity element) is called a *subgroup* if it contains the inverse of any element $x \in G'$. A subgroup G' of G is called *normal* if for each $x' \in G'$ and each $x \in G$ we have $xx'x^{-1} \in G'$.

If G' is a normal subgroup of G, we let $R(G')$ denote the relation

$$x \sim y(R(G')) \Leftrightarrow yx^{-1} \in G'.$$

Thus, $R(G')$ is an equivalence relation on (the category) G; the factor category is denoted by G/G'. It is easy to see that G/G' is again a group; it is called the *factor group* of G by G'.

Conversely, if R is an equivalence relation on the group G, then there is a unique normal subgroup G' of G, namely the class of the identity element, such that $R = R(G')$.

PROPOSITION 5.3. *Let $T: \mathscr{C} \to \mathscr{C}'$ be a functor, and let R_T denote the following relation in $\mathscr{C}$: if $f, g \in \mathscr{C}(X, Y)$, then $f \sim g(R_T(X, Y)) \Leftrightarrow T(f) = T(g)$. Then R_T is an equivalence relation on $\mathscr{C}$, and there is a unique faithful functor $\bar{T}: \mathscr{C}/R_T \to \mathscr{C}'$ such that the diagram of categories and functors*

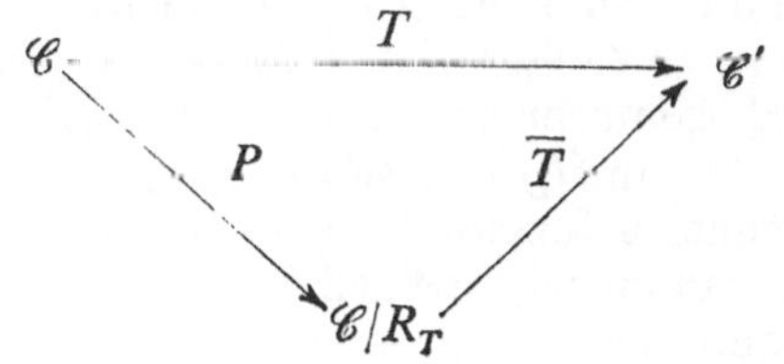

is commutative, where P is the canonical functor.

Moreover, $\bar{T}$ is an equivalence of categories if and only if T is full and representative.

Proof. It is easy to see that R_T is an equivalence relation on $\mathscr{C}$. Now the construction of $\bar{T}$ is obvious. If $X \in \mathscr{C}/R_T$, then there is a unique object X' of $\mathscr{C}$ such that $P(X') = X$. Thus define $\bar{T}(X)$ to be $T(X')$. Also, if $f: X \to Y$ is a morphism in $\mathscr{C}/R_T$, then let X denote $P(X')$ and Y denote $P(Y'')$. There is an $f' \in \mathscr{C}(X', Y')$ such that $P(f') = f$. Then we define $\bar{T}$ by $\bar{T}(f) = T(f')$. The uniqueness of $\bar{T}$ is obvious, since P is full and representative. It is also clear that $\bar{T}$ is faithful.

Now we assume that T is full and representative. Then it is obvious that $\bar{T}$ is also full and representative.

Conversely, if $\bar{T}$ is full and representative then $\bar{T}P = T$ is also a full and representative functor, since it is the composition of full and representative functors. ✳

Note 5.4. Let $T: \mathscr{C} \to \mathscr{C}'$ be a functor; define the *image* of T to be the class $\{T(X)\}_{X \in \mathscr{C}'}$ of objects from $\mathscr{C}'$ together with the class $\{T(f) | f \text{ is a morphism of } \mathscr{C}\}$ of morphisms from $\mathscr{C}'$. Denote the image of T by Im(T). It is easy to see that Im(T) is not necessarily a subategory of $\mathscr{C}'$. The difficulty is due to the fact that T may not be univalent on objects. (This allows situations wherein the composition in $\mathscr{C}'$, of two functions from the image is not in the image of a function from $\mathscr{C}$.) However, if T is univalent with respect to objects, then Im(T) is a subcategory of $\mathscr{C}'$. In this case, the above functor $\bar{T}: \mathscr{C}/R_T \to \text{Im}(T)$ defines an isomorphism of categories. The following proposition shows that in a sense there is no loss in generality by always assuming that T is univalent on objects.

PROPOSITION 5.5. *Let $T: \mathscr{C} \to \mathscr{C}'$ be a functor. Then there is a category $\mathscr{C}''$ an embedding $I: \mathscr{C}' \to \mathscr{C}''$ and a functor $T': \mathscr{C} \to \mathscr{C}''$ such that T' is univalent on objects and is isomorphic with IT.*

Proof. Define a category $\mathscr{C}_1'$ as follows: $\mathcal{O}b(\mathscr{C}_1') = \mathcal{O}b(\mathscr{C}) \coprod \mathcal{O}b(\mathscr{C}')$; a morphism in $\mathscr{C}_1'$ from (X, X') to (Y, Y') is a triple (X, Y, f'), where $f': X' \to Y'$ is a morphism in $\mathscr{C}'$. The composition is defined by the following rule:

$$(Y, Z, g') \cdot (X, Y, f') = (X, Z, g'f').$$

We fix an object $X_0 \in \mathscr{C}$ and define the functor $I_1: \mathscr{C}' \to \mathscr{C}_1'$ by $I_1(X') = (X_0, X')$ and $I_1(f') = (X_0, X_0, f')$. It is clear that I_1 is a full embedding. Furthermore, I_1 defines an equivalence between $\mathscr{C}'$ and $\mathscr{C}_1'$, since I_1 is full, and for every object (X, X') of $\mathscr{C}_1'$ we have an isomorphism $(X, X_0, 1_{X'}): (X, X') \xrightarrow{\sim} (X_0, X') = I_1(X')$. Now we define a functor $T_1: \mathscr{C} \to \mathscr{C}_1'$ by $T_1(X) = (X, T(X))$ and $T_1(f) = (X, Y, T(f))$ for $f: X \to Y$. Then for each object $X \in \mathscr{C}$ we have an isomorphism $u_X: T_1(X) \xrightarrow{\sim} I_1T(X)$ in $\mathscr{C}_1'$ given by $u_X = (X, X_0, 1_{(X)})$. It is easy to check that the morphisms $\{u_X\}_{X \in C}$ give a functorial isomorphism from T_1 to I_1T. Since I_1 is univalent on objects and is faithful, $\mathscr{C}'' = \text{Im}(I_1)$ is a subcategory of $\mathscr{C}_1'$ which is equivalent to $\mathscr{C}'$. It is clear that T_1 defines a functor $T': \mathscr{C} \to \mathscr{C}''$ such that T' is isomorphic to IT where $I: \mathscr{C}' \to \mathscr{C}''$ is canonically defined by I_1. ✳

Let R and R' be equivalence relations on the category $\mathscr{C}$; we say that R' is *finer* than R or that R is *smaller* than R', denoted $R \leqslant R'$, if for each pair of objects (X, Y) of $\mathscr{C}$, the relation $f \sim g(R(X, Y))$ implies $f \sim g(R'(X, Y))$, for $f, g \in \mathscr{C}(X, Y)$. In particular, a functor $T: \mathscr{C} \to \mathscr{C}'$ is *compatible with an equivalence relation R on $\mathscr{C}$ if $R \leqslant R_T$*.

Proposition 5.6. *Let R and R' be equivalence relations on a category $\mathscr{C}$. Then $R \leqslant R'$ if and only if there is functor $T: \mathscr{C}/R \to \mathscr{C}/R'$ such that $TP = P'$, where $P: \mathscr{C} \to \mathscr{C}/R$ and $P': \mathscr{C} \to \mathscr{C}/R'$ are the canonical functors.*

Proof. Assume the existence of T. Then for every $f, g \in \mathscr{C}(X, Y)$ such that $f \sim g(R(X, Y))$ we have that $TP(f) = TP(g) = P'(f) = P'(g)$ so that $f \sim g(R'(X, Y))$, i. e. $R \leqslant R'$.

The converse follows from Proposition 5.3.✳

Proposition 5.7. *Let R and R' be equivalence relations on a category $\mathscr{C}$ and let R' be finer than R. Let R'/R denote the equivalence relation on $\mathscr{C}/R$ defined by: $P(f) \sim P(g)(R'/R(P(X), P(Y))) \Leftrightarrow f \sim g(R'(X, Y))$ for $X, Y \in \mathscr{C}$ and $f, g \in \mathscr{C}(X, Y)$. (Here $P: \mathscr{C} \to \mathscr{C}/R$ is the canonical functor.)*

The assignment

$$R' \rightsquigarrow R'/R$$

defines a bijection between the class of all equivalence relations on $\mathscr{C}$ which are finer than R and the class of equivalence relations on $\mathscr{C}/R$. Moreover, for every R' that is finer than R, we have the following commutative diagram of categories and functors defined in a canonical way:

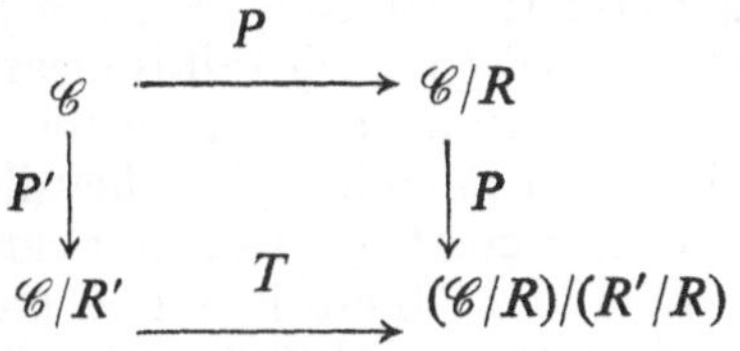

where T is an isomorphism of categories.

Proof. If R_1 is an equivalence relation on $\mathscr{C}/R$, let $P^{-1}(R_1)$ denote the equivalence relation on $\mathscr{C}$ defined by

$$f \sim g(P^{-1}(R_1)(X, Y)) \Leftrightarrow P(f) \sim P(g)(R_1(P(X), P(Y)))$$

for $X, Y \in \mathscr{C}$ and $f, g \in \mathscr{C}(X, Y)$.

It is easy to see that $P^{-1}(R_1)$ is an equivalence relation on $\mathscr{C}$ which is finer than R, and $P^{-1}(R_1)/R = R_1$. Also $P^{-1}(R'/R) = R'$ for every equivalence relation R' on $\mathscr{C}$ such that $R \leqslant R'$.

The last part of Proposition 5.7 follows from Propositions 5.3 and 5.5. ✳

Now we define a kind of category which will occasionally be useful, namely a free category. This is a generalization of the concept of a free monoid.

A morphism $f: X \to Y$ in a category $\mathscr{C}$ is called a *free morphism* if the existence of a commutative diagram

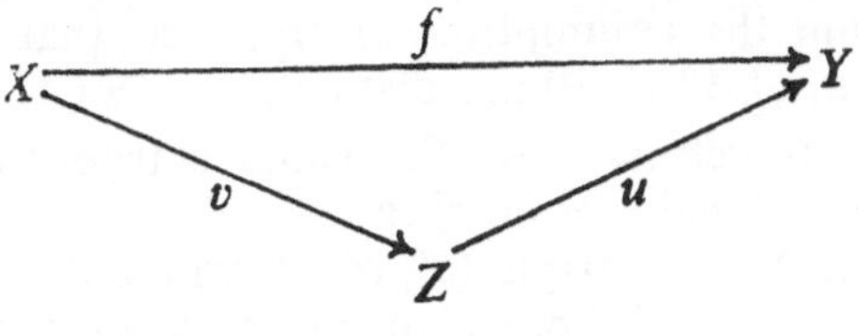

implies that $Z = X$ and $v = 1_X$, or $Z = Y$ and $u = 1_Y$.

Let $f: X \to Y$ be a morphism in $\mathscr{C}$ which is not an identity morphism. We say that f has a *free factorization* if there are free morphisms $f_1: X \to X_1$, $f_2: X_1 \to X_2, \ldots, f_n: X_{n-1} \to X_n = Y$, all different from identities, such that $f = f_n f_{n-1} \ldots f_1$. We say that a category $\mathscr{C}$ is *free* if the following two conditions are satisfied:

a) Every identity morphism of $\mathscr{C}$ is free.

b) Every non-identity morphism $f: X \to Y$ of $\mathscr{C}$ has a unique free factorization, i.e. if $f_n \ldots f_1 = g_m \ldots g_1 = f$ are two free factorizations of f, then $m = n$, and $f_i = g_i$ for $1 \leqslant i \leqslant n$.

The next proposition justifies the terminology.

PROPOSITION 5.8. *Let $\mathscr{C}$ be a free category, and $\mathscr{C}'$ an arbitrary category. Assume that there is a function α which assigns to each object X of $\mathscr{C}$ an object $\alpha(X)$ of $\mathscr{C}'$ and a function β which assigns to each free morphism $f: X \to Y$ in $\mathscr{C}$ a morphism $\beta(f): \alpha(X) \to \alpha(Y)$ in $\mathscr{C}'$, such that $\beta(f)$ is an identity if f is an identity. Then there exists a unique functor $F: \mathscr{C} \to \mathscr{C}'$ such that $F(X) = \alpha(X)$ for all $X \in \mathscr{C}$, and $F(f) = \beta(f)$ for every free morphism in $\mathscr{C}$.*

Proof. The key point is how we define $F(f)$ when f is not necessarily a free morphism. Since f has a unique free factorization, i.e. $f = f_n f_{n-1} \ldots f_1$, where $f_n, \ldots, f_1$ are free morphisms, we define $F(f) = \beta(f_n) \ldots \beta(f_1)$; it is clear that the morphisms $\beta(f_n), \ldots, \beta(f_1)$ are composable in $\mathscr{C}'$. The details are left for the reader.✳

We know from the theory of monoids (see Chevalley [1]) that every monoid is a factor monoid of a free one. Actually, a free monoid is a free category with one object; hence, for every monoid M there is a free monoid L and a full functor (i.e. a surjective morphism of monoids) $f: L \to M$. This suggests the following natural question: If $\mathscr{C}$ is a category, do there exist a free category $\mathscr{L}$ and a full and representative functor $F: \mathscr{L} \to \mathscr{C}$? The following example shows that in general the answer is "no".

Example 5.9. Let $\mathscr{C}$ be the category whose objects are X, Y and $\{X_i\}_{i \in \mathscr{I}}$, where $\mathscr{I}$ is a proper class (not a set) and the X_i's are all distinct. The morphisms are the identity morphisms, together with, for every $i \in \mathscr{I}$, morphisms g_i and f_i:

$$X \xrightarrow{g_i} X_i \xrightarrow{f_i} Y$$

such that $f_i g_i = f_j g_j$ for all $i, j \in \mathscr{I}$. Now assume that there is a free category $\mathscr{L}$ and a full representative functor $F: \mathscr{L} \to \mathscr{C}$. Let A, B and $\{A_i\}_{i \in \mathscr{I}}$ be objects of $\mathscr{L}$ such that $F(A) = X$. $F(B) = Y$ and $F(A_i) = X_i$ for all $i \in \mathscr{I}$. Also let $u_i: A \to A_i$ and $v_i: A_i \to B$ be morphisms such that $F(u_i) = g_i$ and $F(v_i) = f_i$ for every $i \in \mathscr{I}$. Since g_i and f_i are not isomorphisms, it is clear that v_i and u_i are not identity morphisms. It follows from the assumption of freeness that for $i, j \in \mathscr{I}$, $i \neq j$, the morphisms $v_i u_i$ and $v_j u_j$ of $\mathscr{L}(A, B)$ are distinct, i. e. $\mathscr{L}(A, B)$ is not a set.

Nevertheless, the answer is "yes" for small categories. Before showing this, we develop some additional terminology.

A *diagram scheme*, Σ, is a triple $(\mathscr{I}, \mathscr{M}, f)$, where $\mathscr{I}$ is a set whose elements are called *vertices*, $\mathscr{M}$ is a set whose elements are called *arrows* and f is a map from $\mathscr{M}$ to $\mathscr{I} \prod \mathscr{I}$. If $m \in \mathscr{M}$ and $f(m) = (i, j)$, then we call i the *domain* of m, denoted $i = d(m)$, and j the *codomain* of m, denoted $j = c(m)$. A *composite arrow* in Σ is a finite sequence $x = m_p m_{p-1} \ldots m_1$ of arrows such that $c(m_{k+1}) = c(m_k)$ for $1 \leqslant k \leqslant p-1$. The domain of m_1 is called the domain of x, written $d(x)$, and the codomain of m_p is called the codomain of x, written $c(x)$. If $x = m_p \ldots m_1$ and $y = m_s \ldots m_{p+1}$ are two composite arrows such that $d(y) = c(x)$, then we speak of the composite arrow $m_s \ldots m_{p+1} m_p \ldots m_1$, which is usually denoted by yx, and called the *formal composition of y with x*. Clearly, every composite arrow is a formal composition of arrows and every arrow can be viewed as a composite arrow. Now assume that for each $i \in \mathscr{I}$, the set $f^{-1}(i, i)$ contains a fixed element e_i called a *preunity*. An arrow m which is not a preunity is called a *veritable arrow*. If $x = m_p \ldots m_1$ is a composite arrow, the *skeleton* of y is defined to be the set of all veritable arrows which occur in the composition of x; if, for every k, $1 \leqslant \leqslant k \leqslant p$, m_k is a preunity, then the skeleton of x is empty. For every pair $i, j \in \mathscr{I}$, let $S(i,j)$ denote the set of all composite arrows x such that $d(x) = i$ and $c(x) = j$ Let $R(i,j)$ denote the equivalence relation on $S(i,j)$ defined by $x \sim y(R(i, j)) \Leftrightarrow x$ and y have the same skeleton. If $x, y \in S(i,j), x \sim y(R(i,j))$, then for every $z \in S(j, k)$ we have $zx \sim zy(R(i, k))$; also, for every $s \in R(h, i)$ we have $xs \sim ys(R(h, j))$. Hence these equivalence relations are compatible with the formal composition of composite arrows.

Now we define a free category which we shall denote by $\mathscr{L}'(\Sigma)$. The objects of $\mathscr{L}'(\Sigma)$ are the vertices of Σ. Precisely, if i is a vertex of Σ, the corresponding object of $\mathscr{L}'(\Sigma)$ is denoted by X_i. By definition, $\mathscr{L}'(\Sigma)(X_i, X_j) = S(i,j)/R(i,j)$. If $\alpha: X_i \to X_j$ and $\beta: X_j \to X_k$ are morphisms in $\mathscr{L}'(\Sigma)$, then there is an $x \in S(i,j)$ and a $y \in S(j, k)$ such that $\bar{x} = \alpha$, $\bar{y} = \beta$ (here $\bar{x}$ is the equivalence class of x in $S(i,j)/R(i,j)$). Thus $yx \in S(i, k)$ and, by definition, $\beta\alpha = \overline{yx}$. The reader may verify that the composition of morphisms in $\mathscr{L}'(\Sigma)$ is associative and that the morphisms $\bar{e}_i$ are the identities, i.e. $\bar{e}_i = 1_{X_i}$ for every $i \in \mathscr{I}$. It is easy to see that for every arrow m of Σ the morphism $\bar{m}$ of $\mathscr{L}'(\Sigma)$ is a free morphism and every morphism of $\mathscr{L}'(\Sigma)$ is a unique composition of free arrows, i.e. $\mathscr{L}'(\Sigma)$ is a free category.

A *diagram in a category $\mathscr{C}$ over the scheme* Σ is a function D which assigns to each vertex $i \in \mathscr{I}$ and object D_i of $\mathscr{C}$ and to each arrow m of Σ with domain i and codomain j a morphism $D(m): D_i \to D_j$ in $\mathscr{C}$. Now we are able to formulate the following result about diagram schemes and free categories.

THEOREM 5.10. *Let $\Sigma = (\mathscr{I}, \mathscr{M}, f)$ be a diagram scheme. Then a free category $\mathscr{L}(\Sigma)$ and a diagram $K: \Sigma \to \mathscr{L}(\Sigma)$ can be constructed such that for each diagram $D: \Sigma \to \mathscr{C}$, there exists a unique functor $\bar{D}: \mathscr{L}(\Sigma) \to \mathscr{C}$ such that $\bar{D}K = D$. The category $\mathscr{L}(\Sigma)$ is defined up to isomorphism of categories.*

Proof. Consider a new diagram scheme $\bar{\Sigma} = (\bar{\mathscr{I}}, \bar{\mathscr{M}}, \bar{f})$ such that: 1) $\mathscr{I} = \bar{\mathscr{I}}$, 2) for every $i, j \in \bar{\mathscr{I}}, i \neq j, \bar{f}^{-1}(i,j) = f^{-1}(i,j)$, 3) for each $i \in \mathscr{I}, \bar{f}^{-1}(i, i) = f^{-1}(i, i) \cup \{e_i\}$ (disjoint union), i.e. there is a new arrow e_i with domain and codomain i. Denote $\mathscr{L}'(\bar{\Sigma})$, the free category defined above, by $\mathscr{L}(\Sigma)$. Further, let K: $\Sigma \to \mathscr{L}(\Sigma)$ denote the following diagram: For each vertex i of $\bar{\Sigma}$, let $K(i)$ be X_i; if m is an arrow of $\bar{\Sigma}$ then let $K(m)$ be $\bar{m}$. Now let $D: \Sigma \to \mathscr{C}$ be a diagram.

By Proposition 5.8, the functor $\bar{D}: \mathscr{L}(\Sigma) \to \mathscr{C}$ is defined by $\bar{D}(X_i) = D_i$ and $\bar{D}(\bar{m}) = D(m)$ for every arrow m of Σ. In addition, $\bar{D}(e_i) = D(1_{X_i}) = 1_{D_i}$ for all $i \in \mathscr{I}$. It is clear that $\bar{D}$ is the unique functor such that $\bar{D}K = D$. The last part of the theorem is obtained in Exercise 5.8. ✳

We call $\mathscr{L}(\Sigma)$ the *free category generated by* Σ, K the *canonical diagram* (when some ambiguity could otherwise arise, we write K_Σ instead of K), and $\bar{D}$ the *functor induced by the diagram* D.

Example 5.11. Let X be a set. Then X can be viewed as a diagram scheme (also denoted by X) with one vertex i and the set of arrows equal to X.

The category $\mathscr{L}(X)$ defined by Theorem 5.10 is called the *free monoid generated by* X. In particular, if X contains only one element, i.e. $X = \{*\}$, then $\mathscr{L}(\{*\})$ is the free monoid $\mathbf{N}$ of the natural numbers.

Let $\mathscr{C}$ be a small category. Then $\mathscr{C}$ may be regarded as a diagram scheme whose set of vertices is $\mathscr{O}b\,\mathscr{C}$ and whose set of arrows is $\mathscr{M}or\,\mathscr{C}$. Since, for each "vertex" X of $\mathscr{C}$, there is a distinguished arrow $1_X: X \to X$, we can define a free category $\tilde{\mathscr{L}}(\mathscr{C})$, whose objects are the objects of $\mathscr{C}$. The identity morphisms of $\mathscr{C}$ become again identity morphisms in $\tilde{\mathscr{L}}(\mathscr{C})$. If X is an object of $\mathscr{C}$, then the same object regarded in $\tilde{\mathscr{L}}(\mathscr{C})$ will be denoted by $\tilde{X}$; also, if f is a morphism of $\mathscr{C}$, then $\tilde{f}$ will denote the same morphism regarded to be in $\tilde{\mathscr{L}}(\mathscr{C})$. Thus the assignments $X \rightsquigarrow \tilde{X}$, $f \rightsquigarrow \tilde{f}$ define a diagram $\tilde{K}: \mathscr{C} \to \tilde{\mathscr{L}}(\mathscr{C})$. We remark that this free category $\tilde{\mathscr{L}}(\mathscr{C})$ is always different from the category $\mathscr{L}(\mathscr{C})$ defined in Theorem 5.10 and the requirements of this theorem are not satisfied. Nevertheless, we do get the following theorem.

THEOREM 5.12. *For every small category $\mathscr{C}$, there exist a free category $\tilde{\mathscr{L}}(\mathscr{C})$ and a diagram $\tilde{K}: \mathscr{C} \to \tilde{\mathscr{L}}(\mathscr{C})$ such that for every functor $F: \mathscr{C} \to \mathscr{C}'$ there exists a unique functor $\tilde{F}: \tilde{\mathscr{L}}(\mathscr{C}) \to \mathscr{C}'$ with $\tilde{F}\tilde{K} = F$. The category $\tilde{\mathscr{L}}(\mathscr{C})$ is defined up to isomorphism of categories.* ✳

COROLLARY 5.13. *Let $\mathscr{C}$ be a small category. There exists a free category $\mathscr{L}$ and a full functor $T: \mathscr{L} \to \mathscr{C}$ such that T takes distinct objects to distinct objects. Furthermore, every object of $\mathscr{C}$ is of the form $T(X)$ with $X \in \mathscr{L}$ (i.e. T is bijective on objects).* ✳

The following result will be useful later.

PROPOSITION 5.14. *Let $\Sigma = (\mathscr{I}, \mathscr{M}, f)$ be a diagram scheme, and $K: \Sigma \to \mathscr{L}(\Sigma)$ be the canonical diagram, $D: \Sigma \to \mathscr{C}$ be a diagram, and let $F: \mathscr{L}(\Sigma) \to \mathscr{C}$ be a functor. Assume that for each vertex i of Σ a morphism $u_i: F(K_i) \to D_i$ is given such that for every arrow m with domain i and codomain j we have that $u_j F(K(m)) = D(m) u_i$. Then the morphisms $\{u_i\}_{i \in \mathscr{I}}$ define a functorial morphism $u: F \to \bar{D}$, where $\bar{D}$ is the functor associated with D.*

Proof. Let $g: K_i \to K_j$ be a morphism of $\mathscr{L}(\Sigma)$. We must prove that $u_j F(g) = \bar{D}(g) u_i$. By the construction of $\mathscr{L}(\Sigma)$ we have that $g = K(m_p) \ldots K(m_1)$ where $m_1, \ldots, m_p$ are arrows of Σ. The proof follows by induction with respect to p. ✳

We now reexamine equivalence relations on a category. Let $R(\mathscr{C})$ denote the class of all equivalence relations on a category $\mathscr{C}$. It is easy to see that $R(\mathscr{C})$ is not empty and need not be a set. Relative to the relation of fineness defined above $R(\mathscr{C})$ is an ordered class.

LEMMA 5.15. *Let $\mathscr{C}$ be a category. Then every class $\{R_i\}_{i\in\mathscr{I}}$ of elements of $R(\mathscr{C})$ has a greatest lower bound, which is denoted by $\cap_i R_i$ (i.e. $\cap_i R_i \leqslant R_i$ for all i and if $L \in R(\mathscr{C})$ and $L \leqslant R_i$ for all i then $L \leqslant \cap_i R_i$).*

Proof. For every ordered pair (X, Y) of objects of $\mathscr{C}$ denote $\cap_i R_i(X, Y)$ by $(\cap_i R_i)(X, Y)$. We observe that the first intersection is meaningful since $\mathscr{C}(X, Y)$ is a set. The remaining details are left for the reader. ⚹

COROLLARY 5.16. *Let $\mathscr{C}$ be a category. Assume that there is given a class $\{f_i, g_i\}_{i\in\mathscr{I}}$ of pairs of morphisms in $\mathscr{C}$ such that for each $i \in \mathscr{I}$, f_i and g_i have the same domain and the same codomain. Then there is a smallest equivalence relation R on $\mathscr{C}$ such that $f_i \sim g_i(R)$ for every $i \in \mathscr{I}$.*

The proof follows from Lemma 5.15 and is left for the reader.⚹

We note that the equivalence relation defined in Corollary 5.16 is called the *equivalence relation generated by the class of pairs* $\{f_i, g_i\}_{i\in\mathscr{I}}$.

Now let Σ be a diagram scheme. By a *commutativity relation* k on Σ we mean a set $\{f_i, g_i\}_{i\in\mathscr{I}}$ of pairs of morphisms in $\mathscr{L}(\Sigma)$ (or, equivalently, a set of composed arrows of Σ) such that f_i and g_i have the same domain and the same codomain for all $i \in \mathscr{I}$. A diagram $D\colon \Sigma \to \mathscr{C}$ is called *k-commutative* if for every $i \in \mathscr{I}$, $\bar{D}(f_i) = \bar{D}(g_i)$, where $\bar{D}\colon \mathscr{L}(\Sigma) \to \mathscr{C}$ is the functor induced by D.

PROPOSITION 5.17. *Let k be a commutativity relation on a diagram scheme Σ. A diagram $D\colon \Sigma \to \mathscr{C}$ is k-commutative if and only if $\bar{D}\colon \mathscr{L}(\Sigma) \to \mathscr{C}$ (the functor induced by D) may be factored as:*

$$\mathscr{L}(\Sigma) \xrightarrow{P} \mathscr{L}(\Sigma)/R \to \mathscr{C},$$

where R is the equivalence relation on $\mathscr{L}(\Sigma)$ generated by k, and P is the canonical functor.

The proof is left for the reader.⚹

We also note that the category $\mathscr{L}(\Sigma)/R$ is called the *category of commutativity* for k and is also denoted by $\mathscr{L}(\Sigma/k)$.

Exercises

5.1. Let $\mathscr{G}$ be a groupoid. A *normal subgroupoid* $\mathscr{H}$ of $\mathscr{G}$ is given if for each $X \in \mathscr{G}$ there is a subgroup $\mathscr{H}(X)$ of $\mathscr{G}(X, X)$ such that for each $Y \in \mathscr{G}$, for each $g \in \mathscr{G}(X, Y)$ and for each $f \in \mathscr{H}(X)$, we have that $gfg^{-1} \in \mathscr{H}(Y)$.

If $\mathcal{H}$ is a normal subgroupoid of $\mathcal{G}$, let $R(\mathcal{H})$ denote the equivalence relation on $\mathcal{G}$ defined as follows. For $f, g \in \mathcal{G}(X, Y)$, we define

$$f \sim g(R(X, Y)) \Leftrightarrow g^{-1}f \in \mathcal{H}(X).$$

Prove that, for every equivalence relation R on the category $\mathcal{G}$, there is a uniquely defined normal subgroupoid $\mathcal{H}$ of $\mathcal{G}$ such that $R(\mathcal{H}) = R$, and moreover, that the factor category $\mathcal{G}/\mathcal{H} = \mathcal{G}/R(\mathcal{H})$ is a canonical groupoid.

5.2. Let R be an equivalence relation on the category $\mathcal{C}$, $P: \mathcal{C} \to \mathcal{C}/R$ be the canonical functor and let $T: \mathcal{C} \to \mathcal{C}'$ be a functor. Prove that the following are equivalent:

1) T is compatible with R.

2) There is a unique functor $T': \mathcal{C}/R \to \mathcal{C}'$ such that $T'P = T$.

5.3. Let $\mathcal{C} \xrightarrow{F} \mathcal{C}' \xrightarrow{G} \mathcal{C}''$ be functors. Show that $R_{GF} \geq R_F$. Prove: If F is representative, then $R_{GF} = R_F$ if and only if G is faithful.

5.4. Let R be an equivalence relation on a category $\mathcal{C}$, and let $\mathcal{C}'$ be a subcategory of $\mathcal{C}$. Let $R/\mathcal{C}'$ denote the equivalence relation on $\mathcal{C}'$ defined by:

$$f \sim g(R/\mathcal{C}'(X, Y)) \Leftrightarrow f \sim g(R(X, Y))$$

for every $X, Y \in \mathcal{C}'$ and for each pair $f, g \in \mathcal{C}'(X, Y)$. ($R/\mathcal{C}'$ is called the *restriction* of R to $\mathcal{C}'$.) Show that we get the following commutative diagram of categories and functors:

$$
\begin{array}{ccc}
\mathcal{C} & \xrightarrow{\ \ P\ \ } & \mathcal{C}/R \\
{\scriptstyle I'}\big\uparrow & & \big\uparrow{\scriptstyle I'} \\
\mathcal{C}' & \xrightarrow{\ \ P'\ \ } & \mathcal{C}'/(R/\mathcal{C}')
\end{array}
$$

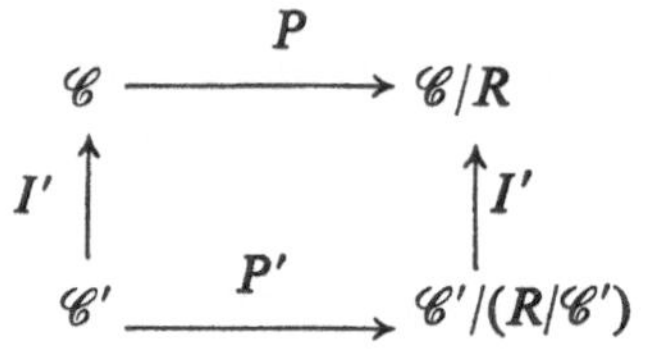

where I is the inclusion functor and I' is canonically constructed, and moreover, that I' is embedding.

5.5. Let R be an equivalence relation on $\mathcal{C}$ and let $\mathcal{C}_1$ be a subcategory of $\mathcal{C}/R$. Prove that there is a subcategory $\mathcal{C}'$ of $\mathcal{C}$ and that the following canonically constructed diagram of categories and functors

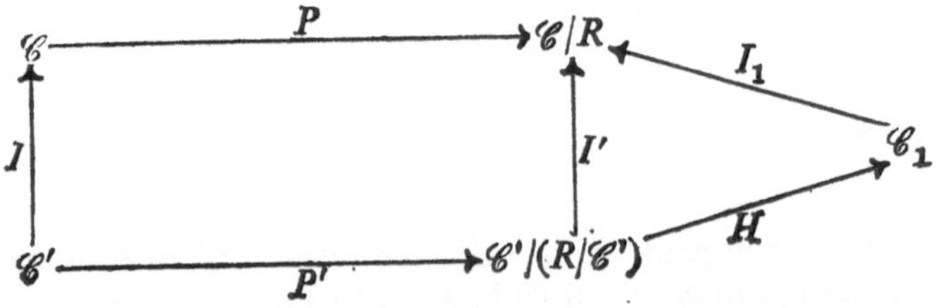

is commutative, where H is an isomorphism of categories.

5.6. Prove that the following are equivalent for a category $\mathscr{C}$:
a) Every identity morphism of $\mathscr{C}$ is free.
b) Every retraction of $\mathscr{C}$ is an identity.
c) Every coretraction of $\mathscr{C}$ is an identity.

5.7. Let $\mathscr{C}$ be a free category, and let $F, G: \mathscr{C} \to \mathscr{C}'$ be any two functors. Assume that for each object X of $\mathscr{C}$ there is a morphism $u_X: F(X) \to G(X)$ of $\mathscr{C}'$ such that for every free morphism $f: X \to Y$ of $\mathscr{C}$, we have that $u_Y F(f) = G(f) u_X$. Prove that the morphisms $\{u_X\}_{X \in \mathscr{C}}$ define a functorial morphism $u: F \to G$.

5.8. Prove that two isomorphic objects of a free category are identical, and that two free categories are equivalent if and only if they are isomorphic.

5.9. Let $\mathscr{C}$ be a category with the following property. For each morphism $f: X \to Y$ in $\mathscr{C}$ there is a set of distinct commutative diagrams:

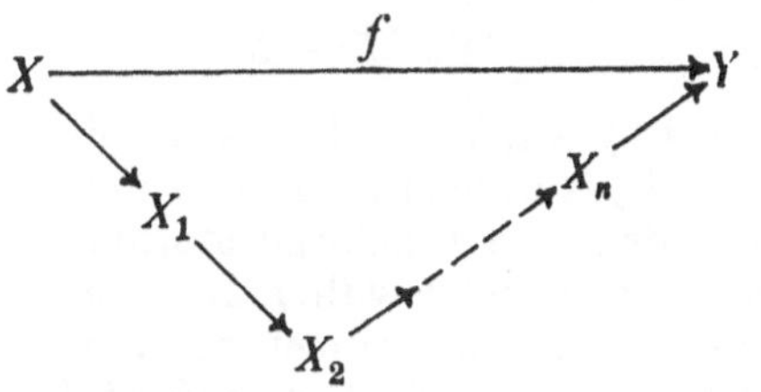

.

Show that there is a free category $\mathscr{L}$ and a full representative functor $F: \mathscr{L} \to \mathscr{C}$.

5.10. Let Σ be the diagram scheme:

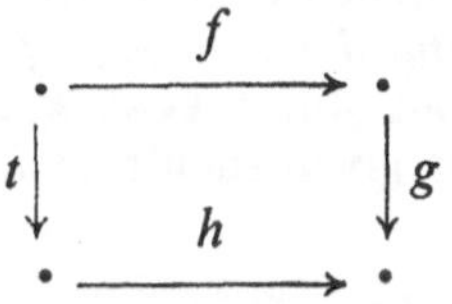

.

a) Give a description of the category $\mathscr{L}(\Sigma)$.

b) Let k be the commutativity relation on Σ given by (gf, ht). Give a description of the category of commutativity for k.

1.6. Limits and colimits

The concepts of limit and colimit play an important role in the theory of categories. These concepts are defined in this section and some general results are obtained.

References: Ekman—Hilton [1], Grothendieck [1], Grothendieck—Verdier [1], MacLane [3], Maranda [1], Mitchell [2], Pareigis [1], Schubert [1].

An object X of a category $\mathscr{C}$ is called *final* (or *terminal*, or *coinitial*) if for each object Y of $\mathscr{C}$ the set $\mathscr{C}(Y, X)$ contains exactly one element. Duality gives us the corresponding concept of an *initial* object.

PROPOSITION 6.1. *Any two final (initial) objects of a category $\mathscr{C}$ are isomorphic. In fact, there is a unique isomorphism between them.* ✳

Example 6.2. In the category $\mathscr{S}et$, the empty set is an initial object and any set that contains exactly one element is a final object.

Example 6.3. In the category $\mathscr{S}^+$, [0] is both a final object and an initial object. However, [0] is only a final object in $\mathscr{S}$.

Note 6.4. For simplicity of notation, the objects of some particular categories will be denoted by lower case italic Latin letters.

Let $\mathscr{I}$ and $\mathscr{C}$ be categories; for each $X \in \mathscr{C}$ let

$$X_{\mathscr{I}} : \mathscr{I} \to \mathscr{C}$$

denote the functor defined by $X_{\mathscr{I}}(i) = X$ for each object i of $\mathscr{I}$ and $X_{\mathscr{I}}(u) = 1_X$ for every morphism u in $\mathscr{I}$. $X_{\mathscr{I}}$ is called the *constant functor* associated with X. Every morphism $f: X \to Y$ in $\mathscr{C}$ defines a functorial morphism $f_{\mathscr{I}}: X_{\mathscr{I}} \to Y_{\mathscr{I}}$, called the *constant functorial morphism* associated with f. It is clear that $X_{\mathscr{I}}$ can be regarded as either a functor or as a cofunctor. Sometimes if no confusion can occur, we shall simply write X instead of $X_{\mathscr{I}}$ and f instead of $f_{\mathscr{I}}$.

Now let $F: \mathscr{I} \to \mathscr{C}$ be a functor. Let $\mathscr{C}/F$ denote the category whose objects are pairs (X, u), where $X \in \mathscr{C}$ and $u: X_{\mathscr{I}} \to F$ is a functorial morphism, and a morphism in $\mathscr{C}/F$ from (X, u) to (X', u') is a morphism $f \in \mathscr{C}(X, X')$ such that $u'f_{\mathscr{I}} = u$.

An object (X, u) of $\mathscr{C}/F$ is called a *cone over* F, X is called the *vertex* of the cone, and u is called the *structural morphism of the cone*. It is easy to see that a cone over F is determined if an object X of $\mathscr{C}$ is specified and if for each object $i \in \mathscr{I}$ a morphism $u_i: X \to F(i)$ is given such that for every morphism $t: i \to j$ of $\mathscr{I}$, the diagram

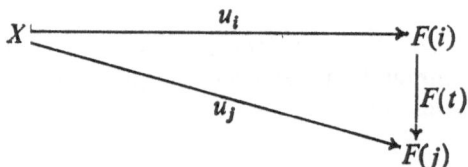

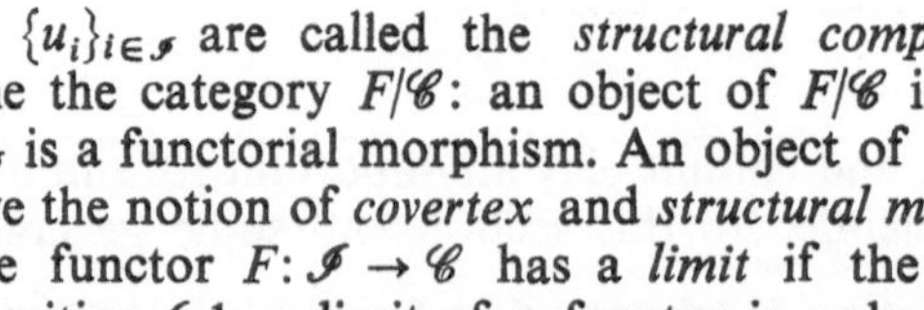

is commutative.

The morphisms $\{u_i\}_{i \in \mathscr{I}}$ are called the *structural components* of the cone.

Dually, we define the category $F/\mathscr{C}$: an object of $F/\mathscr{C}$ is a pair (u, X) where $X \in \mathscr{C}$ and $u: F \to X_{\mathscr{I}}$ is a functorial morphism. An object of $F/\mathscr{C}$ is called a *cocone* over F. Thus we have the notion of *covertex* and *structural morphism of the cocone.*

We say that the functor $F: \mathscr{I} \to \mathscr{C}$ has a *limit* if the category $\mathscr{C}/F$ has a final object. By Proposition 6.1, a limit of a functor is only unique up to isomorphism. Thus a limit of the functor F is a cone (X, u) over F such that for every cone (Y, v) over F there exists a unique morphism $f: Y \to X$ of $\mathscr{C}$ with $uf = v$. Frequently the vertex of a limit of F is denoted by $\varprojlim F$ and the structural components of a limit are called *structural components of the limit*, or the *structural projections of the limit*. In general, a limit of a functor F (if it exists!) is denoted simply by $\varprojlim F$, the structural components of the limits being implied.

Dually, a functor $F: \mathscr{I} \to \mathscr{C}$ has a *colimit* if the category $F/\mathscr{C}$ has an initial object; the colimit of a functor is unique up to isomorphism and its denoted by $\varinjlim F$. Analogously, we have the concepts of the *structural components* of the colimit, which are also called the *structural injections* of the colimit.

Note 6.5. Simple examples, even in the category $\mathscr{S}et$, show that in general the projections of a limit (the injections of a colimit) need not be epimorphisms (monomorphisms). [See Exercise 6.9.]

PROPOSITION 6.6. *Let i be an initial object of a category $\mathscr{I}$. Then every functor $F: \mathscr{I} \to \mathscr{C}$ has $F(i)$ as a limit. Moreover, if for each $j \in \mathscr{I}$, $t_j: i \to j$ is the unique morphism, then $\{F(t_j)\}_{j \in \mathscr{I}}$ are structural projections.* ✱

Example 6.7. Let $\mathscr{I}$ be a set (i.e. a small discrete category) and let $F: \mathscr{I} \to \mathscr{S}et$ be a functor (i.e. a family $\{F(i)\}_{i \in \mathscr{I}}$ of sets). Then the cartesian product $\prod_{i \in \mathscr{I}} F(i)$ is a limit of F and the structural projections of the limit are the natural projections of the cartesian product.

Example 6.8. The identity functor of a category has a limit (colimit) if and only if the category has an initial (a final) object.

Let $F: \mathscr{I} \to \mathscr{C}$ be a functor. For each $X \in \mathscr{C}$, we consider the class $[X, F]$ of all functorial morphisms from $X_{\mathscr{I}}$ into F. If $f: X \to Y$ is a morphism in $\mathscr{C}$, let $[f, F]: [Y, F] \to [X, F]$ denote the map obtained by composition with f. Thus the assignments

$$X \rightsquigarrow [X, F],$$

$$f \rightsquigarrow [f, F]$$

look very much like a cofunctor, usually called a *quasi-functor* and denoted by $[?, F]$. It is clear that $[?, F]$ is a functor whenever $\mathscr{I}$ is small, although $[?, F]$ can be a functor even when $\mathscr{I}$ is not small. An important example is given by the following result.

PROPOSITION 6.9. *Let $F: \mathscr{I} \to \mathscr{C}$ be a functor. The following are equivalent:*

1) *F has a limit.*

2) *$[?, F]$ is a representable cofunctor.*

Proof. 1) $\Rightarrow$ 2). It is easy to see that if (X, u) is a limit of F, then there is a functorial isomorphism:

$$v_Y: [Y, F] \xrightarrow{\sim} \mathscr{C}(Y, X) = H_X(Y)$$

defined by $v_Y(t) = \bar{t}$, where $\bar{t}: Y \to X$ is the unique morphism such that $u\bar{t} = t$.

2) $\Rightarrow$ 1). Let $v: [?, F] \to H_X$ be a functorial isomorphism and let (Y, t) be a cone over F. Then $v_Y(t): Y \to X$ is the unique morphism of $\mathscr{C}$ such that $v_X^{-1}(1_X)v_Y(t) = t$, i.e. $(X, v_X^{-1}(1_X))$ is a limit of F. ✳

The following result is a theoretical (i.e. non-constructive) criterion for determining when a functor has a limit.

THEOREM 6.10. *Let* $F: \mathscr{I} \to \mathscr{C}$ *be a functor and* (X, u) *a cone over* F. *The following are equivalent:*

1) (X, u) *is a limit of* F.

2) *For each object* Y *of* $\mathscr{C}$, $(H^Y(X), H^Y(u))$ *is a limit of* H^YF.

Proof. 1) $\Rightarrow$ 2). Let (M, v) be a cone over H^YF. Thus, for each object i of $\mathscr{I}$, $v_i: M \to H^YF(i) = H^Y(F(i)) = \mathscr{C}(Y, F(i))$ is a map such that if $t: i \to j$ is a morphism in $\mathscr{I}$, then the diagram

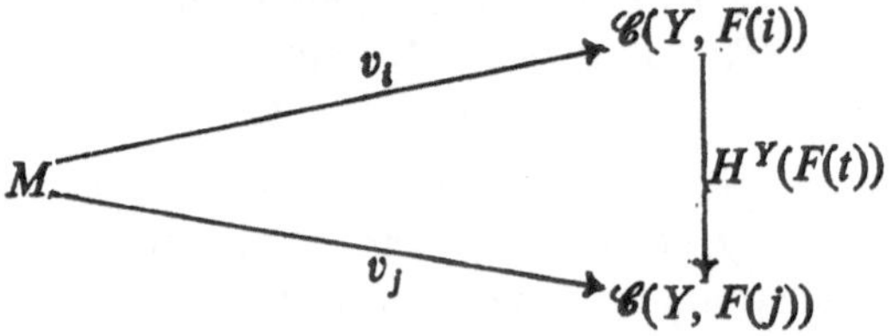

s commutative.

Now, if $x \in M$, $v_i(x) \in \mathscr{C}(Y, F(i))$, and $H^Y(F(t))(v_i(x)) = F(t)v_i(x) = v_j(x)$, then the morphisms $\{v_i(x)\}_{i \in \mathscr{I}}$ give a morphism $v_x: Y \to F$. Thus, by hypothesis, there exists a unique morphism $\bar{v}_x: Y \to X$ such that $u\bar{v}_x = v_x$. It is easy to check that the map

$$\bar{v}: M \to H^Y(X)$$

defined by $\bar{v}(x) = \bar{v}_x$ is the unique map which satisfies $(H^Yu)\bar{v} = v$.

2) $\Rightarrow$ 1). Let (Y, v) be a cone over F, M a set with a single element x and $\bar{v}: M \to H^YF$ the cone defined by

$$\bar{v}_i(x) = v_i,$$

where $v_i: Y \to F(i)$, $i \in \mathscr{I}$, are the structural components of the cone under consideration. It is clear that for every morphism $t: i \to j$ of $\mathscr{I}$,

$$H^Y(F(t))(\bar{v}_i(x)) = F(t)v_i = v_j = \bar{v}_j(x).$$

Thus the maps $\{\bar{v}_i\}_{i \in \mathscr{I}}$ determine a morphism $\bar{v}: M \to H^YF$. Hence, by hypothesis, there is a unique map $r: M \to H^Y(X)$ such that $(H^Yu)r = \bar{v}$. Finally, the morphism $r(x): Y \to X$ is the unique morphism of $\mathscr{C}$ such that $ur(x) = \bar{v}$; i.e., (X, u) is a limit of F. ✳

The above result can be dualized as follows.

THEOREM 6.11. *Let $F: \mathscr{I} \to \mathscr{C}$ be a functor and (u, X) a cocone over F. The following are equivalent:*

1) (u, X) is a colimit of F.

2) For each object Y of $\mathscr{C}$, $(H_Y(X), H_Y(u))$ is a limit of $H_Y F$.

The proof follows from Theorem 6.10 by duality. ✳

A subcategory $\mathscr{I}'$ of a category $\mathscr{I}$ is called an *initial* subcategory if for each object i of $\mathscr{I}$ there is a morphism $s: i' \to i$ where i' is an object of $\mathscr{I}'$ and every diagram

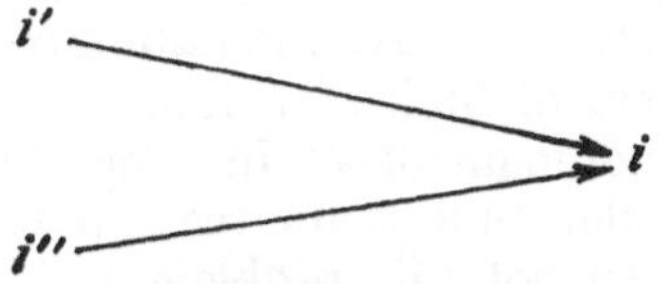

with i' and i'' objects of $\mathscr{I}'$ can be embedded in a commutative diagram

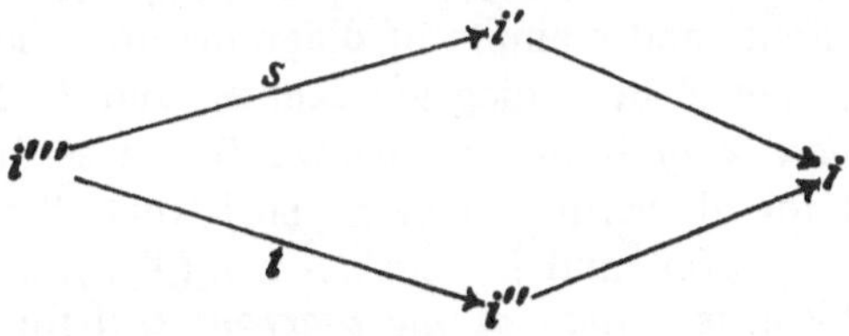

with s, t (and hence i''') in $\mathscr{I}'$. Dually, we have the notion of a *final* subcategory

THEOREM 6.12. *Let $F: \mathscr{I} \to \mathscr{C}$ be a functor, $\mathscr{I}'$ an initial subcategory of $\mathscr{I}$, and $H: \mathscr{I}' \to \mathscr{I}$ the canonical inclusion. Then the functor F has a limit if and only if the functor FH has one. Moreover, there is a canonical isomorphism:*

$$\varprojlim (FH) \xrightarrow{\sim} \varprojlim F.$$

Proof. Assume that (X, u) is a limit of F. Let (X, u') denote the cone over FH whose structural components $u'_{i'}$ are equal to $u_{i'}$ for each object i' of $\mathscr{I}'$. Let (Y, v) be another cone over FH. We define a cone $(Y, \hat{v})$ over F as follows. If i is an object of $\mathscr{I}$, then $\hat{v}_i: Y \to F(i)$ will be defined as the composition $F(s) v_{i'}$ where $s: i' \to i$ is a morphism in $\mathscr{I}$, with $i' \in \mathscr{I}'$. Now it must be checked that $\hat{v}$ does not depend on the morphism s. For this, let $s': i'' \to i$ be another morphism in $\mathscr{I}$, with $i'' \in \mathscr{I}'$. By hypothesis we have the commutative diagram of $\mathscr{I}$:

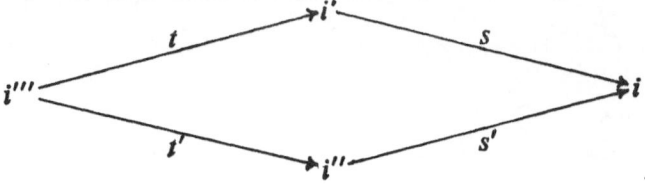

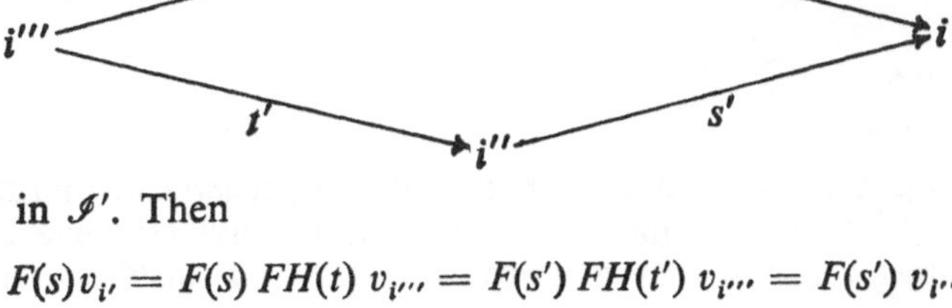

with t and t' in $\mathscr{I}'$. Then

$$F(s) v_{i'} = F(s)\, FH(t)\, v_{i'''} = F(s')\, FH(t')\, v_{i'''} = F(s')\, v_{i''}$$

since $FH(t) = F(t)$ for every morphism of $\mathscr{I}'$, where H is the inclusion functor. In particular, for every object i' of $\mathscr{I}$, we have $\hat{v}_{i'} = v_{i'}$, or equivalently, $\hat{v}$ is the unique "extension" of v to F. Now we can say that there is a unique morphism $f\colon Y \to X$ such that $uf = \hat{v}$ and necessarily $u'f = \hat{v}$. Hence (X, u') is a limit of FH.

Conversely, assume that FH has a limit (X, u). As above we may extend u to a morphism $\hat{u}\colon X \to F$, and it is easy to check that $(X, \hat{u})$ is a limit of F. The details are left for the reader. ✳

We say that a category $\mathscr{C}$ has *$\mathscr{I}$-limits*, where $\mathscr{I}$ is a category, if every functor $F\colon \mathscr{I} \to \mathscr{C}$ has a limit. Dually, we have the notion of a category with *$\mathscr{I}$-colimits.*

By definition, the concept of limit of a functor F involves two essential parameters: the domain and codomain of F. In order to obtain some results about limits, we will make particular choices for these parameters. In the next sections we shall present some aspects of this problem.

Note 6.13. In some works on the theory of categories, instead of limits and colimits of functors, limits and colimits of diagrams are considered. This is done in the following way. Let Σ be a diagram scheme and $D\colon \Sigma \to \mathscr{C}$ a diagram. A *morphism from an object X of $\mathscr{C}$ to the diagram D* is a set $\{u_i\colon X \to D_i\}_i$ of morphisms in $\mathscr{C}$, defined for all vertices i of Σ, such that for every arrow m of Σ, $D(m)u_i = u_j$ where $i = d(m)$ and $j = c(m)$. A pair (X, u), where $X \in \mathscr{C}$ and where $u\colon X \to D$ is a morphism, is a *limit of the diagram D* if for each $Y \in \mathscr{C}$ and each morphism $v\colon Y \to D$ there is a unique morphism $f\colon Y \to X$ of $\mathscr{C}$ such that $uf = v$. The following result shows that the notion of the limit of a diagram is equivalent to the notion of the limit of a suitable functor.

Proposition 6.14. *Let Σ be a diagram scheme, $D\colon \Sigma \to \mathscr{C}$ a diagram, $\mathscr{L}(\Sigma)$ the free category generated by Σ and $K\colon \Sigma \to \mathscr{L}(\Sigma)$ the canonical diagram. The diagram D has a limit if and only if the functor $\bar{D}$ induced by the diagram D has a limit and these limits are equal.*

Proof. Let (X, u) be a limit of D. By Proposition 5.14, the morphisms $u_i\colon X \to D_i$ define a functorial morphism (also denoted by u) from the constant functor $X_{\mathscr{L}(\Sigma)}$ to $\bar{D}$. If $v\colon Y \to \bar{D}$ is a functorial morphism, where $Y \in \mathscr{C}$, then the morphisms $v_i\colon Y \to D_i = D(K_i)$ define a morphism v from Y into the diagram D. Thus there is a unique morphism $f\colon Y \to X$ of $\mathscr{C}$ such that $u_i f = v_i$ for all vertices i of Σ.

We leave it for the reader to check this claim: if (X, u) is a limit of $\bar{D}$, then this pair is also a limit of the diagram D. The dual result may also be verified. ✳

Exercises

6.1. An object X of a category $\mathscr{C}$ is called a *pseudofinal* object if for each Y of $\mathscr{C}$ the set $\mathscr{C}(Y, X)$ has at most one element. Dually, we have the notion of a *pseudoinitial* object. Prove that a morphism $f\colon X \to Y$ in $\mathscr{C}$ is a monomorphism (an epimorphism) if and only if (X,f) $((f, Y))$ is a pseudofinal (pseudoinitial) object of $\mathscr{C}/Y(X/\mathscr{C})$.

6.2. Prove that the following are equivalent for a category $\mathscr{C}$:

1) $\mathscr{C}$ is a preordered category.
2) Every object of $\mathscr{C}$ is pseudoinitial.
3) Every object of $\mathscr{C}$ is pseudofinal.

6.3. An object 0 of a category $\mathscr{C}$ is called a *zero* object if 0 is both a final and an initial object. Then for every object X there are morphisms $0_{0X}: 0 \to X$ and $0_{X0}: X \to 0$. If Y is another object of $\mathscr{C}$, then the morphism $X \xrightarrow{0_{X0}} 0 \xrightarrow{0_{XY}} Y$ is denoted by 0_{XY} or simply by 0 if no confusion can occur and is called the *zero* morphism. Prove that:

a) For every morphism $f: Z \to X$, we have that $0_{XY}f = 0_{ZY}$.
b) For every morphism $g: Y \to T$, we have that $g0_{XY} = 0_{XT}$.

Prove that for any category $\mathscr{C}$ there is an embedding $H: \mathscr{C} \to \mathscr{C}_0$, such that $\mathscr{C}_0$ has a zero object, and for every functor $T: \mathscr{C} \to \mathscr{C}'$, where $\mathscr{C}'$ is a category with zero object, there is a unique functor $T_0: \mathscr{C}_0 \to \mathscr{C}'$ with $T_0H = T$.

6.4. A category $\mathscr{I}$ is called a *connected* category if for any two objects i, j of $\mathscr{I}$ there are objects $i_1 = i, i_2, \ldots, i_n = j$ and a diagram

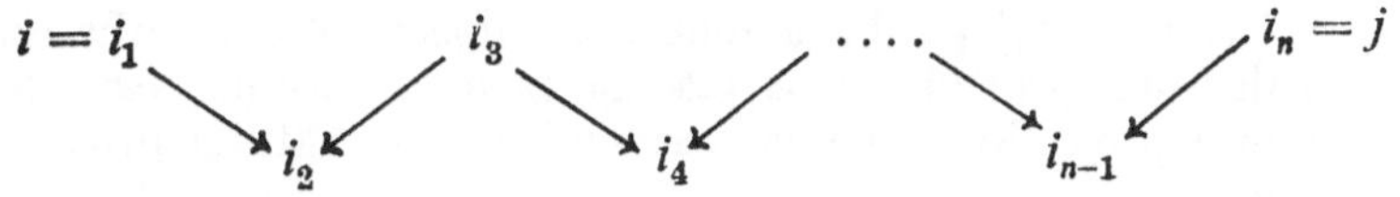

If X and Y are objects of a category $\mathscr{C}$, then for any category $\mathscr{I}$ the assignment $f \rightsquigarrow f_{\mathscr{I}}$ defines an injection from $\mathscr{C}(X, Y)$ into $[X_{\mathscr{I}}, Y_{\mathscr{I}}]$. Assume that $\mathscr{C}(X, Y)$ contains at least two elements. Prove that this map is a bijection if and only if $\mathscr{I}$ is a connected category and, moreover, that $[X_{\mathscr{I}}, Y_{\mathscr{I}}]$ is a set if and only if $\mathscr{I}$ can be written as a coproduct of a set of connected categories.

6.5. Prove that if a category $\mathscr{C}$ has an initial object, then for every functor $F: \mathscr{I} \to \mathscr{C}$, the category $\mathscr{C}/F$ also has an initial object.

6.6. Let $F: \mathscr{I} \to \mathscr{S}et$ be a functor. Show that a cone (M, u) over F is a limit of F if and only if for every cone (P, v), where P is a set with a single element, there is a unique map $f: P \to M$ such that $uf = v$. Can this result be dualized?

6.7. Prove that the identity functor of a category $\mathscr{C}$ has a limit (colimit) if and only if $\mathscr{C}$ has an initial (a final) object.

Thus give an example of a functor which has neither a limit nor a colimit.

6.8. Let $\mathscr{C}$ be a category with a final object. Prove that for every small category $\mathscr{I}$, the category $[\mathscr{I}, \mathscr{C}]$ has a final object.

6.9. Give an example of a limit whose projections are not epimorphisms and an example of a colimit whose injections are not monomorphisms.

1.7. Products and coproducts

The concepts of product and coproduct probably were the first examples of limits and colimits. These concepts are important in the study of limits and colimits in categories.

References: Bourbaki [1], Cartan—Eilenberg [1], Eilenberg—Steenrod [1], Eckman—Hilton [1], Grothendieck [1], MacLane [3], Mitchell [2], Pareigis [1], Samuel [1], Calenko [2].

Let $\mathscr{I}$ be a discrete category. A functor $F: \mathscr{I} \to \mathscr{C}$ is completely determined by a family $\{X_i\}_{i \in \mathscr{I}}$ of objects of $\mathscr{C}$. The limit of this functor is called a *product* and is denoted by $\prod_{i \in \mathscr{I}} X_i$ or simply by $\prod X_i$ when there is no possibility of confusion concerning the index set. Dually, the colimit of this functor is called a *coproduct* and is denoted by $\coprod_{i \in \mathscr{I}} X_i$ or simply by $\coprod X_i$. Usually we shall speak about the product (coproduct) of the family $\{X_i\}_{i \in \mathscr{I}}$ of objects of $\mathscr{C}$.

Let $\mathscr{C}$ be an arbitrary category, X an object of $\mathscr{C}$ and $\mathscr{I}$ a set. We let $X^{\mathscr{I}}$ denote the product (if it exists!) of the family $\{X_i\}_{i \in \mathscr{I}}$ of objects of $\mathscr{C}$ such that $X_i = X$ for all $i \in \mathscr{I}$. Dually, we denote the coproduct of the same family of objects by $X^{(\mathscr{I})}$.

Example 7.1. Let $\{X_i\}_{i \in \mathscr{I}}$ be a family of objects of $\mathscr{S}et$ where $\mathscr{I}$ is a set. A product of this family of objects is just the cartesian product (see Bourbaki [1]) and the structural projections in this case are just the natural projections of the cartesian product.

Example 7.2. Let $\{X_i\}_{i \in \mathscr{I}}$ be a family of objects of $\mathscr{S}et$ where $\mathscr{I}$ is a set. A coproduct is the set whose objects are all the pairs (i, x) where $i \in \mathscr{I}$ and $x \in X_i$ (two pairs (i, x) and (i', x') being equal if and only if $i = i'$ and $x = x'$). In this case the structural injections $X_i \xrightarrow{u_i} \coprod_{i \in \mathscr{I}} X_i$ are given by $u_i(x) = (i, x)$. Hence the coproduct of a family of sets is just the *disjoint union*.

Example 7.3. Let $\{X_i\}_{i \in \mathscr{I}}$ be a family of topological spaces. A product in $\mathscr{T}op$ of this family of objects is the cartesian product $\prod X_i$ of the underlying sets endowed with the product topology (see Bourbaki [3]). Recall that an open set of $\prod X_i$ is a union of finite intersections of subsets of the form $\prod U_i$, where U_i is an open set of X_i for all i and $U_i = X_i$ for all but a finite number of indices i. A coproduct of this family of topological spaces is the disjoint union $\coprod X_i$ of the underlying sets endowed with the union topology (see Bourbaki [3]). Recall that a subset U of $\coprod X_i$ is open if and only if $u_i^{-1}(U)$ is open for all $i \in \mathscr{I}$, where the $u_i: X_i \to \coprod X_i$ are the structural injections.

If each family $\{X_i\}_{i \in \mathscr{I}}$ of objects of $\mathscr{C}$, where $\mathscr{I}$ is a set, has a product, then we call $\mathscr{C}$ a *category with products*. Dually, we have the notion of a *category with coproducts*. It should be observed that the restriction "where $\mathscr{I}$ is a set" is important. Indeed, in Example 7.1 we saw that $\mathscr{S}et$ is a category with products; on the other hand, we have the following result.

PROPOSITION 7.4. *Let $\mathscr{C}$ be a category such that every family $\{X_i\}_{i\in\mathscr{I}}$ of objects of $\mathscr{C}$ ($\mathscr{I}$ is not necessarily a set) has a product. Then $\mathscr{C}$ is a preordered category.*

Proof. Assume to the contrary that there are two distinct morphisms $f, g: X \to Y$ in $\mathscr{C}$. Let $\{Y_i\}_{i\in\mathscr{I}}$ be a family of objects of $\mathscr{C}$ where $Y_i = Y$ for all i, and $\mathscr{I}$ is a proper class. Let Z be a product of the family $\{Y_i\}_{i\in\mathscr{I}}$ and let $p_i: Z \to Y_i$ be the structural projections. For each $i \in \mathscr{I}$, let $f_i: X \to Z$ denote the unique morphism such that $p_i f_i = f$ and $p_j f_i = g$ if $j \neq i$. It is clear that $f_i \neq f_j$ whenever $i \neq j$; in other words, $\mathscr{C}(X, Z)$ is a proper class. That is a contradiction (see the definition of a category in Section 1.1.). Therefore, for each pair (X, Y) of objects of $\mathscr{C}$ the set $\mathscr{C}(X, Y)$ has at most one element. ✳

Note 7.5. In the arrow category, every family of objects (not necessarily indexed by a set) has a product. It may also be proven that in the preordered category of all ordinal numbers every family of objects has a product.

Proposition 7.4 has the following variant if $\mathscr{C}$ is a small category.

PROPOSITION 7.6. *If $\mathscr{C}$ is a small category with products then $\mathscr{C}$ is a preordered category.*

Proof. Choose a set $\mathscr{I}$ whose cardinality is strictly greater than the cardinality of each of the sets $\mathscr{C}(X, Y)$ for all pairs (X, Y) of objects of $\mathscr{C}$. As in the proof of Proposition 7.4, if there are distinct morphisms $f, g: X \to Y$ in $\mathscr{C}$, then the cardinality of the set $\mathscr{C}(X, Y^{\mathscr{I}})$ is at least equal to the cardinality of $\mathscr{I}$, which is a contradiction. Thus again $\mathscr{C}$ is a preordered category. ✳

We say that a category $\mathscr{C}$ has *finite products* if every finite family of objects of $\mathscr{C}$ has a product. Dually, we have the notion of a category with *finite coproducts.*

Example 7.7. The *category of finite sets*, denoted by $\mathscr{S}et_f$, whose objects are finite sets (see Bourbaki [1]) and whose morphisms are the maps, is a category with finite products and finite coproducts (the cartesian product is a product and the disjoint union is a coproduct). However, this category has neither products nor coproducts.

PROPOSITION 7.8. *Category $\mathscr{C}$ has finite products if and only if for every pair (X, Y) of objects of $\mathscr{C}$ the product $X \coprod Y$ exists.*

The proof is by induction on the number of objects and is left for the reader. ✳

A category $\mathscr{I}$ is called a *filtered* category if the following conditions are satisfied:

i) For every pair of morphisms s and t in $\mathscr{I}(i,j)$ there is a morphism $r: j \to k$ such that $rs = rt$.

ii) For any two objects i and j, there is an object k of $\mathscr{I}$ and morphisms

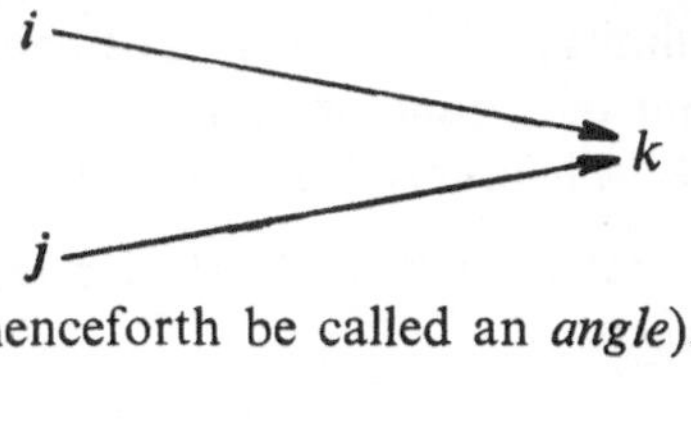

(such a configuration will henceforth be called an *angle*).

A functor (cofunctor) $F: \mathscr{I} \to C$ where $\mathscr{I}$ is a small filtered category is called a *filtered (cofiltered) system* in $\mathscr{C}$. By a *monofiltered (epifiltered) system* we mean a filtered (cofiltered) system such that $F(s)$ is a monomorphism (epimorphism) for all morphisms s in $\mathscr{I}$. We shall say that $\mathscr{C}$ is a *category with filtered colimits (cofiltered limits)* if every filtered (cofiltered) system in $\mathscr{C}$ has a colimit (limit).

By a *directed* set we mean a preordered set $\mathscr{M}$ which, considered as a category, is filtered; i.e. for every pair $i, j \in \mathscr{M}$, there is a $k \in \mathscr{M}$ such that $i \leqslant k$ and $j \leqslant k$. Sometimes a filtered (cofiltered) system $F: \mathscr{M} \to \mathscr{C}$, where $\mathscr{M}$ is a directed set, is called a *direct* (*inverse*) *system*. If $\mathscr{M}$ is a directed set, then a direct (inverse) system in $\mathscr{C}$ is given, if for each $i \in \mathscr{M}$ there are an object X_i of $\mathscr{C}$ and a morphism $u_{ij}: X_i \to X_j$ ($u_{ij}: X_j \to X_i$) whenever $i \leqslant j$, such that the system $\{X_i, u_{ij}\}_{i \in \mathscr{M}}$ meets the following requirements:

$$u_{ii} = 1_{X_i},$$

$$u_{jk}u_{ij} = u_{ik} \text{ (resp. } u_{ij}u_{jk} = u_{ik}\text{) whenever } i \leqslant j \leqslant k.$$

The colimit (limit) of the direct (inverse) system $\{X_i, u_{ij}\}_i$ is denoted by $\varinjlim_{i \in \mathscr{M}} X_i$ ($\varprojlim_{i \in \mathscr{M}} X_i$). We remark that the notions directed set, direct system, and inverse system are also examples of filtered and cofiltered systems. These notions play an important role in mathematics.

Example 7.9. Let $F: \mathscr{I} \to \mathscr{S}et$ be a filtered system. Consider the set $X = \coprod_i F_i$ and denote by R the equivalence relation on X defined by $(i, x) \sim (i', x')$ $(R) \Leftrightarrow$ there exist morphisms $s: i \to j$ and $s': i' \to j$ such that $F(s)(x) = F(s)(x')$. Let $p: X \to X/R$ be the canonical map and let $v_i = pu_i$, where the $u_i: X \to \coprod_i X_i$ are the structural injections of the coproduct. Now it is easy to see that the maps v_i give a morphism $v: D \to X/R$, and the pair $(v, X/R)$ is a colimit of D. Hence $\mathscr{S}et$ is a category with filtered limits. The same construction can be used to see that the categories $\mathscr{T}op$, $\mathscr{R}el$, $\mathscr{P}re$, $\mathscr{O}rd$, $\mathscr{M}on$, $\mathscr{G}r$, $\mathscr{A}b$ and $\mathscr{R}g$ are categories with filtered limits.

Example 7.10. $\mathscr{S}et$ is a category with inverse limits. Indeed, let $F: \mathscr{I} \to \mathscr{S}et$ be a cofiltered system in $\mathscr{S}et$. Consider the cartesian product $\prod_i X_i$ and its subset X' that consists of all families $\{x_i\}_{i \in \mathscr{I}}$ such that $F(s)(x_j) = x_i$ for each $i, j \in \mathscr{I}$ and morphism $s: i \to j$. Let q_i be the restriction of the structural projection p_i of the product to the set X'. We leave it for the reader to verify that the maps q_i define a cone (X', q) over F and to verify that this cone is the limit of F. Again, this construction can be applied to the categories listed above in 7.9.

Note 7.11. The above constructions of the filtered colimit and cofiltered limit were the first examples of limits and colimits in a category. Actually, the construction of the limit and the colimit of a functor in a category is a natural generalization of these constructions (see Theorems 9.1 and 9.2).

PROPOSITION 7.12. *A category $\mathscr{C}$ with filtered colimits and finite coproducts is a category with coproducts.*

Proof. Let $\{X_i\}_{i\in\mathscr{I}}$ be a family of objects of $\mathscr{C}$ and let $\mathscr{J}$ denote the set of all finite subsets of $\mathscr{I}$. It is clear that $\mathscr{J}$ is a directed set with respect to inclusion of subsets. For each $\alpha\in\mathscr{J}$ fix a coproduct $X = \coprod_{i\in\alpha} X_i$ of the family $\{X_i\}_{i\in\alpha}$; if $\alpha\leqslant\beta$, let $u_{\alpha\beta}\colon X_\alpha\to X_\beta$ denote the unique morphism such that $u_{\alpha\beta}u_i=\bar{u}_i$, where $u_i\colon X_i\to X_\alpha$, $i\in\alpha$, and $\bar{u}_j\colon X_j\to X_\beta$, $j\in\beta$ are the structural injections. Now it is easy to see that $\{X_\alpha, u_{\alpha\beta}\}_{\alpha\in\mathscr{J}}$ is a direct system of $\mathscr{C}$. Let $X=\varinjlim_{\alpha\in\mathscr{J}} X_\alpha$ and let $u_\alpha\colon X_\alpha\to X$, $\alpha\in\mathscr{J}$ be the structural injections. For each $i\in\mathscr{I}$, choose $\alpha\in\mathscr{J}$ such that $i\in\alpha$ and let $v_i\colon X_i\to X$ be the composition $u_\alpha u_i$, where the $u_i\colon X_i\to X^\alpha$ are, as above, the structural injections of the coproduct. It is easy to see that the morphism v_i does not depend on the choice of α, and that the object X together with the morphisms v_i give a coproduct of this family of objects. ✱

Example 7.13. We shall now give an example of a category with direct colimits which does not have coproducts. By a *field* we mean a commutative unitary nonzero ring A in which every nonzero element has an inverse. Let $\mathscr{F}d$ denote the category whose objects are fields and whose morphisms are the ring morphisms. Let $D=\{X_i, u_{ij}\}_{i\in\mathscr{I}}$ be a direct system in $\mathscr{F}d$; let X be the direct limit of the underlying system of sets and let $u_i\colon X_i\to X$ be the structural injections. Then X has the canonical structure of a field and the u_i are morphisms of fields. Indeed, by the construction of direct limits in $\mathscr{S}et$, for each $x\in X$, there is an $i\in\mathscr{I}$ and an $x_i\in X_i$ such that $u_i(x_i)=x$ (see Exercise 7.3). By the same exercise, if $x, x'\in X$, we can find an $i\in\mathscr{I}$ and $x_i, x_i'\in X_i$ such that $u_i(x_i)=x$ and $u_i(x_i')=x'$. Therefore we define $x+x'=u_i(x_i+x_i')$, and $xx'=u_i(x_ix_i')$. It can thus be proven that X becomes a field and the u_i become morphisms of fields. The details are left for the reader.

The following result is sometimes useful.

Proposition 7.14. *Let $\{X_i\}_{i\in\mathscr{I}}$ be a family of objects of a category $\mathscr{C}$ such that $\prod_{i\in\mathscr{I}}X_i$ exists and let $i_0\in\mathscr{I}$. Then the structural projection $p_{i_0}\colon\prod_{i\in\mathscr{I}}X_i\to X_{i_0}$ is a retraction if and only if the sets $\mathscr{C}(X_{i_0}, X_j)$ are nonempty for all $j\in\mathscr{I}$.*

Proof. If $\mathscr{C}(X_{i_0}, X_j)\neq\emptyset$ for all $j\in\mathscr{I}$, choose an element $f_j\in\mathscr{C}(X_{i_0}, X_j)$, $j\in\mathscr{I}$, such that $f_{i_0}=1_{X_{i_0}}$. By the definition of the product there is a unique morphism

$$s_{i_0}\colon X_{i_0}\to\prod\nolimits_{i\in\mathscr{I}}X_i$$

such that $p_js_{i_0}=f_j$ for all $j\in\mathscr{I}$; in particular, $p_{i_0}s_{i_0}=1_{X_{i_0}}$. The converse is obvious. ✱

Note 7.15. Let $\{f_i\colon X_i\to Y_i\}_{i\in\mathscr{I}}$ be a family of morphisms of $\mathscr{C}$. We assume that $\prod_iX_i$ and $\prod_iY_i$ exist in $\mathscr{C}$. Then there is a unique morphism $h\colon\prod_iX_i\to\prod_iY_i$ such that $q_ih=f_ip_i$ for all $i\in\mathscr{I}$, where $p_i\colon\prod_iX_i\to X_i$ and $q_i\colon\prod_iY_i\to Y_i$ are the corresponding structural projections. The morphism h is usually denoted by $\prod_if_i$ and is called the *product of the family of morphisms* $\{f_i\}_{i\in\mathscr{I}}$. Dually, we have the notion of the *coproduct of a family of morphisms*, which is denoted by $\coprod_if_i$.

Now let $\{f_i: X \to X_i\}_{i \in \mathscr{I}}$ be a family of morphisms of $\mathscr{C}$. We assume that $\prod_i X_i$ is defined. Then there exists a unique morphism $k: X \to \prod_i X_i$ such that $p_i k = f_i$ for all $i \in \mathscr{I}$. The morphism k is usually denoted by $[f_i]_i$ or simply by $[f_i]$ and is called the *morphism induced by the family* $\{f_i\}_{i \in \mathscr{I}}$ of morphisms. It is clear that $[f_i] = (\prod_i f_i)\Delta$, where $\Delta: X \to X^{\mathscr{I}}$ is the *diagonal morphism* (see Exercise 7.2). Dually, if $\{f_i: X_i \to X\}_{i \in \mathscr{I}}$ is a family of morphisms, then there is a unique morphism, denoted by $\langle f_i \rangle: \coprod_i X_i \to X$, such that $\langle f_i \rangle u_i = f_i$ for all $i \in \mathscr{I}$, where the $u_i: X_i \to \coprod_i X_i$ are the structural injections. It is clear that $\langle f_i \rangle = \nabla \coprod_i f_i$ (see Exercise 7.2).

Exercises

7.1. (Associativity of the product). Let $\{X_i\}_{i \in \mathscr{I}}$ be a family of objects of a category $\mathscr{C}$ and assume that the subsets $\{\mathscr{I}_j\}_{j \in \mathscr{J}}$ form a partition of $\mathscr{I}$. Assume that for every $j \in \mathscr{J}$ the product $\prod_{i \in \mathscr{I}_j} X_i$ of the family $\{X_i\}_{i \in \mathscr{I}_j}$ exists and also that the product $\prod_{j \in \mathscr{J}}(\prod_{i \in \mathscr{I}_j} X_i)$ can be formed in $\mathscr{C}$. Show that the product $\prod_{i \in \mathscr{I}} X_i$ exists in $\mathscr{C}$ and that we have a canonical isomorphism

$$\prod_{i \in \mathscr{I}} X_i \xrightarrow{\sim} \prod_{j \in \mathscr{J}}(\prod_{i \in \mathscr{I}_j} X_i).$$

7.2. Let X be an object of a category $\mathscr{C}$ such that $X^{\mathscr{I}}$ exists, where $\mathscr{I}$ is a set. Let $\Delta: X \to X^{\mathscr{I}}$ denote the unique morphism such that $p_i \Delta = 1_X$ for all $i \in \mathscr{I}$, where the p_i are the structural projections and Δ is the *diagonal morphism*. Prove that the following are equivalent:

a) X is a pseudofinal object of $\mathscr{C}$.

b) For each set $\mathscr{I}$, $X^{\mathscr{I}}$ exists and $\Delta: X \to X^{\mathscr{I}}$ is an isomorphism.

c) There is a set $\mathscr{I}$ having more than one element such that $X^{\mathscr{I}}$ exists and the structural projection $p_i: X^{\mathscr{I}} \to X$ is an isomorphism for some $i \in \mathscr{I}$.

The dual notion of the diagonal morphism is called the *codiagonal morphism* and is denoted by $\nabla: X^{(\mathscr{I})} \to X$. Dualize the above assertions for pseudoinitial objects.

7.3. Let $\mathscr{I}$ be a directed set, let $D = \{X_i, u_{ij}\}_{i \in \mathscr{I}}$ be a direct system in $\mathscr{S}et$ over $\mathscr{I}$ and let $u_i: X_i \to \varinjlim_{i \in \mathscr{I}} X_i$ be the structural morphisms. Prove that the following hold:

a) The u_i are monomorphisms for all $i \in \mathscr{I}$ if and only if the u_{ij} are monomorphisms for all $i \leqslant j$.

b) If $x_1, \ldots, x_n$ are elements of $\varinjlim_{i \in \mathscr{I}} X_i$ then there is an $i \in \mathscr{I}$ and there are elements $x_1^i, \ldots, x_n^i \in X_i$ such that $u_i(x_1^i) = x_1, \ldots, u_i(x_n^i) = x_n$. Thus we infer that $\bigcup_{i \in \mathscr{I}} u_i(X_i) = \varinjlim_{i \in \mathscr{I}} X_i$.

c) If all the u_{ij}, $i \leqslant j$, are surjections, then all the u_i are surjections. The converse is not generally true.

d) If there is an element $i \in \mathscr{I}$ such that the u_{ij} are isomorphisms for all $j \geqslant i$, then the u_j are also isomorphisms for all $j \geqslant i$.

Dualize these assertions for inverse limits in $\mathscr{S}et$.

7.4. Prove that the field $\mathbf{Q}$ of rational numbers and the fields $\mathbf{Z}_p$ (the cosets of $\mathbf{Z}$ modulo a prime number p) are pseudoinitial and pseudofinal objects of $\mathscr{F}d$. Do there exist other pseudoinitial or pseudofinal objects in this category?

7.5. Let $f: X \to Y$ be a morphism of $\mathscr{F}d$. Prove that:

a) f is a monomorphism.

b) If X has characteristic zero, then f is an epimorphism if and only if f is a surjective map.

c) If X has characteristic p (p a prime number), then f is an epimorphism if and only if for each $y \in Y$ there exists a natural number n such that $y^{p^n} = f(x)$ for some $x \in X$.

7.6. Prove that $\mathscr{F}d$ is a coproduct (disjoint union) of a countable number of categories. Also prove that a morphism of fields regarded as a morphism in the category $\mathscr{R}g$ is an epimorphism if and only if it is a surjection.

7.7. Let $\mathscr{C}$ be a category with coproducts and $\{G_i\}_{i \in \mathscr{I}}$ a set of objects of $\mathscr{C}$. Prove that the following are equivalent:

1) The given set of objects is a set of separators.

2) The set $\{\coprod_{i \in \mathscr{I}'} G_i\}_{\mathscr{I}'}$ where $\mathscr{I}'$ runs over the power set of $\mathscr{I}$, is a set of separators.

3) For each object X of $\mathscr{C}$ there are a subset $\mathscr{I}'$ of $\mathscr{I}$, a set $\mathscr{A}$ and an epimorphism:

$$(\coprod_{i \in \mathscr{I}'} G_i)^{(\mathscr{A})} \longrightarrow X.$$

7.8. (The formal completion of a category with products). Let $\mathscr{C}$ be a category. We denote by $\mathscr{C}_p$ the category defined as follows. The objects of $\mathscr{C}_p$ are pairs $(\mathscr{I}, \{X_i\}_{i \in \mathscr{I}})$, where $\mathscr{I}$ is a set and $\{X_i\}_{i \in \mathscr{I}}$ is a family of objects of $\mathscr{C}$ indexed by $\mathscr{I}$. A morphism $\alpha: (\mathscr{I}, \{X_i\}_{i \in \mathscr{I}}) \to (\mathscr{J}, \{Y_j\}_{j \in \mathscr{J}})$ of $\mathscr{C}_p$ is given by a pair $(u, \{f_i\}_{i \in \mathscr{I}})$ where $u: \mathscr{I} \to \mathscr{J}$ is a map and the $f_i: X_i \to Y_{u(i)}$ are morphisms of $\mathscr{C}$. The composition of morphisms in $\mathscr{C}_p$ is defined in an obvious way.

It is clear that the assignment $X \rightsquigarrow (\{*\}, X)$, where $\{*\}$ is a one-point set, determines a functor $S: \mathscr{C} \to \mathscr{C}_p$. Prove that:

1) The functor S is full and faithful.

2) $\mathscr{C}_p$ is a category with products and every object of $\mathscr{C}_p$ is a product of objects of the form $S(X)$, $X \in \mathscr{C}$.

3) If $F: \mathscr{C} \to \mathscr{C}'$ is a functor and $\mathscr{C}'$ is a category with products, then there exists a unique functor $\bar{F}: \mathscr{C}_p \to \mathscr{C}'$ such that $\bar{F}S = F$.

7.9. Let $\{X_n, f_{n,n-1}\}_{n \in \mathbf{N}}$ determine an inverse epifiltered system of sets. Prove that $\varprojlim_n X_n$ is a nonempty set.

7.10. Let $\{\mathscr{C}_i\}$ be a collection of categories. Assume that for every i, $\mathscr{C}_i$ is a category with products. Show that $\coprod_i \mathscr{C}_i$ is a category with products.

1.8. Some special limits and colimits

In this section we shall investigate some special functors in a category and examine their limits and colimits.

References: Bourbaki [1], Bucur—Deleanu [1], Cartan—Eilenberg [1], Gabriel-Zisman [1], MacLane [3], Mitchell [2], Pareigis [1], Schubert [1], Roux [1].

Let $\mathscr{C}$ be a category. A functor $F: \{0 \rightrightarrows 1\} \to \mathscr{C}$ is completely defined by a pair of morphisms $(f, g): X \to Y$ in $\mathscr{C}$. Sometimes such a functor is called a *double arrow*. A limit of a double arrow is usually called a *kernel*. Thus a kernel for a double arrow $(f, g): X \to Y$ is a pair (Z, u) where $u: Z \to X$ is a morphism in $\mathscr{C}$ such that $fu = gu$ and for every morphism $h: T \to X$ such that $fh = gh$ there is a unique morphism $\bar{h}: T \to Z$ such that $u\bar{h} = h$. A colimit of a double arrow is called a *cokernel*.

Sometimes the kernel (cokernel) of a double arrow $(f, g): X \to Y$ is denoted by $\mathrm{Ker}(f, g)$ ($\mathrm{Coker}\,(f, g)$) and the structural morphism $u: \mathrm{Ker}(f, g) \to X$ ($p: Y \to$ $\to \mathrm{Coker}\,(f, g)$) is called the *equalizer* (*coequalizer*) of this double arrow. Generally, a morphism u in $\mathscr{C}$ is called an *equalizer* (*coequalizer*) if there exists a double arrow such that u is the equalizer (coequalizer) of this double arrow. It is clear that every equalizer (coequalizer) is a monomorphism (an epimorphism), but simple examples show that the converse is not true in general.

Example 8.1. Let $(f, g): X \to Y$ be a double arrow in $\mathscr{S}et$; let X' denote the subset of X consisting of all elements x such that $f(x) = g(x)$ and let $u: X' \to X$ denote the inclusion mapping. Then (X', u) is a kernel of the considered double arrow. A cokernel of this double arrow is obtained as follows. Consider in the set Y the smallest equivalence relation R generated by all pairs $(f(x), g(x))$ when x runs through X (see Bourbaki [1]). Let $p: Y \to Y/R$ be the canonical map. It is easy to see that $(p, Y/R)$ is a cokernel of (f, g). The reader may wish to construct in the same manner the kernel of a double arrow in the categories $\mathscr{T}op$, $\mathscr{R}el$, $\mathscr{P}re$, $\mathscr{O}rd$, $\mathscr{M}on$, $\mathscr{G}r$, $\mathscr{A}b$, $\mathscr{R}g$ and $\mathscr{F}d$. The construction of cokernels in $\mathscr{T}op$ is similar to that in $\mathscr{S}et$ (see Bourbaki [3]).

8.2. Let X be a monoid; an equivalence relation R on X is said to be *compatible* with the structure of the monoid if for every pair x, $x' \in X$ such that $x \sim x'(R)$ we have that $yx \sim yx'(R)$ and $xy \sim x'y(R)$ for all $y \in X$. Thus the quotient set X/R has, in a canonical way, the structure of a monoid where the canonical map $p: X \to X/R$ is a morphism of monoids. (Cf. Section 1.5.) Thus, if $(f, g): X \to Y$ is a double arrow in $\mathscr{M}on$, consider the smallest equivalence relation R on Y that is generated by all pairs $(f(x), g(x))$ where x runs through X, and is compatible with the structure of the monoid. Then $(p, Y/R)$ is a cokernel of (f, g), where $p: Y \to Y/R$ is the canonical map. The details are left for the reader.

8.3. Let $(f, g): X \to Y$ be a double arrow in the category $\mathscr{G}r$ and let H denote the smallest normal subgroup of Y that contains all elements $f(x)g(x)^{-1}$, where x

runs through X. Then the pair $(p, Y/H)$, (where Y/H is the quotient group and where $p: Y \to Y/H$ is the canonical morphism), is a cokernel of (f, g).

The cokernels in $\mathscr{A}b$ are defined in a similar manner.

Note 8.4. The construction of cokernels in $\mathscr{R}g$ is somewhat more complicated (see Section 3.3). In general, cokernels need not exist in the category $\mathscr{F}d$.

A category $\mathscr{C}$ in which every double arrow has a kernel is called a *category with kernels*. Dually, we have the notion of a *category with cokernels*.

Let $\mathscr{C}$ be a category with zero morphisms. Let $f: X \to Y$ be a morphism in $\mathscr{C}$. A morphism $g: Z \to X$ is called a *kernel* or *equalizer* of f if and only if

a) $fg = 0$, and

b) if $h: T \to X$ is a morphism with $fh = 0$, then there is a unique morphism $k: T \to Z$ with $h = gk$.

It is easy to see that a morphism $f: X \to Y$ has a kernel if and only if the double arrow $(f, 0)$ has a kernel.

By an *angle category* we mean a category $\mathscr{A}$ with three objects, A, B, C and five morphisms, $f: A \to C$, $g: B \to C$ and the identities. A functor $F: \mathscr{A} \to \mathscr{C}$ is called an *angle* in $\mathscr{C}$. Obviously an angle in $\mathscr{C}$ is in fact a diagram of $\mathscr{C}$:

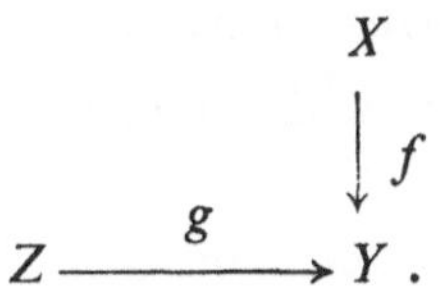

(6)

A limit of the angle (6) actually is an object of $\mathscr{C}$, usually denoted by $X \prod_Y Z$, and two morphisms $f': X \prod_Y Z \to Z$, $g': X \prod_Y Z \to X$ with $gf' = fg'$ such that for each triple (T, h, k) with $fh = gk$ there exists a unique morphism $l: T \to X \prod_Y Z$ such that the diagram

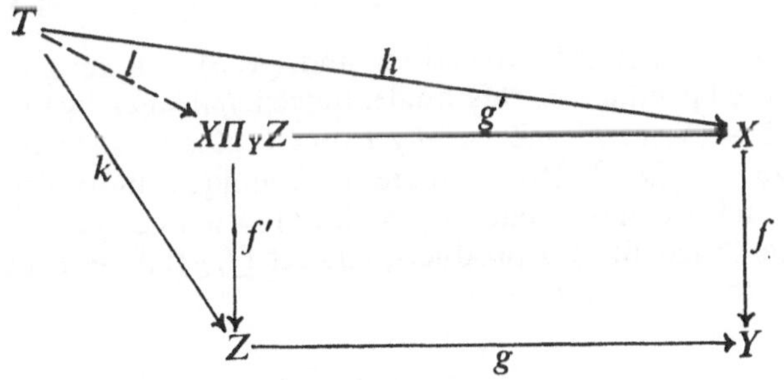

is commutative. The limit of an angle (6) is called a *fibered product of X and Z over Y*. Other common names for this are *cartesian square* and *pullback*.

A cofunctor from the angle category to a category $\mathscr{C}$ is called a *coangle.* In fact a coangle in $\mathscr{C}$ is a diagram:

$$\begin{array}{ccc} Y & \longrightarrow & X \\ \downarrow & & \\ Z & & \end{array} \qquad . \tag{7}$$

The colimit of the coangle (7) is denoted by $X\coprod_Y Z$ and is called a *fibered coproduct.* Other common names are *cocartesian square, pushout* and *fibered sum.* Note that in the category $\mathscr{G}r$, this is just the amalgamated product. By a *bicartesian square* we mean a square that is both cartesian and cocartesian.

A category in which every angle has a fibered product is called a *category with fibered products.* Dually, we have the notion of a *category with fibered coproducts.*

Example 8.5. Assume that (6) is an angle of $\mathscr{S}et$. Let U be the subset of $X\prod Z$ consisting of all pairs (x, z) such that $fp(x) = gq(z)$, where $p: X\prod Z \to X$ and $q: X\prod Z \to Z$ are the structural projections. Then $U = X\prod_Y Z$ and pu, qu are the structural projections of the fibered product, where $u: U \to X\prod Z$ is the natural inclusion.

The above construction suggests the construction of fibered products in a very large class of categories.

PROPOSITION 8.6. *Let $\mathscr{C}$ be a category with finite products. $\mathscr{C}$ is a category with kernels if and only if $\mathscr{C}$ has fibered products.*

Proof. Let $\mathscr{C}$ be a category with kernels. Let $X\xrightarrow{f} Y \xleftarrow{g} Z$ be an angle in $\mathscr{C}$; consider the diagram

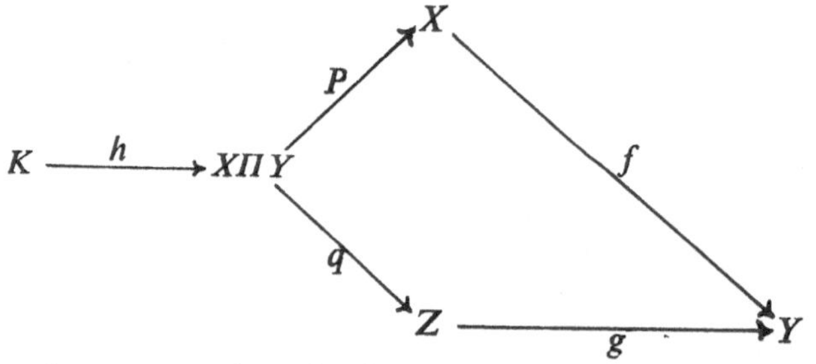

in which p, q are the structural projections and (K, h) is a kernel of (fp, gq). Then (K, ph, qh) is a fibered product of this angle. Indeed, $fph = gqh$ by hypothesis.

Now let $u: T \to X$, $v: T \to Z$ be any two morphisms such that $fu = gv$. Then $fp[u, v] = fu = gv = gq[u, v]$. Hence there is a unique morphism $t: T \to K$ such that $ht = [u, v]$ and we have that $pht = p[u, v] = u$ and $qht = q[u, v] = v$.

Suppose that $\mathscr{C}$ has fibered products, and let $(f, g): X \rightrightarrows Y$ be a double arrow of $\mathscr{C}$. Consider the cartesian square

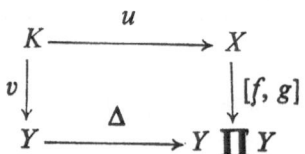

$$\begin{array}{ccc} K & \xrightarrow{\;u\;} & X \\ {\scriptstyle v}\downarrow & & \downarrow{\scriptstyle [f,\, g]} \\ Y & \xrightarrow{\;\Delta\;} & Y\prod Y \end{array}$$

where Δ is the diagonal morphism. We claim that (K, u) is a kernel of (f, g). Indeed, $fu = p[f, g]u = p\Delta v = v$ and $gu = q[f, g]u = q\Delta v = v$ so that $fu = gu$. Let $k: Z \to X$ be such that $fk = gk$. Then $k: Z \to X$ and $gk: Z \to Y$ are such that $[f, g]k = \Delta gk$. In fact, by the definition of the product, the latter morphisms will be equal if and only if their composition with the structural projections are equal. But $p[f, g]k = fk = gk = (p\Delta)gk = p(\Delta gk)$ and $q[f, g]k = gk = (q\Delta)gk = q(\Delta gk)$. Consequently, there is a unique morphism $t: Z \to K$ such that $ut = k$ and $vt = gk$, i.e. (K, u) is a kernel of (f, g). ✳

The next notion is closely related to that of kernel.

Let $f: X \to Y$ be a morphism. An ordered pair $(u, v): Z \to X$ of morphisms is called a *kernel pair* of f if $fu = fv$, and furthermore if for each ordered pair $(u', v'): Z' \to X$ of morphisms, such that $fu' = fv'$, there exists a unique morphism $g: Z' \to Z$ with $ug = u'$ and $vg = v'$.

PROPOSITION 8.7. *Let $f: X \to Y$ be a morphism and let $(u, v): Z \to X$ be a double arrow. The following are equivalent:*

a) (u, v) is a kernel pair of f.

b) The square

$$\begin{array}{ccc} Z & \xrightarrow{u} & X \\ {\scriptstyle v}\downarrow & & \downarrow{\scriptstyle f} \\ X & \xrightarrow{f} & Y \end{array}$$

is cartesian.

Moreover, if the product $X \coprod X$ is defined, these assertions are also equivalent to the following.

c) The morphism $[u, v]: Z \to X \coprod X$ is an equalizer of the double arrow $(fp, fq): X \coprod X \to Y$ where $p, q: X \coprod X \to X$ are structural morphisms of the product.

The proof is left for the reader. ✳

From this proposition it follows that any two kernel pairs of a morphism are canonically isomorphic. As before, for each morphism we fix a kernel pair.

A *category with kernel pairs* means a category $\mathscr{C}$ such that each morphism of $\mathscr{C}$ has a kernel pair. It is clear that each category with fibered products is a category with kernel pairs. On the other hand, the angle category is a category with kernel pairs but not a category with fibered products.

In Example 8.1 we saw that the cokernel $p: Y \to Z$ of a double arrow $(u, v): X \to Y$ is a set of equivalence classes of Y.

Conversely, if $p: Y \to Z$ is an epimorphism in $\mathscr{Set}$ and $(f, g): X \to Y$ is its kernel pair, then $(p, Z) = \mathrm{Coker}\,(f, g)$ and the corresponding equivalence relation on Y is defined as follows: two elements y, y' of Y are equivalent if and only if there is an element $x \in X$ such that $f(x) = y$ and $g(x) = y'$. This situation suggests the notion of equivalence relation on an object of a category.

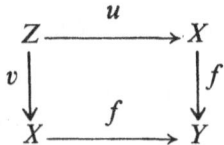

An *equivalence relation on an object* X of a category $\mathscr{C}$ is a double arrow $(f, g): Y \to X$ such that, for each object Z of $\mathscr{C}$, the image of the map

$$[H^Z(f), H^Z(g)]: \mathscr{C}(Z, Y) \to \mathscr{C}(Z, X) \coprod \mathscr{C}(Z, X)$$

is an equivalence relation. If in addition the map $[H^Z(f), H^Z(g)]$ is an injection for every $Z \in \mathscr{C}$, then the equivalence relation is called a *monomorphic equivalence relation*.

LEMMA 8.8. *Let* $(f, g): Y \to X$ *be an equivalence relation on* X. *Assume that the object* $X \coprod X$ *is defined. Then this equivalence relation is a monomorphic equivalence relation if and only if the morphism* $[f, g]: Y \to X \coprod X$ *is a monomorphism.*

The proof is left for the reader. ✳

PROPOSITION 8.9. *Let* $(f, g): Y \to X$ *be an equivalence relation on* X. *Then:*

a) There is a morphism $s: X \to Y$ *such that* $fs = gs = 1_X$. *Moreover, if the relation is monomorphic, then* (X, s) *is a kernel of* (f, g).

b) There is a morphism $\sigma: Y \to Y$ *such that* $f\sigma = g$ *and* $g\sigma = f$. *Moreover, if the relation is monomorphic, then* $\sigma\sigma = 1_Y$.

Proof. a) Consider the canonically defined map

$$\mathscr{C}(X, Y) \xrightarrow{[H^X(f), H^X(g)] = u} \mathscr{C}(X, X) \coprod \mathscr{C}(X, X).$$

By the definition of an equivalence relation on a set, there is an element $s \in \mathscr{C}(X, X)$ such that $u(s) = (1_X, 1_X)$, or equivalently $fs = gs = 1_X$.

Assume that the relation is monomorphic. We claim that (X, s) is a kernel of (f, g). To show that, let $t: Z \to Y$ be a morphism such that $ft = gt$. Then $f(sft) = g(sft) = ft = gt$, so that $sft = t$, since, by hypothesis, the equivalence relation is monomorphic.

b) Let us consider the canonically defined map

$$\mathscr{C}(Y, Y) \xrightarrow{[H^Y(f), H^Y(g)] = v} \mathscr{C}(Y, X) \coprod \mathscr{C}(Y, X).$$

Then (f, g) belongs to the image of v and so (g, f) also belongs to the image of v, i.e. there is an element $\sigma \in \mathscr{C}(Y, Y)$ such that $f\sigma = g$ and $g\sigma = f$. The equality $\sigma\sigma = 1_X$ follows easily if the equivalence relation is assumed to be monomorphic. ✳

THEOREM 8.10. *Let* $(f, g): Y \to X$ *be a double arrow of* $\mathscr{C}$. *Assume that* $X \coprod X$ *is defined and that the morphism* $[f, g]: Y \to X \coprod X$ *is a monomorphism. Assume also that the angle* $Y \xrightarrow{f} X \xleftarrow{g} Y$ *has a fibered product. Then the following are equivalent:*

1) *The double arrow* (f, g) *is a monomorphic equivalence relation on* X.

2) *Conditions a) and b) of Proposition 8.9 are satisfied. If, in addition, the square*

$$\begin{array}{ccc} P & \xrightarrow{u} & Y \\ {\scriptstyle v}\downarrow & & \downarrow{\scriptstyle f} \\ Y & \xrightarrow{g} & X \end{array}$$

is cartesian, then there is a morphism $p\colon P \to Y$ such that $fv = fp$ and $gu = gp$.

Proof. 1)$\Rightarrow$ 2). That conditions a) and b) are satisfied follows from Proposition 8.9. Furthermore, the elements (fv, gv) and (fu, gu) belong to the equivalence relation defined by $[H^P(f), H^P(g)]$ on $\mathscr{C}(P, X)$. Now, since $fu = gv$, we check that (fv, gu) also belongs to this equivalence relation. Hence there is a morphism $p\colon P \to Y$ such that $fp = fv$ and $gp = gu$, as required.

2) $\Rightarrow$ 1). Let Z be an object of $\mathscr{C}$. By the hypothesis and Lemma 8.8, the canonical map

$$\mathscr{C}(Z, Y) \xrightarrow{[H^Z(f),\ H^Z(g)] = t} \mathscr{C}(Z, X) \prod \mathscr{C}(Z, X)$$

is an injection. Now we must show that the image of t (say R) is an equivalence relation on $\mathscr{C}(Z, X)$.

Reflexivity. Let $h \in \mathscr{C}(Z, X)$. Then $f(sh) = h$, $g(sh) = h$, so that $t(sh) = (h, h) \in R$.

Symmetry. Let $h \in \mathscr{C}(Z, Y)$. Then $t(h) = (fh, gh)$ and $t(\sigma h) = (gh, fh)$.

Transitivity. Let $h, k \in \mathscr{C}(Z, Y)$ be such that $gh = fk$. Then there is a unique morphism $m\colon Z \to P$ such that $um = k$ and $vm = h$. But then $fpm = fvm = fh$, and $gpm = gum = gk$, i.e. (fh, gk) belongs to R.

The proof is now complete. ✳

Note 8.11. The assumption "the product $X \prod X$ is defined" in Theorem 8.10 was not used in the proof of the theorem. In fact, we used only the fact that the morphism $[f, g]$ is a monomorphism. It is clear that this last requirement can be changed as follows: "for each $Z \in \mathscr{C}$, the map

$$[H^Z(f),\ H^Z(g)]\colon \mathscr{C}(Z, Y) \to \mathscr{C}(Z, X) \prod \mathscr{C}(Z, X)$$

is an injection". This situation will be encountered in the following result.

COROLLARY 8.12. *Let $(f, g)\colon Y \to X$ be a kernel pair of the a morphism $t\colon X \to Z$. Assume that we have the cartesian square*

$$\begin{array}{ccc} P & \xrightarrow{u} & Y \\ {\scriptstyle v}\downarrow & & \downarrow{\scriptstyle f} \\ Y & \xrightarrow{g} & X\ . \end{array}$$

Then the pair (f, g) defines a monomorphic equivalence relation on X.

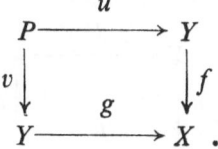

Proof. By Proposition 8.7, for each object Z of $\mathscr{C}$, the map $[H^Z(f), H^Z(g)]$: $\mathscr{C}(Z, Y) \to \mathscr{C}(Z, X) \coprod \mathscr{C}(Z, X)$ is an injection. The proof will be complete if we prove that condition 2) of Theorem 8.10 is satisfied. Since a) and b) follow by Exercise 8.19 we need only show the existence of a morphism $p: P \to Y$ such that $fp = fv$ and $gp = gu$. But $tfv = tgv = tfu = tgu$, so that there is a unique morphism $p: P \to Y$ such that $fp = fv$ and $gp = gu$. ✳

We say that an equivalence relation $(f, g): Y \to X$ is *effective* if the double arrow (f, g) has a cokernel.

PROPOSITION 8.13. *Let $(f, g): Y \to X$ be an equivalence relation. The following are equivalent:*

1) *The equivalence relation (f, g) is effective.*

2) *The coangle $X \overset{f}{\leftarrow} Y \overset{g}{\rightarrow} X$ has a fibered coproduct.*

Proof. 1)⇒2). Let $q: X \to Z$ be a cokernel of (f, g). We claim that the square

$$\begin{array}{ccc} Y & \xrightarrow{f} & X \\ {\scriptstyle g}\downarrow & & \downarrow{\scriptstyle q} \\ X & \xrightarrow[q]{} & Z \end{array}$$

is cocartesian. To show that, let $(u, v): X \to T$ be a pair of morphisms such that $uf = vg$. By Proposition 8.9 a, there is a morphism $s: X \to Y$ such that $fs = gs = 1_X$. But then $ufs = vgs$, so that $u = v$. Hence, by hypothesis, there is a unique morphism $t: Z \to T$ such that $tq = u = v$.

2) ⇒ 1). Consider the cocartesian square

$$\begin{array}{ccc} Y & \xrightarrow{f} & X \\ {\scriptstyle g}\downarrow & & \downarrow{\scriptstyle q} \\ X & \xrightarrow[p]{} & Z \end{array} \quad .$$

As above, the equality $qf = pg$ implies $qfs = pgs$ or equivalently $q = p$. Now it is easy to see that (p, Z) is a cokernel of (f, g). ✳

COROLLARY 8.14. *Let $(f, g): Y \to X$ be a kernel pair. The following assertions are equivalent:*

1) *The equivalence relation (f, g) is effective.*

2) *There is a morphism $p: X \to Z$ such that the square*

$$\begin{array}{ccc} Y & \xrightarrow{f} & X \\ {\scriptstyle g}\downarrow & & \downarrow{\scriptstyle p} \\ X & \xrightarrow{p} & Z \end{array}$$

is bicartesian.

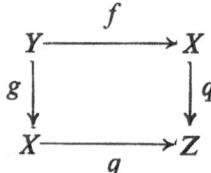

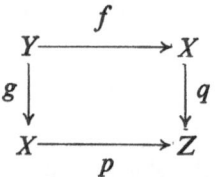

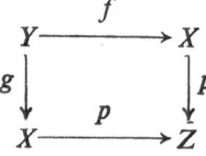

The proof follows from Proposition 8.13 and Exercise 8.18. ✳

Note 8.15. a) A diagram

$$X \xrightarrow{f} Y \underset{v}{\overset{u}{\rightrightarrows}} Z$$

is called an *equalizer diagram* if f is an equalizer of (u, v). Dually, we have the notion of *coequalizer diagram.*

b) The notions of kernel and cokernel of a double arrow can be generalized as follows. Let $\{f_i : X \to Y\}_{i \in \mathcal{I}}$ be a set of morphisms in a category $\mathscr{C}$. A *kernel* of this set of arrows is a pair (Z, u), where $u: Z \to X$ is a morphism such that $f_i u = f_j u$ for every $i, j \in \mathcal{I}$, and furthermore, if $v: Z' \to X$ is a morphism such that $f_i v = f_j v$ for all $i, j \in \mathcal{I}$, then there is a unique morphism $t: Z' \to Z$ such that $ut = v$. Dually, we have the notion of a *cokernel* of a set of arrows.

c) Similarly, the notions of fibered product and fibered coproduct can be generalized. A *pencil* in a category $\mathscr{C}$ is a set $\{f_i : X_i \to Y\}_{i \in \mathcal{I}}$ of morphisms all having the same codomain. Dually, we have the notion of a *copencil.*

A fibered product of a pencil $\{f_i : X_i \to Y\}_{i \in \mathcal{I}}$ is a copencil $\{u_i : Z \to X_i\}_{i \in \mathcal{I}}$ such that $f_i u_i = f_j u_j$ for all $i, j \in \mathcal{I}$; moreover, for every copencil $\{v_i : Z' \to X_i\}_{i \in \mathcal{I}}$, such that $f_i v_i = f_j v_j$ for all $i, j \in \mathcal{I}$, there is a unique morphism $t: Z' \to Z$ such that $u_i t = v_i$ for all $i \in \mathcal{I}$.

d) An epimorphism f of $\mathscr{C}$ is called *effective* if it is the equalizer of its kernel pair.

e) A morphism $f: X \to Y$ is called *universal* if for each morphism $g: Z \to X$, the angle

$$X \xrightarrow{f} Y \xleftarrow{g} Z$$

has a fibered product. Dually, we have the notion of a *couniversal* morphism.

f) An epimorphism $f: X \to Y$ of $\mathscr{C}$ is called *universal* (*effective universal*) if for each morphism $g: Z \to Y$ there is a cartesian square, i.e. it is universal,

$$\begin{array}{ccc} R & \xrightarrow{f'} & Z \\ {\scriptstyle g'}\downarrow & & \downarrow{\scriptstyle g} \\ X & \xrightarrow{f} & Y \end{array}$$

and g' is also an epimorphism (effective epimorphism). Dually we have the notions of a *couniversal monomorphism* and an *effective couniversal monomorphism.*

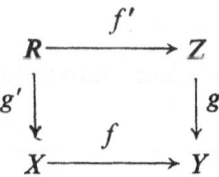

Exercises

8.1. Prove that if (Z, u) is a kernel of a double arrow in a category then u is a monomorphism.

8.2. Prove that the double arrow $(f, f)\colon X \to Y$ always has $(X, 1_X)$ $((1_Y, Y))$ as a kernel (cokernel).

8.3. Let $f\colon X \to Y$ be a retraction in a category $\mathscr{C}$, and let g be a right inverse of f, i.e. $fg = 1_Y$. Show that (Y, g) is a kernel of the double arrow $(gf, 1_X)\colon X \to X$. Dualize this result.

8.4. Let $(f, g)\colon X \to Y$ be a double arrow in a category $\mathscr{C}$. Show that the commutative diagram

$$\begin{array}{ccc} K & \xrightarrow{\quad u \quad} & X \\ {\scriptstyle u}\downarrow & & \downarrow{\scriptstyle [1_X, f]} \\ X & \xrightarrow{\ [1_X, g]\ } & X \coprod Y \end{array}$$

is cartesian if and only if (K, u) is a kernel of (f, g).

8.5. Let $X \to Y \leftarrow Z$ be an angle of $\mathscr{C}$ such that Y is a pseudofinal object. Show that this angle has a fibered product if and only if the product $X \prod Z$ exists in $\mathscr{C}$, and, in this case, we have the natural isomorphism $X \prod_Y Z \simeq X \prod Z$. Consequently, if $\mathscr{C}$ is a category with fibered products and has a final object, then $\mathscr{C}$ has finite products.

8.6. Prove that a morphism $f\colon X \to Y$ of $\mathscr{C}$ is a monomorphism if and only if the square

$$\begin{array}{ccc} X & \xrightarrow{\quad 1_X \quad} & X \\ {\scriptstyle 1_X}\downarrow & & \downarrow{\scriptstyle f} \\ X & \xrightarrow{\quad f \quad} & Y \end{array}$$

is cartesian.

8.7. Consider the following commutative diagram in a category $\mathscr{C}$:

$$\begin{array}{ccccc} X & \xrightarrow{\quad f \quad} & Y & \xrightarrow{\quad g \quad} & Z \\ {\scriptstyle u}\downarrow & & {\scriptstyle v}\downarrow & & \downarrow{\scriptstyle w} \\ X' & \xrightarrow{\quad f' \quad} & Y' & \xrightarrow{\quad g' \quad} & Z' \end{array}$$

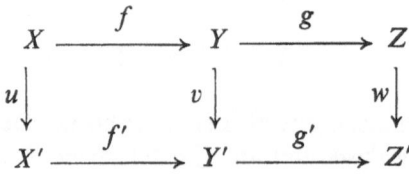

in which the right hand square is cartesian. Show that the left hand square is cartesian if and only if the composite square is cartesian i.e. (X, gf, u) is a fibered product of the angle

$$Z \xrightarrow{w} Z' \xleftarrow{g'f'} X').$$

8.8. If in the cartesian square:

$$\begin{array}{ccc} T & \longrightarrow & X \\ {\scriptstyle f'}\big\downarrow & & \big\downarrow{\scriptstyle f} \\ Z & \longrightarrow & Y \end{array}$$

f is a monomorphism (isomorphism), show that f' is also a monomorphism (isomorphism).

8.9. Consider the cartesian square

$$\begin{array}{ccc} Z & \xrightarrow{u} & X \\ {\scriptstyle v}\big\downarrow & & \big\downarrow{\scriptstyle f} \\ X & \xrightarrow{f} & Y \end{array} \quad .$$

Prove that:

a) u and v are coretractions, having the same retraction.

b) The morphism f is an equalizer if and only if it is an equalizer of its kernel pair, i.e. if and only if $(f, Y) = \text{Coker}(u, v)$.

8.10. A monomorphism $f: X \to Y$ is called *strict* if and only if a morphism $g: Z \to Y$ can be factored through f, provided that, for any double arrow $(u, v): Y \to H$, the condition $uf = vf$ implies that also $ug = vg$.

a) Prove that every equalizer is a strict monomorphism.

b) Consider the cocartesian square

$$\begin{array}{ccc} X & \xrightarrow{f} & Y \\ {\scriptstyle f}\big\downarrow & & \big\downarrow{\scriptstyle u} \\ Y & \xrightarrow{v} & Z \end{array}$$

in which f is a monomorphism. Prove that f is a strict monomorphism if and only if this square is also cartesian.

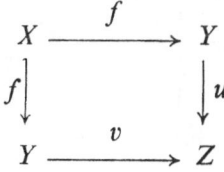

8.11. Consider the commutative diagram:

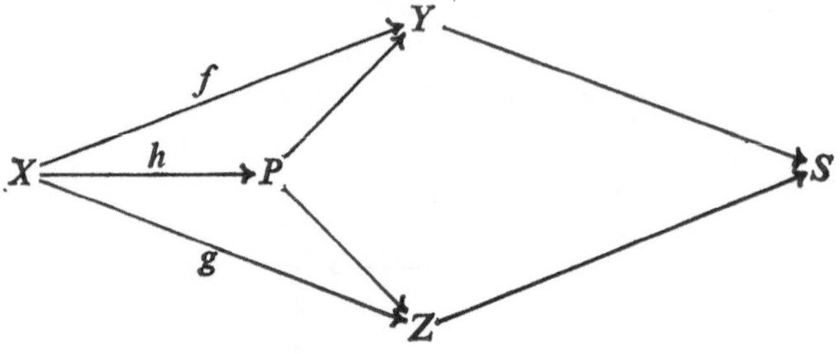

where $S = Y \coprod_X Z$ and $P = Y \coprod_S Z$. Prove that $S = Y \coprod_P Z$, and if $f = g$ and h is an isomorphism, then f is a strict monomorphism.

8.12. Let $(f, g)\colon X \to Y$ be a double arrow of $\mathscr{C}$. If in the cartesian square

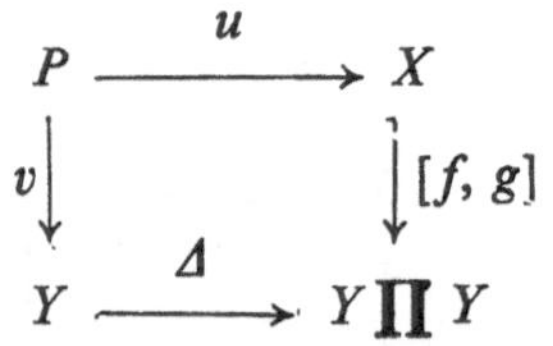

u is an epimorphism, show that $f = g$ and u is an isomorphism.

8.13. An object U of a category $\mathscr{C}$ is called a *quasigenerator* if for every monomorphism $f\colon X \to Y$, the condition that $H^U(f)$ is a surjection implies that f is a bimorphism. Prove that:

a) Every separator is a quasigenerator.

b) Conversely, if $\mathscr{C}$ is a category with fibered products, then every quasigenerator is a separator.

8.14. Let $\mathscr{C}$ be a category with zero object. Consider the commutative diagram

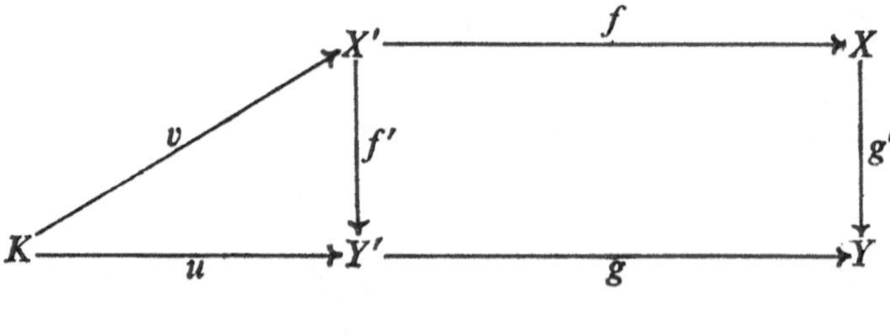

where the right hand square is cartesian, u is the kernel of g, v is the unique morphism induced by u and $0\colon K \to X$. Show that v is the kernel of f.

8.15. Consider the following diagram in a category with zero object:

$$\begin{array}{ccccc} X' & \xrightarrow{f} & X & & \\ & & \downarrow{\scriptstyle h} & & \\ Y' & \xrightarrow{g} & Y & \xrightarrow{t} & Y'', \end{array}$$

where g is the kernel of t. Prove that the diagram can be enlarged to a cartesian square if and only if f is the kernel of th.

8.16. Suppose that $\mathscr{C}$ is a category with zero object. Let $u: X' \to X$ be the kernel of $f: X \to Y$ and let $p: X \to Z$ be the cokernel of u. Show that u is the kernel of p.

8.17. Prove: If $(u, v): Z \to Y$ is a kernel pair of the morphism $f: X \to Y$ and $p: X \to Y'$ is a coequalizer of (u, v), then (u, v) is a kernel pair of p.

8.18. Let $(u, v): Z \to X$ be a kernel pair. Prove that a morphism $p: X \to Y$ is a coequalizer of (u, v) if and only if the square

$$\begin{array}{ccc} Z & \xrightarrow{u} & X \\ {\scriptstyle v}\downarrow & & \downarrow{\scriptstyle p} \\ X & \xrightarrow{p} & Y \end{array}$$

is cocartesian. In that case the square is also cartesian.

8.19. Prove: If $(u, v): Z \to X$ is a kernel pair, then u and v have a common right inverse t. Also prove that (X, t) is a kernel of (u, v).

Show that, moreover, there exists a morphism $s: Z \to Z$ such that $us = v$, $vs = u$ and $ss = 1_Z$.

8.20. Let $\mathscr{C}$ be a category with kernel pairs and cokernels. Dualize the concept of strict monomorphism to the notion of strict epimorphism. Prove:

a) An epimorphism $p: X \to Y$ is strict if and only if it is the coequalizer of its kernel pair.

b) A double arrow is a kernel pair if and only if it is a kernel pair of its coequalizer (cf. Exercise 8.9).

1.9. Existence of limits and colimits

In this section we define important classes of categories: the complete and cocomplete categories. Theorems 9.1 and 9.2 give useful criteria for recognizing whether a category is complete or cocomplete.

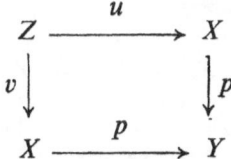

References: Bourbaki [1]. Bucur—Deleanu [1], Grothendieck—Verdier [1], Lambeck [2], MacLane [3], Maranda [1], Mitchell [2], Pareigis [1], Roux [2], Schubert [1].

A category $\mathscr{C}$ is said to be *complete* if every small functor with values in $\mathscr{C}$ has a limit i.e. $\mathscr{C}$ is $\mathscr{I}$-complete for every small category $\mathscr{I}$. Dually, we have the notion of a *cocomplete category*. The purpose of this section is to prove the following result which will allow us to determine whether or not a category is complete.

THEOREM 9.1. *The following are equivalent for a category* $\mathscr{C}$:
1) $\mathscr{C}$ *is complete.*
2) $\mathscr{C}$ *has products and kernels.*
3) $\mathscr{C}$ *has products and fibered products.*

Proof. The implications 1) $\Rightarrow$ 2) and 1) $\Rightarrow$ 3) are obvious, and the equivalence 2) $\Leftrightarrow$ 3) follows from Proposition 8.6.

2) $\Rightarrow$ 1). Let $F\colon \mathscr{I} \to \mathscr{C}$ be a small functor. Consider the object $X = \coprod_{i \in \mathscr{I}} F_i$ (here i runs through all objects of $\mathscr{I}$ and $F_i = F(i)$) and let $\{p_i\colon X \to F_i\}_{i \in \mathscr{I}}$, be the structural projections. Also consider the set $\mathscr{A}$ of all morphisms in $\mathscr{I}$ and the product $Y = \coprod_{s \in \mathscr{A}} F_{c(s)}$ where $c(s)$ is the codomain of s; let $q_s\colon Y \to F_{c(s)}$ denote the structural projections. Let $f\colon X \to Y$ be the unique morphism such that the diagram

$$\begin{array}{ccc} X & \xrightarrow{\quad f \quad} & Y \\ {\scriptstyle p_{c(s)}}\downarrow & & \downarrow{\scriptstyle q_s} \\ F_{c(s)} & \xrightarrow{\quad 1_{F_{c(s)}} \quad} & F_{c(s)} \end{array}$$

is commutative for all $s \in \mathscr{A}$.

Similarly, let $g\colon X \to Y$ be the unique morphism such that the diagram

$$\begin{array}{ccc} X & \xrightarrow{\quad g \quad} & Y \\ {\scriptstyle p_{d(s)}}\downarrow & & \downarrow{\scriptstyle q_s} \\ F_{d(s)} & \xrightarrow{\quad F(s) \quad} & F_{c(s)} \end{array}$$

is commutative for all $s \in \mathscr{A}$, where $d(s)$ is the domain of s. Let (K, v) be a kernel of (f, g). For each object i of $\mathscr{I}$, let u_i denote $p_i v\colon K \to F_i$. Thus, if $s\colon i \to j$ is a morphism of $\mathscr{I}$, then $F(s)u_i = F(s)p_i v = q_s g v = q_s f v = p_j v = u_j$, since $d(s) = i$ and $c(s) = j$. These relations show that the morphisms $\{u_i\}_{i \in \mathscr{I}}$ form a morphism $u\colon K \to F$, i.e. (K, u) is a cone over F. Now let (H, t) be a cone over F. There is a unique morphism $t_1\colon H \to X$ such that $p_i t_1 = t_i$ for every $i \in \mathscr{I}$. Furthermore, $q_s f t_1 = p_{c(s)} t_1 = t_{c(s)}$ and $q_s g t_1 = F(s) p_{d(s)} t_1 = F(s) t_{d(s)} = t_{c(s)}$ for all $s \in \mathscr{A}$, i.e. $f t_1 = g t_1$. Hence there exists a unique morphism $\bar{t}\colon H \to K$ such that $v\bar{t} = t_1$. But then $p_i v \bar{t} = p_i t_1 = t_i$ for all $i \in \mathscr{I}$, or, equivalently, $u\bar{t} = t$. Since the uniqueness of $\bar{t}$ is obvious, we see that (K, u) is a limit of F. ✳

We have the following dual result for colimits.

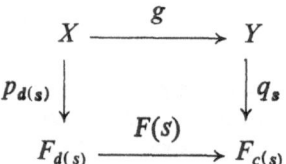

THEOREM 9.2. *The following are equivalent for a category* $\mathscr{C}$:

1) $\mathscr{C}$ *is cocomplete.*

2) $\mathscr{C}$ *has coproducts and cokernels.*

3) $\mathscr{C}$ *has coproducts and fibered coproducts.* ✳

Example 9.3. It follows from Examples 7.1 and 8.1 that the category $\mathscr{S}et$ is both complete and cocomplete. The same assertion holds for the categories $\mathscr{T}op$, $\mathscr{R}el$, $\mathscr{P}re$, $\mathscr{O}rd$, $\mathscr{M}on$, $\mathscr{G}r$, $\mathscr{A}b$ and $\mathscr{R}g$ (also see Section 3.3).

For small categories we have the following result.

THEOREM 9.4. *Let* $\mathscr{C}$ *be a small category. The following are equivalent:*

1) $\mathscr{C}$ *is a category with products and has a final object.*

2) $\mathscr{C}$ *is complete and has a final object.*

3) $\mathscr{C}$ *is a category with coproducts and has an initial object.*

4) $\mathscr{C}$ *is cocomplete and has an initial object.*

5) $\mathscr{C}$ *is a complete preordered category and has a final object.*

6) $\mathscr{C}$ *is a cocomplete preordered category and has an initial object.*

Proof. It follows from Proposition 7.6 that 1) implies that $\mathscr{C}$ is a preordered category. Then, using Exercise 8.5 and Theorem 9.1, we get that 1) ⇒ 5).

5) ⇒ 6). By Exercise 8.5, it will suffice to prove that $\mathscr{C}$ has coproducts. Indeed, if $\{X_i\}_{i\in\mathscr{I}}$ is a family of objects of $\mathscr{C}$, let $\{Y_j\}_{j\in\mathscr{J}}$ denote the family of all objects Y of $\mathscr{C}$ such that $\mathscr{C}(X_i, Y_j)\neq\emptyset$ for all $i\in\mathscr{I}$ (the family is nonempty since $\mathscr{C}$ has a final object). It is easy to verify that $\prod_{j\in\mathscr{J}}Y_j$ is a coproduct of the family $\{X_i\}_{i\in\mathscr{I}}$. The other implications are left for the reader. ✳

Note 9.5. The angle category is a cocomplete category but is not complete. Therefore, the condition "has a final or initial object" in Theorem 9.4 is not superfluous. There are also complete categories which are not cocomplete (see Section 2.1).

We say that a category $\mathscr{C}$ is *finitely complete* if every functor $F: \mathscr{I}\to\mathscr{C}$, where $\mathscr{I}$ is a finite category, has a limit. It follows from Theorem 9.1 that a category is finitely complete if and only if it has finite products and kernels or, equivalently, if and only if it has finite products and fibered products. Dually, we have the notion of a *finitely cocomplete* category. The category $\mathscr{S}etf$ is a finitely complete and finitely cocomplete category, but it is neither complete nor cocomplete.

The following considerations will be useful later.

Let $F: \mathscr{I}\to\mathscr{S}et$ be a functor and, for each $i\in\mathscr{I}$, let x_i be an element of F_i. We say that $\{x_i\}_{i\in\mathscr{I}}$ is a *coherent* family of elements over F if, for every morphism $s: i\to j$ of $\mathscr{I}$, $F(s)(x_i) = x_j$. The notion of a coherent family of elements is closely related to the notion of a limit.

PROPOSITION 9.6. *Let* (X, u) *be a cone over the functor* $F: \mathscr{I}\to\mathscr{S}et$. *The following are equivalent:*

a) (X, u) *is a limit of* F.

b) *If* $\{x_i\}_{i\in\mathscr{I}}$ *is a coherent family of elements over* F, *then there exists a unique element* $x\in X$ *such that* $u_i(x) = x_i$ *for all* $i\in\mathscr{I}$.

Proof. a) $\Rightarrow$ b). Assume there exists a coherent family $\{x_i\}_{i\in\mathscr{I}}$ such that for each $x \in X$ there is an $i(x) \in \mathscr{I}$ such that $u_{i(x)}(x) \neq x_{i(x)}$. Let Y be the set $X \coprod \{a\}$ where a is an element which does not belong to X. For each $i \in \mathscr{I}$, let $v_i \colon Y \to F_i$ denote the map defined by

$$v_i(y) = \begin{cases} u_i(y) & \text{if } y \in X, \\ x_i & \text{if } y = a. \end{cases}$$

It is easy to check that the maps $\{v_i\}_{i\in\mathscr{I}}$ give a functorial morphism $v \colon Y \to F$. Hence there is a unique map $f \colon Y \to X$ such that $uf = v$. But then $u_i f(a) = v_i(a) = x_i$ for all $i \in \mathscr{I}$, which is a contradiction. Therefore, if $\{x_i\}_{i\in\mathscr{I}}$ is a coherent family of elements over F, then there is an element $x \in X$ such that $u_i(x) = x_i$ for all $i \in \mathscr{I}$. Now we check the uniqueness of x. To show this let $x, x' \in X$ be such that $x \neq x'$ and $u_i(x) = u_i(x')$ for all $i \in \mathscr{I}$. Let Z denote $X - \{x'\}$ and let $v_i \colon Z \to F_i$ be the map defined by $v_i(z) = u_i(z)$, $z \in Z$. It is clear that the maps $\{v_i\}_{i\in\mathscr{I}}$ give a functorial morphism $v \colon Z \to F$. Hence there is a unique map $f \colon Z \to X$ such that $uf = v$, which is a contradiction.

b) $\Rightarrow$ a). Let (Y, v) be a cone over F. Then, for each $y \in Y$, $\{v_i(y)\}_{i\in\mathscr{I}}$ is a coherent family of elements over F. Hence there is a unique element $f(y) \in X$ such that $u_i(f(y)) = v_i(y)$ for all $i \in \mathscr{I}$. Then the assignment $y \rightsquigarrow f(y)$ gives a map $f \colon Y \to X$ such that $uf = v$. The uniqueness of f is obvious. ✳

Let $F \colon \mathscr{I} \to \mathscr{S}et$ be a functor, $i, i' \in \mathscr{I}$, $x_i \in F_i$, and $x_{i'} \in F_{i'}$. We say that x_i is *equivalent to* x_i *relative to* F if there exists a diagram

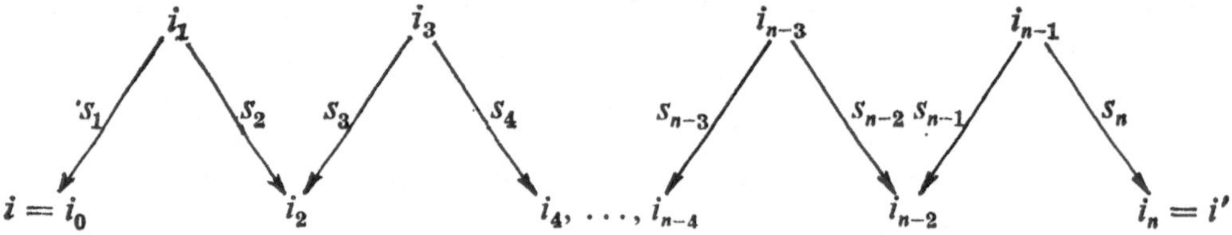

in $\mathscr{I}$ and elements $x_1 \in F_{i_1}, x_2 \in F_{i_2} \ldots x_{n-1} \in F_{i_{n-1}}$ such that $F(s_1)(x_1) = x_i$, $F(s_2)(x_1) = x_2, \ldots, F(s_{n-1})(x_{n-1}) = x_{n-2}$, $F(s_n)(x_{n-1}) = x_{i'}$. Using this, Proposition 9.6 can thus be dualized as follows.

PROPOSITION 9.7. *If* (p, X) *is a cocone over the functor* $F \colon \mathscr{I} \to \mathscr{S}et$, *then the following are equivalent:*

a) (p, X) *is a colimit of* F.

b) *For each* $x \in X$, *there are an* $i \in \mathscr{I}$ *and an element* $x_i \in F_i$ *such that* $p_i(x_i) = x$; *if, in addition,* $x_i \in F_i$ *and* $x_{i'} \in F_{i'}$, *then* $p_{i'}(x_i) = p_i(x_{i'})$ *if and only if* x_i *and* $x_{i'}$ *are equivalent relative to* F.

The proof is left for the reader. ✳

Note 9.8. Let $F \colon \mathscr{I} \to \mathscr{S}et$ be a small functor. By Theorem 9.1 and Proposition 9.6, we have the following "practical" procedure for the construction of the limit of F. Consider the product $\prod_{i\in\mathscr{I}} F_i = P$ and let K be the subset of P whose elements are the coherent families of elements over F. Let $t: K \to P$ be the canonical inclusion and $p_i: P \to F_i$ the structural projections. Then the morphisms $\{u_i = p_i t: K \to F_i\}$ define a functorial morphism $u: K \to F$ and (K, u) is a limit of F.

For the construction of the colimit of F, consider the coproduct $S = \coprod_{i \in \mathcal{I}} F_i$ and let R be the equivalence relation on S defined as follows. If $x, x' \in S$, then $x \in F_i$, $x' \in F_{i'}$ and we say that $x \sim x'(R)$ if and only if x and x' are equivalent relative to F. Now, if $h: S \to S/R$ is the canonical surjection and $v_i: F_i \to S$ are the structural injections, then the morphisms $\{q_i = hv_i\}_{i \in \mathcal{I}}$ define a functorial morphism $q: F \to S/R$, and $(q, S/R)$ is a colimit of F. All details are left for the reader.

Example 9.9. The category $\mathcal{C}at$ is complete and cocomplete.

To show this, let $\{\mathcal{C}_i\}_{i \in \mathcal{I}}$ be a set of objects of $\mathcal{C}at$. Consider the product category $\mathcal{C} = \prod_{i \in \mathcal{I}} \mathcal{C}_i$ and let $P_i: \mathcal{C} \to \mathcal{C}_i$ be the projection functors, defined in the obvious way. Then $\mathcal{C}$ and $\{P_i\}_{i \in \mathcal{I}}$ yield a product for the given set of objects in $\mathcal{C}at$. Moreover, if $F, G: \mathcal{C} \to \mathcal{C}'$ are morphisms in $\mathcal{C}at$, then denote by K the subcategory of $\mathcal{C}$ whose objects are the objects of $\mathcal{C}$ and whose morphisms are all the morphisms f of $\mathcal{C}$ such that $F(f) = G(f)$. If $U: K \to \mathcal{C}$ is the inclusion functor, then it is easy to see that (K, U) is a kernel of the pair (F, G). Hence, by Theorem 9.1, the category $\mathcal{C}at$ is complete.

The construction of colimits in $\mathcal{C}at$ is somewhat more difficult. Let $\{\mathcal{C}_i\}_{i \in \mathcal{I}}$ be a set of objects in $\mathcal{C}at$. Denote by $\mathcal{C} = \coprod_{i \in \mathcal{I}} \mathcal{C}_i$ the disjoint union of categories, i.e. the category whose objects are $\mathcal{O}b\,\mathcal{C} = \coprod_{i \in \mathcal{I}} \mathcal{O}b\,\mathcal{C}_i$. Furthermore, if $X, Y \in \mathcal{O}b\,\mathcal{C}$, then $X \in \mathcal{O}b\,\mathcal{C}_i$, $Y \in \mathcal{O}b\,\mathcal{C}_j$ and we define $\mathcal{C}(X, Y) = \mathcal{C}_i(X, Y)$ if $i = j$ and $\mathcal{C}(X, Y) = \emptyset$ otherwise. Then $\mathcal{C}$ is the coproduct of this set of objects of $\mathcal{C}at$. The structural injections are defined in the obvious way.

Now let $(F, G): \mathcal{C}' \to \mathcal{C}$ be a double arrow in $\mathcal{C}at$. Consider the following diagram in $\mathcal{S}et$:

$$
\begin{array}{ccccc}
\mathcal{M}or(\mathcal{C}') & \underset{\beta}{\overset{\alpha}{\rightrightarrows}} & \mathcal{M}or(\mathcal{C}) & \xrightarrow{p} & \mathcal{M}, \\
{\scriptstyle d'}\downarrow\downarrow{\scriptstyle c'} & & {\scriptstyle d}\downarrow\downarrow{\scriptstyle c} & & \\
\mathcal{O}b(\mathcal{C}') & \underset{g}{\overset{f}{\rightrightarrows}} & \mathcal{O}b(\mathcal{C}) & \xrightarrow{q} & \mathcal{I}
\end{array}
\qquad
\begin{array}{c}
\alpha(t) = F(t), \quad \beta(t) = G(t), \\
t \in \mathcal{M}or(\mathcal{C}'), \\
f(X) = F(X), \quad g(X) = G(X), \\
X \in \mathcal{O}b(\mathcal{C}'),
\end{array}
$$

where d and c (resp. d$'$ and c') are the map domain and codomain, i.e. $d(t) =$ domain of t and $c(t) =$ codomain of t. Let $(p, \mathcal{M})$ be a cokernel of (α, β) and $(q, \mathcal{I})$ be a cokernel of (f, g). Since $d\alpha = fd'$, $d\beta = gd'$, $c\alpha = fc'$, and $c\beta = gc'$, there are morphisms $u, v: \mathcal{M} \to \mathcal{I}$ such that $qd = u$ and $qv = c$. Let $h: \mathcal{M} \to \mathcal{I} \coprod \mathcal{I}$ be the map: $h(m) = (u(m), v(m))$, i.e. $h = [u, v]$. Furthermore, consider the diagram scheme $\Sigma = (\mathcal{I}, \mathcal{M}, h)$. If $i \in \mathcal{I}$, then $i = q(X)$, where $X \in \mathcal{O}b\,\mathcal{C}$ and $up(1_X) = vp(1_X) = i$. The element $p(1_X) = e_i$ does not depend on the choice of the element $X \in \mathcal{O}b\,\mathcal{C}$. These elements are the preunities of the diagram scheme Σ. Let $\mathcal{L}$ be the free category generated by the diagram scheme Σ (such that the e_i become identities) and $D: \Sigma \to \mathcal{L}$ the canonical diagram.

Let R denote the smallest equivalence relation on $\mathscr{C}$ defined such that for any morphisms $X \xrightarrow{s} Y \xrightarrow{t} Z$ of $\mathscr{C}$, we have that

$$p(ts) \sim p(t)p(s)\,(R(q(X), q(Z)).$$

Let $\mathscr{D} = \mathscr{L}/R$ and $P: \mathscr{L} \to \mathscr{D}$ be the canonical functor. Let $K: \mathscr{C} \to \mathscr{D}$ denote the functor defined by $K(X) = P(q(X))$ for every $X \in \mathscr{O}b\,\mathscr{C}$ and $K(t) = P(p(t))$ for every $t \in \mathscr{M}or(\mathscr{C})$. We leave it for the reader to check that $(K, \mathscr{D})$ is a cokernel of (F, G) in $\mathscr{C}at$. Finally $\mathscr{C}at$ is cocomplete, by Theorem 9.2.

Note 9.10. Let $\mathscr{C}'$ be a subcategory of a category $\mathscr{C}$, and let $F: \mathscr{I} \to \mathscr{C}'$ be a functor. There are simple examples which show that the functor F may have a limit, even when the functor IF (where $I: \mathscr{C}' \to \mathscr{C}$ is the inclusion functor) does not. On the other hand, other examples show that the functor IF may have a limit, even when F does not. Hence, we make the following conventions. A subcategory $\mathscr{C}'$ of $\mathscr{C}$ is said to be *closed under limits* if, given a functor $F: \mathscr{I} \to \mathscr{C}'$ such that the functor IF has a limit (X, u), then there exists an isomorphism $f: X' \to X$ such that $X' \in \mathscr{C}'$ and the morphism $u_i f: X' \to F_i$ also belong to $\mathscr{C}'$. We define the concepts of *subcategory closed under products* (*subcategory closed under coproducts*, *colimits*, etc.) analogously.

Note 9.11. An ordered set that is a finitely complete category is called a *left semilattice*. Dually, a finitely cocomplete ordered set is called a *right semilattice*. Finally, an ordered set which is simultaneously a left and a right semilattice is called a *lattice*. If $\{x_i\}_{i \in \mathscr{I}}$ is a family of elements in a left (right) semilattice, then we denote the product (coproduct) of this family by $\inf(x_i)$ $(\sup(x_i))$ and $\inf(x_i)$ $(\sup(x_i))$ is called a *lower bound* (an *upper bound*) of the family $\{x_i\}_{i \in \mathscr{I}}$.

Note 9.12. The considerations of Example 9.9 can be formally applied to show that every (not necessarily small) "functor" $F: \mathscr{I} \to \mathscr{C}\mathscr{A}\mathscr{T}$ has a limit and a colimit.

Note 9.13. A category $\mathscr{I}$ is called *left bounded* if it contains a small subcategory $\mathscr{J}$ such that for each $i \in \mathscr{I}$ there exists a morphism $s: j \to i$ with $j \in \mathscr{J}$. $\mathscr{I}$ is called *right bounded* if $\mathscr{I}^0$, the dual of $\mathscr{I}$, is left bounded. The subcategory $\mathscr{J}$ which appears in the definition of the notion of a left (right) bounded category is sometimes called a *left* (*right*) *bounding subcategory*.

A category $\mathscr{C}$ has *bounded limits* (*bounded colimits*) if every functor $F: \mathscr{I} \to \mathscr{C}$, where $\mathscr{I}$ is a left (right) bounded category, has a limit (colimit). Exercise 9.5 shows that $\mathscr{S}et$, $\mathscr{G}r$, etc. are categories with bounded limits and colimits (see Section 1.19).

Note 9.14. An infinite cardinal number α is called *regular* if for every set $\{\alpha_i\}_{i \in \mathscr{I}}$ of cardinal numbers such that $\alpha_i < \alpha$ for all i and $\mathrm{Card}(\mathscr{I}) < \alpha$, we have that $\Sigma_i \alpha_i < \alpha$ (the sum of cardinal numbers — see Bourbaki [1]). Let α be a regular cardinal number. We say that a small category $\mathscr{I}$ is *α-small* if $\mathrm{Card}(\mathscr{O}b\,\mathscr{I}) < \alpha$ and if $\mathrm{Card}(\mathscr{I}(i, j)) < \alpha$ for all $i, j \in \mathscr{I}$. A category $\mathscr{I}$ is called *α-right bounded* if it has an α-small right bounding subcategory. By an *α-limit* (*α-colimit*) in a category $\mathscr{C}$, we mean the limit (colimit) of a functor $F: \mathscr{I} \to \mathscr{C}$, where $\mathscr{I}$ is α-small.

We say that $\mathscr{C}$ is *α-cocomplete* (has *α-bounded colimits*) if every functor $F: \mathscr{I} \to \mathscr{C}$ with $\mathscr{I}$ α-small (α-right bounded) has a colimit. Dually, we have the notions *α-left bounded category*, *α-complete category* and *category with α-bounded limits.*

If α is the first regular cardinal number, then the α-complete (α-cocomplete) categories are just the finitely complete (finitely cocomplete) categories.

These notions will be discussed and made use of in the next sections.

Exercises

9.1. Let $\mathscr{C}$ be a complete category and $\mathscr{I}$ a category having an initial small subcategory. Prove that every functor $F: \mathscr{I} \to \mathscr{C}$ has a limit.

9.2. Let $\mathscr{C}$ be a complete category, $F: \mathscr{I} \to \mathscr{C}$ a functor and $\mathscr{J}$ a small final subcategory of $\mathscr{I}$. Assume that $i \coprod j$ exists in $\mathscr{I}$ for all $i, j \in \mathscr{J}$. Prove that $\varprojlim F$ exists in $\mathscr{C}$ and that there is an isomorphism

$$\varprojlim F \simeq \operatorname{Ker}(u, v)$$

where u and v are the morphisms arising from the following commutative diagrams:

$$\begin{array}{ccc} \prod_{i\in\mathscr{J}} F(i) & \xrightarrow{\ u\ } & \prod_{i,j\in\mathscr{J}} F(i\coprod j); \\ {\scriptstyle p_i}\downarrow & & \downarrow{\scriptstyle p_{ij}} \\ F(i) & \xrightarrow{\ F(u_i)\ } & F(i\coprod j) \end{array} \qquad \begin{array}{ccc} \prod_{i\in\mathscr{J}} F(i) & \xrightarrow{\ v\ } & \prod_{i,j\in\mathscr{J}} F(i\coprod j) \\ {\scriptstyle p_j}\downarrow & & \downarrow{\scriptstyle p_{ij}} \\ F(j) & \xrightarrow{\ F(u_j)\ } & F(i\coprod j) \end{array}$$

(here $u_i: i \to i\coprod j$, $u_j: j \to i \coprod j$ are the structural injections, and p_i, p_j and p_{ij} are the structural projections).

9.3. Prove that a category with fibered products and a final object is finitely complete.

9.4. A functor $T: \mathscr{C} \to \mathscr{C}'$ *reflects* a property of a diagram D in $\mathscr{C}$ if D has the property whenever TD has the property. Prove that every faithful functor reflects monomorphisms, epimorphisms and commutative diagrams. Prove, in addition: If $\mathscr{C}$ and $\mathscr{C}'$ are categories with zero objects, then T reflects zero objects; every full and faithul functor reflects limits and colimits (i.e. if $F: \mathscr{I} \to \mathscr{C}$ is a functor and for every $i \in \mathscr{I}$ there is a morphism $u_i: F_i \to X$ such that $\{T(u_i)\}_{i\in\mathscr{I}}$ and $T(X)$ give a colimit for TF, then $\{u_i\}_{i\in\mathscr{I}}$ and X define a colimit of F).

9.5. Prove that every functor of $F: \mathscr{I} \to \mathscr{S}et$ where $\mathscr{I}$ is left (right) bounded has a limit (colimit). Show that the same result holds in $\mathscr{T}op$, $\mathscr{R}el$, $\mathscr{O}rd$, $\mathscr{P}re$, $\mathscr{M}on$, $\mathscr{G}r$ and $\mathscr{A}b$.

9.6. Prove that a cocomplete category $\mathscr{C}$ has a final object if and only if it is right bounded.

9.7. Let $T: \mathscr{C} \to \mathscr{C}$ be a functor. With T we associate a category $(1, T)$ whose objects are all morphisms $f: X \to T(X)$ where $X \in \mathscr{C}$ and a morphism $\alpha: (X \to T(X)) \to (X' \xrightarrow{f'} T(X'))$ in $(1, T)$ is defined by a morphism $u: X \to X'$ such that $f'u = T(u)f$. Prove: a) There is a canonical functor $T': (1, T) \to \mathscr{C}$ where $T'(X \xrightarrow{f} T(X)) = X$ and $T'(\alpha) = u$. b) A functor $F: \mathscr{I} \to (1, T)$ has a colimit if $T'F$ has a colimit. c) If $\mathscr{C}$ is cocomplete then $(1, T)$ is also cocomplete. d) If $(1, T)$ has a final object $X \xrightarrow{f} T(X)$ then f is an isomorphism.

9.8. Let $T: \mathscr{C} \to \mathscr{C}$ be a functor and assume that $\mathscr{C}$ is cocomplete. Using the notation of Exercise 9.7, show that there exists an object X of $\mathscr{C}$ such that $T(X) \simeq X$ if the category $(1, T)$ is right bounded. In particular, if $\mathscr{C}$ is a small cocomplete category and $T: \mathscr{C} \to \mathscr{C}$ is a functor, then there exists an object X of $\mathscr{C}$ such that $T(X) \simeq X$. In particular, show that if $\mathscr{C}$ is a cocomplete lattice then every nondecreasing map $T: \mathscr{C} \to \mathscr{C}$ has fixed points, i.e. points x such that $T(x) = x$.

9.9. (The formal completion of a category). Let $\mathscr{C}$ be a category and denote by $\mathscr{C}'$ the category whose objects are small functors in $\mathscr{C}$. If $F: \mathscr{I} \to \mathscr{C}$ and $G: \mathscr{J} \to \mathscr{C}$ are two small functors in $\mathscr{C}$, define $\mathscr{C}'(F, G) = \varprojlim_j (\varinjlim_i \mathscr{C}(F(i), G(j)))$. Let $\overleftarrow{\mathscr{C}}$ be the dual of $\mathscr{C}'$. Since every object of $\overleftarrow{\mathscr{C}}$ can be viewed as a small functor defined on the point category, there exists a canonical functor $S: \mathscr{C} \to \overleftarrow{\mathscr{C}}$ which is full and faithful. Prove that:

a) The category $\overleftarrow{\mathscr{C}}$ is complete and the functor S commutes with finite products.

b) If $\mathscr{C}$ is a category with finite coproducts, then $\overleftarrow{\mathscr{C}}$ is also a category with finite coproducts, and S commutes with colimits.

c) If $\mathscr{C}'$ is a complete category and $T: \mathscr{C} \to \mathscr{C}'$ is a functor, then there exists a unique functor $\overleftarrow{T}: \overleftarrow{\mathscr{C}} \to \mathscr{C}'$ such that $\overleftarrow{T}S = T$. Moreover, the functor $\overleftarrow{T}$ commutes with limits.

9.10. Let $\{\mathscr{C}_i\}_i$ be a set of categories and assume that $\mathscr{C}_i$ is complete for every i. Show that $\prod_i \mathscr{C}_i$ is also complete.

1.10. Limits and colimits in the category of functors

The notions of limit and colimit are defined for functors; in particular, we examine them in the category of functors.

References: Bucur—Deleanu [1], Cartan—Eilenberg [1], Gabriel—Zisman [1], MacLane [3], Mitchell [2], Pareigis [1], Schubert [1].

Let $\mathscr{C}, \mathscr{C}'$ and $\mathscr{I}$ be categories. By an *$\mathscr{I}$-diagram F* of functors from $\mathscr{C}$ into $\mathscr{C}'$ we mean the following: For each object i of $\mathscr{I}$ there is a functor $F_i: \mathscr{C} \to \mathscr{C}'$ and for each morphism $s: i \to j$ of $\mathscr{I}$ there is a functorial morphism $u_s: F_i \to F_i$ such that

$$u_{1_i} = 1_{F_i} \quad \text{for all } i \in \mathscr{I}, \text{ and}$$

$$u_{s's} = u_{s'} u_s \quad \text{for } i \xrightarrow{s} j \xrightarrow{s'} k.$$

If $\mathscr{C}$ is a small category, then an $\mathscr{I}$-diagram of functors from $\mathscr{C}$ into $\mathscr{C}'$ is in fact a functor $F: \mathscr{I} \to [\mathscr{C}, \mathscr{C}']$ where $[\mathscr{C}, \mathscr{C}']$ is the category of all functors from $\mathscr{C}$ into $\mathscr{C}'$ (see Section 1.3). A *cone over an $\mathscr{I}$-diagram* is a class $\{G, v_i\}_{i\in\mathscr{I}}$ where $G: \mathscr{C} \to \mathscr{C}'$ is a functor and the $v_i: G \to F_i$ are functorial morphisms such that $u_s v_i = v_j$ for every morphism $s: i \to j$ of $\mathscr{I}$. A cone $\{G, v_i\}_{i\in\mathscr{I}}$ over an $\mathscr{I}$-diagram F is a *limit* of F if for every cone $\{H, t_i\}_{i\in\mathscr{I}}$ over F_i there is a unique functorial morphism $t: H \to G$ such that $v_i t = t_i$ for all $i \in \mathscr{I}$. If $\mathscr{C} = \{0\}$ is the point category, then an $\mathscr{I}$-diagram of functors from $\{0\}$ into $\mathscr{C}'$ is in fact a functor from $\mathscr{I}$ into $\mathscr{C}'$. Thus we obtain again as a special case, the concepts "cone over a functor" and "limit of a functor".

THEOREM 10.1. *Let $\mathscr{C}$ be a complete category, $\mathscr{D}$ be a category and $\mathscr{I}$ be a small category. If F is an $\mathscr{I}$-diagram of functors from $\mathscr{D}$ into $\mathscr{C}$ then F has a limit.*

Proof. Let F be an $\mathscr{I}$-diagram of functors from $\mathscr{D}$ into $\mathscr{C}$. For each object D of $\mathscr{D}$ let $P_D: \mathscr{I} \to \mathscr{C}$ denote the functor defined by $P_D(i) = F_i(D)$ for all objects i of $\mathscr{I}$, and $P_D(s) = (u_s)_D: P_D(i) \to P_D(j)$ for each morphism $s: i \to j$ of $\mathscr{I}$. Here $u_s: F_i \to F_j$ is the functorial morphism which appears in the definition of an $\mathscr{I}$-diagram and $(u_s)_D$ is the D-component of this functorial morphism (see Section 1.3). Choose a limit $(G(D), v_D)$ of the functor P_D. Now let $h: D \to D'$ be a morphism of $\mathscr{D}$. Let $\bar{h}: P_D \to P_{D'}$ denote the functorial morphism defined as follows. If $i \in \mathscr{I}$, then $\bar{h}_i = F_i(h): P_D(i) \to P_{D'}(i)$. It is routine to check that the morphisms $\{\bar{h}_i(v_D)_i: G(D) \to P_{D'}(i)\}_{i\in\mathscr{I}}$ give a morphism $\bar{h}v_D: G(D) \to P_{D'}$. Thus, there is a unique morphism $G(h): G(D) \to G(D')$ such that $v_{D'}G(h) = \bar{h}v_D$. We leave it for the reader to prove that the assignments

$$D \rightsquigarrow G(D), \quad h \rightsquigarrow G(h)$$

give a functor $G: \mathscr{D} \to \mathscr{C}$ and the morphisms $\{(v_D)_i\}_{D\in\mathscr{D}}$ give a functorial morphism $v_i: G \to F_i$ such that $u_s v_i = v_j$ for every morphism $s: i \to j$ of $\mathscr{I}$, i.e. such that $\{G, v_i\}_{i\in\mathscr{I}}$ is a cone over the given $\mathscr{I}$-diagram F. We claim that this cone is in fact a limit of F. Indeed, let $\{H, t_i\}_{i\in\mathscr{I}}$ be another cone over F. For each object D of $\mathscr{D}$, the morphisms $\{(t_i)_D: H(D) \to F_i(D) = P_D(i)\}_{i\in\mathscr{I}}$ define a functorial morphism $t_D: H(D) \to P_D$. Hence a unique morphism $\bar{t}_D: H(D) \to G(D)$ can be found such that $v_D\bar{t}_D = t_D$. Finally the morphisms $\{\bar{t}_D\}_{D\in\mathscr{D}}$ define a unique functorial morphism $\bar{t}: H \to G$ such that $v_i\bar{t} = t_i$ for all $i \in \mathscr{I}$, i.e. (G, v_i) is a limit of F, as claimed. ⁎

COROLLARY 10.2. *If $\mathscr{C}$ is a complete category, then for each small category $\mathscr{I}$ the category $[\mathscr{I}, \mathscr{C}]$ is also complete. Also if $\mathscr{C}$ is cocomplete, then $[\mathscr{I}, \mathscr{C}]$ is cocomplete.* ⁎

COROLLARY 10.3. *If $\mathscr{C}$ is a finitely complete category, then for every category $\mathscr{D}$ and every finite category $\mathscr{I}$ any $\mathscr{I}$-diagram of functors from $\mathscr{D}$ into $\mathscr{C}$ has a limit.*

Proof. Suppose that $\mathscr{I}$ is a finite category; then in the proof of Theorem 10.1 the functors P_D are finite functors. ✳

In referring to the construction of a limit of an $\mathscr{I}$-diagram of functors in a category we shall say that they are *formed argumentwise.*

Let $\mathscr{C}$ be a complete category. If $\{F_i: \mathscr{D} \to \mathscr{C}\}_{i \in \mathscr{I}}$ is a family of functors, then we let $\prod_{i \in \mathscr{I}} F_i$ denote the limit of this diagram of functors and call this functor the *product of the given functors.* Dually, we have the concept of *coproduct of functors.* Similarly, if $F \to G \leftarrow H$ is an angle of functors, then its limit will be denoted by $F \prod_G H$ and called the *fibered product of the functors F and H over G.* Dually, we have the notion of *fibered coproduct of functors.*

Let $\mathscr{C}$ be a complete category. Then for every small category $\mathscr{I}$, we may consider the functor

$$\varprojlim_{\mathscr{I}} : [\mathscr{I}, \mathscr{C}] \to \mathscr{C} \tag{8}$$

(usually denoted by $\varprojlim \mathscr{I}$ if there is no danger of confusion) which is defined as follows: For each functor $F: \mathscr{I} \to \mathscr{C}$ choose a limit $\varprojlim F$ of F in $\mathscr{C}$ and define $\varprojlim (F) = \varprojlim F$. If $u: F \to G$ is a functorial morphism, i.e. a morphism of $[\mathscr{I}, \mathscr{C}]$, then $\varprojlim (u): \varprojlim F \to \varprojlim G$ is the unique morphism (usually denoted by $\varprojlim u$) which makes the following diagram commutative:

$$\begin{array}{ccc}
\varprojlim F & \xrightarrow{\;f\;} & F \\
{\scriptstyle \varprojlim u}\downarrow & & \downarrow{\scriptstyle u} \\
\varprojlim G & \xrightarrow[\;g\;]{} & G
\end{array}$$

where f and g are the structural morphisms. The morphism $\varprojlim u$ is called the *limit of the functorial morphism u.* The functor (8) is called the *limit functor.* In particular, if $\mathscr{I}$ is a set, then F and G are respectively families $\{X_i\}_{i \in \mathscr{I}}$ and $\{Y_i\}_{i \in \mathscr{I}}$ of objects of $\mathscr{C}$, and u is a family $\{u_i: X_i \to Y_i\}_{i \in \mathscr{I}}$ of morphisms. In this case, the limit of the functorial morphism u is in fact $\prod_{i \in \mathscr{I}} u_i$, the product of this family of morphisms, and the functor (8) is called the *product functor.* Dually, we have the notions of *colimit functor, colimit of a functorial morphism, coproduct of a family of morphisms* and *coproduct functor.* (See Note 7.15.)

We shall now prove a result which is a variation of Theorem 6.5.

THEOREM 10.4. *Let: $F \mathscr{I} \to \mathscr{C}$ be a functor and let (X, u) be a cone over F. Let D denote the $\mathscr{I}$-diagram of cofunctors from $\mathscr{C}$ into $\mathscr{S}et$ defined as follows: $D_i = H_{F_i}$ for each $i \in \mathscr{I}$ and $D(s) = H(F(s))$ for each morphism s in $\mathscr{I}$. Then (X, u) is a limit of F if and only if $\{H_X, D(u_i)\}_{i \in \mathscr{I}}$ is a limit of the diagram D.*

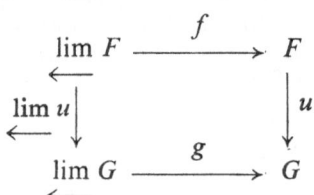

Proof. Assume that (X, u) is a limit of F and let $(G, t_i)_{i\in\mathscr{I}}$ be a cone over the diagram D (here $G: \mathscr{C} \to \mathscr{S}et$ is a cofunctor). Then for each $Y \in \mathscr{C}$, the maps $\{(t_i)_Y: G(Y) \to H_{F_i}(Y) = H^Y(F_i)\}_{i\in\mathscr{I}}$ give a functorial morphism $\bar{t}_Y: G(Y) \to \to H^H F$. Hence by Theorem 6.10 there is a unique map $\bar{t}_Y: G(Y) \to H^Y(X)$ such that $(H^Y u)\bar{t}_Y = t_Y$. Now it is easy to check that the maps $\{\bar{t}_Y\}_{Y\in\mathscr{C}}$ define a functorial morphism $\bar{t}: G \to H_X$ such that $H(u_i)\bar{t} = t_i$ for all $i \in \mathscr{I}$. The proofs of the remaining assertions are left for the reader. ✳

The following result is simple but useful.

PROPOSITION 10.5. *Let $\mathscr{C}$ be a category with fibered products, $G, F: \mathscr{C}' \to \mathscr{C}$ be functors and $u: F \to G$ be a functorial morphism. Then u is a functorial monomorphism if and only if u is an argumentwise monomorphism.*

Proof. Assume that u is a functorial monomorphism. Then the diagram

$$\begin{array}{ccc} F & \xrightarrow{1_F} & F \\ {\scriptstyle 1_F}\downarrow & & \downarrow{\scriptstyle u} \\ F & \xrightarrow{u} & G \end{array}$$

is cartesian.

On the other hand, by Theorem 10.1, for each $X \in \mathscr{C}'$ the diagram

$$\begin{array}{ccc} F(X) & \xrightarrow{1_{F(X)}} & F(X) \\ {\scriptstyle 1_{F(X)}}\downarrow & & \downarrow{\scriptstyle u_X} \\ F(X) & \xrightarrow{u_X} & G(X) \end{array}$$

is a cartesian diagram of sets, i.e. u_X is a monomorphism. Consequently u is an argumentwise monomorphism. ✳

We now give an example of a limit of a diagram of functors. Let $F: \mathscr{C} \to \mathscr{S}et$ be a functor.

The category whose objects are pairs (X, x), where $X \in \mathscr{C}$ and $x \in F(X)$ and a morphism $f: (X, x) \to (Y, y)$ of which is by definition a morphism $f: X \to Y$ of $\mathscr{C}$ such that $F(f)(x) = y$, will be denoted by $\mathscr{P}_F$; $\mathscr{P}_F$ is called the *pointed category associated with F*.

THEOREM 10.6. *Let $F: \mathscr{C} \to \mathscr{S}et$ be a functor, $\mathscr{P}_F$ the pointed category associated with F and D the $\mathscr{P}_F^0$-diagram of functors from $\mathscr{C}$ into $\mathscr{S}et$, which associates with each $(X, x) \in \mathscr{P}_F$ the functor $D((X, x)) = H_X: \mathscr{C} \to \mathscr{S}et$. Also, if $f: (X, x) \to \to (Y, y)$ is a morphism of $\mathscr{P}_F$, then $D(f^0) = F(f): D(Y, y) = H_Y \to H_X = D(X, x)$. For each object (X, x) of $\mathscr{P}_F$,* let *$\varphi_{(X,x)}: D(X, x) \to F$ denote the morphism u^x associated with x (see Section 1.3). Then the morphisms $\varphi = \{\varphi_{(X,x)}\}_{(X,x)\in\mathscr{P}_F}$ define a cocone over the diagram D whose vertex is F and (φ, F) is a colimit of the diagram D.*

Proof. We first check that for every morphism $f: (X, x) \to (Y, y)$ of $\mathscr{P}_F$ we have $\varphi_{(X,x)}D(f) = \varphi_{(Y,y)}$. But that follows from Theorem 3.19 since $\varphi_{(X,x)} = u^x$, $D(f) = H^0(f)$ and $\varphi_{(Y,y)} = u^y$.

Now let (t, G) be another cocone over D. For each $X \in \mathscr{C}$ denote by $\bar{t}_X: F(X) \to G(X)$ the map which takes the element $x \in F(X)$ to the element $[t_{(X,x)}]_X(1_X)$. We leave it for the reader to prove that the maps $\{\bar{t}_X\}_{X\in\mathscr{C}}$ define a functorial morphism $\bar{t}: F \to G$ such that $\bar{t}\varphi = t$. The uniqueness of $\bar{t}$ is obvious, by its construction and by Theorem 3.19. ⁎

Theorem 10.6 can be dualized for cofunctors.

Exercises

10.1. Let $\{f_i: X_i \to Y_i\}_{i\in\mathscr{I}}$ be a set of morphisms in a category $\mathscr{C}$. Assume that for each $i \in \mathscr{I}$, f_i is a monomorphism (epimorphism). Show that $\prod f_i(\coprod f_i)$ is also a monomorphism (an epimorphism). Also show that if $\mathscr{C}$ has a zero object then the converse is also true.

10.2. Let $\{f_i: X_i \to Y_i\}_{i\in\mathscr{I}}$ be a set of epimorphisms (monomorphisms) in $\mathscr{S}et$. Show that $\prod f_i(\coprod f_i)$ is also an epimorphism (a monomorphism). Show that the same result also holds in $\mathscr{R}el$, $\mathscr{P}re$, $\mathscr{O}rd$, $\mathscr{M}on$, $\mathscr{G}r$, $\mathscr{A}b$ and $\mathscr{T}op$.

10.3. Let $F_1: \mathscr{D}_1 \to \mathscr{C}$, $F_2: \mathscr{D}_2 \to \mathscr{C}$ and $G: \mathscr{D}_2 \to \mathscr{D}_1$ be functors. Assume that (X_1, u^1) and (X_2, u^2) are limits of F_1 and F_2 respectively. Finally, let $f: F_1G \to F_2$ be a functorial morphism. Prove that there is a unique morphism $v: X_1 \to X_2$ such that the diagram

$$\begin{array}{ccc} X_1 & \xrightarrow{\quad v \quad} & X_2 \\ {\scriptstyle u^1_{G(X)}}\downarrow & & \downarrow{\scriptstyle u^2_X} \\ F_1G(X) & \xrightarrow{\quad f_X \quad} & F_2(X) \end{array}$$

is commutative for all objects X of $\mathscr{D}_2$.

10.4. Let $F, G: \mathscr{C} \to \mathscr{S}et$ be functors, and $u: F \to G$ be a functorial epimorphism. Prove that for each object X of $\mathscr{C}$ and each functorial morphism $v: H^X \to G$ there is a functorial morphism $\bar{v}: H^X \to F$ such that $u\bar{v} = v$. In particular, every functorial epimorphism $F \to H^X$ has a left inverse.

10.5. Let $\mathscr{C}$ be a complete category and $F: \mathscr{I} \to \mathscr{C}$ a functor. Assume there is a set $\{u_j, X_j\}_{j\in\mathscr{J}}$ of cocones over F such that if (u, X) is a cocone over F, then there is an index $j \in \mathscr{J}$ and a morphism $f: X_j \to X$ such that $fu_j = u$. Prove that F has a colimit in $\mathscr{C}$. (Hint: Since $\mathscr{C}$ is complete, the category $F/\mathscr{C}$ has an initial object.)

10.6. Let $F: \mathscr{I} \to \mathscr{S}et$ be a functor such that $\mathscr{P}_F$, the pointed category associated with F, is left bounded. Show that F has a limit and a colimit.

1.11. Adjoint functors

One of the most important notions in the theory of categories is that of an adjoint functor. This concept was introduced by Kan [1] although many examples of adjoint functors can be found in earlier papers (see Cartan—Eilenberg [1] and Grothendieck [1]). In the following sections we shall consider some general results, and some examples of adjoint functors. In Section 2.6, Ch. 2, we give some criteria for the existence of adjoint functors. Other examples of adjoint functors are given in the subsequent sections.

References: Bucur—Deleanu [1], Cartan—Eilenberg [1], Gabriel [1], Kan [1], Mitchell [2], Pareigis [1], Schubert [1], Linton [1], Grothendieck—Verdier [1], Kleisly [1], Eilenberg—Moore [1], MacLane [3].

Let $F:\mathscr{C}\to\mathscr{C}'$ and $G:\mathscr{C}'\to\mathscr{C}$ be functors. A functorial morphism $u:\mathrm{Id}\,\mathscr{C}\to GF$ is called a *functorial arrow*. Likewise, a functorial morphism $v: FG\to\mathrm{Id}\,\mathscr{C}'$ is called a *functorial coarrow*. The functors considered above enable us to define the functors

$$\begin{aligned}&\mathscr{C}'(F(?),?):\mathscr{C}^0\textstyle\prod\mathscr{C}'\to\mathscr{S}et,\\&\mathscr{C}(?,G(?)):\mathscr{C}^0\textstyle\prod\mathscr{C}'\to\mathscr{S}et\end{aligned}\qquad(9)$$

in the following manner: If (X^0,X') is an object of $\mathscr{C}^0\prod\mathscr{C}'$, then $\mathscr{C}'(F(?),?)(X^0,X')=\mathscr{C}'(F(X),X')$ and $\mathscr{C}(?,G(?))(X^0,X')=\mathscr{C}(X,G(X'))$. If $(f^0,f')\colon(X^0,X')\to(Y^0,Y')$ is a morphism in $\mathscr{C}^0\prod\mathscr{C}'$, then $\mathscr{C}'(G(?),?)(f^0,f')=\mathscr{C}'(F(f),f')$ is the map that associates with each element $g\in\mathscr{C}'(F(X),X')$, the morphism $f'gF(f)$, i.e. $\mathscr{C}'(F(f),f')(g)=f'gF(f)$. (Observe that $f\colon Y\to X$ and $f'\colon X'\to Y'$ are morphisms of $\mathscr{C}$ and $\mathscr{C}'$, respectively, so that $\mathscr{C}'(F(f),f')$ associates with the morphism $g\colon F(X)\to X'$ the composition of morphisms: $F(Y)\xrightarrow{F(f)}F(X)\xrightarrow{g}X'\xrightarrow{f'}Y'$.) Also, $\mathscr{C}(?,G(?))(f^0,f')=\mathscr{C}(f,G(f'))$ is the map which associates with a morphism $h\colon X\to G(X')$ the morphism $G(f')hf\colon Y\to G(Y')$, i.e. $\mathscr{C}(f,G(f'))(h)=G(f')hf$. We leave it for the reader to verify that the morphisms in (9) are functors.

PROPOSITION 11.1. *Let $F:\mathscr{C}\to\mathscr{C}'$ and $G:\mathscr{C}'\to\mathscr{C}$ be functors and let $u:\mathrm{Id}\,\mathscr{C}\to GF$ be a functorial arrow. For each object (X^0,X') of $\mathscr{C}^0\prod\mathscr{C}'$ let $\tilde{u}_{(X,X')}\colon\mathscr{C}'(F(X),X')\to\mathscr{C}(X,G(X'))$ denote the map defined by*

$$u_{(X,X')}(g)=G(g)u_X.$$

Then:

a) *The class of maps $\{\tilde{u}_{(X,X')}\}_{(X^0,X')\in\mathscr{C}^0\prod\mathscr{C}'}$ define a functorial morphism $\tilde{u}\colon\mathscr{C}'(F(?),?)\to\mathscr{C}(?,G(?))$.*

b) *The assignment $u\rightsquigarrow\tilde{u}$ gives a bijection of the class $[Id\,\mathscr{C},GF]$ onto the class $[\mathscr{C}'(F(?),?),\ \mathscr{C}(?,G(?))]$.*

Proof. a) We prove that for each morphism $(f^0, f'): (X^0, X') \to (Y^0, Y')$ of $\mathscr{C}^0 \coprod \mathscr{C}'$ the diagram

$$\begin{array}{ccc} \mathscr{C}'(F(X), X') & \xrightarrow{\tilde{u}_{(X, X')}} & \mathscr{C}(X, G(X')) \\ {\scriptstyle \mathscr{C}'(F(f), f')}\downarrow & & \downarrow{\scriptstyle \mathscr{C}(f, G(f'))} \\ \mathscr{C}'(G(Y), Y') & \xrightarrow{\tilde{u}_{(Y, Y')}} & \mathscr{C}(Y, G(Y')) \end{array} \tag{10}$$

is commutative.

On the one hand,

$$\mathscr{C}(f, G(f'))\, \tilde{u}_{(X, X')}(g) = \mathscr{C}(f, G(f'))\, (G(g) u_X) = G(f')\, G(g) u_X f.$$

On the other hand

$$\tilde{u}_{(Y, Y')} \mathscr{C}'(F(f), f')\, (g) = \tilde{u}_{(Y, Y')}(f' g F(f)) = G(f' g F(f)) u_Y = G(f') G(g) GF(f) u_Y.$$

Thus the commutativity of diagram (10) follows from the fact that the diagram

$$\begin{array}{ccc} Y & \xrightarrow{u_Y} & GF(Y) \\ {\scriptstyle f}\downarrow & & \downarrow{\scriptstyle GF(f)} \\ X & \xrightarrow{u_X} & GF(X) \end{array}$$

is commutative.

b) Assume that $\tilde{u}: \mathscr{C}'(F(?), ?) \to \mathscr{C}(?, G(?))$ is a functorial morphism and for each $X \in \mathscr{C}$ consider the morphism

$$u_X = \tilde{u}_{(X, F(X))}(1_{F(X)}): X \to GF(X)$$

(here $\tilde{u}_{(X, F(X))}: \mathscr{C}'(F(X), F(X)) \to \mathscr{C}(X, GF(X))$). We shall prove that the morphisms $\{u_X\}_{X \in \mathscr{C}}$ define a functorial arrow. To show that, it will suffice to prove that the diagram

$$\begin{array}{ccc} X & \xrightarrow{u_X} & GF(X) \\ {\scriptstyle f}\downarrow & & \downarrow{\scriptstyle GF(f)} \\ Y & \xrightarrow{u_Y} & GF(Y) \end{array} \tag{11}$$

is commutative for all morphisms $f: X \to Y$ of $\mathscr{C}$. By hypothesis, the diagram

$$\begin{array}{ccc} \mathscr{C}'(F(X), F(X)) & \xrightarrow{\tilde{u}_{(X, F(X))}} & \mathscr{C}(X, GF(X)) \\ {\scriptstyle \mathscr{C}'(1_{F(X)}, F(f))}\downarrow & & \downarrow{\scriptstyle \mathscr{C}(1_X, GF(f))} \\ \mathscr{C}'(F(X), F(Y)) & \xrightarrow{\tilde{u}_{(X, F(Y))}} & \mathscr{C}(X, GF(Y)) \end{array}$$

is commutative, i.e.

$$\mathscr{C}(1_X, GF(f))\tilde{u}_{(X,F(X))}(1_{F(X)}) = \mathscr{C}(1_X, GF(f))(u_X) = GF(f)u_X$$
$$= \tilde{u}_{(X,F(Y))}\mathscr{C}'(1_{F(X)}, F(f))(1_{F(X)}) = \tilde{u}_{(X,F(Y))}(F(f)). \tag{12}$$

From the hypothesis, the diagram

$$\begin{array}{ccc} \mathscr{C}'(F(Y), F(Y)) & \xrightarrow{\tilde{u}_{(Y,F(Y))}} & \mathscr{C}(Y, GF(Y)) \\ {\scriptstyle \mathscr{C}'(F(f), 1_{F(Y)})} \downarrow & & \downarrow {\scriptstyle \mathscr{C}(f, 1_{GF(Y)})} \\ \mathscr{C}'(F(X), F(Y) & \xrightarrow{\tilde{u}_{(X,F(Y))}} & \mathscr{C}(X, GF(Y)) \end{array}$$

must also be commutative. Hence

$$\mathscr{C}(f, 1_{GF(Y)})\tilde{u}_{(Y,F(Y))}(1_{F(Y)}) = \mathscr{C}(f, 1_{GF(Y)})(u_Y) = u_Y f \tag{13}$$
$$= \tilde{u}_{(X,F(Y))}\mathscr{C}'(F(f), 1_{F(Y)})(1_{F(Y)}) = \tilde{u}_{(X,F(Y))}(F(f)),$$

Now a glance at relations (12) and (13) shows that diagram (11) is commutative.

In the above we have associated with each functorial arrow $u\colon \mathrm{Id}\,\mathscr{C} \to GF$ a functorial morphism $\tilde{u}\colon \mathscr{C}(F(?), ?) \to \mathscr{C}(?, G(?))$ and conversely. We leave it for the reader to check that the assignment $u \rightsquigarrow \tilde{u}$ is a bijection. ✳

We note that $\tilde{u}$ is called the *functorial morphism associated with the functorial arrow* u. On the other hand, if the functorial morphism $\tilde{u}$ is given, then the functorial arrow u defined in Proposition 11.1 is called the *functorial arrow associated with the functorial morphism* $\tilde{u}$.

We also have the following dual result.

PROPOSITION 11.2. *Let* $F\colon \mathscr{C} \to \mathscr{C}'$ *and* $G\colon \mathscr{C}' \to \mathscr{C}$ *be functors, and let* $v\colon FG \to \mathrm{Id}\,\mathscr{C}'$ *be a functorial coarrow. For each object* (X^0, X') *of* $\mathscr{C}^0 \prod \mathscr{C}'$ *let* $\tilde{v}_{(X,X')}\colon \mathscr{C}(X, G(X')) \to \mathscr{C}'(F(X), X')$ *denote the map defined by*

$$\tilde{v}_{(X,X')}(g) = v_{X'}F(g).$$

Then:

a) *The class of maps* $\{\tilde{v}_{(X^0,X')}\}_{(X^0,X') \in \mathscr{C}^0 \prod \mathscr{C}'}$ *defines a functorial morphism* $\tilde{v}\colon \mathscr{C}(?, G(?)) \to \mathscr{C}'(F(?), ?)$.

b) *The assignment* $v \rightsquigarrow \tilde{v}$ *gives a bijection of the class* $[FG, \mathrm{Id}\,\mathscr{C}']$ *onto the class* $[\mathscr{C}(?, G(?)), \mathscr{C}'(F(?), ?)]$. ✳

We thus have the notions of *functorial morphism associated with a functorial coarrow* and *functorial coarrow associated with a functorial morphism.*

PROPOSITION 11.3. *Let* $F\colon \mathscr{C} \to \mathscr{C}'$ *and* $G\colon \mathscr{C}' \to \mathscr{C}$ *be functors,* $u\colon \mathrm{Id}\,\mathscr{C} \to GF$ *a functorial arrow and* $v\colon FG \to \mathrm{Id}\,\mathscr{C}'$ *a functorial coarrow. Also, let*

$$\tilde{u}\colon \mathscr{C}'(F(?), ?) \to \mathscr{C}(?, G(?))$$

and

$$\tilde{v}: \mathscr{C}(?, G(?)) \to \mathscr{C}'(F(?), ?)$$

be their associated functorial morphisms. Then:

a) *$\tilde{u}\tilde{v}$ is the identity functorial morphism of the functor $\mathscr{C}(?, G)?))$ if and only if the composition of functorial morphisms*

$$G \xrightarrow{uG} GFG \xrightarrow{Gv} G$$

is the identity of G, i.e. $(Gv)(uG) = 1_G$.

b) *$\tilde{v}\tilde{u}$ is the identity functorial morphism of the functor $\mathscr{C}'(F(?), ?)$ if and only if the composition of functorial morphisms*

$$F \xrightarrow{Fu} FGF \xrightarrow{vF} F$$

is the identity of F, i.e. $(vF)(Fu) = 1_F$.

Proof. a) Assume that $\tilde{u}\tilde{v}$ is the identity morphism of the functor $\mathscr{C}(?, G(?))$, i.e. for each object (X^0, X') of $\mathscr{C}^0 \prod \mathscr{C}'$ and each element f of $\mathscr{C}(X, G(X'))$ we have that

$$\tilde{u}_{(X, X')}\tilde{v}_{(X, X')}(f) = f.$$

Then for every $X' \in \mathscr{C}'$ we have that

$$[(Gv)(uG)]_{X'} = (Gv)_{X'}(uG)_{X'} = G(u_{X'})u_{G(X')}.$$

But by Proposition 11.1,

$$G(v_{X'})u_{G(X')} = \tilde{u}_{(G(X'), X')}(v_{X'}) = \tilde{u}_{(G(X'), X')}\tilde{v}_{(G(X'), X')}(1_{G(X')}) = 1_{G(X')}.$$

Conversely, assume that $(Gv)(uG) = 1_G$ and let $(X^0, X') \in \mathscr{C}^0 \prod \mathscr{C}'$ and $f \in \mathscr{C}(X, G(X'))$. By Proposition 11.2, $\tilde{v}_{(X, X')}(f) = v_{X'}F(t)$ so that $\tilde{u}_{(X, X')}\tilde{v}_{(X, X')}(f) = \tilde{u}_{(X, X')}(v_{X'}F(f)) = G(v_{X'})\, GF(f)u_X$ (see Proposition 11.1). But we have the commutative diagram

$$\begin{array}{ccc} X & \xrightarrow{\quad f \quad} & G(X') \\ {\scriptstyle u_X}\downarrow & & \downarrow{\scriptstyle u_{G(X')}} \\ GF(X) & \xrightarrow{\quad GF(f) \quad} & GFG(X') \end{array}$$

and thus $G(v_{X'})GF(f)u_X = G(v_{X'})u_{G(X')}f = 1_{G(X')}f = f$. Part b) of the proposition is obtained in a similar manner. ✳

We say that a functor $F: \mathscr{C} \to \mathscr{C}'$ is *adjoint* to the functor $G: \mathscr{C}' \to \mathscr{C}$ or that G is *coadjoint* to F if there is a functorial arrow $u: \mathrm{Id}\,\mathscr{C} \to GF$ and a functorial coarrow $v: FG \to Id\,\mathscr{C}'$ such that the following conditions are satisfied:

(Ad) $(Gv)(uG) = 1_G$ and $(vF)(Fu) = 1_F$.

These equalities are called the *conditions of adjointness*. The functorial arrow u is said to be an *arrow of adjunction quasi-inverse* to the coarrow v and v is said to be a *coarrow of adjunction quasi-inverse* to the arrow u. Furthermore, we say that F is left adjoint to G, and that G is right adjoint to F.

PROPOSITION 11.4. *A functor* $F: \mathscr{C} \to \mathscr{C}'$ *is adjoint to the functor* $G: \mathscr{C}' \to \mathscr{C}$ *if and only if there is a functorial isomorphism*

$$\varphi: \mathscr{C}'(F(?), ?) \to \mathscr{C}(?, G(?)),$$

i.e. if and only if for each pair of objects (X, X'), $X \in \mathscr{C}$ *and* $X' \in \mathscr{C}'$, *there is a bijection*

$$\varphi_{(X, X')}: \mathscr{C}'(F(X), X') \to \mathscr{C}(X, G(X'))$$

that is functorial in X *and* X' *simultaneously.*

The proof is a direct consequence of Proposition 11.3. ✳

The isomorphism φ is called the *isomorphism of adjointness.*

Example 11.5. Let $S: \mathcal{T}op \to \mathcal{S}et$ be the forgetful functor and let $D: \mathcal{S}et \to \mathcal{T}op$ be the functor defined in Example 3.2. Then D is adjoint to S. Indeed, since for each set X, we have $SD(X) = X$, so that $SD = \mathrm{Id}\,\mathcal{S}et$, the identity of SD is an arrow of adjunction u. Likewise, if Y is a topological space, then the identity map $1_Y: DS(Y) \to Y$ is continuous and this map defines the coarrow of adjunction quasi-inverse to u.

Example 11.6. Let $I: \mathcal{A}b \to \mathcal{G}r$ be the inclusion functor. Let $A: \mathcal{G}r \to \mathcal{A}b$ denote the functor defined as follows. If G is a group, consider the smallest normal subgroup $c(G)$ that contains the elements $xyx^{-1}y^{-1}$ for all $x, y \in G$, i.e. the *commutator subgroup* of G. Define $A(G)$ to be $G/c(G)$. If $f: G \to G'$ is a morphism of groups then $f(xyx^{-1}y^{-1}) = f(x)f(y)f(x)^{-1}f(y)^{-1}$ so that $f(c(G)) \subseteq c(G')$. Thus there is a unique morphism $A(f): A(G) \to A(G')$ such that the diagram

$$\begin{array}{ccc} G & \xrightarrow{\;u_G\;} & A(G) \\ {\scriptstyle f}\downarrow & & \downarrow{\scriptstyle A(f)} \\ G' & \xrightarrow{\;u_{G'}\;} & A(G') \end{array}$$

is commutative (here u_G and $u_{G'}$ are the canonical morphisms). It is easy to check that $A(G)$ is an abelian group for all G. The functor A is coadjoint to I. To prove this we observe that $AI \equiv \mathrm{Id}\,\mathcal{A}b$; hence the identity morphism of $\mathrm{Id}\,\mathcal{A}b$ can be considered to be a coarrow of adjunction. Because $IA(G) = A(G)$, the above morphisms $\{u_G\}_{G \in \mathcal{G}r}$ define an arrow of adjunction.

Example 11.7. Let M be a set. Consider the functor

$$M \prod ?: \mathcal{S}et \to \mathcal{S}et$$

defined by $(M \prod ?)(X) = M \prod X$ for each set X and $(M \prod ?)(f) = M \prod f = 1_M \prod f$ for each map $f: X \to Y$ (recall that $1_M \prod f: M \prod X \to M \prod Y$ is the product of the family of morphisms $(1_M, f)$). Thus the functor $M\coprod ?$ is adjoint to the functor H^M (recall that $H^M(X) = \mathscr{S}et(M, X)$). Indeed, if X and Y are sets, define a map

$$\varphi_{(X,Y)}: \mathscr{S}et(M\coprod X, Y) \to \mathscr{S}et(X, H^M(Y))$$

as follows: If $f: M\coprod X \to Y$ is a map, then $\varphi_{(X,Y)}(f): X \to H^M(Y) = \mathscr{S}et(M, Y)$ is the map that associates with each element x of X, the map $\varphi_{(X,Y)}(f)(x): M \to Y$ defined by $[\varphi_{(X,Y)}(f)(x)](m) = f(m, x)$. We invite the reader to prove that the maps $\{\varphi_{(X,Y)}\}_{(X,Y)}$ give a functorial isomorphism between $\mathscr{S}et(M\coprod ?, ?)$ and $\mathscr{S}et(?, H^M(?))$.

Let $F: \mathscr{C} \to \mathscr{C}'$ be a functor. For each object X' of $\mathscr{C}'$, let X'/F denote the category whose objects are pairs (f, X) where $X \in \mathscr{C}$ and $f: X' \to F(X)$ is a morphism of $\mathscr{C}'$ and a morphism $t: (f, X) \to (g, Y)$ in X'/F is a morphism $t: X \to Y$ of $\mathscr{C}$ such that $F(t)f = g$. Dually, we may define the category F/X'. In particular, $X'/\mathscr{C}'$ is canonically isomorphic to $X'/\mathrm{Id}\,\mathscr{C}'$.

THEOREM 11.8. *Let $G: \mathscr{C}' \to \mathscr{C}$ be a functor. The following are equivalent:*
i) G has an adjoint $F: \mathscr{C} \to \mathscr{C}'$.
ii) For each object X of $\mathscr{C}$ the functor $H^X G: \mathscr{C}' \to \mathscr{S}et$ is representable.
iii) For each $X \in \mathscr{C}$ the category X/G has an initial object.

Proof. i) $\Rightarrow$ ii). Since F is adjoint to G, there is a functional isomorphism $\varphi: \mathscr{C}'(F(?), ?) \to \mathscr{C}(?, G(?))$ and hence we have the functorial isomorphism

$$\varphi_{(X,?)}: \mathscr{C}'(F(X), ?) \to \mathscr{C}(X, G(?)) = H^X G$$

or, equivalently, $H^X G$ is representable, a universal element being $(F(X), 1_{F(X)})$.

ii) $\to$ iii). Assume that $H^X G$ is representable for each $X \in \mathscr{C}$ and let (X', x) be a universal element. where $X' \in \mathscr{C}'$ and $X \in H^X G(X') = \mathscr{C}(X, G(X'))$. Thus the functorial morphism

$$u^x : H^{X'} \to H^X G$$

associated with x is an isomorphism, and hence for each $Y' \in \mathscr{C}'$ the map $u^x_{Y'}: H^{X'}(Y') \to H^X G(Y')$ is a bijection. We claim that the object (f, X') (where $f = u^x_{X'}(1_{X'})$) is an initial object in the category X/G. Indeed, if (t, Y') is another object of X/G then $t: X \to G(Y')$ is a morphism, so that there is a unique element $s \in \mathscr{C}'(X', Y')$ such that $u^x_{Y'}(s) = t$. From the commutative diagram

$$\begin{array}{ccc}
H^{X'}(X') = \mathscr{C}'(X', X') & \xrightarrow{\;u^x_{X'}\;} & \mathscr{C}(X, G(X')) = H^X G(X') \\
{\scriptstyle H^{X'}(s)}\downarrow & & \downarrow{\scriptstyle H^X(G(s))} \\
H^{X'}(Y') = \mathscr{C}'(X', Y') & \xrightarrow{\;u^x_{Y'}\;} & \mathscr{C}(X, G(Y')) = H^X G(Y'))
\end{array}$$

we have that $H^X(G(s))u^x_{X'}(1_{X'}) = H^X(G(s))(f) = G(s)f = u^x_{Y'}H^{X'}(s)(1_{X'}) = u^x_{Y'}(s) = t$.

iii) ⇒ *i*). For each $X \in \mathscr{C}$, let $(u_X, F(X))$ be an initial object of X/G. Here, $F(X) \in \mathscr{C}'$ and $u_X: X \to GF(X))$ is a morphism of $\mathscr{C}'$. If $f: X \to Y$ is a morphism of $\mathscr{C}$, let $F(f): F(X) \to F(Y)$ denote the unique morphism of $\mathscr{C}'$ such that the diagram

$$\begin{array}{ccc} X & \xrightarrow{u_X} & GF(X) \\ {\scriptstyle f}\downarrow & & \downarrow{\scriptstyle GF(f)} \\ Y & \xrightarrow{u_Y} & GF(Y) \end{array}$$

is commutative. Now it is clear that the assignments

$$X \rightsquigarrow F(X),$$
$$f \rightsquigarrow F(f)$$

give a functor $F: \mathscr{C} \to \mathscr{C}'$. Finally, we shall prove that F is adjoint to G. To this end, define the functorial morphism

$$\varphi: \mathscr{C}(?, G(?)) \to \mathscr{C}'(F(?), ?)$$

as follows. If $(X^0, X') \in \mathscr{C}^0 \coprod \mathscr{C}'$, then $\varphi_{(X,X')}: \mathscr{C}(X, G(X')) \to \mathscr{C}'(F(X), X')$ is the map that associates with each element $f \in \mathscr{C}(X, G(X'))$ the unique morphism $\varphi_{(X,X')}(f) \in \mathscr{C}(F(X), X')$ that makes the diagram

$$\begin{array}{ccc} X & \xrightarrow{u_X} & GF(X) \\ & {\scriptstyle f}\searrow \quad \swarrow{\scriptstyle G(\varphi_{(X,X')}(f))} & \\ & G(Y') & \end{array}$$

commutative.

We leave it for the reader to check that φ is indeed a functorial isomorphism. ✻

By duality, we have the following result:

THEOREM 11.9. *Let $F: \mathscr{C} \to \mathscr{C}'$ be a functor. The following are equivalent:*

i) The functor F has a coadjoint.

ii) For each object $X' \in \mathscr{C}'$, the functor $H_{X'}F$ is representable.

iii) For each $X' \in \mathscr{C}'$, the category F/X' has a final object. ✻

A subcategory $\mathscr{D}$ of a category $\mathscr{C}$ is called *reflective* if the inclusion functor $U: \mathscr{D} \to \mathscr{C}$ has an adjoint. By Theorem 11.8, the subcategory $\mathscr{D}$ is reflective if and only if for each object X of $\mathscr{C}$ the category X/U has an initial object. But this means that for each object X of $\mathscr{C}$ there is a morphism $u_X: X \to \bar{X}$, where $\bar{X} \in \mathscr{D}$, such that for each morphism $f: X \to D$, where $D \in \mathscr{D}$, there is a unique morphism $\bar{f}: \bar{X} \to D$ of $\mathscr{D}$ with $\bar{f}u_X = f$. Usually the morphism u_X is called an *arrow of reflection*. Dually, we have the concepts of *coreflective subcategory* and *coarrow of coreflection*. An adjoint of the inclusion functor is called a *reflector* (*coreflector*).

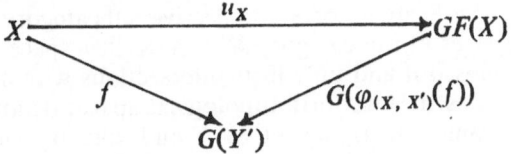

Example 11.10. Let X be a preordered set, $\overline{X}$ be the ordered set defined in Example 4.2, and $p_X: X \to \overline{X}$ the canonical map. It is easy to see that the pair $(p_X, \overline{X})$ is an initial object in the category $X/\mathcal{O}rd$. Consequently $\mathcal{O}rd$ is a reflective subcategory of $\mathcal{P}re$.

Example 11.11. The category $\mathcal{H}d$ is a reflective subcategory of $\mathcal{T}op$. We first make two observations:

a) In $\mathcal{T}op$ a product of Hausdorff spaces is also a Hausdorff space. Indeed, let $\{X_i\}_i$ be a family of Hausdorff spaces and let $x = \{x_i\}_i$ and $y = \{y_i\}_i$ be two distinct points of $\prod_i X_i$. Then there exists an index j such that $x_j \neq y_j$. Let U be a neighborhood of x_j and V be neighborhood of y_j such that $U \cap V = \emptyset$. The subsets $U\prod(\prod_{i\neq j}X_i)$ and $V\prod(\prod_{i\neq j}X_i)$ are disjoint neighborhoods of x and y respectively.

b) If $f: X' \to X$ is an injective continuous map and X is Hausdorff, then X' is also Hausdorff.

By a) and b), and by the construction of limits in $\mathcal{T}op$ (see Theorem 9.1), we conclude that a limit of Hausdorff spaces is again a Hausdorff space. Now, let X be a topological space. Consider the category $X/\mathcal{A}$ to be the full subcategory of $X/\mathcal{H}d$ whose objects are pairs (f, Y) such that f is a surjective continuous map. It is clear that the codomain functor $\mathrm{K}: X/\mathcal{A} \to \mathcal{T}op$ has a limit $\overline{X}$ and that there is a canonical morphism $q_X: X \to \overline{X}$. By the above considerations, $\overline{X}$ is a Hausdorff space and q_X is an arrow of reflection. Sometimes $\overline{X}$ is called the "largest Hausdorff quotient" of X.

Example 11.12. An example of a coreflective subcategory is given by the subcategory of *Kelley spaces* in the category $\mathcal{H}d$. A Kelley space is a Hausdorff space in which a subset is closed if and only if its intersections with all compact subspaces are closed. Now, if X is a Hausdorff topological space, denote by X^k the topological space with the same underlying set as X and take the closed sets to be those whose intersections with the compact subsets of the topology on X are closed. The identity map of the set X gives an coarrow of coreflection $X^k \to X$.

Exercises

11.1. Prove that the functor $G: \mathcal{S}et \to \mathcal{T}op$ defined in Example 3.12 is coadjoint to the forgetful functor $S: \mathcal{T}op \to \mathcal{S}et$.

11.2. Let $\mathscr{C} \xrightarrow{F} \mathscr{C}' \xrightarrow{F'} \mathscr{C}''$, $\mathscr{C}'' \xrightarrow{G} \mathscr{C}' \xrightarrow{G'} \mathscr{C}$ be functors such that F is adjoint to G and F' is adjoint to G'. Show that $F'F$ is adjoint to GG'.

11.3. Find the functorial arrow and the functorial coarrow associated with the functorial isomorphism in Exercise 11.2.

11.4. Let $F: \mathscr{C} \to \mathscr{C}'$ be a functor. Prove that if F is representative and full, then X'/F is equivalent to $X'/\mathscr{C}'$ for all $X' \in \mathscr{C}'$.

11.5. Let $F, F': \mathscr{C} \to \mathscr{C}'$ and $G, G': \mathscr{C}' \to \mathscr{C}$ be functors such that F is adjoint to G and F' is adjoint to G'. Assume that there is a functorial morphism $t: G \to G'$. Prove that:

a) There is a unique functorial morphism $s: F' \to F$ such that for each $X \in \mathscr{C}$, and each $X' \in \mathscr{C}'$, the diagram

$$\begin{array}{ccc} \mathscr{C}'(F(X), X') & \xrightarrow{\varphi_{(X,X')}} & \mathscr{C}(X, G(X')) \\ \downarrow{\scriptstyle H_{X'}(s_X)} & & \downarrow{\scriptstyle H^X(t_{X'})} \\ \mathscr{C}'(F'(X), X') & \xrightarrow{\varphi'_{(X,X')}} & \mathscr{C}(X, G'(X')) \end{array}$$

is commutative, where φ and φ' are the isomorphisms of adjointness.

b) If t is an isomorphism then s is also an isomorphism. In particular, if a functor F had two coadjoints, G, G', then $G \simeq G'$.

c) Let $u: Id\,\mathscr{C} \to GF$, $v: FG \to Id\,\mathscr{C}'$, $u': Id\,\mathscr{C} \to G'F'$, $v': F'G' \to Id\,\mathscr{C}'$ be the arrows and coarrows of adjunction respectively. Then we have the following equalities:

$$(tF)u = (G's)u' \text{ and } v(sG) = v'(F't).$$

11.6. Let $F: \mathscr{C} \to \mathscr{C}'$ be a functor and let G be coadjoint to F. Show that for each small category $\mathscr{I}$, the functor $[\mathscr{I}, F]: [\mathscr{I}, \mathscr{C}] \to [\mathscr{I}, \mathscr{C}']$ is adjoint to the functor $[\mathscr{I}, G]$. Moreover, if $\mathscr{C}$ and $\mathscr{C}'$ are small categories, then, show that for each category $\mathscr{D}$, the functor $[F, \mathscr{D}]: [\mathscr{C}', \mathscr{D}] \to [\mathscr{C}, \mathscr{D}]$ is coadjoint to the functor $[G, \mathscr{D}]$.

11.7. Describe the adjoint functors between preordered categories. What are the conditions (Ad) in this case?

11.8. Assume that the functor $F: \mathscr{C} \to \mathscr{C}'$ has a coadjoint G. Let $u: Id\,\mathscr{C} \to GF$ be an arrow of adjunction. Prove that:

a) For each $X \in \mathscr{C}$, the assignment $(f, Y) \rightsquigarrow (X, G(f)u_X)$ gives an equivalence between $F(X)/\mathscr{C}'$ and X/G. Dually, F/X' and $\mathscr{C}/G(X')$ are equivalent for each $X' \in \mathscr{C}'$.

b) For each object X' of $\mathscr{C}'$, the functor G defines a functor $G': \mathscr{C}'/X' \to \mathscr{C}/G(X')$ which has an adjoint.

11.9. a) Let X be an object of $\mathscr{C}$. Show that the functor $H^X: \mathscr{C} \to \mathscr{S}et$ has an adjoint if and only if the coproduct $X^{(\mathscr{I})}$ is defined for each set $\mathscr{I}$.

b) If $\mathscr{C}$ is cocomplete, show that a functor $F: \mathscr{C} \to \mathscr{S}et$ had an adjoint if and only if it is representable. Prove that, moreover, a functor $G: \mathscr{S}et \to \mathscr{C}$ has a coadjoint if and only if it commutes with coproducts.

11.10. Suppose that $\mathscr{C}$ is a category with coproducts and X is an object of $\mathscr{C}$ and that $X^{(\mathscr{I})}$ is defined for every set $\mathscr{I}$. Show that the functor $F: \mathscr{C} \to \mathscr{S}et$ has an adjoint if and only if it is representable. Show that, moreover, a functor $G: \mathscr{S}et \to \mathscr{C}$ has a coadjoint if and only if G commutes with coproducts.

11.11. A functor $F: \mathscr{C} \to \mathscr{C}$ is called *selfadjoint* if it is its own adjoint. Show that every selfadjoint functor $F: \mathscr{S}et \to \mathscr{S}et$ is isomorphic to the identity functor. The same result holds for $\mathscr{T}op$, $\mathscr{G}r$ and $\mathscr{A}b$.

11.12. Let $\mathscr{C}$ be a small category; denote by $[\mathscr{C}, ?]: \mathscr{C}at \to \mathscr{C}at$ the functor which associates with every category $\mathscr{D}$ the functor category $[\mathscr{C}, \mathscr{D}]$. Prove that this functor has an adjoint.

11.13. By a *monad* (or *triple, triad*) over a category $\mathscr{C}$, we mean a triple $T = (F, u, v)$ where $F: \mathscr{C} \to \mathscr{C}$ is a functor and $u: \mathrm{Id}\,\mathscr{C} \to F$, $v: FF \to F$ are functorial morphisms which satisfy the following conditions:

t_1) The functorial morphisms

$$F \xrightarrow{Fu} FF \xrightarrow{v} F,$$

and

$$F \xrightarrow{uF} FF \xrightarrow{v} F$$

are both equal to 1_F, i.e. $vFu = vuF = 1_F$.

t_2) The functorial morphisms

$$FFF \xrightarrow{Fv} FF \xrightarrow{v} F,$$

and

$$FFF \xrightarrow{vF} FF \xrightarrow{v} F$$

are equal, i.e. $vFv = vvF$.

Let $F: \mathscr{C} \to \mathscr{C}'$ be a functor, G be a coadjoint of F, $u: \mathrm{Id}\,\mathscr{C} \to GF$ be an arrow of adjunction and w be a quasi-inverse coarrow of u. Put $T = GF$ and $v = GwF$. Show that (T, u, v) is a monad over $\mathscr{C}$, called the *monad induced by the adjoint functors F and G*.

Conversely, let $F = (T, u, v)$ be a monad over a category $\mathscr{C}$. Show that there is a category $\mathscr{C}_T$ and functors $F_T: \mathscr{C} \to \mathscr{C}_T$ and $G_T: \mathscr{C}_T \to \mathscr{C}$ such that F_T is adjoint to G_T and T is the monad induced by these functors (see Exercise 5.12.a, Ch. 3).

11.14. Let $T: \mathscr{C} \to \mathscr{C}'$ be a functor between small categories and $\mathscr{D}$ be a cocomplete (complete) category. Prove that the restriction functor $[T, \mathscr{D}]: [\mathscr{C}', \mathscr{D}] \to [\mathscr{C}, \mathscr{D}]$ has an adjoint (coadjoint). (Hint: Let $F: \mathscr{C} \to \mathscr{D}$ be a functor, $X' \in \mathscr{C}'$ and $r_{X'}: T/X' \to \mathscr{C}$ the underlying functor, i.e. $r_{X'}(X, h) = X$, and $r_{X'}(f) = f$, where $f: (X, h) \to (Y, g)$ is a morphism of T/X'.) Let $T^*(F)(X')$ denote $\varinjlim (Fr_{X'})$. If $u: X' \to Y'$ is a morphism of $\mathscr{C}'$, then u defines a functor $u^*: T/X' \to T/Y'$, defined by $u^*(X, h) = (X, uh)$. It is easy to check that the functor u^* defines in a canonical way a morphism $T^*(F)(u): T^*(F)(X') \to T^*(F)(Y')$. Then the assignments $X' \rightsquigarrow T^*(F)(X')$ and $u \rightsquigarrow T^*(F)(u)$ give a functor $T^*(F): \mathscr{C}' \to \mathscr{D}$, and the assignment $F \rightsquigarrow T^*(F)$ defines a functor $T^*: [\mathscr{C}, \mathscr{D}] \to [\mathscr{C}', \mathscr{D}]$, the adjoint to $[T, \mathscr{D}]$. (A coadjoint to $[T, \mathscr{D}]$ is denoted by T^+.)

11.15. Let $F: \mathscr{C} \to \mathscr{C}'$ be a functor and let G be coadjoint to F. Show that for each small category $\mathscr{I}$, the functor $[\mathscr{I}, F]$ is an adjoint of $[\mathscr{I}, G]$.

11.16. Let $F, G: \mathscr{C} \to \mathscr{C}'$ be two functors. Show that the class $[F, G]$ is a set provided that any one of the following conditions is satisfied:

i) $\mathscr{C}'$ has a separator and F has an adjoint.
ii) $\mathscr{C}'$ has a coseparator and G has a coadjoint.
iii) $\mathscr{C}$ has a separator and G has a coadjoint.
iv) $\mathscr{C}$ has a coseparator and G has an adjoint.

11.17. By a *contractible coequalizer diagram* we mean a diagram of the form:

$$X \underset{\overset{t}{\dashleftarrow}}{\overset{\overset{u}{\longrightarrow}}{\underset{v}{\longrightarrow}}} Y \underset{\overset{s}{\dashleftarrow}}{\overset{d}{\longrightarrow}} Z$$

in which the following equalities hold: $du = dv$, $ds = 1_Z$, $ut = 1_Y$ and $vt = sd$.

a) Show that every contractible coequalizer diagram is a coequalizer diagram.

b) Let $F: \mathscr{C} \to \mathscr{C}'$ be a functor, G be coadjoint to F, and u and v the corresponding arrow and coarrow of adjunction. Prove that for each object X of $\mathscr{C}$, the canonically defined diagram:

$$FGFGF(X) \underset{\overset{(FGFu)_X}{\dashleftarrow}}{\overset{\overset{(vFGF)_X}{\longrightarrow}}{\underset{(FGv)_X}{\longrightarrow}}} FGF(X) \underset{\overset{(Fu)_X}{\dashleftarrow}}{\overset{(vF)_X}{\longrightarrow}} F(X)$$

is a contractible coequalizer diagram.

1.12. Commutation of functors with limits and colimits

Here we consider functors that commute with limits or colimits. We also discuss a particular class of colimits in $\mathscr{S}et$ that commute with certain limits.

References: Bourbaki [1], Gabriel—Zisman [1], Grothendieck—Verdier [1], Gabriel—Ulmer [1].

Let $\mathscr{C}$ be a category. For each small category $\mathscr{I}$ let $C_{\mathscr{I}}: \mathscr{C} \to [\mathscr{I}, \mathscr{C}]$ denote the *constant functor*, i.e. the functor that associates with each object X of $\mathscr{C}$ the constant functor $X_{\mathscr{I}}: \mathscr{I} \to \mathscr{C}$. We shall write C instead of $C_{\mathscr{I}}$ if there is no danger of confusion.

PROPOSITION 12.1. *Let $\mathscr{C}$ be a category, $\mathscr{I}$ a small category and $C: \mathscr{C} \to$ $\to [\mathscr{I}, \mathscr{C}]$ the constant functor. Then $\mathscr{C}$ is a category with $\mathscr{I}$-limits if and only if the functor C has a coadjoint.*

Proof. The proof follows from Theorem 11.9 and Proposition 6.9. The limit functor $\overleftarrow{\lim}: [\mathscr{I}, \mathscr{C}] \to \mathscr{C}$ is coadjoint to C.✳

We have the following dual result.

PROPOSITION 12.2. *Let $\mathscr{C}$ be a category, $\mathscr{I}$ a small category and $C: \mathscr{C} \to \to [\mathscr{I}, \mathscr{C}]$ the constant functor. Then $\mathscr{C}$ is a category with $\mathscr{I}$-colimits if and only if $\mathscr{C}$ has an adjoint.* ✳

The colimit functor is adjoint to C.

Let $F: \mathscr{C} \to \mathscr{C}'$ and $G: \mathscr{D} \to \mathscr{C}$ be functors, and let (X, u) be a limit of G. We say that F *commutes with a limit* of G if $(F(X), Fu)$ is a limit of FG. Dually, we have the notion of a functor that *commutes with a colimit* of a functor. If $F: \mathscr{C} \to \to \mathscr{C}'$ is a cofunctor and (u, X) is a colimit of the functor $G: \mathscr{D} \to \mathscr{C}$, then we say that F *commutes with the limit* of G, if $(F(X), Fu)$ is a limit of FG, i.e. F takes the colimit of G to a limit of FG. Similarly, we have the notion of cofunctor that *commutes with the colimit* of a functor. A functor $F: \mathscr{C} \to \mathscr{C}'$ *commutes with limits* (*commutes with small limits*) if it commutes with the limit of each (small) functor $G: \mathscr{D} \to \mathscr{C}$. Also we have the notion of a functor that *commutes with colimits* (*commutes with small colimits*). The same terminology is used for cofunctors.

PROPOSITION 12.3. *For each $X \in \mathscr{C}$, the functors H^X and $H_X: \mathscr{C} \to \mathscr{S}et$ commute with limits. Likewise, a representable functor commutes with limits.*

The proof follows from Theorems 6.10 and 6.11. ✳

The following result is an important consequence of Proposition 12.3.

THEOREM 12.4. *Let $F: \mathscr{C} \to \mathscr{C}'$ be a functor and let G be coadjoint to F. Then F commutes with colimits and G commutes with limits.*

Proof. Let $K: \mathscr{D} \to \mathscr{C}$ be a functor and (u, X) a colimit of K. By Theorem 6.11, $(Fu, F(X))$ is a colimit of FK if and only if for each $X' \in \mathscr{C}'$, $(H_{X'}F(X), H_{X'}Fu)$ is a limit of $H_{X'}FK$. But by Theorem 11.9, the functor $H_{X'}F$ is representable so that the result follows from Proposition 12.3.

The proof of the last part of the theorem is analogous. ✳

In Section 2.6 we shall see that in some cases a functor that commutes with limits has an adjoint (see also Exercise 11.9).

Note 12.5. If a functor $F: \mathscr{C} \to \mathscr{C}'$ commutes with the limit of a functor $G: \mathscr{D} \to \mathscr{C}$, we denote this by

$$F(\underset{\longleftarrow}{\lim}\, G) \simeq \underset{\longleftarrow}{\lim}\, FG.$$

If F is a cofunctor, then we denote it by

$$F(\underset{\longrightarrow}{\lim}\, G) \simeq \underset{\longleftarrow}{\lim}\, FG.$$

The analogous notions will be used for functors which commute with colimits

A special class of functors that commute with limits is described next.

Let $K: \mathscr{I} \prod \mathscr{J} \to \mathscr{C}$ be a functor. For each $i \in \mathscr{I}$, let $K(i, ?): \mathscr{J} \to \mathscr{C}$ denote the functor defined by $K(i, ?)(j) = K(i, j)$ for each $j \in \mathscr{J}$ and $K(i, ?)(s) = K(1_i, s)$ for each $s: j \to j'$ a morphism of $\mathscr{J}$. If $t: i \to i'$ is a morphism of $\mathscr{I}$, then the morphisms $\{K(t, 1_j)\}_{j \in \mathscr{J}}$ determine a functorial morphism $\bar{t}: K(i, ?) \to K(i', ?)$. In an analogous manner, for each $j \in \mathscr{J}$ we get the functor $K(?. j): \mathscr{I} \to \mathscr{C}$ and for each morphism $s: j \to j'$ of $\mathscr{J}$, we get the functorial morphism $\bar{s}: K(?, j) \to K(?, j')$.

LEMMA 12.6. *Let* $K: \mathcal{I}\prod\mathcal{J} \to \mathscr{C}$ *be a functor.*

a) *If for each* $i \in \mathcal{I}$, *the functor* $K(i, ?): \mathcal{J} \to \mathscr{C}$ *has a limit, denoted by* X_i, *then the assignment* $i \rightsquigarrow X_i$ *gives a functor* $\varprojlim_{\mathcal{J}} K(?, j): \mathcal{I} \to \mathscr{C}$. *The functor* K *has a limit if and only if the functor* $\varprojlim_{\mathcal{I}} \varprojlim_{\mathcal{J}} K(?, j)$ *has a limit.*

b) *If for each* $j \in \mathcal{J}$ *the functor* $K(?, j)$ *has a limit, denoted by* X_j, *then the assignment* $j \rightsquigarrow X_j$ *gives a functor* $\varprojlim_{\mathcal{I}} K(i, ?): \mathcal{J} \to \mathscr{C}$. *The functor* K *has a limit if and only if the functor* $\varprojlim_{\mathcal{I}} K(i, ?)$ *has a limit.*

Proof. a) Let $u_i: X_i \to K(i, ?)$ be the structural projections. For each morphism $t: i \to i'$ of $\mathcal{I}$ there is a unique morphism $\bar{t}: X_i \to X_{i'}$ such that $\bar{t}u_i = u_{i'}\tilde{t}$. Let us denote $[\varprojlim_{\mathcal{J}} K(?, j)](t)$ by $\bar{t}$.

Now assume that the functor $\varprojlim_{\mathcal{J}} K(?, j)$ has a limit X and let $u^i: X \to X_i$ be structural projections. For each $(i, j) \in \mathcal{I}\prod\mathcal{J}$ let $u_{(i,j)}$ denote the composition

$$X \xrightarrow{u^i} X_i \xrightarrow{(u_i)_j} K(i, ?)(j) = K(i, j).$$

It is easy to see that for every morphism $(t, s): (i, j) \to (i', j')$ of $\mathcal{I}\prod\mathcal{J}$ we have $u_{(i',j')} = K(t, s)u_{(i,j)}$, i.e. the morphisms $\{u_{(i,j)}\}_{(i,j) \in \mathcal{I}\prod\mathcal{J}}$ define a functorial morphism $u: X \to K$. We leave it for the reader to check that (X, u) is a limit of K.

Conversely, assume that (X, u) is a limit of K. Thus for a fixed i, the structural morphisms $\{u_{(i,j)}\}_{j \in \mathcal{J}}$ give a functorial morphism $v_i: X \to K(i, ?)$
Thus there is a unique morphism $u^i: X \to X_i$ such that $u_i u^i = v_i$. We leave it for the reader to check that the morphisms $\{u^i\}_{i \in \mathcal{I}}$ give a functorial morphism $w: X \to \varprojlim_{\mathcal{J}} K(?, j)$ and that (X, w) is a limit of $\varprojlim_{\mathcal{J}} K(?, j)$. ✳

COROLLARY 12.7. *With the conditions and notation as in the previous lemma, we have the following canonical isomorphisms:*

$$\varprojlim_{\mathcal{I}} (\varprojlim_{\mathcal{J}} K(?, j)) \xrightarrow{\sim} \varprojlim_{\mathcal{I}\prod\mathcal{J}} K \xrightarrow{\sim} \varprojlim_{\mathcal{J}} (\varprojlim_{\mathcal{I}} K(i, ?)). \; ✳$$

The corollary says that limits commute with limits. This is also the conclusion of the following result:

THEOREM 12.8. *Let* $\mathscr{C}$ *be a complete category,* $\mathcal{I}$ *and* $\mathcal{J}$ *be small categories and* $F: \mathcal{J} \to [\mathcal{I}, \mathscr{C}]$ *be a functor. Denote by* $G: \mathcal{I} \to [\mathcal{J}, \mathscr{C}]$ *the functor that associates with* $i \in \mathcal{I}$ *the functor* $G(i): \mathcal{J} \to \mathscr{C}$ *defined by* $G(i)(j) = F(j)(i)$. *If* $\varprojlim_{\mathcal{I}}: [\mathcal{I}, \mathscr{C}] \to \mathscr{C}$ *and* $\varprojlim_{\mathcal{J}}: [\mathcal{J}, \mathscr{C}] \to \mathscr{C}$ *are the respective limit functors, then we have the isomorphisms*

$$\varprojlim_{\mathscr{I}} (\varprojlim_{\mathscr{J}} F) \xrightarrow{\sim} \varprojlim_{\mathscr{J}} (\varprojlim_{\mathscr{I}} G) \simeq \varprojlim_{\mathscr{J}\amalg\mathscr{I}} K$$

where $K: \mathscr{I} \amalg \mathscr{J} \to \mathscr{C}$ *is the functor defined by* $K(i, j) = F(j)(i)$.

The result follows from Proposition 12.1, Theorem 12.4 and Lemma 12.6. ✳

We suggest that the reader dualize the above results for colimits.

An interesting problem is the study of those colimits which commute with certain limits. We will consider this problem after having first developed some preliminary results.

PROPOSITION 12.9. *Let* $F: \mathscr{I} \to \mathscr{S}et$ *be a functor. Assume that* $\mathscr{I}$ *is a filtered category and that* (u, M) *is a cocone over* F. *The following are equivalent:*

1) (u, M) *is a colimit of* F.

2) *The following two conditions are satisfied:*

a) *For each* $x \in M$ *there exist an object* i *of* $\mathscr{I}$ *and an element* $y \in F_i$ *such that* $u_i(y) = x$.

(b) *If* i *and* i' *are objects of* $\mathscr{I}$, *and* $y \in F_i$ *and* $y' \in F_{i'}$ *are such that* $u_i(y) = u_{i'}(y')$, *then there is an angle* $i \xrightarrow{s} k \xleftarrow{s'} i'$ *such that* $F(s)(y) = F(s')(y')$.

Proof. 1) ⇒ 2). a) Assume, to the contrary, that there is an element $x \in M$ such that $x \notin u_i(F_i)$ for every $i \in \mathscr{I}$. Let $M' = M - \{x\}$. It is easy to see that the maps $\{u_i\}_{i \in \mathscr{I}}$ define a cocone (u', M') over F, and that there are at least two maps $f, g: M \to M'$ such that $fu = u' = gu$, which is a contradiction.

b) For each element $x \in M$ fix an object $j(x)$ of $\mathscr{I}$ and an element $x' \in F_{j(x)}$ such that $u_{j(x)}(x') = x$.

We claim that for each $i \in \mathscr{I}$, and for each $y \in F_i$, there is an angle $i \xrightarrow{s} k \xleftarrow{t} j(x)$, $x \in M$, such that $F(s)(y) = F(t)(x')$. To show this, assume, to the contrary, that this condition does not hold. Then for each $i \in \mathscr{I}$ let A_i denote $\{y \in F_i$ such that there is an angle $i \xrightarrow{s} k \xleftarrow{t} j(x)$, with $F(s)(y) = F(t)(x')\}$ and let $B_i = F_i - A_i$, i.e. the complement of A_i. Since $\mathscr{I}$ is filtered, we have for each morphism $s: i \to i'$ of $\mathscr{I}$, that $F(s)(A_i) \subseteq A_i$ and $F(s)(B_i) \subseteq B_i$. Furthermore, let $\overline{M}$ denote $M \cup \{*\}$, where $*$ is a symbol which does not belong to M, and for each $i \in \mathscr{I}$ let $v_i: F_i \to \overline{M}$ denote the map defined by

$$v_i(y) = \begin{cases} u_i(y), & \text{if } y \in A_i \\ * & \text{otherwise.} \end{cases}$$

Thus the maps $\{v_i\}_{i \in \mathscr{I}}$ define a cocone $(v, \overline{M})$ over F, so that there is a unique map $f: M \to \overline{M}$ with $fu = v$. If $B_i \neq \emptyset$ then for some $i, y \in B_i$, and $x = u_i(y)$, we have $v_i(y) = * = fu_i(y) = f(x) = f(u_{j(x)}(x')) = v_{j(x)}(x')$, which is a contradiction.

Hence, if $y \in F_i$ and $y' \in F_{i'}$ are such that $u_i(y) = u_{i'}(y') = x$, then we have the diagram

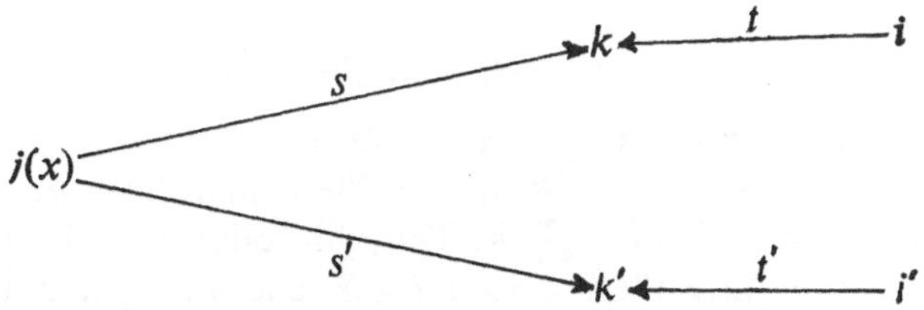

in $\mathscr{I}$, such that $F(t)(y) = F(s)(x')$ and $F(t')(y') = F(s')(x')$. Since $\mathscr{I}$ is filtered there are morphisms $h\colon k \to k''$ and $h'\colon k' \to k''$ such that $h's' = hs$. Thus $F(ht)(y) = F(h't')(y')$, as claimed.

The proof of the implication 2) $\Rightarrow$ 1) is left for the reader. $\ast$

The following lemma asserts that filtered limits commute with fibered products.

LEMMA 12.10. *Let $\mathscr{I}$ be a filtered category. Also, let $F, G, K\colon \mathscr{I} \to \mathscr{S}et$ be functors, and let $u\colon F \to G$ and $v\colon K \to G$ be functorial morphisms. Asume that $\varinjlim F$, $\varinjlim G$ and $\varinjlim K$ exist. Then the functor $F \prod_G K\colon \mathscr{I} \to \mathscr{S}et$ also has a colimit and the square*

$$\begin{array}{ccc} \varinjlim (F \prod_G K) & \longrightarrow & \varinjlim F \\ \downarrow & & \downarrow \varinjlim u = \tilde{u} \\ \varinjlim K & \xrightarrow{\varinjlim v = \tilde{v}} & \varinjlim G \end{array}$$

is cartesian.

Proof. Consider the cartesian square:

$$\begin{array}{ccc} M = (\varinjlim F) \prod_{(\varinjlim G)} (\varinjlim K) & \xrightarrow{m} & \varinjlim F \\ {\scriptstyle n}\downarrow & & \downarrow{\scriptstyle \tilde{u}} \\ \varinjlim K & \xrightarrow{\tilde{v}} & \varinjlim G \end{array} .$$

By Corollary 10.2, the functor $F \prod_G K$ is defined argumentwise, i.e. for each object i of $\mathscr{I}$ we have the cartesian square:

$$\begin{array}{ccc} (F \prod_G K)_i = F_i \prod_{G_i} K_i & \xrightarrow{\alpha_i} & F_i \\ {\scriptstyle \beta_i}\downarrow & & \downarrow{\scriptstyle u_i} \\ K_i & \xrightarrow{v_i} & G_i \end{array} .$$

Now we denote the structural morphisms by $f\colon F \to \varinjlim F$, $g\colon G \to \varinjlim G$ and $k\colon K \to \varinjlim K$. Thus, for each $i \in \mathscr{I}$,

$$\tilde{u} f_i \alpha_i = g_i u_i \alpha_i = g_i v_i \beta_i = \tilde{v} k_i \beta_i$$

so that there is a unique map $h_i\colon (F \prod_G K)_i \to M$ such that $mh_i = f_i\alpha_i$ and $nh_i = k_i\beta_i$. It is easy to show that the maps $\{h_i\}_{i\in\mathscr{I}}$ define a morphism $h\colon F\prod_G K \to M$. We claim that (h, m) is a colimit of $F\prod_G K$. This will follow from Proposition 12.9. Let $x \in M$; by Proposition 12.9 there exist $i, i' \in \mathscr{I}$ and $y \in F_i$ and $y' \in K_i$ such that $f_i(y) = m(x)$ and $k_{i'}(y') = n(x)$. From Exercise 12.4 we can assume $i = i'$. Then $g_i u_i(y) = g_i v_i(y') = um(x) = vn(x)$. Using Proposition 12.9 again and also Exercise 12.4, we may assume that $u_i(y) = v_i(y')$. Hence there is an element $z \in F_i \prod_{G_i} K_i$ such that $\alpha_i(z) = y$ and $\beta_i(z) = y'$. Obviously, $h_i(z) = x$.

Now let $z \in (F\prod_G K)_i$ and $z' \in (F\prod_G K)_{i'}$ be such that $h_i(z) = h_{i'}(z')$. Since $\mathscr{I}$ is filtered we can assume that $i = i'$. Then consider the elements $\alpha_i(z)$ and $\alpha_i(z')$ of F_i. A trivial computation proves that $f_i(\alpha_i(z)) = f_i(\alpha_i(z'))$, so that, by Proposition 12.9, there is a morphism $s\colon i \to j$ in $\mathscr{I}$ such that $F(s)(\alpha_i(z)) = F(s)(\alpha_i(z')) = a$. Also since $k_i(\beta_i(z)) = k_i(\beta_i(z'))$ and $\mathscr{I}$ is filtered, we may assume that $K(s)(\beta_i(z)) = K(s)(\beta_i(z')) = b$ and $u_j(a) = v_j(b)$. From the construction of $F\prod_G K$, we must have that $(F\prod_G K)(s)(z) = (F\prod_G K)(s)(z')$. Hence the proof follows from Proposition 12.9. ✳

THEOREM 12.11. *Let $\mathscr{I}$ be a filtered category. Also let $\mathscr{J}$ be a finite category and F a $\mathscr{J}$-diagram of functors from $\mathscr{I}$ into $\mathscr{S}et$. Assume that for every object j of $\mathscr{J}$ the functor $F_j\colon \mathscr{I} \to \mathscr{S}et$ has a colimit. Then the functor $\varprojlim F\colon \mathscr{I} \to \mathscr{S}et$ (the limit of the diagram F) has a colimit and this colimit is a limit of the functor $F\colon \mathscr{J} \to \mathscr{S}et$, defined by $F(j) = \varinjlim F_j$, for every object j and $F(s) = \varinjlim F(s)$ for every morphism s of $\mathscr{J}$.*

Proof. Since the limits of functors are defined argumentwise, the proof follows from Lemma 12.10 and Exercise 9.3. Indeed, the existence of a ‘final functor”, i.e. the constant functor associated with a final object of $\mathscr{S}et$, proves that a product of functors can be considered as a fibered product. ✳

A functor $F\colon \mathscr{C} \to \mathscr{C}'$ is called *left exact* if it commutes with finite limits. Dually, we have the notion of a *right exact* functor.

THEOREM 12.12. *Let $\mathscr{I}$ be a small filtered category. Then the colimit functor*

$$\varinjlim\colon [\mathscr{I}, \mathscr{S}et] \to \mathscr{S}et$$

is left exact.

The proof follows from Theorem 12.11. ✳

Finally, we define a class of colimits which commute with limits. We say that a category $\mathscr{I}$ is *strongly filtered,* if for each set $\{i_\alpha\}_{\alpha\in K}$ of objects there exists a set of morphisms $\{t_\alpha\colon i_\alpha \to j\}_{\alpha\in K}$ with the same codomain j.

LEMMA 12.13. *Let $\mathcal{I}$ be a strongly filtered category. Also let $\{F_\alpha\}_{\alpha \in K}$ be a set of functors from $\mathcal{I}$ into $\mathcal{S}et$. Assume that for each α, the functor F_α has a colimit (u^α, X_α). Then the functor $F = \prod_\alpha F_\alpha$ (which associates with each $i \in \mathcal{I}$, the set $F_i = \prod_\alpha F_\alpha(i)$ and $F(t) = \prod_\alpha F_\alpha(t)$ for each morphism t in $\mathcal{I}$) also has a colimit X and $X = \prod_\alpha X_\alpha$.*

Proof. For each $\alpha \in K$, let $p^\alpha \colon F \to F_\alpha$ denote the structural projections. Also, let $X = \prod_\alpha X_\alpha$ and let $q_\alpha \colon X \to X_\alpha$ be the structural projections. For each $i \in \mathcal{I}$ let $u_i \colon F_i \to X$ denote the unique morphism such that $q_\alpha u_i = u_i^\alpha p_i^\alpha$ for all $\alpha \in K$. A slight computation proves that the morphisms $\{u_i\}_{i \in \mathcal{I}}$ define a functorial morphism $u \colon F \to X$. We leave it for the reader to check that (u, X) is a colimit of F, and suggest that he use Proposition 12.9 to do so. ✳

THEOREM 12.14. *Let $\mathcal{I}$ be a strongly filtered category. Also let $\mathcal{J}$ be a small category and D a $\mathcal{J}$-diagram of functors from $\mathcal{I}$ into $\mathcal{S}et$. Assume that for all $j \in \mathcal{J}$ the functor D_j has a colimit. Then the functor $\varprojlim D$ also has a colimit and we have the isomorphism*

$$\varinjlim (\varprojlim D) \simeq \varprojlim_{j \in \mathcal{J}} (\varinjlim D_j).$$

In particular, if $\mathcal{I}$ is a small category, then the functor

$$\varinjlim \colon [\mathcal{I}, \mathcal{S}et] \to \mathcal{S}et$$

commutes with limits. ✳

There exist other colimits which commute with certain limits.

PROPOSITION 12.15. *Let $\mathcal{I}$ be a discrete small category, i.e. a set. Then the functor*

$$\textstyle\coprod_i = \varinjlim_{\mathcal{I}} \colon [\mathcal{I}, \mathcal{S}et] \to \mathcal{S}et$$

commutes with kernels and filtered products.

Proof. Consider in $[\mathcal{I}, \mathcal{S}et]$ the exact diagram

$$F \xrightarrow{f} G \underset{v}{\overset{u}{\rightrightarrows}} H.$$

Then, by Proposition 10.5, for each $i \in \mathcal{I}$, we have the following exact diagram in $\mathcal{S}et$:

$$F_i \xrightarrow{f_i} G_i \underset{v_i}{\overset{u_i}{\rightrightarrows}} H_i.$$

Furthermore, it is easy to see that the diagram

$$\textstyle\coprod_i F_i \xrightarrow{\coprod f_i} \coprod_i G_i \underset{\coprod v_i}{\overset{\coprod u_i}{\rightrightarrows}} \coprod_i H_i$$

in $\mathcal{S}et$ is exact, that is, the functor $\coprod_i$ commutes with kernels. In the same way, it can be checked that this functor commutes with fibered products. ✳

Note 12.16. An easy counterexample shows that the functor $\coprod_i$ in Proposition 12.15 does not commute with products although the products in $[\mathcal{I}, \mathcal{S}et]$ are a special case of fibered products. The problem arises since the functor $\coprod_i$ does not preserve final objects.

Note 12.17. In general, if for a category $\mathcal{C}$ the functor $\varinjlim_{\mathcal{I}} : [\mathcal{I}, \mathcal{C}] \to \mathcal{C}$ commutes with certain limits, then we shall say that *$\mathcal{I}$-colimits in $\mathcal{C}$ commute with $\mathcal{J}$-limits*. In particular, the filtered colimits in $\mathcal{S}et$ commute with finite limits, the coproducts commute with kernels and fibered products, etc.

Note 12.18. Let α be a regular cardinal number. A category $\mathcal{I}$ is called *α-filtered* if the following conditions hold:

i) For each set $\{i_m\}_{m \in \mathcal{M}}$ of objects of $\mathcal{I}$ such that $\mathrm{Card}(\mathcal{M}) < \alpha$, there exists a *pencil*, i.e. $\{s_m : i_m \to j\}_{m \in \mathcal{M}}$.

ii) If $\{s_m : i \to j\}_{m \in \mathcal{M}}$ is a set of morphisms of $\mathcal{I}$ such that $\mathrm{Card}(\mathcal{M}) < \alpha$, then there is a morphism $t: j \to k$ with $ts_m = ts_n$ for all $m, n \in \mathcal{M}$.

Since the first infinite cardinal number, α_0, is regular, see that a category $\mathcal{I}$ is α_0-filtered if and only if it is filtered. It is also clear that every α-filtered category is also filtered.

Let α be a regular cardinal number. A functor $F: \mathcal{I} \to \mathcal{C}$, where $\mathcal{I}$ is a small α-filtered category, is called an *α-filtered system in $\mathcal{C}$*. By an *α-monofiltered system* we mean a filtered system $F: \mathcal{I} \to \mathcal{C}$ such that $F(s)$ is a monomorphism for all morphisms s of $\mathcal{I}$.

The notions *α-cofiltered category*, *α-cofiltered system* and *α-epicofiltered system* are obtained by duality.

Exercises

12.1. Let $\mathcal{I}$ be a small category. Show that for every category $\mathcal{C}$, the constant functor $C: \mathcal{C} \to [\mathcal{I}, \mathcal{C}]$ commutes with limits and colimits.

Show that if $F: \mathcal{C} \to \mathcal{C}'$ is a functor, then for every small category $\mathcal{I}$, the functor $[\mathcal{I}, F]: [\mathcal{I}, \mathcal{C}] \to [\mathcal{I}, \mathcal{C}']$ commutes with limits and colimits.

12.2. Let $F: \mathcal{C} \to \mathcal{C}'$ be a functor. Assume that $\mathcal{C}$ is finitely complete. Prove that the following are equivalent:

a) F is left exact.

b) F commutes with finite products and kernels.

c) F commutes with finite products and fibered products.

Show that moreover if $\mathscr{C}$ is complete, then F commutes with small limits if and only if it commutes with products and kernels or, alternately, F commutes with small limits if and only if it commutes with products and fibered products.

12.3. Let $F, G: \mathscr{I} \to \mathscr{S}et$ be functors, where $\mathscr{I}$ is a small filtered category. Prove: If $u: F \to G$ is a functorial morphism such that $u_i: F(i) \to G(i)$ is an injection for every $i \in \mathscr{I}$, then $\varinjlim u: \varinjlim F \to \varinjlim G$ is also an injection.

12.4. Let $\mathscr{I}$ be a filtered category and $\mathscr{M}$ be a set of objects of $\mathscr{I}$. Show there is a small, filtered and full subcategory of $\mathscr{I}$ that contains $\mathscr{M}$, and that if $\mathscr{I}$ is right bounded and $\mathscr{C}$ is cocomplete, then every functor $F: \mathscr{I} \to \mathscr{C}$ has a colimit.

12.5. Let $\mathscr{C}$ be a category with fibered products, $F: \mathscr{I} \to \mathscr{C}$ a functor having a colimit, $f: X \to Y$ a morphism of $\mathscr{C}$ and $v: F \to Y$ a functorial morphism. Denote by $F\prod_Y X: \mathscr{I} \to \mathscr{C}$ the functor $i \rightsquigarrow F(i)\prod_Y X$ (in fact $F\prod_Y X$ is a fibered product of functors). We say that colimit of F is *universal* in $\mathscr{C}$ if for every morphism $f: X \to Y$ and every $v: F \to Y$ the functor $F\prod_Y X$ has a colimit and the canonical morphism

$$\varinjlim (F\prod_Y X) \to (\varinjlim F)\prod_Y X$$

is an isomorphism. Prove that the colimit of every functor $F: \mathscr{I} \to \mathscr{S}et$ is universal.

12.6. Prove that a functor which commutes with fibered products carries monomorphisms into monomorphisms.

12.7. Let $\mathscr{I}$ be a small category such that the functor $\varinjlim: [\mathscr{I}, \mathscr{S}et] \to \mathscr{S}et$ commutes with products. Show that $\mathscr{I}$ is connected.

12.8. Let $\mathscr{C}$ be a small category. For each $X \in \mathscr{C}$ and each category $\mathscr{D}$ denote by $A^X_{\mathscr{D}}: [\mathscr{C}, \mathscr{D}] \to \mathscr{D}$ the *argument functor* defined as follows: $A^X_{\mathscr{D}}(F) = F(X)$ for every functor F, and for every functorial morphism u, $A^X_{\mathscr{D}}(u) = u_X$. Prove that the argument functor commutes with limits and colimits.

We denote $A^X_{\mathscr{S}et}$ by A^X.

12.9. Let $\mathscr{I}$ be a preordered category. Prove: If $F: \mathscr{I} \to \mathscr{S}et$ is a functor that commutes with limits, then for each $i \in \mathscr{I}$ the set $F(i)$ contains no more than one element.

12.10. Let X be a set. Prove that the functor $H^X: \mathscr{S}et \to \mathscr{S}et$ commutes with coproducts if and only if X contains no more than one element. Also prove that H^X commutes with filtered colimits if and only if X is a finite set.

12.11. Let $\mathscr{C}$ be a category with products. Prove that the following are equivalent:

a) Every functor $F: \mathscr{C} \to \mathscr{C}'$ that commutes with products, also commutes with limits.

b) Every functor $G: \mathscr{C} \to \mathscr{S}et$, which commutes with products, also commutes with limits.

12.12. Let $\{X_i\}_i$ be a set of sets. Prove that, for every set M, there is a canonical bijection

$$M \prod (\coprod_i X_i) \simeq \coprod_i (M \prod X_i).$$

(Hint: See Example 11.7.)

12.13. Let $F: \mathscr{C} \to \mathscr{S}et$ be a functor that commutes with limits. Let F' be a function from the class of objects of $\mathscr{C}$ to the class of sets. Assume $F'(X) \subseteq F(X)$ for all $X \in \mathscr{C}$. Show that F' may be regarded as a functor if and only if for each morphism $f: X \to Y$ of $\mathscr{C}$ we have that $F(f)(F'(X)) \subseteq F'(Y)$. Prove that, moreover, F' commutes with limits if and only if whenever (X, u) is a limit of a functor $H: \mathscr{I} \to \mathscr{C}$, then $F(X) \supseteq \bigcap_{i \in \mathscr{I}} u_i^{-1}(F(X_i))$.

12.14. Let $\mathscr{C}$ be a cocomplete category such that the $\mathscr{I}$-colimits in $\mathscr{C}$ commute with $\mathscr{J}$-limits. Show that, for every small category $\mathscr{A}$, the category $[\mathscr{A}, \mathscr{C}]$ has the same property.

12.15. With the notation and hypothesis as in Exercise 11.14 assume that $\mathscr{C}$ and $\mathscr{D}$ are finitely complete, T is left exact and in $\mathscr{D}$ the filtered colimits commute with finite limits. Prove that the functor T^* is left exact.

12.16. Let α be a regular cardinal number. Prove that, for a small category $\mathscr{I}$, the following are equivalent:

a) $\mathscr{I}$ is α-filtered.

b) In the category $\mathscr{S}et$, $\mathscr{I}$-colimits commute with $\mathscr{J}$-limits, where $\mathscr{J}$ is an α-small category.

In particular, $\mathscr{I}$ is a filtered category if and only if the $\mathscr{I}$-colimits in $\mathscr{S}et$ commute with finite limits.

12.17. Let $\mathscr{C}$ be a small category and $F: \mathscr{C} \to \mathscr{S}et$ a cofunctor. Prove that the following are equivalent:

a) The pointed category associated with F is α-filtered.

b) F is the colimit of an α-filtered system of representable cofunctors.

12.18. Let X be a set. Prove that the functor H^X commutes with α-filtered colimits if and only if $\mathrm{Card}(X) < \alpha$.

1.13. Categories of fractions

The categories of fractions are generalizations of ordinary fractions. In this section we define the basic notions of the theory of categories of fractions and also examine some results about these notions. A special case is closely related to the idea of adjoint functors in which one of the two functors is full and faithful.

References: Bănică—Popescu [2], Bourbaki [2], Gabriel—Zisman [1], Schubert [1].

A functor $F: \mathscr{C} \to \mathscr{C}'$ is said to *make a morphism s of $\mathscr{C}$ invertible* if $F(s)$ is invertible. Let Σ be a class of morphisms of a category $\mathscr{C}$. By a *category of fractions* of $\mathscr{C}$ relative to Σ we mean a pair $(P_\Sigma, \mathscr{C}(\Sigma^{-1}))$, where $\mathscr{C}(\Sigma^{-1})$ is a category and $P_\Sigma: \mathscr{C} \to \mathscr{C}(\Sigma^{-1})$ is a functor satisfying the following conditions:

i) P_Σ makes the morphisms of Σ invertible.

ii) If a functor $F: \mathscr{C} \to \mathscr{D}$ makes the morphisms of Σ invertible, then there is a unique functor $\bar{F}: \mathscr{C}(\Sigma^{-1}) \to \mathscr{D}$ such that $\bar{F}P_\Sigma = F$.

The functor P_Σ (or simply P, if there is no danger of confusion) is called the *canonical functor*. It is easy to see that the category $\mathscr{C}(\Sigma^{-1})$ is defined up to equivalence of categories.

Note 13.1. The following example shows that the category $\mathscr{C}(\Sigma^{-1})$ need not exist. Let $\mathscr{I}$ be a class (which is not a set) and let $\mathscr{C}$ denote the category whose objects are X, Y and $\{X_\alpha\}_{\alpha \in \mathscr{I}}$. The morphisms of $\mathscr{C}$ are the identity morphisms, and the morphisms $f_\alpha: X \to X_\alpha$ and $s_\alpha: Y \to X_\alpha$, $\alpha \in \mathscr{I}$. It is clear that $\mathscr{C}$ is a free category. Consider $\Sigma = \{s_\alpha\}_{\alpha \in \mathscr{I}}$ and suppose the category $\mathscr{C}(\Sigma^{-1})$ exists. Since for each α, $P(s_\alpha)^{-1}P(f_\alpha): P(X) \to P(Y)$, there are necessarily two distinct elements α and α' of $\mathscr{I}$ such that $P(s_\alpha)^{-1}P(f_\alpha) = P(s_{\alpha'})^{-1}P(f_{\alpha'})$. Now let $\mathscr{C}'$ be the category whose objects are X', Y', Z, Z', and whose morphisms are the identity morphisms and $f: X' \to Z$, $s: Y' \to Z$, $f': X' \to Z'$ and $s': Y' \to Z'$. Assume that s and s' are isomorphisms and $s^{-1}f \neq s'^{-1}f'$. Let $G: \mathscr{C} \to \mathscr{C}'$ be the functor defined by $G(X) = X'$, $G(Y) = Y'$, $G(X_\alpha) = Z$, and $G(X_\beta) = Z'$ for $\beta \neq \alpha$; also, $G(f_\alpha) = f$, $G(s_\alpha) = s$ and $G(f_\beta) = f'$, $G(s_\beta) = s'$ for $\beta \neq \alpha$. It is clear that G is a functor which makes the morphisms of Σ invertible, so that there is a unique functor $\bar{G}: \mathscr{C}(\Sigma^{-1}) \to \mathscr{C}'$such that $\bar{G}P = G$. But then necessarily $\bar{G}(P(s_\alpha)^{-1}) = G(s_\alpha)^{-1}$ and $\bar{G}(P(s_{\alpha'})^{-1}) = G(s_{\alpha'})^{-1}$ and so $\bar{G}(P(s_\alpha)^{-1}P(f_\alpha)) = \bar{G}(P(s_\alpha))^{-1}\,\bar{G}P(f_\alpha) = G(s_\alpha)^{-1}G(f_\alpha) = \bar{G}(P(s_{\alpha'})^{-1}P(f_{\alpha'})) = \bar{G}(P(s_{\alpha'}))^{-1}\bar{G}P(f_{\alpha'}) = G(s_{\alpha'})^{-1}G(f_{\alpha'})$, which is a contradiction. Hence the category $\mathscr{C}(\Sigma^{-1})$ does not exist (see Section 1.1) since $\mathscr{C}(\Sigma^{-1})(P(X), P(Y))$ would not be a set.

13.2. By Proposition 5.5 and Exercise 13.1 we may assume that the canonical functor $P: \mathscr{C} \to \mathscr{C}(\Sigma^{-1})$ defines a bijection between the objects of $\mathscr{C}$ and the objects of $\mathscr{C}(\Sigma^{-1})$. We assume that this condition is always satisfied.

THEOREM 13.3. *Let $\mathscr{C}$ be a small category and Σ a set of morphisms of $\mathscr{C}$. Then the category $\mathscr{C}(\Sigma^{-1})$ exists.*

Proof. For each pair $X, Y \in \mathscr{C}$, denote $\Sigma \cap \mathscr{C}(X, Y)$ by $\Sigma(X, Y)$, and let Γ be the diagram scheme whose vertices are the objects of $\mathscr{C}$. For all vertices X, Y of Γ, the set of all arrows with domain X and codomain Y is $\mathscr{C}(X, Y) \coprod \Sigma(Y, X)$ (the coproduct of sets). If a morphism s belongs to $\Sigma(Y, X)$, then, considered as an arrow with domain X and codomain Y, it will be denoted by s^{-1}. For each vertex X, 1_X is considered to be a preunity. Let $\mathscr{L}$ be the free category generated by the diagram scheme Γ (see Section 1.5). We make the convention that the objects of $\mathscr{L}$ are denoted by the same symbols as the vertices of Γ, i.e. as the objects of $\mathscr{C}$. Also we denote by $D: \mathscr{C} \to \mathscr{L}$ the diagram which assigns to each object X of $\mathscr{C}$ the same object of $\mathscr{L}$, and to each morphism $f: X \to Y$ in $\mathscr{C}$, the morphism

$D(f): X \to Y$ in $\mathscr{L}$ (in fact, $D(f) = K(f)$ where $K: \Gamma \to \mathscr{L}$ is the canonical diagram). Now denote by R the smallest equivalence relation defined on $\mathscr{L}$ such that:

a) For each pair of morphisms $f: X \to Y$ and $g: Y \to Z$ of $\mathscr{C}$, $D(g)D(f) \sim \sim D(gf)\ (R(X, Z))$.

b) For every $s: X \to Y$, $s \in \Sigma$, we have $K(s^{-1})\,D(s) \sim 1_X(R(X, X))$ and $D(s)K(s^{-1}) \sim 1_Y(R(Y, Y))$.

Let $\mathscr{C}(\Sigma^{-1}) = \mathscr{L}/R$ and let $T: \mathscr{L} \to \mathscr{C}(\Sigma^{-1})$ be the canonical functor. It is easy to see that the diagram TD is in fact a functor (actually denoted by $P_\Sigma: \mathscr{C} \to \mathscr{C}(\Sigma^{-1})$). If $s \in \Sigma$, then $P_\Sigma(s) = T(D(s))$ and it is obvious that $T(K(s^{-1}))$ is an inverse of $P_\Sigma(s)$. Finally, let $F: \mathscr{C} \to \mathscr{C}'$ be a functor that makes every morphism s of Σ invertible. Let $F_1: \Gamma \to \mathscr{C}'$ denote the diagram $F_1(X) = F(X)$, $F_1(f) = F(f)$ and $F_1(s^{-1}) = F(s)^{-1}$. Also, let $F_2: \mathscr{L} \to \mathscr{C}'$ denote the unique functor such that $F_2K = F_1$. It is easy to see that $R_{F_2} \geqslant R$, where R_{F_2} is the equivalence relation on $\mathscr{L}$ associated with F_2, so that by Proposition 5.6 there is a unique functor $\overline{F}: \mathscr{C}(\Sigma^{-1}) \to \mathscr{C}'$ such that $\overline{F}T = F_2$. Thus it is obvious that $\overline{F}TD = \overline{F}P_\Sigma = F$. Hence $(P_\Sigma, \mathscr{C}(\Sigma^{-1}))$ is a category of fractions of $\mathscr{C}$ relative to Σ. ⁎

Let $\mathscr{C}$ be a category; we denote by $U(\mathscr{C})$ the subcategory of $\mathscr{C}$ whose objects are the objects of $\mathscr{C}$ and the morphisms are all the isomorphisms in $\mathscr{C}$. It is easy to see that $U(\mathscr{C})$ is a groupoid — it is called the *groupoid of unities* of $\mathscr{C}$.

PROPOSITION 13.4. *Let $I: \mathscr{G}rp \to \mathscr{C}at$ be the inclusion functor. Then I has an adjoint and a coadjoint.*

Proof. Let $\mathscr{C}$ be an object of $\mathscr{C}at$, i.e. a small category, and $\Sigma = \mathscr{M}or(\mathscr{C})$. The category $\mathscr{C}(\Sigma^{-1})$ is denoted by $T(\mathscr{C})$. If $F: \mathscr{C} \to \mathscr{C}'$ is a morphism of $\mathscr{C}at$, i.e. a functor, then there is a unique functor $T(F): T(\mathscr{C}) \to T(\mathscr{C}')$ such that $T(F)P_{\mathscr{C}} = P_{\mathscr{C}'}F$, where $P_{\mathscr{C}}: \mathscr{C} \to T(\mathscr{C})$ is the canonical functor. We leave it for the reader to check that $T(\mathscr{C})$ is a groupoid and that the assignments $\mathscr{C} \rightsquigarrow T(\mathscr{C})$ and $F \rightsquigarrow T(F)$ define a functor $T: \mathscr{C}at \to \mathscr{G}rp$, having the inclusion functor $I: \mathscr{G}rp \to \mathscr{C}at$ as a coadjoint. An adjoint of I is the functor $U: \mathscr{C}at \to \mathscr{G}rp$ which assigns to each category $\mathscr{C}$ its groupoid of unities. ⁎

PROPOSITION 13.5. *Let Σ be a class of morphisms in $\mathscr{C}$ such that $\mathscr{C}(\Sigma^{-1})$ exists. Let $F, G: \mathscr{C}(\Sigma^{-1}) \to \mathscr{D}$ be functors and let $v: FP \to GP$ be a functorial morphism where $P: \mathscr{C} \to \mathscr{C}(\Sigma^{-1})$ is the canonical functor. Then there is a unique functorial morphism $u: F \to G$ such that $uP = v$.*

Proof. Let X be an object of $\mathscr{C}(\Sigma^{-1})$. By Note 13.2, $X = P(Y)$ with $Y \in \mathscr{C}$, so we define $u_X = v_{P(Y)}$. Now let $f: X \to Y$ be a morphism of $\mathscr{C}(\Sigma^{-1})$; by Exercise 13.1, we may assume that $f = P(s')^{-1}P(g)P(s)^{-1}$, where $A \xleftarrow{s} B \xrightarrow{g} C \xleftarrow{s'} D$ are morphisms of $\mathscr{C}$ and $s, s' \in \Sigma$. Thus the commutativity of the diagram

$$\begin{array}{ccc} F(X) & \xrightarrow{F(f)} & F(Y) \\ {\scriptstyle u_X}\downarrow & & \downarrow{\scriptstyle u_Y} \\ G(X) & \xrightarrow{G(f)} & G(Y) \end{array}$$

is obtained from the commutativity of the diagram

$$\begin{array}{ccccccc}
FP(A) & \xleftarrow{FP(s)} & FP(B) & \xrightarrow{FP(g)} & FP(C) & \xleftarrow{FP(s')} & FP(D) \\
\downarrow{\scriptstyle v_A} & & \downarrow{\scriptstyle v_B} & & \downarrow{\scriptstyle v_C} & & \downarrow{\scriptstyle v_D} \\
GP(A) & \xleftarrow{GP(s)} & GP(B) & \xrightarrow{GP(g)} & GP(C) & \xleftarrow{GP(s')} & GP(D)
\end{array}$$

since $P(s)$ and $P(s')$ are invertible. ✳

COROLLARY 13.6. *Let Σ be a class of morphisms from a category $\mathscr{C}$ and assume that the category $\mathscr{C}(\Sigma^{-1})$ exists. Also, let $F, G: \mathscr{C}(\Sigma^{-1}) \to \mathscr{C}'$ be functors such that $FP \simeq GP$. Then $F \simeq G$.* ✳

Now we shall give some examples of categories of fractions. Before doing this, we make some useful observations.

PROPOSITION 13.7. *Let $F: \mathscr{C} \to \mathscr{C}'$ be a functor, G a coadjoint of F, $u: \mathrm{Id}\,\mathscr{C} \to GF$ an arrow of adjunction, v a coarrow quasi-inverse of u, $\tilde{v}: \mathscr{C}(?, G(?)) \to \mathscr{C}'(F(?), ?)$ the functorial isomorphism associated with v and $\tilde{u}$ the inverse of $\tilde{v}$* (see Proposition 11.2). *Then for each object Y of $\mathscr{C}$:*

i) *The following are equivalent:*

a) *For each object X of $\mathscr{C}$, the map*

$$F(X, Y): \mathscr{C}(X, Y) \to \mathscr{C}'(F(X), F(Y))$$

is an injection.

b) *The morphism $u_Y: Y \to GF(Y)$ is a monomorphism.*

c) *If $f: Y \to X$ is a morphism of $\mathscr{C}$ such that $F(f)$ is a monomorphism, then f is a monomorphism.*

d) *The canonical bijection*

$$\tilde{u}(Y, X'): \mathscr{C}'(F(Y), X') \to \mathscr{C}(Y, G(X'))$$

preserves monomorphisms for all $X' \in \mathscr{C}'$.

ii) *The following are equivalent:*

a) *For each object X of $\mathscr{C}$, the map*

$$F(X, Y): \mathscr{C}(X, Y) \to \mathscr{C}'(F(X), F(Y))$$

is a bijection.

b) *The morphism $u_Y: Y \to GF(Y)$ is an isomorphism.*

Proof. i) The equivalence a) $\Leftrightarrow$ b) holds because $F(X, Y)$ is equal to the composition

$$\mathscr{C}(X, Y) \xrightarrow{H^X(u_Y)} \mathscr{C}(X, GF(Y)) \xrightarrow{\tilde{v}(X, F(Y))} \mathscr{C}'(F(X), F(Y)).$$

b) $\Rightarrow$ c). Assume that $F(f)$ is a monomorphism. Then $GF(f)$ is also a monomorphism, and hence it follows from the commutativity of the diagram

$$
\begin{array}{ccc}
Y & \xrightarrow{\quad f \quad} & X \\
{\scriptstyle u_Y}\Big\downarrow & & \Big\downarrow{\scriptstyle u_X} \\
GF(Y) & \xdashrightarrow{\quad GF(f) \quad} & GF(X)
\end{array}
$$

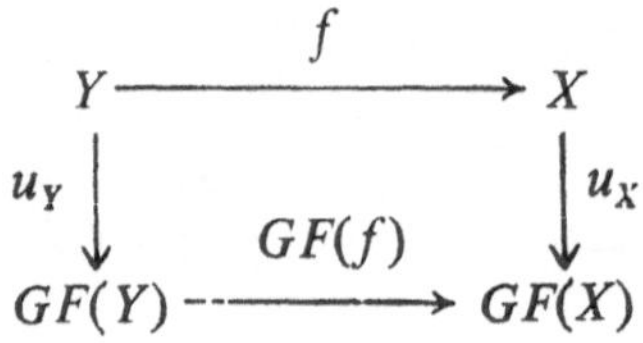

that f is a monomorphism.

c) $\Rightarrow$ b). This follows from the fact that $F(u_Y)$ is a monomorphism (see conditions (Ad) in Section 1.11).

b) $\Rightarrow$ d). Let $f: F(Y) \to X'$ be a monomorphism of $\mathscr{C}'$. Then $\tilde{u}(Y, X')(f) = G(f)u_Y$, so that $\tilde{u}(Y, X')$ is a monomorphism (G preserve monomorphisms!).

d) $\Rightarrow$ c). To show this let $f: Y \to X$ be a morphism such that $F(f)$ is a monomorphism. Then $\tilde{u}(Y, F(X))(F(f)) = GF(f)u_Y$ is a monomorphism, so that f is a monomorphism (see the previous diagram).

Assertion ii) can be obtained similarly. $\ast$

COROLLARY 13.8. *Assume the hypotheses of Proposition* 13.7. *Then the following are equivalent:*

a) *The functor* $F: \mathscr{C} \to \mathscr{C}'$ *is faithful.*

b) *The functorial arrow* $u: \mathrm{Id}\,\mathscr{C} \to GF$ *is an argumentwise functorial monomorphism.*

c) F *reflects monomorphisms.*

d) *For each* $X \in \mathscr{C}$ *and* $X' \in \mathscr{C}'$, *the map*

$$\tilde{u}(X, X'): \mathscr{C}'(F(X), X') \to \mathscr{C}(X, G(X'))$$

preserves monomorphisms. $\ast$

We suggest that the reader dualize Proposition 13.7 and Corollary 13.8.

LEMMA 13.9. *Consider the following diagram of categories and functors*

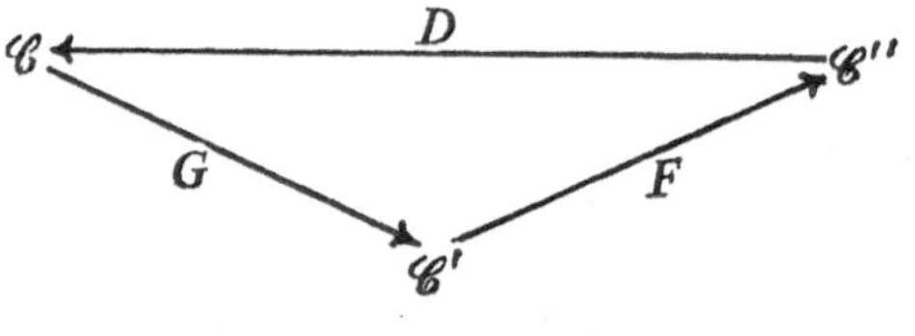

and let $v: (FG)D \to \mathrm{Id}\mathscr{C}''$ *be a coarrow of adjunction of D with FG. Assume that for every pair of functors* $T, T': \mathscr{C}' \to \mathscr{D}$, *and for every functorial morphism* $k: TG \to T'G$, *there is a unique functorial morphism* $t: T \to T'$ *such that* $tG = k$. *Then GD is coadjoint to F and v is a coarrow of adjunction (of F with GD).*

Proof. Let $u: \mathrm{Id}\mathscr{C} \to D(FG)$ be an arrow of adjunction quasi-inverse to v. Thus the functorial morphism $Gu: G \to GD(FG)$ can be considered as a functorial morphism $Gu: (\mathrm{Id}\,\mathscr{C}')G \to (GDF)G$, so that there is a unique functorial morphism

$u' : \mathrm{Id}\,\mathscr{C}' \to GDF$ such that $u'G = Gu$. We claim that u' is an arrow of adjunction of GD with F, quasi-inverse to v. By Proposition 11.3 we consider the composition of arrows:

$$GD \xrightarrow{u'GF} GDFGD \xrightarrow{GDv} GD.$$

Since $u'G = Gu$ and since D is coadjoint to FG we have that

$$(GDv)(u'GD) = (GDv)(GuD) = G(DvuD) = G(1_D).$$

But then $G(1_D) = 1_{GD}$, i.e. $(GDv)(u'GD) = 1_{GD}$.

We also consider the composition

$$F \xrightarrow{Fu'} FGDF \xrightarrow{vF} F$$

which is a functorial morphism $t: F \to F$. We have that $tG = (vFFu')G = vFGFu'G = (vFG)(FGu) = 1_{FG}$, because $u'G = Gu$ and v is quasi-inverse to u. But we also have $1_F G = 1_{FG}$, that is, by hypothesis, $vFFu' = 1_F$. ✳

THEOREM 13.10. *Let $G: \mathscr{C}' \to \mathscr{C}$ be a functor, F adjoint to G and $v: FG \to \mathrm{Id}\mathscr{C}'$ a coarrow of adjunction. We denote by Σ the class of all morphisms s of $\mathscr{C}$ such that $F(s)$ is invertible. The following are equivalent:*

1) *G is full and faithful.*

2) *The functorial morphism $v: FG \to \mathrm{Id}\,\mathscr{C}'$ is invertible.*

3) *The category $\mathscr{C}(\Sigma^{-1})$ exists and is equivalent to $\mathscr{C}'$.*

4) *If $T, T': \mathscr{C}' \to \mathscr{D}$ are functors and $t: TF \to T'F$ is a functorial morphism, then there is a unique functorial morphism $s: T \to T'$ such that $t = sF$.*

Proof. The equivalence 1) ⇔ 2) follows from Proposition 13.7, ii).

2) ⇒ 3). Let $H: \mathscr{C} \to \mathscr{D}$ be a functor that makes the morphisms of Σ invertible and let $\bar{H}: \mathscr{C}' \to \mathscr{D}$ be the functor HG. It is easy to check that $\bar{H}F = HGF \simeq H$. The uniqueness of $\bar{H}$ follows since F is representative. Hence $(F, \mathscr{C}')$ is a category of fractions of $\mathscr{C}$ relative to Σ. We finish the implication by using Exercise 13.1.

3) ⇒ 4). This follows from Proposition 13.5.

4) ⇒ 2). Let $u: \mathrm{Id}\,\mathscr{C} \to GF$ be an arrow of adjunction quasi-inverse to v. Then, by Proposition 11.3, the composition of arrows

$$F \xrightarrow{Fu} FGF \xrightarrow{vF} F$$

is the identity of F. By hypothesis there is a unique morphism $t: \mathrm{Id}\,\mathscr{C}' \to FG$ such that $Fu = tF$. Thus, $1_F = vFFu = vFtF = (vt)F$, i.e. vt is the identity morphism of $\mathrm{Id}\,\mathscr{C}'$. In the same way we see that tv is the identity morphism of FG. Hence v is an isomorphism. ✳

It is easy to check that Theorem 13.10 dualizes in the following way.

THEOREM 13.11. *Let* $F: \mathscr{C} \to \mathscr{C}'$ *be a functor,* G *an adjoint of* F *and* $u: \mathrm{Id}\mathscr{C}' \to FG$ *an arrow of adjunction. We denote by* Σ *the class of all morphisms* s *of* $\mathscr{C}$ *such that* $F(s)$ *is invertible. The following are equivalent:*

1) G *is full and faithful.*
2) *The functorial morphism* $u: Id\mathscr{C}' \to FG$ *is invertible.*
3) *The category* $\mathscr{C}(\Sigma^{-1})$ *exists and is equivalent to* $\mathscr{C}'$.
4) *If* $T, T': \mathscr{C}' \to D$ *are functors and* $t: TF \to T'F$ *is a functorial morphism, there is a unique functorial morphism* $s: T \to T'$ *such that* $t = sF$. ✳

PROPOSITION 13.12. *Let* $G: \mathscr{C}' \to \mathscr{C}$ *be a full and faithful functor and* F *adjoint to* G. *A functor* $K: \mathscr{I} \to \mathscr{C}'$ *has a limit in* $\mathscr{C}'$ *if and only if the functor* GK *has a limit in* $\mathscr{C}$. *Hence we have the canonical isomorphism* $\varprojlim K \simeq F(\varprojlim GK)$. *Also the functor* K *has a colimit in* $\mathscr{C}'$ *if and only if the functor* GK *has a colimit in* $\mathscr{C}$ *and there is the canonical isomorphism* $\varinjlim K \simeq F(\varinjlim GK)$. *Therefore, if* $\mathscr{C}$ *is a complete (finitely complete, cocomplete or finitely cocomplete) category, then so is* $\mathscr{C}'$.

Proof. Assume that GK has a limit (X, w) in $\mathscr{C}$. Then in $\mathscr{C}'$ we have the diagram

$$F(X) \xrightarrow{\ Fw\ } FGK \xrightarrow{\ vK\ } K$$

where $v: FG \to \mathrm{Id}\,\mathscr{C}'$ is a coarrow of adjunction. Let $t: Y \to K$ be a cone; then $Gt: G(Y) \to GK$ is again a cone and hence there is a unique morphism $f: G(Y) \to X$ such that $wf = Gt$. Thus in $\mathscr{C}$ we have the following commutative diagram:

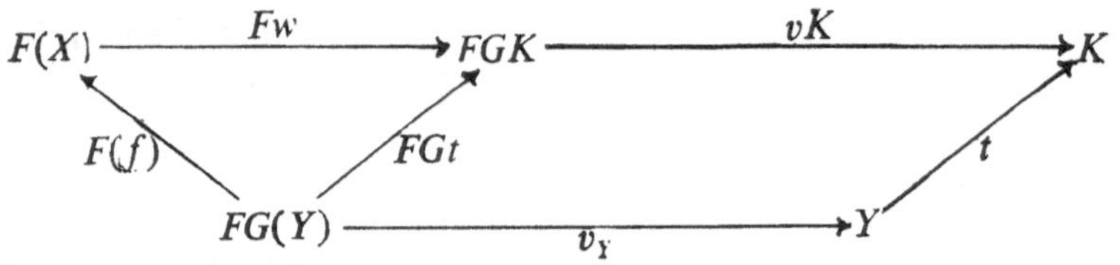

By Theorem 13.10, v is a functorial isomorphism so that v_Y and vK are also isomorphisms. Thus $t = F(f)v_Y^{-1}$ is the unique morphism such that $vKFwt = t$, i.e. $(F(X), vKw)$ is a limit of K. The remaining assertions can be checked in the same way, and we leave them for the reader. ✳

Exercises

13.1. Let Σ be a system of morphisms of $\mathscr{C}$ such that $\mathscr{C}(\Sigma^{-1})$ exists.

a) Prove that the canonical functor $P: \mathscr{C} \to \mathscr{C}(\Sigma^{-1})$ is representative.

b) Prove that every morphism of $\mathscr{C}(\Sigma^{-1})$ is a composition of morphisms of the form $P(s)^{-1}P(f)P(s')^{-1}$ with $s, s' \in \Sigma$, and $f \in \mathscr{C}$. (Hint: Let $P: \mathscr{C} \to \mathscr{C}(\Sigma^{-1})$ be the canonical functor; denote by $\mathscr{C}_1$ the subcategory of $\mathscr{C}(\Sigma^{-1})$ whose objects are $\{P(X)\}_{X \in \mathscr{C}}$ and whose morphisms are all compositions of morphisms of the form $P(s)^{-1}P(f)\,P(s')^{-1}$, where f, s, s' are morphisms of $\mathscr{C}$ and $s, s' \in \Sigma$. Thus $\mathscr{C}_1$ is canonically a category of fractions of $\mathscr{C}$ relative to Σ.)

13.2. Let $F: \mathscr{C} \to \mathscr{C}'$ be a functor, G coadjoint to F, and $u: \mathrm{Id}\mathscr{C} \to GF$ and $v: FG \to \mathrm{Id}\mathscr{C}'$ the corresponding arrow and coarrow of adjunction. Assume that F is full and faithful. Prove that the following hold:

a) The functorial morphisms $Gv: GFG \to G$ and $vF: FGF \to F$ are isomorphisms.

b) The functorial morphisms $FGv: FGFG \to FG$ and $vFG: FGFG \to FG$ are equal and are isomorphisms.

13.3. Prove that the inclusion functor $\mathscr{G}\imath \to \mathscr{M}on$ has an adjoint and a coadjoint.

13.4. Let Σ be a class of morphisms of $\mathscr{C}$ such that $\mathscr{C}(\Sigma^{-1})$ exists. Let Σ_s denote the *saturation* of Σ, i.e. the class of all morphisms s of $\mathscr{C}$ which are taken to isomorphisms by $P: \mathscr{C} \to \mathscr{C}(\Sigma^{-1})$. Prove that:

a) The category $\mathscr{C}(\Sigma_s^{-1})$ exists and is equivalent to $\mathscr{C}(\Sigma^{-1})$.

b) If Σ' is another class of morphisms of $\mathscr{C}$ such that $\mathscr{C}(\Sigma'^{-1})$ exists, then $\mathscr{C}(\Sigma^{-1})$ and $\mathscr{C}(\Sigma'^{-1})$ are equivalent if and only if $\Sigma_s = \Sigma'_s$.

13.5. A functor $F: \mathscr{C} \to \mathscr{C}'$ is called *conservative* if $F(f)$ being an isomorphism implies that f is an isomorphism.

a) Show that the canonical functor $P: \mathscr{C} \to \mathscr{C}(\Sigma^{-1})$ is conservative if and only if every morphism of Σ is an isomorphism. In this case P is an equivalence of categories.

b) Prove that a conservative functor having a full and faithful coadjoint is an equivalence of categories.

1.14. Calculus of fractions

A particular, but interesting, type of category of fractions is the left (right) fractional category of fractions. They are closely related to certain special systems of morphisms, namely left (right) calculable systems. In this section we describe left fractional categories and give some of their simple properties.

References: Almkvist [1], Bănică—Popescu [2], Gabriel [1], Gabriel—Zisman [1], Ore [1], Mal'cev [1], Schubert [1].

A *multiplicative system* of morphisms in a category $\mathscr{C}$ is a class Σ of morphisms of $\mathscr{C}$ such that:

a) all identity morphisms of $\mathscr{C}$ are in Σ;

b) if $s: X \to Y$ and $s': Y \to Z$ are in Σ, then $s's \in \Sigma$.

A multiplicative system Σ is called *left permutable* if:

c) each coangle $X' \overset{s}{\leftarrow} X \overset{f}{\rightarrow} Y$ with $s \in \Sigma$ can be embedded in a commutative square

$$\begin{array}{ccc} X & \xrightarrow{f} & Y \\ {\scriptstyle s}\downarrow & & \downarrow{\scriptstyle s'} \\ X' & \xrightarrow{f'} & Y' \end{array}$$

with $s' \in \Sigma$.

A multiplicative system Σ is called *left simplifiable* if:

d) for each pair $f, g\colon X \to Y$ of morphisms in $\mathscr{C}$ and morphisms $s\colon X' \to X$, $s \in \Sigma$, such that $fs = gs$, there is a morphism $s'\colon Y \to Y'$, $s' \in \Sigma$, such that $s'f = s'g$.

By a *left calculable system* we mean a multiplicative system which is also left permutable and left simplifiable. Dually, we have the notions of *right permutable*, *right simplifiable* and *right calculable systems.* The terminology "left calculable system" is justified by the simple description of the category $\mathscr{C}(\Sigma^{-1})$ which is given in the following theorem. Before stating it we build some useful terminology. For each object X of $\mathscr{C}$, let X/Σ denote the full subcategory of $X/\mathscr{C}$ whose objects are pairs (s, Y), with $s \in \Sigma$. Also let $\sigma_X\colon X/\Sigma \to \mathscr{C}$ denote the forgetful functor, i.e. the functor that assigns to each object (s, Y) of X/Σ the object Y of $\mathscr{C}$ and to each morphism $f\colon (s, Y) \to (s', Y')$ of X/Σ, the morphism $\sigma_X(f) = f$.

Let Σ be a multiplicative system of $\mathscr{C}$. We say that the *left fractional category* of $\mathscr{C}$ with respect to Σ is defined if:

l_1) The category $\mathscr{C}(\Sigma^{-1})$ exists.

l_2) For each morphism $\alpha\colon U \to V$ of $\mathscr{C}(\Sigma^{-1})$ there is an angle $X \to Y \overset{s}{\leftarrow} Z$ with $s \in \Sigma$ such that $\alpha = P(s)^{-1}P(f)$.

l_3) If $f, g\colon X \to Y$ are morphisms of $\mathscr{C}$ such that $P(f) = P(g)$, then there is an $s\colon Y \to Z$, $s \in \Sigma$, such that $sf = sg$.

Dually, we have the notion of a *right fractional category.*

Let $\mathscr{C}^l(\Sigma^{-1})$ ($\mathscr{C}^r(\Sigma^{-1})$) denote the left (right) fractional category of $\mathscr{C}$ with respect to Σ.

THEOREM 14.1. *Let Σ be a multiplicative system of morphisms in a category $\mathscr{C}$. The following are equivalent:*

1) *The left fractional category of $\mathscr{C}$ with respect to Σ is defined.*

2) *Σ is left calculable and for every pair (Y, X) of objects of $\mathscr{C}$, the functor $H^Y\sigma_X\colon X/\Sigma \to \mathscr{S}et$ has a colimit.*

Proof. 1) $\Rightarrow$ 2). First we prove that Σ is left calculable. Indeed, if $X \overset{f}{\leftarrow} Y \overset{s}{\rightarrow} Y'$ is a coangle of $\mathscr{C}$, with $s \in \Sigma$, then $P(f)P(s)^{-1}$ is of the form $P(s')^{-1}P(f')$, where $s' \in \Sigma$ (see l_2)). Thus $P(s')P(f) = P(f')P(s)$, i.e. $P(s'f) = P(f's)$. But then by l_3), there is an $s'' \in \Sigma$ such that $s''s'f = s''f's$. Hence Σ is left permutable. Now if $fs = gs$, with $s \in \Sigma$, then obviously $P(f) = P(g)$, i.e. $s'f = s'g$ for some $s' \in \Sigma$ (see l_3)). Hence Σ is also left simplifiable.

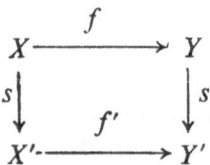

Now we prove that the functor $H^Y\sigma_X$ has a colimit. For that, let M denote $\mathscr{C}(\Sigma^{-1})(P(Y), P(X))$. If (s, Z) is an object of X/Σ, let $u_{(s, Z)}: H^Y\sigma_X(s, Z) = H^Y(Z) = \mathscr{C}(Y, Z) \to M$ denote the map that associates with $f \in \mathscr{C}(Y, Z)$ the element $P(s)^{-1}P(f)$ of M. If $g: (s, Z) \to (s', Z')$ is a morphism of X/Σ, then $H^Y\sigma_X(g) = H^Y(g)$. If $f \in \mathscr{C}(Y, Z)$ then we have the diagram of $\mathscr{C}$

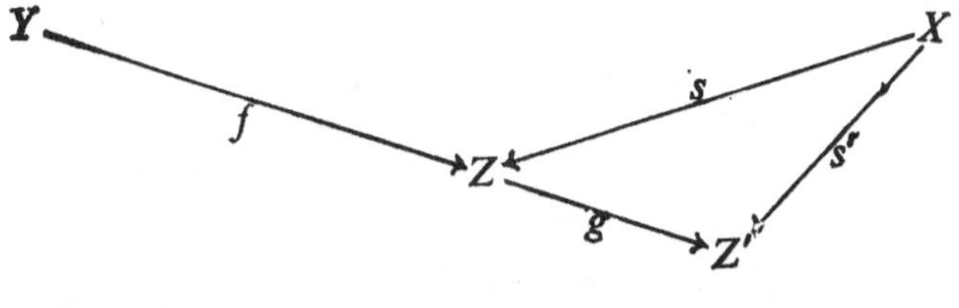

where $gs = s'$. Hence $=u_{(s, Z)}(f) = P(s)^{-1}P(f) = P(s')^{-1}P(g)P(f) = P(s')^{-1}P(gf) = u_{(s', Z')}(gf) = u_{(s', Z')}H^Y(g)(f)$. Finally, we have that the maps $\{u_{(s, Z)}\}_{(s, Z) \in X/\Sigma}$ define a cocone (u, M) over $H^Y\sigma_X$. This cocone is the colimit of $H^Y\sigma_X$. To show that, let (v, N) be another cocone over $H^Y\sigma_X$. Define a map $\bar{v}: M \to N$ in the following way. If $\alpha \in M$, then $\alpha = P(s)^{-1}P(f)$, where $s: X \to Z$ is an element of Σ, and $f: Y \to Z$. Then we define $\bar{v}(\alpha) = v_{(s, Z)}(f)$. We now show that the map $\bar{v}$ is well defined. For that let $s': X \to Z'$ be another element of Σ such that $\alpha = P(s')^{-1}P(f')$, where $f': Y \to Z'$. Since Σ is left calculable, we may construct the following diagram

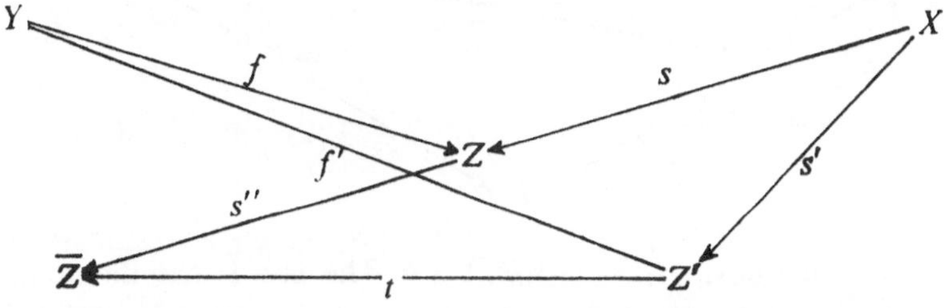

where $s'' \in \Sigma$ and $s''s = ts' = \bar{s}$.

Then $P(\bar{s})P(s)^{-1}P(f) = P(s'')P(f) = P(\bar{s})P(s')^{-1}P(f') = P(t)P(f')$, i.e. by l_3) we can assume that $s''f = tf'$. But then $v_{(\bar{s}, \bar{Z})}(s''f) = v_{(\bar{s}, \bar{Z})}H^Y(s'')(f) = v_{(s, Z)}(f) = \bar{v}(\alpha)$ and similarly $v_{(\bar{s}, \bar{Z})}(t, f') = v_{(\bar{s}, \bar{Z})}H^Y(t)(f') = v_{(s', Z')}(f') = \bar{v}(\alpha)$, i.e. the map $\bar{v}$ is well defined. In the same way we see that $\bar{v}u = v$. The uniqueness of $\bar{v}$ follows from the fact that for each $\alpha \in M$ there is an object (s, Z) of X/Σ such that $u_{(s, Z)}(f) = \alpha$ (see l_2)). Another proof of this part of the theorem can be obtained from Proposition 12.9.

2) ⇒ 1). Define a category $\overline{\mathscr{C}}$ in the following way. The objects of $\overline{\mathscr{C}}$ are the objects of $\mathscr{C}$. If X is an object of $\mathscr{C}$, then the same object considered in $\overline{\mathscr{C}}$ will be denoted by $\bar{X}$. If $\bar{X}$ and $\bar{Y}$ are objects of $\overline{\mathscr{C}}$ denote by $\overline{\mathscr{C}}(\bar{Y}, \bar{X})$ a colimit of $H^Y\sigma_X$, and by $\{u_{(s, Z)}\}_{(s, Z) \in X/\Sigma}$ the structural injections of that colimit. If $f \in H^Y\sigma_X(s, Z) = \mathscr{C}(Y, Z)$, denote $u_{(s, Z)}(f)$ by (s/f).

Now we define the composition of morphisms of $\overline{\mathscr{C}}$. For that let $\alpha: \overline{Y} \to \overline{X}$ and $\beta: \overline{X} \to \overline{Z}$ be morphisms of $\overline{\mathscr{C}}$. By Exercise 14.1 and Proposition 12.9, there is an object (s, K) of X/Σ and an $f \in H^Y\sigma_X(s, K) = \mathscr{C}(Y, K)$ such that $u_{(s,K)}(f) = (s/f) = \alpha$. Also there exist $t: Z \to K'$ and $g: X \to K'$ such that $t \in \Sigma$ and $(t/g') = \beta$. Since Σ is left calculable we have the commutative diagram of $\mathscr{C}$:

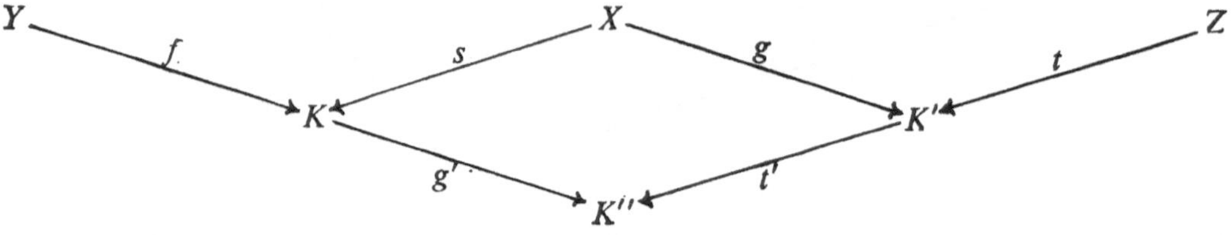

where $t' \in \Sigma$.

Let us define composition by $\beta\alpha = (t/g)(s/f) = (t't/g'f)$. We prove that composition defined this way does not depend on representatives. To this end, assume that $\alpha = (s_1/f_1)$ where $s_1: X \to K_1$, $f_1: Y \to K_1$ and $s_1 \in \Sigma$. Again by Exercise 14.1 and Proposition 12.9 we can check that we have the commutative diagram of $\mathscr{C}$

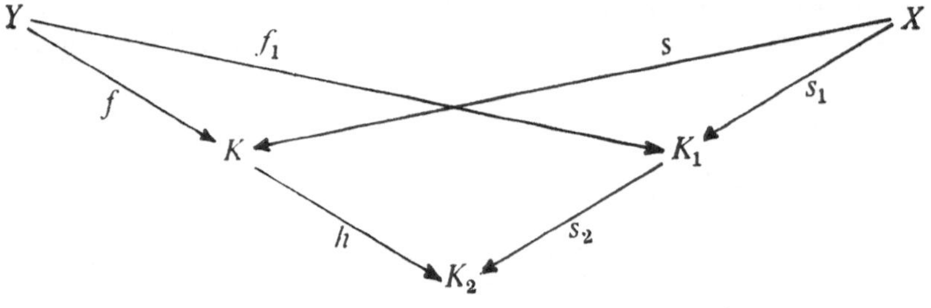

where $s_2 \in \Sigma$, such that $(s_2s_1/s_2f_1) = (s_2s_1/hf) = \alpha$. The proof that the composition of morphisms in $\overline{\mathscr{C}}$ does not depend on the representatives used is straightforward and is left for the reader. Also the composition of morphisms is associative and for each object X of $\mathscr{C}$, the element $(1_X/1_X)$ of $\overline{\mathscr{C}}(\overline{X}, \overline{X})$ is the identity of $\overline{X}$.

Furthermore, let $P: \mathscr{C} \to \overline{\mathscr{C}}$ denote the functor defined as follows: for each object X, $P(X) = \overline{X}$ and for each morphism $f: Y \to X$ of $\mathscr{C}$, $P(f) = (1_X/f)$. We claim that $(P, \overline{\mathscr{C}})$ is a category of fractions of $\mathscr{C}$ relative to Σ. To show this, let $s: Y \to X$ be a morphism of $\mathscr{C}$ such that $s \in \Sigma$. Then we may consider the morphisms $P(s) = (1_X/s)$ of $\overline{\mathscr{C}}(\overline{Y}, \overline{X})$ and $(s/1_X)$ of $\overline{\mathscr{C}}(\overline{X}, \overline{Y})$. From the diagram of $\mathscr{C}$

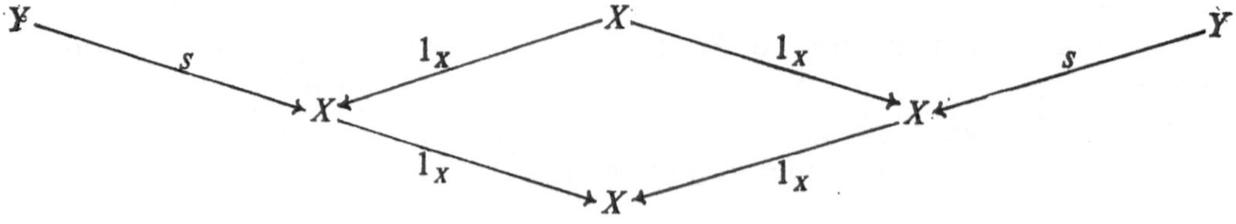

we see that $(s/1_X)(1_X/s) = (s/s)$. Also from the diagram

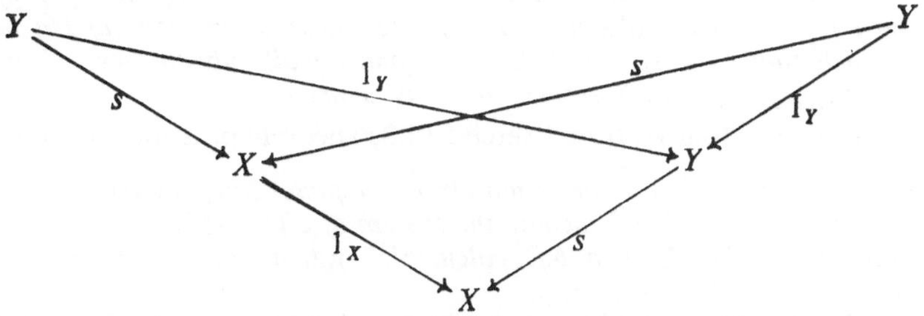

we see that $(s/s) = (1_Y/1_Y) = 1_{\bar{Y}}$. It can be proved similarly that $(1_X/s)(s/1_X) = 1_{\bar{X}}$, i.e. $(s/1_X) = P(s)^{-1}$. Hence P makes every morphism of Σ invertible.

Let $(s/f): \bar{X} \to \bar{Y}$ be a morphism of $\bar{\mathscr{C}}$ where $s: Y \to Z$ and $f: X \to Z$ are morphisms of $\mathscr{C}$ and $s \in \Sigma$. A simple computation proves that $(s/f) = (s/1_Z)(1_Z/f) = P(s)^{-1}P(f)$.

Now suppose that $T: \mathscr{C} \to \mathscr{C}'$ is a functor that makes the morphisms of Σ invertible. Define the functor $\bar{T}: \bar{\mathscr{C}} \to \mathscr{C}'$ as follows: $\bar{T}(\bar{X}) = T(X)$ for each object $\bar{X}$ and $\bar{T}(s/f) = \bar{T}(P(s)^{-1}P(f)) = T(s)^{-1}T(f)$ for each morphism (s/f) of $\bar{\mathscr{C}}$. We leave it for the reader to check that $\bar{T}$ is well defined, i.e., that $(s/f) = (s'/f')$ implies $T(s)^{-1}T(f) = T(s')^{-1}T(f')$. The uniqueness of $\bar{T}$ is obvious. Hence $\bar{\mathscr{C}}$ is a category of fractions of $\mathscr{C}$ with respect to Σ.

Every morphism in $\bar{\mathscr{C}}$ can be written as $P(s)^{-1}P(f)$, with $s \in \Sigma$. So in order to prove that $\bar{\mathscr{C}}$ is the left fractional category of $\mathscr{C}$ with respect to Σ, we will consider morphisms $f, g: X \to Y$ in $\mathscr{C}$ such that $P(f) = P(g)$. Then we have the commutative diagram in $\mathscr{C}$

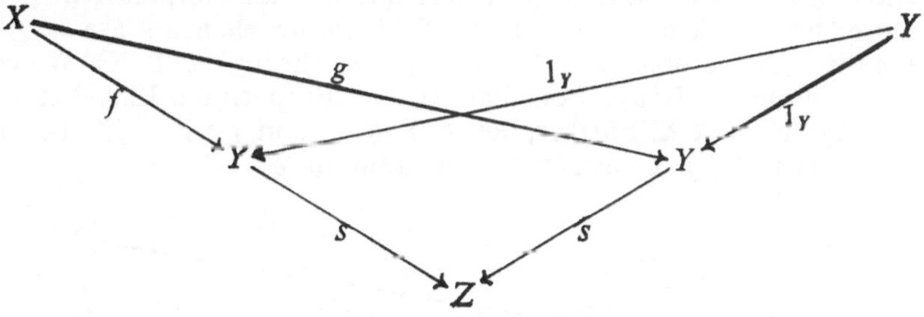

or, equivalently, $sf = sg$, where $s \in \Sigma$.

By the construction of $\bar{\mathscr{C}}$, we may define $\mathscr{C}^l(\Sigma^{-1}) = \bar{\mathscr{C}}$, and P is the canonical functor. ✳

Observe that in proving the above theorem we used only the assumption that some functors have a colimit. Now we show a situation where this condition is satisfied.

Corollary 14.2. *Let Σ be a left calculable system of morphisms in a category $\mathscr{C}$. Assume that for each object $X \in \mathscr{C}$, there is a set $\{(s_i\, X_i)\}_{i \in \mathscr{I}}$ of objects of X/Σ such that for each object (s, Y) of X/Σ there is an $i \in \mathscr{I}$ and a morphism $f\colon Y \to X_i$ such that $fs = s_i$, i.e. X/Σ has a final small subcategory. Then the left fractional category of $\mathscr{C}$ with respect to Σ is defined.*

Proof. The proof follows from Exercise 14.3, Theorem 6.12 and Theorem 14.1. ✠

Proposition 14.3. *Let $\mathscr{C}$ be a finitely cocomplete category and let $T\colon \mathscr{C} \to \mathscr{C}'$ be a right exact functor. Let Σ denote the system of all morphisms s in $\mathscr{C}$ such that $T(s)$ is invertible. Then Σ is a left calculable system and the category $\mathscr{C}^l\,(\Sigma^{-1})$ is defined.*

Proof. Conditions a) and b) are clearly satisfied. In order to check c) consider a coangle $X' \overset{s}{\leftarrow} X \overset{f}{\rightarrow} Y$ of $\mathscr{C}$. Since the canonically defined square of $\mathscr{C}'$:

$$\begin{array}{ccc} T(X) & \xrightarrow{\;T(s)\;} & T(X') \\ {\scriptstyle T(f)}\downarrow & & \downarrow{\scriptstyle T(f')} \\ T(Y) & \xrightarrow{\;T(s')\;} & T(X' \textstyle\coprod_X Y) \end{array}$$

is cocartesian and $T(s)$ is an isomorphism, we get that $T(s')$ is also an isomorphism, i.e. $s' \in \Sigma$. For d) let $(f, g)\colon X \to Y$ be a double arrow of $\mathscr{C}$ and let $s\colon X' \to X$ be such that $fs = gs$, where $s \in \Sigma$. Also let $t\colon Y \to Y'$ be a cokernel of (f, g). Then $T(t)\colon T(Y) \to T(Y')$ is a cokernel of $(T(f), T(g))$ and since $T(s)$ is invertible we have that $T(f) = T(g)$, i.e. $T(t)$ is an isomorphism. Hence $t \in \Sigma$.

Now we wish to prove the existence of the category $\mathscr{C}^l(\Sigma^{-1})$. For this we shall use Theorem 14.1. Hence, let X and Y be objects of $\mathscr{C}$, and let (s, Z) be an object of X/Σ. Consider the set $M = \mathscr{C}'(T(Y), T(X))$ and let $u_{(s,Z)}\colon H^Y\sigma_X(s, Z) = \mathscr{C}(Y, Z) \to$ $\to M$ be the map that associates with $f\colon Y \to Z$ the element $T(s)^{-1}T(f)$ of M. A computation proves that these maps define a functorial morphism $u\colon H^Y\sigma_X \to M$. Furthermore, let M' denote the subset of M whose elements are $u_{(s,Z)}(f)$ where (s, Z) runs through the objects of X/Σ and f runs through $\mathscr{C}(Y, Z)$. We claim that (u, M') is a colimit of $H^Y\sigma_X$. For that we use Proposition 12.9. Let (s, Z) and (s', Z') be objects of X/Σ. Further, let $f\colon Y \to Z$ and $f'\colon Y \to Z'$ be such that $T(s)^{-1}T(f) = T(s')^{-1}T(f')$. Consider the diagram in $\mathscr{C}$:

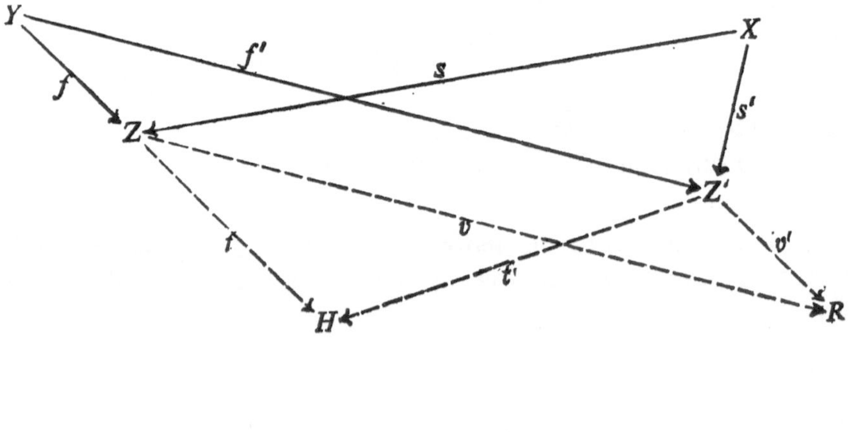

where (t, t', H) is a colimit of the diagram drawn with the unbroken arrows and (v, v', R) is a fibered coproduct of the coangle $Z \overset{s}{\leftarrow} X \overset{s'}{\rightarrow} Z'$. Since $ts = t's'$, there is a unique morphism $p: R \to H$ such that $pv = t$ and $pv' = t'$. Furthermore, the equality $T(s)^{-1}T(f) = T(s')^{-1}T(f')$ implies that $T(v)T(f) = T(v')T(f')$. Hence, since T commutes with colimits, there is a unique morphism $h: T(H) \to T(R)$ such that $T(v) = hT(t)$ and $T(v') = hT(t')$. Again a computation proves that h is an inverse of $T(p)$, i.e. $p \in \Sigma$. But then t and t' also belong to Σ since v and v' are obviously elements of Σ. Therefore (u, M') is a colimit of the functor $H^Y\sigma_X$. ✳

Now we examine limits and colimits in the category of fractions.

PROPOSITION 14.4. *Let Σ be a multiplicative system of morphisms in $\mathscr{C}$ such that $\mathscr{C}^l(\Sigma^{-1})$ is defined. Then the canonical functor $P: \mathscr{C} \to \mathscr{C}^l(\Sigma^{-1})$ commutes with finite colimits, i.e. it is right exact. Moreover, if $\mathscr{C}$ is finitely cocomplete then $\mathscr{C}^l(\Sigma^{-1})$ is also finitely cocomplete.*

Proof. Let $\mathscr{I}$ be a finite category, $F: \mathscr{I} \to \mathscr{C}$ a functor and (u, X) a colimit of F. By Theorem 6.11 in order to prove that $(Pu, P(X))$ is a colimit of PF it will suffice to check that for each object $\bar{Y}$ of $\mathscr{C}^l(\Sigma^{-1})$ we have that $(H_{\bar{Y}}(P(X)), H_{\bar{Y}}(Pu))$ is a limit of $H_{\bar{Y}}PF$. Indeed, since P defines a bijection between the objects of $\mathscr{C}$ and the objects of $\mathscr{C}^l(\Sigma^{-1})$, we deduce that $\bar{Y} = P(Y)$. Hence, by Theorem 14.1, $H_Y(P(Z)) = \mathscr{C}^l(\Sigma^{-1})(P(Z), P(Y)) = \varinjlim_{Y/\Sigma} (H^Z\sigma_Y)$, for all $Z \in \mathscr{C}$. Now the assignment

$$i \rightsquigarrow H^{Fi}\sigma_Y, \quad i \in \mathscr{I}$$

defines an $\mathscr{I}^0$-diagram of functors from Y/Σ into $\mathscr{S}et$ and $H^X\sigma_Y$ is a limit of this diagram. Hence the assertion follows from Theorem 12.11 since Y/Σ is a filtered category.

Now let us assume that $\mathscr{C}$ is finitely cocomplete. According to Theorem 9.2 and Note 9.5, we must check that $\mathscr{C}^l(\Sigma^{-1})$ is a category with fibered coproducts and finite coproducts. To show this, let $P(X) \overset{\alpha}{\leftarrow} P(Y) \overset{\beta}{\rightarrow} P(Z)$ be a coangle of $\mathscr{C}^l(\Sigma^{-1})$. Then we can define the diagram

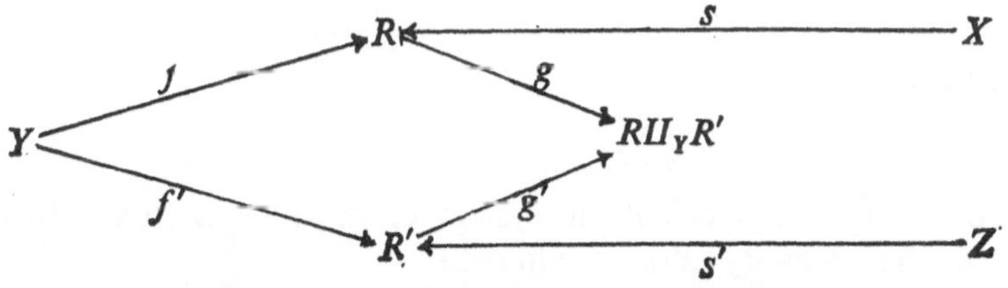

in $\mathscr{C}$, where $s, s' \in \Sigma$, such that $\alpha = P(s)^{-1}P(f)$, $\beta = P(s')^{-1}P(f')$. Let $u = P(g)P(s)$, $v = P(g')P(s')$. We claim that $(u, v, P(R\coprod_Y R'))$ is a fibered coproduct of the given coangle. To show this, let $\gamma: P(X) \to A$ and $\delta: P(Z) \to A$ be morphisms in $\mathscr{C}^l(\Sigma^{-1})$ such that $\gamma\alpha = \delta\beta$, i.e. $\gamma P(s)^{-1}P(f) = \delta P(s')^{-1}P(f')$. Since P is right exact, there is a unique morphism $\varepsilon: P(R\coprod_Y R') \to A$ such that $\varepsilon P(g) = \gamma P(s)^{-1}$ and $\varepsilon P(g') = \delta P(s')^{-1}$. But then $\varepsilon P(g)P(s) = \varepsilon u = \gamma$ and $\varepsilon P(g')P(s') = \varepsilon v = \delta$. The uniqueness of ε is obvious.

The fact that every pair of objects of $\mathscr{C}^l(\Sigma^{-1})$ has a coproduct is obvious, since P is bijective on objects and right exact. ✳

PROPOSITION 14.5. *Let Σ be a multiplicative system of morphisms of $\mathscr{C}$. The following are equivalent:*

a) *The category $\mathscr{C}^l(\Sigma^{-1})$ is defined and Σ is right calculable.*

b) *The category $\mathscr{C}^r(\Sigma^{-1})$ is defined and Σ is left calculable.*

If a) *or* b) *holds then $\mathscr{C}^l(\Sigma^{-1}) = \mathscr{C}^r(\Sigma^{-1}) = \mathscr{C}(\Sigma^{-1})$, and the canonical functor $P: \mathscr{C} \to \mathscr{C}(\Sigma^{-1})$ is left and right exact. Moreover, if $\mathscr{C}$ is a finitely complete and cocomplete category, then so is $\mathscr{C}(\Sigma^{-1})$.*

Proof. a) $\Rightarrow$ b). By Exercise 14.5, the dual of condition l_2) holds. Now let us assume that $f, g: X \to Y$ are morphisms in $\mathscr{C}$ such that $P(f) = P(g)$. Then there is an $s \in \Sigma$ such that $sf = sg$. Now since Σ is right calculable there is a $t \in \Sigma$ such that $ft = gt$, i.e. the dual of l_3) also holds. The remaining assertions are obvious and their verification is left for the reader. ✳

A system Σ of morphisms in $\mathscr{C}$ is called *strong left calculable* if it is left calculable and if in addition the following condition holds:

e) For every copencil $\{s_i: Y \to Y_i\}_{i \in \mathscr{I}}$ of morphisms of Σ there is a set $\{f_i: Y_i \to Z\}_{i \in \mathscr{I}}$ of morphisms in $\mathscr{C}$ such that $f_i s_i \in \Sigma$ for all $i \in \mathscr{I}$.

As a consequence of Theorems 12.4 and 14.1 we have the following result.

PROPOSITION 14.6. *Let Σ be a strong left calculable system of $\mathscr{C}$. Assume that the category $\mathscr{C}^l(\Sigma^{-1})$ is defined. Then the canonical functor $P: \mathscr{C} \to \mathscr{C}^l(\Sigma^{-1})$ commutes with coproducts. In particular, if $\mathscr{C}$ is a category with coproducts or a complete category then $\mathscr{C}^l(\Sigma^{-1})$ is also.* ✳

Note 14.7. In this section we have dealt with categories of fractions relative to left calculable systems. Of course these notions and results can be dualized to right calculable systems. We suggest that the reader do this.

Exercises

14.1. Let Σ be a left calculable system of morphisms in $\mathscr{C}$. Prove that, for every $X \in \mathscr{C}$, the category X/Σ is filtered.

14.2. Prove that the system Σ of all isomorphisms of a category $\mathscr{C}$ is both left and right calculable and, moreover, $\mathscr{C}^l(\Sigma^{-1}) = \mathscr{C}^r(\Sigma^{-1}) = \mathscr{C}$.

14.3. Let $\mathscr{I}$ be a filtered category. Assume that there exists a set $\mathscr{J}$ of objects of $\mathscr{I}$ such that, for each object i of $\mathscr{I}$, there is an object j of $\mathscr{J}$ for which $\mathscr{I}(i, j) \neq \emptyset$. Show that the full subcategory of $\mathscr{I}$ generated by $\mathscr{J}$ is a filtered category and it is a final subcategory of $\mathscr{I}$.

14.4. Let Σ be a class of morphisms of $\mathscr{C}$ such that $\mathscr{C}(\Sigma^{-1})$ is defined. Prove:

a) If the canonical functor $P: \mathscr{C} \to \mathscr{C}(\Sigma^{-1})$ is faithful, then Σ consists of bimorphisms.

b) If Σ consists of monomorphisms and if Σ is left calculable, then P is faithful.

14.5. (Obtaining a common denominator for fractions.) Let Σ be a left calculable system of morphisms in $\mathscr{C}$ such that $\mathscr{C}^l(\Sigma^{-1})$ is defined. Prove: If $\{\alpha_i: P(X_i) \to P(Y)\}_{1 \leqslant i \leqslant n}$ are morphisms in $\mathscr{C}^l(\Sigma^{-1})$, then there exist an $s \in \Sigma$ and morphisms $\{g_i\}_{1 \leqslant i \leqslant n}$ in $\mathscr{C}$ such that $\alpha_i = P(s)^{-1}P(g_i)$, $i = 1, \ldots, n$.

14.6. Let Σ be a left calculable system of morphisms in $\mathscr{C}$ such that $\mathscr{C}^l(\Sigma^{-1})$ is defined. Prove that the following hold:

a) A morphism f in $\mathscr{C}$ belongs to Σ_s, the saturation of Σ, if and only if there are morphisms u, v such that uf and fv exist and belong to Σ.

b) If $f \in \Sigma_s, f = up$, where p is an epimorphism, then u and p also belong to Σ_s.

c) Σ_s is a left calculable system.

d) If Σ is strong left calculable then Σ_s is also strong left calculable.

14.7. Let Σ be a left calculable system of morphisms in $\mathscr{C}$ such that the category of fractions $\mathscr{C}(\Sigma^{-1})$ is defined. Show that this category is a left fractional category of $\mathscr{C}$ with respect to Σ if and only if condition l_3) is satisfied. Give a simple example of a left calculable system Σ of a category $\mathscr{C}$ such that the category $\mathscr{C}(\Sigma^{-1})$ is defined, but which is not a left fractional category.

14.8. Let $\{\Sigma_i\}_{i \in \mathscr{I}}$ be a class of left calculable saturated systems in $\mathscr{C}$ such that $\mathscr{C}^l(\Sigma_i^{-1})$ is defined for all $i \in \mathscr{I}$. Prove that the system $\Sigma = \cap_i \Sigma_i$ is also left calculable and saturated, and, moreover, that if $\mathscr{I}$ is a set, then the category $\mathscr{C}^l(\Sigma^{-1})$ is defined.

14.9. Assume the hypothesis of Proposition 14.3. Show that the unique functor $\overline{T}: \mathscr{C}(\Sigma^{-1}) \to \mathscr{C}'$ such that $\overline{T}P = T$ is conservative and right exact.

14.10. Let $F: \mathscr{C} \to \mathscr{C}'$ be a functor and denote by Σ the class of all morphisms s in $\mathscr{C}$ such that $F(s)$ is an isomorphism. Let P be the canonical functor, $P: \mathscr{C} \to \mathscr{C}(\Sigma^{-1})$.

a) Give an example for which the category $\mathscr{C}(\Sigma^{-1})$ is not defined. (Hint: Use Example 13.1.)

b) Prove: If the category $\mathscr{C}(\Sigma^{-1})$ is defined, then the unique functor $\overline{F}: \mathscr{C}(\Sigma^{-1}) \to \mathscr{C}'$ such that $\overline{F}P = F$ is conservative. In general, $\overline{F}$ is not faithful.

c) Assume that $\mathscr{C}$ is a category with cokernels and that F commutes with cokernels and is full. Prove that Σ is left calculable and $\mathscr{C}^l(\Sigma^{-1})$ is defined, and moreover, that $\overline{F}$ is full and faithful and P is full.

d) Prove: If $\mathscr{C}$ is cocomplete and if F commutes with colimits, then Σ is strongly left calculable and F commutes with colimits.

14.11. Let $F: \mathscr{C} \to \mathscr{C}'$ be a functor and G a full and faithful coadjoint of F. Denote by Σ the class of all morphisms of $\mathscr{C}$ that are made invertible by F. Show that Σ is left calculable and $\mathscr{C}^l(\Sigma^{-1})$ is defined and equivalent to $\mathscr{C}'$ (see Theorem 13.10).

14.12. Let α be an infinite cardinal number and Σ the class of all morphisms of $\mathscr{S}et$ consisting of all maps $f\colon X \to Y$ such that:

i) $Y \neq \emptyset \Rightarrow X \neq \emptyset$.

ii) $\operatorname{Card}(Y - f(X)) < \alpha$.

iii) There exists a subset X' of X such that the restriction of f to X' is injective and $\operatorname{Card}(X - X') < \alpha$.

Show that Σ is a left calculable system and the category $\mathscr{S}et^l(\Sigma^{-1})$ is defined, and that the same holds for the system consisting of all maps which satisfy only ii) and iii).

14.13. Let Σ be a left calculable system of a category $\mathscr{C}$. Assume that the category $\mathscr{C}^l(\Sigma^{-1})$ is defined. Prove that a morphism $f\colon X \to Y$ of $\mathscr{C}$ belongs to Σ, if and only if it can be inserted in a commutative diagram of the form:

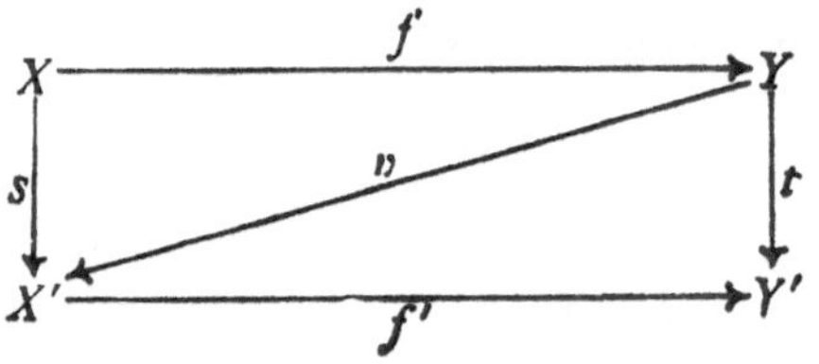

where $s, t \in \Sigma$. (Compare with Exercise 14.6.a).)

14.14. Let $\mathscr{C}$ be a finitely complete category such that all epimorphisms are universal and all monomorphisms are couniversal. Prove that the system Σ of all bimorphisms in $\mathscr{C}$ is bicalculable and saturated, and, moreover, that if the category $\mathscr{C}(\Sigma^{-1})$ is defined, then it is balanced.

1.15. Existence of a coadjoint to the canonical functor $P\colon \mathscr{C} \to \mathscr{C}(\Sigma^{-1})$

In this section we describe some situations where the canonical functor $P\colon \mathscr{C} \to \mathscr{C}(\Sigma^{-1})$ has a coadjoint.

References: Almkvist [1], Gabriel [1], Gabriel—Zisman [1], Schubert [1], N. Popescu [3].

LEMMA 15.1. *Let Σ be a left calculable system of morphisms of a category $\mathscr{C}$ and let X be an object of $\mathscr{C}$. The following are equivalent:*

1) *For every morphism $s \in \Sigma$, the map $H_X(s)$ is a bijection.*

2) *For every every morphism $s \in \Sigma$, the map $H_X(s)$ is a surjection.*

3) *Every morphism $s\colon X \to Y$, $s \in \Sigma$, is left invertible.*

Proof. 3) $\Rightarrow$ 1). Let $s: Y \to Z$ and $f, g: Z \to X$ be such that $s \in \Sigma$ and $fs = gs$. By d) in the definition of a multiplicative system of morphisms, Section 1.14, there exists a $t: X \to X'$ such that $tf = tg$. Now, since t is left invertible we get that $f = g$. Hence $H_X(s)$ is an injection. But it is also a surjection, since for every morphism $f: Y \to X$ we may define the commutative diagram (see Section 1.14.c)):

$$\begin{array}{ccc} Y & \xrightarrow{\ s\ } & Z \\ {\scriptstyle f}\downarrow & & \downarrow{\scriptstyle f'} \\ X & \xrightarrow{\ s'\ } & Z' \end{array}$$

and $f = tf's$, where t is a left inverse of $s' \in \Sigma$.

The proofs of the other implications are left for the reader. ✳

An object X as in the previous lemma is said to be *left closed for* Σ, or simply *left closed.*

PROPOSITION 15.2. *Let Σ be a multiplicative system of morphisms of $\mathscr{C}$ such that $\mathscr{C}^l(\Sigma^{-1})$ is defined. Then an object X of $\mathscr{C}$ is left closed for Σ if and only if the canonical map*

$$P(Y, X): \mathscr{C}(Y, X) \to \mathscr{C}^l(\Sigma^{-1})(P(Y), P(X))$$

is a bijection for each object Y of $\mathscr{C}$.

Proof. Assume that X is left closed for Σ. Using the notation of Theorem 14.1, we have that $\mathscr{C}^l(\Sigma^{-1})(P(Y), P(X)) = \varinjlim H^Y\sigma_X$. Now if $f, g: Y \to X$ are such that $P,(Y, X)(f) = (1_X/f) = (1_X/g) = P(Y, X)(g)$, then there is a morphism $s: X \to Z$, $s \in \Sigma$ such that $sf = sg$. But then $f = g$, since s has a left inverse (see Lemma 15.1). Furthermore, if (s/f) is an element of $\mathscr{C}^l(\Sigma^{-1})(P(Y), P(X))$ where $f: Y \to Z$, $s: X \to Z$, $s \in \Sigma$, then $(s/f) = P(Y, X)(s'f)$ where s' is a left inverse of s.

Conversely, assume that $P(Y, X)$ is a bijection for all $Y \in \mathscr{C}$. Then for each $s: Z \to Y$, $s \in \Sigma$, the diagram

$$\begin{array}{ccc} \mathscr{C}(Y, X) & \xrightarrow{\ H_X(s)\ } & \mathscr{C}(Z, X) \\ {\scriptstyle P(Y, X)}\downarrow & & \downarrow{\scriptstyle P(Z, X)} \\ \mathscr{C}^l(\Sigma^{-1})(P(Y), P(X)) & \xrightarrow{\ H_{P(X)}(P(s))\ } & \mathscr{C}^l(\Sigma^{-1})(P(Z), P(X)) \end{array}$$

is commutative.

Hence, $H_X(s)$ is a bijection since $P(s)$ is invertible. ✳

THEOREM 15.3. *Let Σ be a multiplicative system of $\mathscr{C}$ such that $\mathscr{C}^l(\Sigma^{-1})$ is defined. The following are equivalent:*

1) *The canonical functor $P: \mathscr{C} \to \mathscr{C}^l(\Sigma^{-1})$ has a coadjoint (which is full and faithful).*

2) *For each object X of $\mathscr{C}$ there is a morphism $u_X : X \to \tilde{X}$ such that $\tilde{X}$ is left closed for Σ and $u_X \in \Sigma_s$.*

3) *The category X/Σ_s has a final object for each $X \in \mathscr{C}$.*

Proof. 1) $\Rightarrow$ 2). Let G be a coadjoint of P, u: Id $\mathscr{C} \to GP$ an arrow of adjunction and v a coarrow quasi-inverse to u. We claim that for each $X \in \mathscr{C}$, $GP(X)$ is left closed for Σ and $P(u_X)$ is invertible. Indeed, let $s: Y \to Z$ be an element of Σ. By the canonically defined commutative diagram:

$$\begin{array}{ccc} \mathscr{C}(Z, GP(X)) & \xrightarrow{H_{GP(X)}(s)} & \mathscr{C}(Y, GP(X)) \\ \wr\downarrow & & \wr\downarrow \\ \mathscr{C}^l(\Sigma^{-1})(P(Z), P(X)) & \xrightarrow{H_{P(X)}(P(s))} & \mathscr{C}^l(\Sigma^{-1})(P(Y), P(X)) \end{array}$$

we see that $GP(X)$ is left closed for Σ. Also, from the commutative diagram

$$\begin{array}{ccc} \mathscr{C}(Y, GP(X)) & \xrightarrow{\hspace{3em}} & \mathscr{C}^l(\Sigma^{-1})(P(Y), P(X)) \\ & {\scriptstyle P(Y, GP(X))} \searrow \quad \swarrow {\scriptstyle H^{P(Y)}(P(u_X))} & \\ & \mathscr{C}^l(\Sigma^{-1})(P(Y), PGP(X)) & \end{array}$$

where the horizontal map is given by the morphism of adjointness, we see that $P(u_X)$ is an isomorphism, since Y was arbitrarily chosen and P is a representative functor. Now by the equality $vPPu = 1_P$, we get that vP is an isomorphism and so v is an isomorphism since P is representative. Hence, by Theorem 13.10 G is full and faithful.

2) $\Rightarrow$ 1). By Theorem 11.9, we must prove that the cofunctor $H_Y P$ is representable for each object Y of $\mathscr{C}^l(\Sigma^{-1})$. Since P is representative we may assume that $Y = P(X)$. Now choose a morphism $u_X : X \to \tilde{X}$ such that $\tilde{X}$ is left closed for Σ and $P(u_X)$ is invertible. For each $Z \in \mathscr{C}$, let $r_Z : H_X(Z) \to H_{P(X)}P(Z)$ denote the map defined by $r_Z(f) = P(u_X)^{-1} P(f)$, $f: Z \to X$. We leave it for the reader to check that the maps $\{r_Z\}_{Z \in \mathscr{C}}$ define a functorial isomorphism.

1) $\Rightarrow$ 3). It is easy to see that $(u_X, GP(X))$ is a final object of X/Σ_s.

3) $\Rightarrow$ 2). Let $u_X : X \to \tilde{X}$ be a final object of X/Σ_s. We claim that $\tilde{X}$ is left closed for Σ. To see this, let $s: X \to Y$ be an element of Σ. By hypothesis, there is a morphism $t: Y \to X$ such that $tsu_X = u_X$. But then $ts = 1_X$; hence, by Lemma 15.1, $\tilde{X}$ is left closed for Σ. ✳

Let Σ be a class of morphisms of $\mathscr{C}$. We say that Σ is a (*finitely*) *cocomplete system* if two (finite) small functors $F, G: \mathscr{I} \to \mathscr{C}$ and a functorial morphism $u: F \to G$ such that $u_i \in \Sigma$ for all $i \in \mathscr{I}$ are given, then the morphism $\varinjlim u$ (if it exists!) also belongs to Σ.

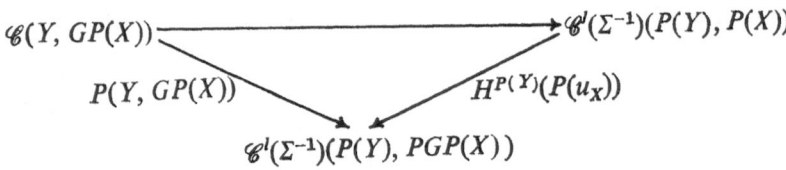

THEOREM 15.4. *Let $\mathscr{C}$ be a cocomplete category and Σ a left calculable system of $\mathscr{C}$. The following are equivalent:*

1) *The category $\mathscr{C}^l(\Sigma^{-1})$ is defined and the canonical functor $P: \mathscr{C} \to \mathscr{C}^l(\Sigma^{-1})$ has a coadjoint.*

2) *For each object X of $\mathscr{C}$, the category X/Σ has a final small subcategory and the system Σ_s is cocomplete.*

Proof. The implication 1) $\Rightarrow$ 2) follows from Theorem 15.3 and from the fact that P commutes with colimits.

2) $\Rightarrow$ 1). By Corollary 14.2, the category $\mathscr{C}^l(\Sigma^{-1})$ is defined. Now let X be an object of $\mathscr{C}$. We prove that for all $X \in \mathscr{C}$, the category X/Σ_s has a small final subcategory. For that purpose let $\mathscr{A}$ be a small final subcategory of X/Σ and $(t, Y) \in X/\Sigma_s$. By Exercise 14.6. a, there exists a morphism $v: Y \to Z$ such that $vt \in \Sigma_s$ so that $(vt, Z) \in X/\Sigma$. Now there is an object (s, R) of $\mathscr{A}$ and a morphism $f: Z \to R$ such that $fvt = s$. Hence it is clear that $\mathscr{A}$ is also a small final subcategory of X/Σ_s.

Furthermore, by Theorem 6.12, the forgetful functor $\sigma_X: X/\Sigma_s \to \mathscr{C}$ has a colimit which is $\varinjlim (\sigma_X I)$, where $I: \mathscr{A} \to X/\Sigma_s$ is the inclusion functor. Denote this colimit by $\bar{X}$ and for each $(t, Y) \in X/\Sigma_s$ denote the structural morphism by $u_t: Y \to \bar{X}$. Now, since Σ_s is cocomplete, the unique morphism $v: X \to \bar{X}$ that is defined so that $v = u_s s$ for all $(s, Y) \in \mathscr{A}$, belongs to Σ_s; furthermore, for every $(t, Z) \in X/\Sigma_s$ we have $u_t t = v$.

Let D be the family of all endomorphisms h of $\bar{X}$ such that $hv = v$. It is clear that $1_X \in D$ and $D \subseteq \Sigma_s$. Now let $(p, \tilde{X})$ be a cokernel of this family of arrows. By Proposition 14.6, the functor P commutes with colimits, so $P(p)$ will be an isomorphism, or equivalently $p \in \Sigma_s$. Now, we claim that the object $\tilde{X}$ is left closed for Σ. To show this, let $s: \tilde{X} \to Y$ be an element of Σ. From the definition of $\bar{X}$, there is a morphism $h: Y \to \bar{X}$ such that $hspv = v$. But then $hsp \in D$, so that $phsp = p$, or equivalently $phs = 1_X$, since p is an epimorphism. Hence s is left invertible. Now an application of Lemma 15.1 and Theorem 15.3.2) completes the proof. ⋇

Exercises

15.1. Show that the conditions of Lemma 15.1 are also equivalent to:

4) For each $s: X \to Y$ such that $s \in \Sigma$, $H_X(s)$ is a surjection.

15.2. Let $\mathscr{C}$ be a finitely cocomplete category. Prove that every finitely cocomplete multiplicative system Σ of $\mathscr{C}$ is left calculable, and moreover, that if Σ is a left calculable system such that $\mathscr{C}^l(\Sigma^{-1})$ is defined, then Σ_s is finitely cocomplete.

15.3. Let $F: \mathscr{C} \to \mathscr{C}'$ be a functor and let G be coadjoint to F. Let Σ be a class of morphisms of $\mathscr{C}$ which are made invertible by F. Prove that for each $X \in \mathscr{C}'$ the object $G(X)$ is left closed for Σ.

15.4. Let $\mathscr{A}$ be a class of objects of a category $\mathscr{C}$ and let $\Sigma(\mathscr{A})$ denote the class of all morphisms s in $\mathscr{C}$ such that $H_X(s)$ is a bijection for every $X \in \mathscr{A}$. Also, let $\mathscr{M}$ be a class of morphisms of $\mathscr{C}$, and let $O(\mathscr{M})$ denote the class of all objects X of $\mathscr{C}$ which are left closed for $\mathscr{M}$. Prove:

a) If $\mathscr{A} \subseteq \mathscr{A}'$, then $\Sigma(\mathscr{A}) \supseteq \Sigma(\mathscr{A}')$.

b) If $\mathscr{M}_1 \subseteq \mathscr{M}_2$, then $O(\mathscr{M}_1) \supseteq O(\mathscr{M}_2)$.

c) $\mathscr{A} \subseteq O(\Sigma\ (\mathscr{A}))$ and $\mathscr{M} \subseteq \Sigma(O(\mathscr{M}))$.

d) $\Sigma(\mathscr{A}) = \Sigma(O(\Sigma(\mathscr{A})))$ and $O(\mathscr{M}) = O(\Sigma(O, \mathscr{M})))$.

e) If $\mathscr{C}$ is finitely cocomplete then $\Sigma(\mathscr{A})$ is a left calculable finitely cocomplete system.

f) If $\mathscr{C}$ is cocomplete, then $\Sigma(\mathscr{A})$ is a strongly left calculable cocomplete system.

15.5. Let Σ_1 and Σ_2 be two classes of morphisms from a category. Denote by $\Sigma_1\Sigma_2$ the class of all morphisms of the form $s_{11}s_{12}s_{21}s_{22} \ldots s_{n1}s_{n2}$, where n is a natural number, $s_{i1} \in \Sigma_1$ and $s_{i2} \in \Sigma_2$ for all $1 \leqslant i \leqslant n$.

a) Prove: If both Σ_1 and Σ_2 are left calculable systems, then $\Sigma_1\Sigma_2$ is also a left calculable system, and moreover, $\Sigma_1\Sigma_2 = \Sigma_2\Sigma_1$.

b) Assume that Σ_1 and Σ_2 are both left calculable and that $\mathscr{C}^l(\Sigma_1^{-1})$ is defined. Denote $\{P(s) \mid s \in \Sigma_2\}$ by $\bar{\Sigma}_2$, where $P: \mathscr{C} \to \mathscr{C}^l(\Sigma_1^{-1})$ is the canonical functor. Prove that $\bar{\Sigma}_2$ is a left calculable system, and, moreover, that the category $\mathscr{C}^l(\Sigma_2^{-1})$ is defined if and only if the category $(\mathscr{C}^l(\Sigma_1^{-1}))(\bar{\Sigma}_2^{-1})$ is defined if and only if the category $\mathscr{C}^l((\Sigma_1\Sigma_2)^{-1})$ is defined, and we have the following equivalences of categories:

$$\mathscr{C}^l((\Sigma_1\Sigma_2)^{-1}) \simeq (\mathscr{C}^l(\Sigma_1^{-1}))(\bar{\Sigma}_2^{-1}) \simeq (\mathscr{C}^l(\Sigma_2^{-1}))(\bar{\Sigma}_1^{-1}).$$

15.6. Let Ψ be the class of all epimorphisms of a finitely cocomplete category.

1) Prove that the system Ψ is left calculable.

2) Describe the category $\mathscr{S}et(\Psi^{-1})$. Prove that the canonical functor $P: \mathscr{S}et \to \mathscr{S}et\, l(\Psi^{-1})$ has a coadjoint and that Ψ is saturated.

3) Prove that if Σ is another left calculable system of $\mathscr{S}et$ such that the canonical functor $P': \mathscr{S}et \to \mathscr{S}et^l(\Sigma^{-1})$ has a coadjoint then $\Sigma_s = \Psi$.

4) Describe the categories $\mathscr{T}op(\Psi^{-1})$, $\mathscr{G}r(\Psi^{-1})$ and $\mathscr{A}b(\Psi^{-1})$.

15.7. Let Φ be the class of all monomorphisms of a finitely complete category.

1) Prove that the system Φ is right calculable.

2) Describe the category $\mathscr{S}et^r(\Phi^{-1})$. Prove that this category is canonically equivalent to the category $\mathscr{S}et^l(\Psi^{-1})$ and that the canonical functor $P: \mathscr{S}et \to \mathscr{S}et^l(\Psi^{-1})$ also has an adjoint.

3) Prove that $\mathscr{A}b^l(\Psi^{-1})$ and $\mathscr{A}b^r(\Phi^{-1})$ are isomorphic whereas $\mathscr{G}r^l(\Psi^{-1})$ and $\mathscr{G}r^r(\Phi^{-1})$ are not equivalent.

15.8. Let Σ be a multiplicative system of $\mathscr{C}$ such that $\mathscr{C}(\Sigma^{-1})$ is defined and the canonical functor $P: \mathscr{C} \to \mathscr{C}(\Sigma^{-1})$ has a coadjoint G. Prove that the following hold:

a) If $(f, g): R \to X$ is an equivalence relation on $\mathscr{C}(\Sigma^{-1})$, then $(G(f), G(g)): G(R) \to G(X)$ is an equivalence relation on $\mathscr{C}$.

b) If $\mathscr{C}$ is finitely complete and every equivalence relation is effective in $\mathscr{C}$, then every equivalence relation is effective in $\mathscr{C}(\Sigma^{-1})$.

1.16. Subobjects and quotient objects

Subobjects and quotient objects are the natural generalization of the well-known notions of subset, subgroup, etc. In this section we define these notions and give some of their simple properties. We refer the reader to the following references for other aspects of these notions:

References: Bucur—Deleanu [1], Kuroš [2], Isbell [5], Grothendieck [1], Mac-Lane [3], Mitchell [2], Pareigis [1], Schubert [1], Šulgeifer [1], [2].

A *right ideal* of a category $\mathscr{C}$ is a full subcategory $\mathscr{C}'$ of $\mathscr{C}$ satisfying the condition that, if $f: X \to X'$ is a morphism in $\mathscr{C}$ and $X' \in \mathscr{C}'$, then $X \in \mathscr{C}'$. A *right ideal over an object* X of $\mathscr{C}$ is a right ideal of the category $\mathscr{C}/X$. Dually, we have the notions of a *left ideal* of a category, and a *left ideal over an object.*

Example 16.1. Let $F: \mathscr{C} \to \mathscr{S}et$ be a cofunctor, $X \in \mathscr{C}$ and $u: F \to H_X$ a functorial morphism. For every $Y \in \mathscr{C}$ and $y \in F(Y)$, $u_Y(y): Y \to X$ is a morphism in $\mathscr{C}$. Let $\mathrm{Im}(u)$ denote the full subcategory of $\mathscr{C}/X$ generated by all objects $(Y, u_Y(y))$ when Y runs through the objects of $\mathscr{C}$ and $y \in F(Y)$. Thus $\mathrm{Im}(u)$ is a right ideal over X which is called the *image of the functorial morphism* u. Indeed, if $(Y, u_Y(y))$ is an object of $\mathrm{Im}(u)$ and $g: Z \to Y$ is a morphism in $\mathscr{C}$, then we have the commutative diagram

$$\begin{array}{ccc} F(Y) & \xrightarrow{\ u_Y\ } & H_X(Y) \\ {\scriptstyle F(g)}\downarrow & & \downarrow{\scriptstyle H_X(g)} \\ F(Z) & \xrightarrow{\ u_Z\ } & H_X(Z) \end{array}$$

so that $(Z, u_Y(y)g) = (Z, u_Z F(g)(y))$.

Note 16.2. Let $\Phi: \mathscr{C} \to \mathscr{S}et$ be the cofunctor associated with the empty set, i.e. $\Phi(X) = \emptyset$ for all $X \in \mathscr{C}$. Then for each cofunctor $F: \mathscr{C} \to \mathscr{S}et$ there is a unique functorial morphism $\Phi \to F$. In particular, the image of the unique functorial morphism $\Phi \to H_X$ is also denoted by Φ. Φ is the right ideal over X which does not contain any object.

LEMMA 16.3. *Let* $F, G: \mathscr{C} \to \mathscr{S}et$ *be two cofunctors and let* $u: F \to H_X$ *and* $v: G \to H_X$ *be functorial monomorphisms. Then* $\mathrm{Im}(u) = \mathrm{Im}(v)$ *if and only if there is a functorial isomorphism* $w: F \to G$ *such that* $vw = u$.

Proof. By Proposition 10.5, u and v are argumentwise functorial monomorphisms. Assume that $\mathrm{Im}(u) = \mathrm{Im}(v)$ and let Z be an object of $\mathscr{C}$. If $x \in F(Z)$, then $(Z, u_Z(x)) \in \mathrm{Im}(u)$, so that there necessarily is a unique element $x' \in G(Z)$ such that $v_Z(x') = u_Z(x)$. Define $w_Z: F(Z) \to G(Z)$ by $w_Z(x) = x'$. We leave it for the reader to check that w_Z is a bijection and that the maps $\{w_Z\}_{Z \in \mathscr{C}}$ give a functorial isomorphism $w: F \to G$ such that $vw = u$. ✳

LEMMA 16.4. *Let X be an object of $\mathscr{C}$ and $\mathscr{A}$ a right ideal over X. Then there is a cofunctor $\mathscr{A}^c: \mathscr{C} \to \mathscr{S}et$ and a functorial monomorphism $u_{\mathscr{A}}: \mathscr{A}^c \to H_X$ such that* $\mathrm{Im}(u_{\mathscr{A}}) = \mathscr{A}$.

Proof. The cofunctor $\mathscr{A}^c$ is defined as follows: For each object Y of $\mathscr{C}$ define $\mathscr{A}^c(Y) = \{f \in \mathscr{C}(Y, X) \mid (Y, f) \in \mathscr{A}\}$. If $g: Y \to Z$ is a morphism in $\mathscr{C}$, then $\mathscr{A}^c(g)$ takes $h \in \mathscr{A}^c(Z)$ to $hg \in \mathscr{A}^c(Y)$. It is easy to check that $\mathscr{A}^c$ is a cofunctor and that the functorial monomorphism $u_{\mathscr{A}}$ is defined by $(u_{\mathscr{A}})_Y(f) = f$ for every $Y \in \mathscr{C}$ and $f \in \mathscr{A}^c(Y)$. ✻

We call $\mathscr{A}^c$ the *cofunctor associated with the right ideal* $\mathscr{A}$ and $u_{\mathscr{A}}$ the *canonical inclusion.*

For an object X of $\mathscr{C}$, let $\mathrm{Idr}(X)$ denote the class of all right ideals over X. If $\mathscr{A}, \mathscr{B} \in \mathrm{Idr}(X)$, then we write $\mathscr{A} \subseteq \mathscr{B}$ if every object of $\mathscr{A}$ is also an object of $\mathscr{B}$. Thus $\mathrm{Idr}(X)$ become a preordered category.

LEMMA 16.5. *Let X be an object of $\mathscr{C}$. Then every functor $F: \mathscr{I} \to \mathrm{Idr}(X)$ has a limit and a colimit. In particular,* $\mathrm{Idr}(X)$ *is a complete and cocomplete category.*

Proof. Let $F: \mathscr{I} \to \mathrm{Idr}(X)$ be a functor. Consider the object $\bigcap_{i \in \mathscr{I}} F(i)$ of $\mathrm{Idr}(X)$ to consist of all pairs (Y, f) such that $(Y, f) \in F(i)$ for all $i \in \mathscr{I}$. It is easy to see that $\bigcap_{i \in \mathscr{I}} F(i)$ is an object of $\mathrm{Idr}(X)$ and for every $j \in \mathscr{I}$ we have $\bigcap_{i \in \mathscr{I}} F(i) \subseteq F(j)$. Also, it is straightforward to check that $\bigcap_{i \in \mathscr{I}} F(i)$ is a limit of F. We note that in this case we may have $\bigcap_{i \in \mathscr{I}} F(i) = \emptyset$. Furthermore, let $\mathscr{J}$ be the class of all objects $\mathscr{A}$ of $\mathrm{Idr}(X)$ such that $F(i) \subseteq \mathscr{A}$ for all $i \in \mathscr{I}$. Observe that there are such objects since $\mathscr{C}/X$ is an object of $\mathrm{Idr}(X)$. Thus $\bigcap_{\mathscr{A} \in \mathscr{J}} \mathscr{A}$ is a colimit of F. This colimit will be denoted by $\bigcup_{i \in \mathscr{I}} F(i)$. ✻

Let $f: X \to Y$ be a morphism in $\mathscr{C}$. If $\mathscr{A} \in \mathrm{Idr}(Y)$, we denote by $f^*(\mathscr{A})$ the right ideal over X whose objects are all pairs (Z, g) such that $(Z, fg) \in \mathscr{A}$. $f^*(\mathscr{A})$ is called the *inverse image of the right ideal $\mathscr{A}$ under f.*

If $\mathscr{A} \subseteq \mathscr{A}'$ are two objects of $\mathrm{Idr}(Y)$, then $f^*(\mathscr{A}) \subseteq f^*(\mathscr{A}')$. In this way we have a functor $f^*: \mathrm{Idr}(Y) \to \mathrm{Idr}(X)$ called the *functor inverse image associated with f.* Also, if $\mathscr{B}$ is an object of $\mathrm{Idr}(X)$, let $f_*(\mathscr{B})$ denote the right ideal over Y consisting of all couples (Z, fg) where $(Z, g) \in \mathscr{B}$. $f_*(\mathscr{B})$ is called the *direct image of the right ideal $\mathscr{B}$ under f.* If $\mathscr{B} \subseteq \mathscr{B}'$, then $f_*(\mathscr{B}) \subseteq f_*(\mathscr{B}')$ so that we have a functor $f_*: \mathrm{Idr}(X) \to \mathrm{Idr}(Y)$ called the *functor direct image associated with f.*

PROPOSITION 16.6. *Let $f: X \to Y$ be a morphism in $\mathscr{C}$. Then for all $\mathscr{A} \in \mathrm{Idr}(X)$ we have:*

i) $\mathscr{A} \subseteq f^* f_*(\mathscr{A})$.

ii) $f_*(\mathscr{A}) = f_* f^* f_*(\mathscr{A})$.

Also, for each $\mathscr{B} \in \mathrm{Idr}(Y)$ we have:

iii) $f_* f^*(\mathscr{B}) \subseteq \mathscr{B}$.

iv) $f^* f_* f^*(\mathscr{B}) = f^*(\mathscr{B})$.

Proof. i) Let $(Z, g) \in \mathscr{A}$; then $(Z, fg) \in f_*(\mathscr{A})$ so that $(Z, g) \in f^* f_*(\mathscr{A})$.

ii) Clearly, $f_* f^* f_*(\mathscr{A}) \subseteq f_*(\mathscr{A})$. Now if $(Z, g) \in f_*(\mathscr{A})$, then $g = fh$, where $h: Z \to X$ and $(Z, h) \in \mathscr{A}$, i.e. $(Z, h) \in f^* f_*(\mathscr{A})$ and $(Z, fh) \in f_* f^* f_*(\mathscr{A})$.

The proofs of the other two assertions are left for the reader. ✳

COROLLARY 16.7. *Let $f: X \to Y$ be a morphism in $\mathscr{C}$. Then the functor $f^*: \mathrm{Idr}(X) \to \mathrm{Idr}(X)$ is coadjoint to the functor f_*. In particular, if $\{\mathscr{A}_i\}_{i \in \mathscr{I}}$ is a class of objects of $\mathrm{Idr}(X)$, then $f_*(\bigcup_{i \in \mathscr{I}} \mathscr{A}_i) = \bigcup_{i \in \mathscr{I}} f_*(A_i)$. Also, if $\{\mathscr{B}_i\}_{i \in \mathscr{I}}$ is a class of objects of $\mathrm{Idr}(Y)$, then $f^*(\bigcap_{i \in \mathscr{I}} \mathscr{B}_i) = \bigcap_{i \in \mathscr{I}} f^*(\mathscr{B}_i)$.*

The proof follows from Lemma 16.5, Proposition 16.6 and Theorem 12.4. ✳

Note 16.8. Of course, the above results about right ideals over an object can be dualized to left ideals over an object. We suggest that the reader do this dualization.

We say that a right ideal $\mathscr{A}$ over X is a *principal ideal* if there is a morphism $f: Y \to X$ such that $\mathscr{A} = \mathrm{Im}(H(f))$, where $H(f): H_Y \to X_X$ is the functorial morphism associated with f. In this case f is called a *generator* of $\mathscr{A}$. In general, a principal right ideal over X may have several generators. It is clear that $\mathscr{A} \in \mathrm{Idr}(X)$ is principal if and only if there is a morphism $f: Y \to X$ such that every $(Z, g) \in \mathscr{A}$ has the factorization $Z \xrightarrow{g'} Y \xrightarrow{f} X$, i.e. $fg' = g$.

If $f: X \to Y$ is a morphism then the right ideal $\mathrm{Im}(H(f))$ is usually denoted by $\langle f \rangle$.

PROPOSITION 16.9. *Let $\mathscr{A}$ be a right ideal over X. The following are equivalent:*

1) *$\mathscr{A}^c$ is a representable cofunctor.*

2) *There is a monomorphism $f: X' \to X$ such that $(X', f) \in \mathscr{A}$, and $\mathscr{A}$ is generated by f.*

Proof. 1) ⇒ 2). Let $v: H_{X'} \xrightarrow{\sim} \mathscr{A}^c$ be a functorial isomorphism. Then $f = (v_{X'})(1_{X'}) \in \mathscr{A}^c(X')$ is in fact a morphism $f: X' \to X$ and $(X', f) \in \mathscr{A}$. If $(Y, g) \in \mathscr{A}$, then $g \in \mathscr{A}^c(Y)$ and thus there is a unique morphism $g': Y \to X'$ such that $\mathscr{A}^c(g')(f) = fg' = g$. Hence $\mathscr{A}$ is generated by f, and by hypothesis it necessarily is a monomorphism.

2) ⇒ 1). It is easy to check that $\mathscr{A}^c$ is isomorphic to $H_{X'}$. The canonical inclusion $u_{\mathscr{A}}: \mathscr{A}^c \to H_X$ is in fact induced by $H(f): H_{X'} \to H_X$. ✳

A right ideal $\mathscr{A}$ over X as in the previous proposition is called a *subobject* of X. Specifically, a subobject of X is a right ideal over X that is generated by a monomorphism. In general a subobject may be generated by several monomorphisms, but by use of Proposition 3.25 we obtain the following result.

COROLLARY 16.10. *Two monomorphisms $f': X' \to X$ and $f'': X'' \to X$ generate the same subobject of X if and only if there is an isomorphism $u: X' \to X''$ such that $f''u = f'$.* ✳

If $f: X' \to X$ is a monomorphism, then the principal right ideal over X generated by f is a subobject of X, which generally will be denoted by X', and we refer to f as the inclusion of X' into X. Sometimes we shall write $f: X' \to X$, or simply $X' \subseteq X$, when we want to indicate that X' is a subobject of X, and we shall say that X' is *contained in* X, or that X *contains* X'. In particular, $1_X: X \to X$ defines X as a subobject of X. A subobject $\mathscr{A}$ of X is called *proper* if it does not coincide with $\mathscr{C}/X$, i.e. if it is not generated by 1_X. If $f: X' \to X$ is a monomorphism and the subobject of X generated by f is proper, then we shall write $f: X' \subset X$. However, we again mention that in general there may be more than one monomorphism $f: X' \to X$ which generates the same subobject, and that whenever we are speaking of a subobject of X, we shall have a specific monomorphism f in mind. In this language the statement that the composition of two monomorphisms is a monomorphism becomes: If X' is a subobject of X and $g: X \to Y$ is a monomorphism, then X' is a subobject of Y. In fact, if $\mathscr{A}$ is the right ideal over X generated by f, then $g_*(\mathscr{A})$ is also a subobject of Y, generated by gf. The composition of a morphism $g: X \to Y$ with a monomorphism $f: X' \to X$ is denoted by g/X' and is called the *restriction* of g to X'. If $X \in \mathscr{C}$, we denote the class of all subobjects of X by $\mathscr{S}(X)$. It is clear that $\mathscr{S}(X)$ is a subcategory of Idr(X). The object X is called *wellpowered* if $\mathscr{S}(X)$ is a set. The category $\mathscr{C}$ is called a *wellpowered category* if every object of $\mathscr{C}$ is wellpowered.

As mentioned before, the dual of a right ideal is the notion of a left ideal. Actually, a left ideal of a category $\mathscr{C}$ is a full subcategory $\mathscr{C}'$ such that for every $X' \in \mathscr{C}'$ and every morphism $f: X' \to X$ in $\mathscr{C}$ we have that $X \in \mathscr{C}'$. Also, a left ideal over an object X of $\mathscr{C}$ is a left ideal of the category $X/\mathscr{C}$. Denote the class of all left ideals over X by Idl(X). It is clear that Idl(X) is a preordered category and that Lemma 16.5 holds for this category. If $\mathscr{A}$ is an object of Idl(X), let ${}^c\mathscr{A}: \mathscr{C} \to \mathscr{S}et$ denote the functor defined by ${}^c\mathscr{A}(Y) = \{f: X \to Y,\ (f, Y) \in \mathscr{A}\}$ and let $u^{\mathscr{A}}: {}^c\mathscr{A} \to H^X$ denote the canonical functorial monomorphism. It is clear that Im$(u^{\mathscr{A}}) = \mathscr{A}$. If $f: X \to Y$ is a morphism in $\mathscr{C}$, let f^*: Idl$(X) \to$ Idl(Y) denote the functor that assigns to $\mathscr{A} \in$ Idl(X) the left ideal $f^*(\mathscr{A})$ over Y that consists of all objects (h, Z) of $Y/\mathscr{C}$ such that $(hf, Z) \in \mathscr{A}$. Furthermore, denote by f_*: Idl$(Y) \to$ Idl(X) the functor that assigns to the left ideal $\mathscr{B}$ over Y the left ideal $f_*(\mathscr{B})$ over X that is generated by all objects (gf, Z) when (g, Z) runs through $\mathscr{B}$. It is clear that for every $\mathscr{A} \in$ Idl(X) we have $\mathscr{A} \supseteq f_*f^*(\mathscr{A})$ and, for all $\mathscr{B} \in$ Idl(Y), $\mathscr{B} \subseteq f^*f_*(\mathscr{B})$. Hence the functor f^* is coadjoint to f_*.

We say that a left ideal $\mathscr{A}$ over X is *principal* if there is a morphism $f: X \to Y$ such that $\mathscr{A} = \text{Im}(H^0(f))$, where $H^0(f): H^Y \to H^X$ is the functorial morphism associated with f, concordantly f is called a *generator* of $\mathscr{A}$. A representable left ideal over X is called a *quotient object* of X. It is clear that $\mathscr{A} \in$ Idl(X) is a quotient object if and only if there is an epimorphism $p: X \to Y$ such that $(p, Y) \in \mathscr{A}$, and for every $(g, Z) \in \mathscr{A}$ there is a (necessarily unique) morphism $g': Y \to Z$ such that $g'p = g$.

A quotient object of an object X is usually interpreted as an epimorphism $p: X \to Y$; two epimorphisms $p: X \to Y$ and $q: X \to Z$ define the same quotient object if and only if there is an isomorphism $t: Y \to Z$ such that $tp = q$. The class of all quotient objects of X is denoted by $Q(X)$. The object X is *wellcopowered* if $Q(X)$ is a set and the category $\mathscr{C}$ is called a *wellcopowered category* if every object of $\mathscr{C}$ is wellcopowered.

We say that a family $\{U_i\}_{i \in \mathscr{I}}$ of objects of $\mathscr{C}$ is a *family of generators*, if for each object X there exists a subset $\mathscr{I}(X)$ of $\mathscr{I}$ such that, for each monomorphism $f: X \to Y$ that is not an isomorphism, there are an index $i \in \mathscr{I}(X)$ and a morphism $g: U_i \to Y$ such that g can not be factored through f, i.e. for every $h: U_i \to X$ $i \in \mathscr{I}(X)$ we have that $fh \neq g$. An object U of $\mathscr{C}$ is a *generator* if the family $\{U\}$ is a family of generators.

PROPOSITION 16.11. *If $\mathscr{C}$ is a category with fibered products that has a set of generators, then $\mathscr{C}$ is wellpowered.*

Proof. Let $\{U_i\}_{i\in\mathscr{I}}$ be a set of generators of $\mathscr{C}$ and let $f: X' \to X$ be a monomorphism. Let M be the set $\coprod_i \mathscr{C}(U_i, X)$ and let $M(X')$ be the subset of M defined as follows. An element $(i, g) \in M$ belongs to $M(X')$ if $g: U_i \to X$ is a morphism such that g can be factored through f, i.e. there is a morphism $g': U_i \to X$ such that $g = fg'$. If $f_1: X_1' \to X$ is another monomorphism, then $M(X') = M(X_1')$ if and only if f and f_1 define the same subobject of X. Indeed, assume that $M(X') = M(X_1')$ and let $g: U_i \to X$ be such that $(i, g) \in M(X')$. Then $g = fg' = f_1 g_1'$ and by the commutative diagram

$$\begin{array}{ccc} X'\coprod_X X_1' & \xrightarrow{\;u\;} & X' \\ {\scriptstyle v}\downarrow & & \downarrow{\scriptstyle f} \\ X_1' & \xrightarrow{\;f_1\;} & X \end{array}$$

we get that there is a unique morphism $\bar{g}: U_i \to X'\coprod_X X_1'$ such that $u\bar{g} = g'$ and $v\bar{g} = g_1'$. Then g can be factored through fu and $f_1 v$. Hence every morphism $h: U_j \to X'$ can be factored through u, so that u is an isomorphism. In the same way we see that v is an isomorphism, and so we have that (X', f) and (X_1', f_1) define the same subobject of X. We have thus proven that there is an injection $X' \rightsquigarrow M(X')$ from the set of subobjects of X into the set of all subsets of M, i.e. X is wellpowered. ✳

Dually, we have the notions of *family of cogenerators*, and *cogenerator*.

Now we give examples in some familiar categories of subobjects and quotient objects.

Example 16.12. Let X be a set, i.e. an object of $\mathscr{S}et$, and let $f: X' \to X$ be a monomorphism. Let $f(X')$ denote the image of X', i.e. the subset of X consisting of all elements $f(x')$, where $x' \in X'$. Also, let $u: f(X') \to X$ be the natural inclusion. It is obvious that the assignment $x' \rightsquigarrow f(x')$ gives a bijection $p: X' \to f(X')$ such that $up = f$, or, in other words, the monomorphisms f and u define the same subobject of X. Conversely, each subset of X gives rise to a subobject of X, and distinct subsets generate distinct subobjects. Consequently, a subobject of an object X of $\mathscr{S}et$ is completely determined by a subset of X.

Dually, according to Example 2.1, an epimorphism $f: X \to Y$ of $\mathscr{S}et$ is in fact a surjection. Denote by $R(f)$ the equivalence relation on X defined by $x \sim x'(R(f)) \Leftrightarrow f(x) = f(x')$ and let $X/R(f)$ be the quotient set, and $p: X \to X/R(f)$ the canonical surjection. Now if $y \in Y$, then there is an $x \in X$ such that $f(x) = y$. Let us define $t(y)$ to be $p(x)$. It is clear that we have defined in this way a bijection $t: Y \to X/R(f)$, i.e. p and f define the same quotient objects of X. Thus, a quotient object of X

is completely determined by an equivalence relation on X, or precisely by the quotient set of X determined by the equivalence relation.

Example 16.13. Let $f: X' \to X$ be a monomorphism in $\mathcal{Top}$; then f is an injection. Denote by $f(X')$ the image of f endowed with the relative topology from X, i.e. a subset D of $f(X')$ is open if and only if there is an open subset D' of X such that $D = f(X') \cap D'$. Since f is a continuous function, for every open subset U of X, we thus that have $f^*(U) = f^*(U \cap f(X'))$; in other words, the canonical map $p: X' \to f(X')$ defined such that $up = f$ (where $u: f(X) \to X$ is the canonical inclusion) is continuous. Since p is a bijection, we can assume that it is the identity map; finally we observe that the original topology of X' is finer than the topology induced by X. However, the induced topology is the smallest topology on $f(X')$ for which the inclusion map is continuous. Hence a subobject of X is defined by a subset X' of X endowed with a topology finer than the relative topology. It is clear that a subset of X may yield several distinct subobjects.

Concerning quotient objects in $\mathcal{Top}$ we have the same situation: in general, a quotient set of a topological space X may determine several quotient objects. Indeed, if $f: X \to Y$ is an epimorphism in $\mathcal{Top}$, then f is a surjection. Assume that $Y = X/R(f)$ and f is the canonical surjection (see Example 16.11). We endow Y with the quotient topology: a subset U of Y is open in the quotient topology if and only if $f^*(U)$ is an open subset of X. Since f is a continuous map, we see that the quotient topology is as fine as the original topology of Y.

Example 16.14. If $f: X' \to X$ is a monomorphism in $\mathcal{Gr}$, then $f(X')$ is a subgroup of X, and every subgroup of X defines a subobject. Thus the subobjects of X are in one to one correspondence with the subgroups of X.

Now if $p: X \to Y$ is a surjective morphism of groups, it is clear that p defines a quotient object. Conversely, we shall prove that every epimorphism in $\mathcal{Gr}$ is a surjective map. Indeed, if $f: X \to Y$ is an epimorphism in $\mathcal{Gr}$, denote $f(X)$ by H and let $u: H \to Y$ be the canonical inclusion. We have that $f = up$, where $p: X \to f(X)$ is also canonical. Thus u is an epimorphism in $\mathcal{Gr}$, and we shall show that it is surjective. For each $y \in Y$ denote $\{yh \mid h \in H\}$ by yH. Thus the subsets yH partition Y; in particular, if e is the neutral element of Y, we have $eH = H$. Let $*$ be a symbol which does not belong to the set A of all distinct subsets yH of Y, and let G be the group $\mathrm{Aut}_{\mathcal{Set}}(A \coprod *)$. (Here the coproduct is considered in $\mathcal{Set}$.) Let g be the element of G which interchanges $eH = H$ and $\{*\}$ and leaves all other elements fixed. Then g^2 is the neutral element of G. Let $t: Y \to G$ be the map defined by $t(y)(y'H) = yy'H$ and $t(y)(*) = *$. Then t is a morphism of groups. Likewise, let $s: Y \to G$ be defined by $s(y) - gt(y)g$. Then g is also a morphism of groups. Now if $y \in H$, then it is easy to see that $t(h) = s(h)$. Since $u: H \to Y$ is an epimorphism, we get $t = s$. Thus, for all $y \in Y$,

$$yH = t(y)(eH) = s(y)(eH) = gt(g)g(eH) = gt(g)(*) = g(*) = eH,$$

i.e. $y \in H$, or equivalently $H = Y$.

Finally, since every quotient object of a group X is determined by a surjective morphism of groups and every surjective morphism is determined by a unique normal subgroup of Y, we see that the set of quotient objects of X is in one-to-one correspondence with the set of all normal subgroups of X, i.e., with the set of all quotient groups X/H where H is a normal subgroup of X.

Notions such as subobject and quotient object can also be carried over to functors. Let $F: \mathscr{C} \to \mathscr{C}'$ be a functor. Let $\mathscr{M}(F)$ denote the category whose objects are pairs (G, u) where $G: \mathscr{C} \to \mathscr{C}'$ is also a functor and $u: G \to F$ is a functorial monomorphism. A morphism $v: (G, u) \to (G', u')$ of $\mathscr{M}(F)$ is a functorial morphism $v: G \to G'$ such that $u'v = u$. It is obvious that $\mathscr{M}(F)$ is a preordered category. Denote the skeleton of $\mathscr{M}(F)$ by $\mathscr{S}(F)$. An object of $\mathscr{S}(F)$ is in fact the class of all objects that are isomorphic to a given object. An object of $\mathscr{S}(F)$ is called a *subfunctor* of F. Usually, a subfunctor of F is determined by a monomorphism $u: F' \to F$ and we shall refer to u as the inclusion of F' in F. Sometimes we write $u: F' \to F$, or simply $F' \subseteq F$, when we want to indicate that F' is a subfunctor of F.

If $u: F' \to F$ and $v: F'' \to F$ are subfunctors of F, then we write $F'' \subseteq F'$ if and only if there is a functorial morphism (necessarily a functorial monomorphism) $t: F'' \to F'$ such that $ut = v$. Thus t defines F'' as a subfunctor of F'. If $\mathscr{C}$ is a small category, then the notion of a subfunctor coincides with the notion of a subobject in the category $[\mathscr{C}, \mathscr{C}']$.

Dually, denote by $\mathscr{E}(F)$ the category whose objects are all pairs (p, G) where $p: F \to G$ is a functorial epimorphism, and by $Q(F)$ the skeleton of $\mathscr{E}(F)$. An object of $Q(F)$ is called a *quotient functor* and is usually represented by a pair (p, G), where $p: F \to G$, is a functorial epimorphism.

Exercises

16.1. Let $f: X \to Y$ be a morphism of $\mathscr{C}$. Prove: For every $\mathscr{A} \in \mathrm{Idr}(Y)$ the following diagram of functors (where v is canonically defined) is a cartesian square:

$$\begin{array}{ccc} f^*(\mathscr{A})^c & \xrightarrow{u_{f(\mathscr{A})}} & H_X \\ {\scriptstyle v}\downarrow & & \downarrow{\scriptstyle H(f)} \\ \mathscr{A}^c & \xrightarrow{u} & H_Y \end{array} .$$

16.2. Prove that a morphism $f: X \to Y$ is a monomorphism if and only if the functor $f_*: \mathrm{Idr}(X) \to \mathrm{Idr}(Y)$ is full and faithful, i.e. $\mathscr{A} = f^* f_*(\mathscr{A})$ for all $\mathscr{A} \in \mathrm{Idr}(X)$, and also that f is an isomorphism if and only if f_* is an equivalence of categories.

16.3. Prove: If $f: X \to Y$ is a universal morphism, then the functor $f^*: \mathrm{Idr}(Y) \to \mathrm{Idr}(X)$ carries a principal right ideal over Y to a principal right ideal over X. When does the converse hold?

16.4. Let $f: X \to Y$ be a morphism of $\mathscr{C}$. Prove that Im$(H(f))$ is a subobject if and only if there is a factorization $X \xrightarrow{p} Y' \xrightarrow{u} Y$, $up = f$, where u is a monomorphism and p has a right inverse. Show that for every morphism $f: X \to Y$ in $\mathscr{S}et$, Im$H(F)$) is a subobject.

16.5. Prove that $\mathscr{S}et$, $\mathscr{T}op$, $\mathscr{P}re$, $\mathscr{O}rd$, $\mathscr{R}el$, $\mathscr{M}on$, $\mathscr{G}r$ and $\mathscr{A}b$ are wellpowered categories.

16.6. Let $\mathscr{C}$ be a finitely complete wellpowered category. Prove that the category $[\mathscr{I}, \mathscr{C}]$ is wellpowered for each small category $\mathscr{I}$.

16.7. Prove: If $\mathscr{C}$ has a final object, then there is a one-to-one correspondence between the set of right ideals of $\mathscr{C}$ and the set of right ideals over a final object.

16.8. Let $F: \mathscr{C} \to \mathscr{S}et$ be a functor and let $u: F \to H^X$ be a functorial morphism. Let Im(u) denote the full subcategory of $X/\mathscr{C}$ generated by the objects $(u_Y(y), Y)$ when Y runs through $\mathscr{C}$ and $y \in F(Y)$. Show that Im(u) is a left ideal over X; it is also called the image of the functorial morphism u. Show that Lemma 16.4 holds for left ideals.

16.9. Let $\mathscr{I}$ be a class (not a set). Let $\mathscr{C}$ denote the category whose objects are $\{X_i\}_{i \in \mathscr{I}}$ together with another object X such that $X \neq X_i$ for all $i \in \mathscr{I}$. The morphisms of $\mathscr{C}$ are the identity morphisms and the morphisms $f_i: X_i \to X$, defined for each $i \in \mathscr{I}$. Show that X is not a wellpowered object.

16.10. Show that if $f: X \to Y$ is an epimorphism and Y is a generator, then X is a generator. Let $\{U_i\}_{i \in \mathscr{I}}$ be a set of objects of $\mathscr{C}$. Assume that $U = \coprod_i U_i$ exists in $\mathscr{C}$. Show that if $\{U_i\}_{i \in \mathscr{I}}$ is a family of generators, then U is a generator. Show that the converse holds if $\mathscr{C}$ has a zero object.

16.11. Show that a family $\{U_i\}_{i \in \mathscr{I}}$ of objects of $\mathscr{S}et$ is a family of generators if and only if it is a family of separators. The same result holds in $\mathscr{R}el$, $\mathscr{M}on$, $\mathscr{G}r$ and $\mathscr{A}b$. Every set that contains at least two elements is a cogenerator of $\mathscr{S}et$. It is easy to see that the additive group $\mathbf{Q}/\mathbf{Z}$ of rational numbers modulo 1 is a cogenerator of $\mathscr{A}b$ (see Exercise 6.4, Ch. 4).

16.12. (Tietze). Show that the closed interval [0, 1] of the real line is a cogenerator of $\mathscr{C}om$, the category of compact spaces (see J. L. Kelley [1], Schubert [2]).

16.13. Let $\mathscr{C}$ be a category with zero object. A monomorphism $f: X' \to X$ is called *normal* if every morphism $g: Z \to X$ such that $uf = 0$ implies $ug = 0$ can be factored through f. Prove:

a) If $f: X \to Y$ is the kernel of some morphism g, then f is a normal monomorphism.

b) Conversely, if $\mathscr{C}$ is wellpowered and complete, then every normal monomorphism is the kernel of a suitable morphism.

A subobject $f: X' \to X$ is called a *normal (strict) subobject* if f is a normal (strict) monomorphism. By duality we have the notions of *normal (strict) epimorphism* and *normal (strict) quotient object.*

16.14. Prove that every monomorphism in $\mathcal{G}r$ is strict. A monomorphism $f: X \to Y$ in $\mathcal{G}r$ is normal if and only if $f(X)$ is a normal subgroup of Y. Prove that there exist strict subobjects that are not normal and that in the category $\mathcal{A}b$ every subobject is simultaneously normal and strict.

16.15. Describe all subobjects and quotient objects of an object in $\mathcal{P}re$, $\mathcal{O}rd$ and $\mathcal{R}el$.

1.17. Intersections and unions of subobjects

The notions of intersection and union of a family of subobjects of an object X in a category $\mathscr{C}$ are in a certain sense the generalization of the well-known notions of intersection of subsets, subgroups, etc. and respectively of union of subsets, the subgroup generated by a set of subgroups, etc.

References: Isbell [8], MacLane [3], Mitchell [2], Pareigis [1], D. Popescu [2], Schubert [1].

Let $\{f_i: X_i \to X\}_{i\in\mathscr{I}}$ be a family (not necessarily a set) of subobjects of X. For each $i \in \mathscr{I}$, let $\mathscr{A}_i$ denote the right ideal over X generated by f_i. We say that such a family of subobjects has an *intersection* if the ideal $\bigcap_{i\in\mathscr{I}}\mathscr{A}_i$ is also a subobject of X, i.e. it is generated by a monomorphism $f: X' \to X$. We denote X' by $\bigcap_{i\in\mathscr{I}}X_i$ or simply by $\bigcap X_i$ when there is no doubt concerning the index set.

PROPOSITION 17.1. *Let $\{f_i: X_i \to X\}_{i\in\mathscr{I}}$ be a set of subobjects of an object X. This set of subobjects has an intersection if and only if there exist a monomorphism $f: X' \to X$ and for each $i \in \mathscr{I}$ a morphism $u_i: X' \to X_i$ such that $f_iu_i = f_ju_j$ for all $i, j \in \mathscr{I}$. Furthermore, every morphism $g: Y \to X$ which factors through each f_i factors (necessarily in a unique way) through f.* ✳

COROLLARY 17.2. *The family $\{f_i: X_i \to X\}_{i\in\mathscr{I}}$ of subobjects of X has an intersection if and only if the family $\{X_i, f_i\}_{i\in\mathscr{I}}$ of objects of $\mathscr{C}/X$ has a product. If $\mathscr{C}/X$ is a category with products then every set of subobjects of X has an intersection.* ✳

If the intersection exists for every set of subobjects of each object of $\mathscr{C}$, we say that $\mathscr{C}$ *has intersections.* If intersections exist only for finite sets of subobjects then we say that $\mathscr{C}$ has *finite intersections.*

The following results show that intersections are related to limits.

PROPOSITION 17.3. *A category with fibered products has finite intersections. A category with finite products is finitely complete if and only if it has finite intersections.*

Proof. The first part of the proposition follows from Corollary 17.2. Now assume that a category $\mathscr{C}$ has finite products and finite intersections. Then $\mathscr{C}$ is a category with kernels. Indeed, if $(f, g): X \to Y$ is a double arrow in $\mathscr{C}$, then consider $u = [1_X, f]$ and $v = [1_X, g]$, both from X into $X\prod Y$. It is clear that u and v are monomorphisms; hence, (X, u) and (X, v) are subobjects of $X\prod Y$. Let

X' be their intersection and $u': X' \to X$, $v': X' \to X$ the inclusions, such that $uu' = vv'$. Then $puu' = u' = pvv' = v'$ and (X', u') is a kernel of (f, g). Next we see that $fu' = quu' = qvu' = gu'$. Now if $h: Z \to X$ is a morphism such that $fh = gh$, then $quh = qvh$. Hence $uh = vh$, and thus a unique morphism $h': Z \to X'$ can be found such that $u'h' = h$. ⁎

COROLLARY 17.4. *For a category $\mathscr{C}$ the following are equivalent:*

i) *$\mathscr{C}$ has products and finite intersections.*

ii) *$\mathscr{C}$ has products and intersections.*

iii) *$\mathscr{C}$ is complete.*

The proof follows from the above proposition and Theorem 9.1. ⁎

Another notion closely related to the intersection can be defined. A family $\{f_i: X_i \to X\}_{i \in \mathscr{I}}$ of subobjects of X has a *lower bound* if there is a subobject $f: X' \to X$ such that $X' \subseteq X_i$ for all $i \in \mathscr{I}$, and, furthermore, if X'' is another subobject of X which is contained in X for all $i \in \mathscr{I}$, then $X'' \subseteq X'$. Let $\inf_{i \in \mathscr{I}}(X_i, f_i)$ or simply $\inf(X_i)$ denote a lower bound of the family of subobjects under consideration. The following example shows that the lower bound need not in general coincide with the intersection.

Example 17.5. Let $\mathscr{C}$ be the category whose objects are $X_1, X_2, X_3, X_4, X_5, X_6$; the morphisms in $\mathscr{C}$ are the arrows drawn in the diagram

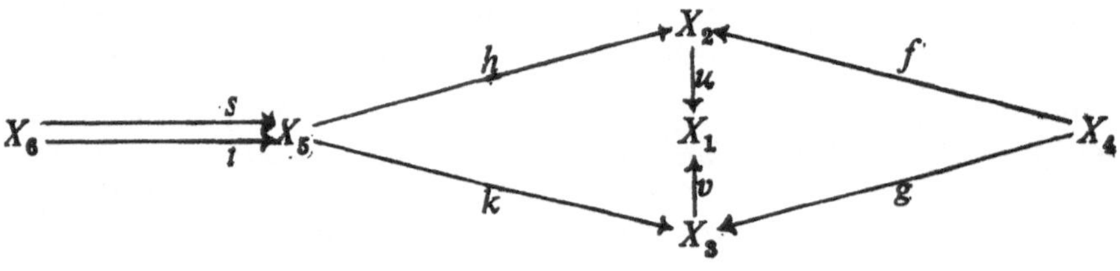

and the identity morphisms.

Assume that $uf = vg$, $uh = vk$, $hs = ht$, and $ks = kt$. Then $X_4 = \mathrm{Inf}(X_2, X_3)$ (because u and v are monomorphisms) but it is not an intersection, since h and k are not monomorphisms (assuming that $s \neq t$).

It is easy to obtain the following result.

PROPOSITION 17.6. *If $\{f_i: X_i \to X\}_{i \in \mathscr{I}}$ is a family of subobjects of X such that $\cap X_i$ and* $\inf(X_i)$ *exist, then $\cap X_i = \inf(X_i)$.* ⁎

This result can be applied, for instance, in $\mathscr{Set}$, $\mathscr{Top}$, $\mathscr{Rel}$, $\mathscr{Pre}$, $\mathscr{Ord}$, $\mathscr{Mon}$, $\mathscr{Gr}$, $\mathscr{Ab}$ and $\mathscr{Rg}$.

For quotient objects, the notion of *cointersection* may be defined. Of course, the duals of the above results are valid.

Let (p, Y) and (q, Z) be quotient objects of X. We write $(p, Y) \leqslant (q, Z)$, or simply $Y \leqslant Z$, if there is a morphism $h: Y \to Z$ such that $hp = q$.

The notion dual to lower bound of subobjects is called an *upper bound of quotient objects*. Specifically, if $\{p_i: X \to X_i\}_{i \in \mathscr{I}}$ is a family of quotient objects of X, then by an upper bound of this family we mean a quotient object (p, Y) such that $X_i \leqslant Y$ for all $i \in \mathscr{I}$, and furthermore if (q, Z) is a quotient object of X such that $X_i \leqslant Z$ for all $i \in \mathscr{I}$, then $Y \leqslant Z$.

Let $\{f_i \colon X_i \to X\}_{i \in \mathscr{I}}$ be a family of subobjects of X and let $\mathscr{A}_i$ be the right ideal over X generated by f_i. We say that this family of subobjects has a *σ-union* if the functor associated with the right ideal $\cup \mathscr{A}_i$ is representable, and denote this σ-union by $\overset{\sigma}{\bigcup}_{i \in \mathscr{I}} X_i$ or simply by $\overset{\sigma}{\cup} X_i$ when the index family is clear from context.

Consider the diagram

$$\begin{array}{ccc} X' & & Y' \\ {\scriptstyle u}\big\downarrow & & \big\downarrow{\scriptstyle v} \\ X & \xrightarrow{\;\;f\;\;} & Y \end{array}$$

where f is a morphism and the vertical morphisms are monomorphisms. We say that the subobject X' of X is *carried into* the subobject Y' if there is a morphism $f' \colon X' \to Y'$ such that $vf' = fu$. The *union* of a family $\{X_i\}_{i \in \mathscr{I}}$ of subobjects of an object X is defined to be a subobject X' of X such that $X_i \subseteq X'$ for every $i \in \mathscr{I}$, and which also has the following property. If $f \colon X \to Y$ and each X_i is carried into some subobject Y' of Y by f, then X' is also carried into Y' by f. The object X' will be denoted by $\cup_{i \in \mathscr{I}} X_i$ or simply by $\cup X_i$.

Proposition 17.7. *Let $\{f_i \colon X_i \to X\}_i$ be a family of subobjects of X and let $f \colon X' \to X$ be the σ-union of this family. If $g \colon X \to Y$ and each X_i is carried into the same subobject Y' of Y by g, then X' is also carried into Y' by g.*

Proof. Let $\mathscr{A}_i$ be the right ideal generated by f_i and $\mathscr{A}$ the right ideal generated by f. The condition "X_i is carried into" a subobject $h \colon Y' \to Y$ by g implies that the right ideal $g_*(\mathscr{A}_i)$ is contained in $\langle h \rangle$, the right ideal generated by h. Thus, by Corollary 16.7, we have that $g_*(\cup \mathscr{A}_i) = \cup g_*(\mathscr{A}_i) \subseteq \langle h \rangle$, so that $g_*(\mathscr{A}) \subseteq \langle h \rangle$. Hence the object (X', gf) belongs to $\langle h \rangle$, so there is a morphism $t \colon X' \to Y'$ such that $ht = gf$, i.e. g carries X' into Y'. ✳

Corollary 17.8. *Let $\{X_i\}_{i \in \mathscr{I}}$ be a family of subobjects of X. Assume that $\overset{\sigma}{\cup} X_i$ and $\cup X_i$ exist. Then these two subobjects coincide.*

Proof. By the definition of $\cup X_i$, if X' is a subobject of X such that $X_i \subseteq X'$ for all $i \in \mathscr{I}$, then $\cup X_i \subseteq X'$. In particular, we get an inclusion $u \colon \cup X_i \subseteq \overset{\sigma}{\cup} X_i$. Now, since the right ideal $\mathscr{A}$ generated by the inclusion $f \colon \cup X_i \to X$ contains the right ideals generated by the inclusions $X_i \subseteq X$ for all $i \in \mathscr{I}$, we see that the right ideal generated by the inclusion $g \colon \overset{\sigma}{\cup} X_i \to X$ is contained in $\mathscr{A}$, i.e. there is a morphism $v \colon \overset{\sigma}{\cup} X_i \to X$ such that $fv = g$. It is easy to see that u and v are isomorphisms and $v = u^{-1}$. ✳

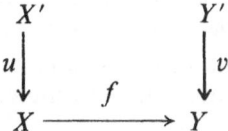

Note 17.9. Let $\{f_i \colon X_i \to X\}_{i \in \mathscr{I}}$ be a family of subobjects of X. Assume that $\overset{\sigma}{\cup} X_i$ is defined and that the coproduct $\coprod X_i$ exists. Let $u_i \colon X_i \to \coprod X_i$ be the structural morphisms. Then there is a unique morphism $g: \coprod X_i \to X$ such that $gu_i = f_i$ for every $i \in \mathscr{I}$. Also, there is a unique morphism $h: \coprod X_i \to \overset{\sigma}{\cup} X_i$ such that $hu_i = v_i$, where $v_i: X_i \to \overset{\sigma}{\cup} X_i$ is an inclusion for each $i \in \mathscr{I}$. It is clear that the right ideal generated by g contains the right ideal generated by f_i for each $i \in \mathscr{I}$. Thus, by the definition of $\overset{\sigma}{\cup} X_i$ there is a morphism $k: \overset{\sigma}{\cup} X_i \to X$ such that $kv_i = u_i$ for all $i \in \mathscr{I}$. An easy computation shows that k is a right inverse of g. In this way we get an example of a category $\mathscr{C}$ with coproduct in which the union $\cup X_i$ of a family $\{X_i\}_{i \in \mathscr{I}}$ of subobjects of an object X exists, whereas the σ-union of this family of subobjects is not defined (see Section 4.4).

Example 17.10. Let $\mathscr{C}$ be the category whose objects and morphisms (except the identity morphisms) are pictured in the following diagram:

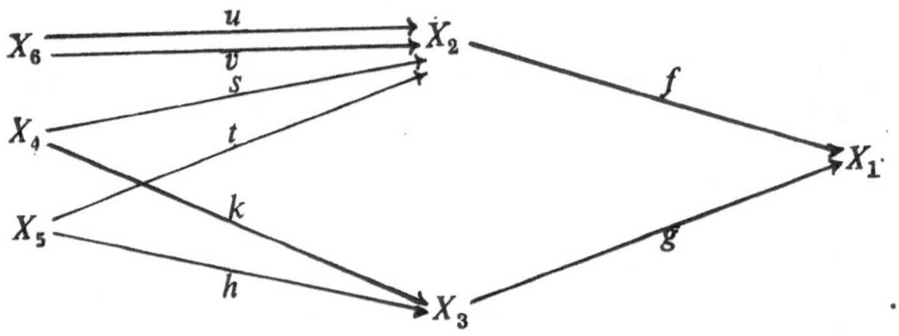

Assume that $fu = fv$ and $u \neq v$. Assume also that $fs = gk$ and $ft = gh$. It is clear that g, h and k are monomorphisms, and $X_3 = X_4 \cup X_5$. Nevertheless, X_3 is not the σ-union of X_4 and X_5.

A family $\{X_i \subseteq X\}_{i \in \mathscr{I}}$ of subobjects has an *upper bound* if the following hold: a) there is a subobject X' of X such that $X_i \subseteq X'$ for each $i \in \mathscr{I}$ and b) if X'' is another subobject of X which contains all subobjects X_i, $i \in \mathscr{I}$, then X'' also contains X'.

An upper bound of this family of subobjects is denoted by $\sup_{i \in \mathscr{I}}(X_i, f_i)$ or simply by $\sup(X_i)$. The reader is invited to find an easy example of a category in which an upper bound of a family of subobjects is defined whereas the union or the σ-union of this family is not defined.

Let $\mathscr{C}$ be a category. We say that $\mathscr{C}$ *has σ-unions* if the σ-union exists for every set of subobjects of each object in $\mathscr{C}$. Analogously, we get the notion of a *category with unions.*

The notion dual to union (σ-union) of subobjects is that of *counion (σ-counion)* of quotient objects. Also the dual to upper bound of subobjects is *lower bound of quotient objects.*

THEOREM 17.11. *Let $\mathscr{C}$ be a complete category such that $\mathscr{C}$ has a set of coseparators. Assume that every family of subobjects of each object X of $\mathscr{C}$ has an intersection. Then $\mathscr{C}$ is a category with bounded colimits. In particular, $\mathscr{C}$ is a cocomplete category.*

Proof. Let $\mathscr{S}$ be the full subcategory of $\mathscr{C}$ whose objects form a set of coseparators of $\mathscr{C}$ (i.e. $\mathscr{S}$ is a small category). Further, let $\mathscr{I}$ be a right bounded category and $F: \mathscr{I} \to \mathscr{C}$ a functor. Let $F/\mathscr{S}$ denote the full subcategory of $F/\mathscr{C}$ that consists of the category of cocones over F, whose covertices belong to $\mathscr{S}$. By Exercise 17.10, for each object X of $\mathscr{S}$, there is a set (possibly empty) of objects of $F/\mathscr{S}$ whose vertex is X. Hence the category $F/\mathscr{S}$ is small. Let $U: F/\mathscr{S} \to \mathscr{C}$ be the underlying functor, i.e. the functor which assigns to each object (α, A) of $F/\mathscr{S}$ the object A, and to each morphism $f: (\alpha, A) \to (\beta, B)$ of $F/\mathscr{S}$ the same morphism $f: A \to B$ of $\mathscr{C}$. Furthermore, since $\mathscr{C}$ is complete, the functor U has a limit (X, φ).

For each object (α, A) of $F/\mathscr{S}$ and $i \in \mathscr{I}$ let $h^i_{(\alpha, A)}$ denote the morphism $\alpha_i: F_i \to A$. It is easy to see that the morphisms $\{h^i_{(\alpha, A)}\}_{(\alpha, A) \in F/\mathscr{S}}$ define a functorial morphism $h^i: F(i) \to U$, so that there is a unique morphism $t^i: F(i) \to X$ such that $h^i = \varphi t^i$ for every $i \in \mathscr{I}$. Now, using the uniqueness of the morphisms t^i, we see that these morphisms define a functorial morphism $t: F \to X$, or equivalently that (t, X) is a cocone over F.

Let $\mathscr{M}$ denote the class of all subobjects (X', f) of X such that there exists a functorial morphism $g: F \to X'$ with $fg = t$. Let (V, v) be an intersection of this family of subobjects. By Proposition 17.1 there is a functorial morphism $w: F \to V$ such that $vw = t$. We claim that (w, V) is a colimit of F. To show this, let (g, G) be a cocone over F. By Exercise 7.7 there are a set $\{A_k\}_{k \in \mathscr{K}}$ of objects of $\mathscr{S}$ and a monomorphism $j: G \to \prod_k A_k$. Let $s: X \to \prod_k A_k$ be the unique morphism such that

$$\varphi_{(p_k jg,\, A_k)} = p_k s, \quad \text{for each } k \in \mathscr{K},$$

where the $p_k: \prod_k A_k \to A_k$ are the structural projections and $(p_k jg, A_k)$ is obviously an object of $F/\mathscr{S}$.

Now consider the canonically defined cartesian square in $\mathscr{C}$:

$$\begin{array}{ccc} P & \xrightarrow{\ \bar{j}\ } & X \\ {\scriptstyle \bar{s}}\big\downarrow & & \big\downarrow{\scriptstyle s} \\ G & \xrightarrow{\ j\ } & \prod_k A_k \end{array} .$$

It is clear that $\bar{j}$ is a monomorphism. Also, for each $k \in \mathscr{K}$ and for each $i \in \mathscr{I}$,

$$p_k j g_i = h^i_{(p_k jg,\, A_k)} = \varphi_{(p_k jg,\, A_k)} t^i = p_k s t^i.$$

But then there is a unique morphism $p_i: F_i \to P$ such that $\bar{j} p_i = t^i$ and $\bar{s} p_i = g_i$. It is easy to see that the morphisms $\{p_i\}_{i \in \mathscr{I}}$ define a functorial morphism $p: F \to P$ such that $\bar{j} p = t$. Thus $(P, \bar{j}) \in \mathscr{M}$. Since (V, v) is the intersection of all the subobjects of $\mathscr{M}$, there is a morphism $r: V \to P$ such that $\bar{j} r = v$. Therefore, the morphism $\bar{s} r: V \to G$ has the property that $j(\bar{s} r)w = s\bar{j} rw = svw = st = jg$ and thus $(\bar{s} r)w = g$ since j is a monomorphism.

To verify that (w, V) is a colimit of F it will suffice to show that $\bar{s} r$ is determined by the property that $(\bar{s} r)w = g$.

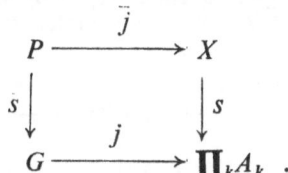

Let $g_1, g_2: V \to G$ be morphisms such that $g_1w = g_2w$. Let (M, m) be a kernel of the double arrow (g_1, g_2); then for each $i \in \mathcal{I}$ there is a unique morphism $q_i: F_i \to M$ such that $mq_i = w_i$. The morphisms $\{q_i\}_{i\in\mathcal{I}}$ determine a functorial morphism $q: F \to M$ such that $mq = w$. Then $(vm)q = vw = t$ and therefore $(M, vm) \in \mathcal{M}$.

Since (V, v) is the intersection of all the objects of $\mathcal{M}$, we see that (M, vm) and (V, v) are equal, or, equivalently, that m is an isomorphism. Therefore $g_1 = g_2$.

Hence (w, V) is a colimit of F and the proof is complete. ✳

COROLLARY 17.12. *Let $\mathcal{C}$ be a wellpowered complete category. Assume that $\mathcal{C}$ has a set of coseparators. Then $\mathcal{C}$ is also cocomplete.* ✳

The above results can be dualized.

Let $F: \mathcal{C} \to \mathcal{C}'$ be a functor. The *intersection* of a class $\{u^i: F_i \to F\}_{i\in\mathcal{I}}$ of subfunctors of F is a subfunctor $u: F' \to F$ such that $F' \subseteq F_i$ for each index $i \in \mathcal{I}$ and, moreover, such that every functorial morphism $v: G \to F$ which factors through each u_i also factors uniquely through u. The functor F' is denoted by $\cap F_i$. In an analogous way we can define the notion of the *lower bound* of a family of subfunctors (which is also denoted by $\inf(F_i)_{i\in\mathcal{I}}$). Dually, we have the notions of *cointersection* and *upper bound* of quotient functors.

PROPOSITION 17.13. *Let $\{u^i: F_i \to F\}_{i\in\mathcal{I}}$ be a family of subfunctors of $F: \mathcal{C} \to \mathcal{C}'$. Assume that for each $i \in \mathcal{I}$, u^i is an argumentwise functorial monomorphism and that for each $X \in \mathcal{C}$, the family $\{u^i_X: F_i(X) \to F(X)\}_{i\in\mathcal{I}}$ of subobjects of $F(X)$ has an intersection. Then this family of subfunctors has an intersection. In particular, if $\mathcal{C}'$ is a complete category, then every set of subfunctors of F has an intersection.*

Proof. Let $v_X: G(X) \to F(X)$ be an intersection of the family $\{u^i_X: F_i(X) \to F(X)\}_{i\in\mathcal{I}}$ of subobjects and $\{v^i_X: G(X) \to F_i(X)\}_{i\in\mathcal{I}}$ the family of corresponding canonical morphisms defined such that $u^i_X v^i_X = v_X$ for each $i \in \mathcal{I}$. Let $G: \mathcal{C} \to \mathcal{C}'$ denote the functor that assigns to each object X of $\mathcal{C}$ the object $G(X)$ of $\mathcal{C}'$. If $f: X \to Y$ is a morphism of $\mathcal{C}$, then for each $i \in \mathcal{I}$ we have the commutative diagram:

$$\begin{array}{ccccc} G(X) & \xrightarrow{v^i_X} & F_i(X) & \xrightarrow{u^i_X} & F(X) \\ \downarrow & & \downarrow{\scriptstyle F_i(f)} & & \downarrow{\scriptstyle F(f)} \\ G(Y) & \xrightarrow{v^i_Y} & F_i(Y) & \xrightarrow{u^i_Y} & F(Y) \end{array} .$$

It can be checked that there is a unique morphism $G(f): G(X) \to G(Y)$ such that the left square of the diagram becomes commutative. It is easy to check that $\{v^i_X\}_{X\in\mathcal{C}}$ defines a functorial monomorphism $v^i: G \to F_i$ and that $v = u^i v^i = u^j v^j$ for all $i, j \in \mathcal{I}$. Consequently (G, v) is the intersection of the family of subfunctors considered. The last part of the proposition follows from Corollary 17.4. ✳

Clearly, the dual result also holds.

The above proposition gives a condition sufficient to ensure that a family of subfunctors of a functor $F: \mathcal{C} \to \mathcal{C}'$ have an intersection. There is a broad class of categories $\mathcal{C}$ such that every class of subfunctors of a functor with values in $\mathcal{C}$ has an intersection. This class of categories is exposed by the following result.

PROPOSITION 17.14. *Let $\mathscr{C}$ be a wellpowered complete category and let $F: \mathscr{C}' \to \mathscr{C}$ be a functor. Then every family $\{u^i: F_i \to F\}_{i \in \mathscr{I}}$ of subfunctors of F has an intersection.*

Proof. Let $X \in \mathscr{C}'$. By Proposition 10.5, for each $i \in \mathscr{I}$, $u^i_X: F_i(X) \to F(X)$ is a monomorphism. Since the family of all subobjects of $F(X)$ forms a set, we can speak of the subobject $\bigcap_{i \in \mathscr{I}} F_i(X) = G(X)$. As in the proof of the previous proposition, we may check that the assignment $X \rightsquigarrow G(X)$ defines a functor $G: \mathscr{C}' \to \mathscr{C}$ that is canonically a subfunctor of F, and moreover $G = \bigcap_{i \in \mathscr{I}} F_i$. ✳

COROLLARY 17.15. *Let $\mathscr{C}$ be a wellpowered complete category and let $F: \mathscr{C}' \to \mathscr{C}$ be a functor. Then every class of objects of $\mathscr{S}(F)$ has a lower bound and an upper bound.*

The proof follows directly from Proposition 17.14. ✳

PROPOSITION 17.16. *Let $F: \mathscr{C} \to \mathscr{S}et$ be a functor. Then every class of quotient functors of F has a lower bound and an upper bound.*

Proof. Let $\{p_i: F \to P_i\}_{i \in \mathscr{I}}$ be a class of quotient functors of F. By the dual of Proposition 10.5, for each $i \in \mathscr{I}$, p_i is an argumentwise functorial epimorphism, i.e. for each $X \in \mathscr{C}$, p^i_X is a surjection. Now let $X \in \mathscr{C}$ and let R_X denote the equivalence relation on $F(X)$ defined as follows: $x \sim x'(R_X)$ if and only if $p^i_X(x) = p^i_X(x')$ for every $i \in \mathscr{I}$. Let $f: X \to Y$ be a morphism in $\mathscr{C}$ and let $x, x' \in F(X)$ be such that $x \sim x'(R_X)$. Then, for every $i \in \mathscr{I}$ we have that $p^i_Y F(f)(x) = P_i(f) p^i_Y(x) = P_i(f) p^i_Y(x') = p^i_Y F(f)(x')$. i.e. $F(f)(x) \sim F(f)(x')(R_Y)$.

Let $P(X) = F(X)/R_X$ and let $p_X: F(X) \to P(X)$ be the canonical surjection. If $f: X \to Y$ is a morphism in $\mathscr{C}$, then, using the above relations, it can be deduced that f defines a unique map $P(f): P(X) \to P(Y)$ such that $P(f)p_X = p_Y F(f)$. In fact, the assignments $X \rightsquigarrow P(X), f \rightsquigarrow P(f)$ define a functor $P: \mathscr{C} \to \mathscr{S}et$, and the surjections $\{p_X\}_{X \in \mathscr{C}}$ define a functorial epimorphism $p: F \to P$. From the construction of (p, P) it is easy to see that $(p, P) = \sup_{i \in \mathscr{C}}(p^i, P_i)$. The proof is completed by an application of Theorem 9.4. ✳

Exercises

17.1. Consider a coproduct $\mathscr{I} = \coprod_{j \in \mathscr{J}} \mathscr{I}_j$ of sets and let $\{X_i \subseteq X\}_{i \in \mathscr{I}}$ be a family of subobjects. Prove: $\bigcap_{i \in \mathscr{I}_j} X_i$ is defined for each $j \in \mathscr{J}$ and if $\bigcap_{j \in \mathscr{J}} (\bigcap_{i \in \mathscr{I}_j} X_i)$ is also defined, then $\bigcap_{i \in \mathscr{I}} X_i$ is defined and

$$\bigcap_{j \in \mathscr{J}} (\bigcap_{i \in \mathscr{I}_j} X_i) = \bigcap_{i \in \mathscr{I}} X_i.$$

A similar result holds for the lower bound.

17.2. Prove: If a family of subobjects has a σ-union then this σ-union is a union and an upper bound. If a family of subobjects has a union then this union is also an upper bound.

17.3. Prove that every set of subobjects of an object in $\mathscr{S}et$ has a σ-union (hence a union and an upper bound).

17.4. Let X be an object of a category $\mathscr{C}$. Denote by $\mathscr{M}(X)$ the full subcategory of $\mathscr{C}/X$ generated by all the objects (Y, f) such that f is a monomorphism. Assume that $\mathscr{M}(X)$ is equivalent to a small category (i.e. X is wellpowered). Prove that the following are equivalent:

a) Every family of subobjects of X has a lower bound.

b) $\mathscr{M}(X)$ has an initial object and every family of subobjects has an upper bound.

c) $\mathscr{M}(X)$ is complete and cocomplete.

17.5. Let $(f, g): X \to Y$ be a double arrow in a category with kernels. Suppose that for each member X_i of a family of subobjects of X, $f|X_i = g|X_i$. Prove that if $X' = \cup_i X_i$ then $f|X' = g|X'$.

17.6. Let $\mathscr{C}$ be a category. Prove:

a) If $\mathscr{C}$ is a category with fibered products, then the intersection of any two strict subobjects is also a strict subobject.

b) If $f: X' \to X$ and $f_1: X_1 \to X$ are strict monomorphisms, and $g: X' \to X$ is an epimorphism such that $f_1 g = f$, then g is an isomorphism.

17.7. Prove: If $\mathscr{C}$ is a category with fibered products and has a set of quasi-generators, then the class of all strict subobjects of any object X of $\mathscr{C}$ is a set.

17.8. Let $\mathscr{C}$ be a category with kernels and finite intersections. Prove that if $f, g, h \in \mathscr{C}(X, Y)$, then

$$\operatorname{Ker}(f, g) \cap \operatorname{Ker}(g, h) \subseteq \operatorname{Ker}(f, h).$$

17.9. Let $\mathscr{C}$ be a category and $X \in \mathscr{C}$. Assume that every family of subobjects of X has a lower bound. Show that every family of subobjects of X has an upper bound, and, in particular, that if $\mathscr{C}$ is complete and X is wellpowered, then $\mathscr{S}(X)$, the set of all subobjects of X, is a complete and cocomplete lattice.

17.10. Let $\mathscr{I}$ be a right bounded category and $F: \mathscr{I} \to \mathscr{C}$ a functor. Show that, for every $X \in \mathscr{C}$, the class $[F, X]$ is a set.

17.11. Let $F: \mathscr{C} \to \mathscr{C}'$ be a functor; the notions of lower bound, upper bound, union and σ-union of subfunctors are defined in the same way as the corresponding notions for subobjects. Prove: If $\mathscr{C}'$ is a category with intersections (unions, σ-unions, lower bounds, upper bounds) for subobjects, then every set of subfunctors has an intersection (union, σ-union, lower bound, upper bound).

1.18. Images and inverse images

The notions of image and inverse image are usually defined relative to a map (morphism of groups, rings, etc.) $f: X \to Y$. These notions are now defined for a morphism in a category.

References: Grothendieck [1], Kuroš [2], MacLane [3], Mitchell [2].

The *image* of a morphism $f: X \to Y$ is a subobject $g: Y' \to Y$ of Y such that f factors through g and, furthermore, if $g_1: Y_1 \to Y$ is another subobject of Y such that f factors through g_1, then $Y' \subseteq Y_1$, i.e. there exists a morphism $h: Y' \to Y_1$, such that $g_1 h = g$.

PROPOSITION 18.1. *The image of a morphism $f: X \to Y$ is unique in the following sense. If in the diagram*

$$\begin{array}{ccccc} X & \xrightarrow{p_1} & Y_1 & \xrightarrow{g_1} & Y \\ \| & & \downarrow h & & \| \\ X & \xrightarrow{p_2} & Y_2 & \xrightarrow{g_2} & Y \end{array}$$

(Y_1, g_1) and (Y_2, g_2) are the images of f, and $g_1 p_1 = g_2 p_2 = f$, then there is a unique isomorphism $h: Y_1 \to Y_2$ such that $g_2 h = g_1$ and $h p_1 = p_2$.

The proof is left for the reader. ⁎

The image of a morphism $f: X \to Y$ (if it exists) is denoted by $(\mathrm{Im}(f), u_f)$ or simply by u if there is no possibility of confusion. The canonical morphism $p: X \to \mathrm{Im}(f)$ such that $up = f$ is denoted by p_f or simply by p. In general, p need not be an epimorphism.

If every morphism in a category $\mathscr{C}$ has an image, then we say that $\mathscr{C}$ is a *category with images.*

PROPOSITION 18.2. *Assume that X is an object such that every set of subobjects of X has an intersection. Then every morphism $f: Y \to X$ has an image. Moreover, if $\mathscr{C}$ is a wellpowered complete category, then $\mathscr{C}$ is a category with images.*

Proof. Let $f: Y \to X$ and let $\{f_i: X_i \to X\}_{i \in \mathscr{I}}$ be the family of all subobjects through which f factors. This family is not empty because it contains $1_X: X \to X$. It is easy to see that $\cap_i X_i = \mathrm{Im}(f)$. The last part of this proposition follows from Corollary 17.4. ⁎

If $f: X \to Y$ and $h: X' \to X$ is a subobject of X, then the image of fh (if it exists) is denoted by $f(X')$ and is called the *image of X' under f*. In particular, the image of $1_X: X \to X$ under f is equal to $\mathrm{Im}(f)$. Therefore, we shall often denote $f(X)$ by $\mathrm{Im}(f)$.

The notion dual to image is that of *coimage*. The coimage of a morphism $f: X \to Y$ is a quotient object $q_f: X \to \mathrm{Coim}(f)$, such that $f = v_f q_f$; furthermore, for every epimorphism $h: X \to X'$ such that $f = gh$ for some g, we have that $th = q_f$. Later we shall see that in some special but important cases the image and coimage of a morphism are related. However, in general, there need not be any relation between these two notions.

Let $f: X \to Y$ be a morphism. We say that the subobject $g: Y' \to Y$ has an *inverse image* under f if the ideal $f^*(\langle g \rangle)$ is principal.

A *category with inverse images* is a category $\mathscr{C}$ such that for every morphism $f: X \to Y$ and every subobject Y' of Y, the inverse image $f^*(Y')$ of Y' under f is defined. By Exercise 16.3, we have the following result.

Proposition 18.3. *The subobject $g: Y' \to Y$ has an inverse image under $f: X \to Y$ if and only if there exists a cartesian square:*

$$\begin{array}{ccc} X' & \xrightarrow{\;g'\;} & X \\ {\scriptstyle f'}\big\downarrow & & \big\downarrow{\scriptstyle f} \\ Y' & \xrightarrow{\;g\;} & Y \end{array} \qquad (13)$$

. ✳

The subobject X' is denoted by $f^*(Y')$ and the inclusion $f^*(Y') \to X$ by g'. $f^*(Y')$ is the largest subobject of X that is carried into Y' by f. However, a simple example shows that the existence of such a maximal subobject does not guarantee the existence of the inverse image.

Proposition 18.4. *Let $f: X \to Y$ and consider the inclusions $X_1 \subseteq X_2 \subseteq X$ and $Y_1 \subseteq Y_2 \subseteq Y$. Then the following relations hold whenever both sides are defined:*

(i) $f(X_1) \subseteq f(X_2)$,

(ii) $f^*(Y_1) \subseteq f^*(Y_2)$,

(iii) $X_1 \subseteq f^*(f(X_1))$,

(iv) $Y_1 \supseteq f(f^*(Y_1))$,

(v) $f(X_1) = f(f^*(f(X_1)))$,

(vi) $f^*(Y_1) = f^*(f(f^*)Y_1)))$.

Proof. (i) and (ii) are direct consequences of the definition of image and inverse image. Assertions (ii) and (iv) follows from Proposition 16.4 and the definitions of the image and inverse image. Assertion (v) results by use of (i), (iii) and (iv). Finally (vi) follows from (ii), (iv) and (iii). ✳

Proposition 18.5. *Let $f: X \to Y$ be a morphism and let $\{Y_i\}_{i \in \mathscr{J}}$ be a family of subobjects of Y. Assume that $\cap Y_i$ and $f^*(\cap Y_i)$ exist as well as $f^*(Y_i)$ for each $i \in \mathscr{J}$.*

Then $\cap_i f^*(Y_i)$ also exists and is equal to $f^*(\cap Y_i)$.

The proof follows from Proposition 18.3 and the definition of inverse image. ✳

Proposition 18.6. *Let $\mathscr{C}$ be a category with images and inverse images, $f: X \to Y$ a morphism in $\mathscr{C}$ and $\{X_i\}_i$ a family of subobjects of X such that $\cup X_i$ exists. Then $\cup f(X_i)$ exists and is equal to $f(\cup X_i)$.*

Proof. Let $g: Y \to Z$ be a morphism and suppose that each $f(X_i)$ is carried by g into some subobject Z' of Z. Then each X_i is carried into Z' by gf. Thus, by the definition of union, $\cup X_i$ is carried into Z' by gf. Hence $\cup X_i$ is carried into $g^*(Z')$ by f, so that $f(\cup X_i)$ is a subobject of $g^*(Z')$. But this means that $f(\cup X_i)$ is carried into Z' by g. Since by Proposition 18.4, $f(\cup X_i)$ contains all of the $f(X_i)$, then $f(\cup X_i)$ is the union of the family $\{f(X_i)\}_{i \in \mathscr{I}}$. ✳

Exercises

18.1. Assume that in the cartesian square (13) the morphism f' factors through a subobject $h: Y_1 \to Y'$. Show that $X' = f^*(Y_1)$, and that, in particular, if $k: Y'' \to Y$ is a subobject through which f factors and the intersection $Y'' \cap Y'$ exists, then $f^*(Y'' \cap Y') = f^*(Y')$.

18.2. Let $f: X \to Y$ be a morphism and let $g: X' \to X$ be a subobject. We shall say that X' has a *σ-image* under f if the right ideal $f_*(\langle f \rangle)$ is a subobject of Y. This subobject is denoted by $f_*(X')$.

a) Prove that in the category $\mathscr{S}et$ every subobject has a σ-image. A similar result is valid in the category $\mathscr{S}$ defined in Example 1.8.

b) Let $\{X_i\}_i$ be a family of subobjects of X and let $f: X \to Y$ be a morphism. Assume that $\overset{\sigma}{\cup} X_i$, $f_*(\overset{\sigma}{\cup} X_i)$ and $f_*(X_i)$ exist for each i. Show that $\overset{\sigma}{\cup} f_*(X_i)$ also exists and is equal to $f_*(\overset{\sigma}{\cup} X_i)$.

18.3. Consider $f: X \to Y$ in a category and let $\{X_i\}_i$ be a family of subobjects of X. If $\cup X_i$ and $\cup f(X_i)$ are defined, show that $f(\cup X_i)$ is also defined and is equal to $\cup f(X_i)$. Show that, likewise, if $\{Y_i\}_i$ is a family of subobjects of Y and if $\cap Y_i$ and $\cap f^*(Y_i)$ are defined then $f^*(\cap Y_i)$ is also and is equal to $\cap f^*(Y_i)$.

18.4. Prove that in Proposition 18.4 it suffices to assume that $f^*(f(X_1))$ exists in (v) and that $f(f^*(Y_1))$ exists in (vi).

18.5. A *normal category* is a category with zero object in which every monomorphism is the kernel of some morphism. Prove that a normal category in which every morphism has a kernel is a category with inverse images and in particular with finite intersections.

18.6. Let $\mathscr{C}$ be a normal category in which every morphism has a cokernel. Prove that then there is an injection from the class of subobjects of an object X into the class of quotient objects of X, and that, in particular, if $\mathscr{C}$ is wellcopowered, then $\mathscr{C}$ is wellpowered. Prove: If $\mathscr{C}$ is normal and conormal and every morphism in $\mathscr{C}$ has a kernel and a cokernel, then the above function is a bijection.

18.7. Prove: If $f: X \to Y$ is a monomorphism with cokernel zero in a normal category, then f is an isomorphism.

18.8. Let $\mathscr{C}$ be a category with zero object. Let $f: X \to Y$ be a morphism and suppose that $p: Y \to Z$ is its cokernel. Finally, suppose that $v: Y' \to Y$ is the kernel of p. Prove that there is a unique morphism $q: X \to Y'$ such that $vq = f$, and that if $\mathscr{C}$ is normal and every morphism has a cokernel, then (Y', v) is the image of f. If $\mathscr{C}$ has kernels, show further that (X, q) is the coimage of f.

18.9. Let $\mathscr{C}$ be a category with zero object. Consider the commutative diagram

$$\begin{array}{ccccc} Z & \xrightarrow{\ v\ } & Y & \xrightarrow{\ f'\ } & X_1 \\ \Vert & & {\scriptstyle g'}\downarrow & & \downarrow{\scriptstyle g} \\ Z & \xrightarrow{\ u\ } & X_2 & \xrightarrow{\ f\ } & X \end{array}$$

where the right hand square is cartesian, u is the kernel of f and v is defined such that $g'v = u$ and $f'v = 0$. Show that (Z, v) is the kernel of f'.

18.10. Consider the diagram in a category with zero object:

$$\begin{array}{ccc} X & \overset{v}{\dashrightarrow} & Y \\ & & \downarrow{\scriptstyle f} \\ T & \xrightarrow{\ u\ } & Z \end{array}$$

where u is the kernel of some morphism $Z \xrightarrow{g} Z'$. Show that then there is a morphism $X \to T$ such that the square obtained in this way is cartesian if and only if v is the kernel of gf.

18.11. In a category with zero object assume that the morphism $u: X \to Y$ is the kernel of $f: Y \to Z$. Also let $p: Y \to Z'$ be the cokernel of u. Show that then u is the kernel of p.

18.12. Let $\mathscr{C}$ be a category with fibered products and images. Let $X' \subseteq X$, $Y' \subseteq Y$ and $f: X \to Y$ be given. Let $g: X' \to f(X')$ be the morphism induced by f. Prove that $g^*(f(X') \cap Y') = X' \cap f^*(Y')$.

18.13. For functorial morphisms we can define the notions images and coimages. Prove that if $\mathscr{C}'$ is a finitely complete and finitely cocomplete category with images (coimages) then for arbitrary functors $F, G: \mathscr{C} \to \mathscr{C}'$, every functorial morphism $u: F \to G$ has an image (a coimage).

1.19. Triangular decompositions of morphisms

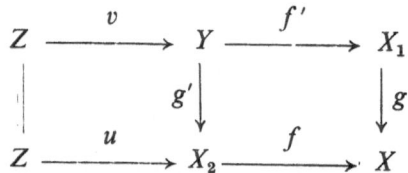

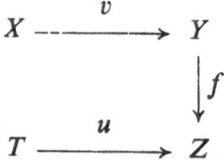

It is well known that every morphism f of $\mathscr{S}et$, $\mathscr{M}on$, $\mathscr{G}r$, $\mathscr{R}g$, etc. can be decomposed as a composition $f = up$ where u is a monomorphism and p an epimorphism. Such a decomposition is frequently used in the study of objects and morphisms of these categories. In this section we give a general definition of the so-called triangular decomposition and develop some aspects of this notion. Some applications to the existence of limits and colimits are also given.

References: Bourbaki [1], Kuroš [2], Mitchell [2], D. Popescu [2], Pupier [1], G. M. Kelly [2], Gabriel-Ulmer [1], Isbell [8], Schubert [1].

We say that a morphism $f: X \to Y$ has a *triangular decomposition* if there is a factorization $f = up$ where u is a monomorphism and p an epimorphism. We say that $\mathscr{C}$ is a *category with triangular decomposition* if every morphism in $\mathscr{C}$ has a triangular decomposition. Frequently a morphism has several triangular decompositions. Nevertheless, in some categories a morphism has in the following sense a unique triangular decomposition. Two triangular decompositions $X \xrightarrow{p_1} X_1 \xrightarrow{u_1} Y$ and $X \xrightarrow{p_2} X_2 \xrightarrow{u_2} Y$ of a morphism f are *isomorphic* if there is an isomorphism $v: X_1 \to X_2$ such that $vp_1 = p_2$ and $u_2 v = u_1$.

PROPOSITION 19.1. *Let $\mathscr{C}$ be a balanced category with fibered products. Then any two triangular decompositions of a morphism f in $\mathscr{C}$ are isomorphic.*

Proof. Let $X \xrightarrow{p_i} X_i \xrightarrow{u_i} Y$, $i = 1, 2$, be two triangular decompositions of f. Consider the cartesian square

$$\begin{array}{ccc} P & \xrightarrow{m_1} & X_1 \\ {\scriptstyle m_2}\downarrow & & \downarrow{\scriptstyle u_1} \\ X_2 & \xrightarrow{u_2} & Y \end{array}\quad .$$

Since $u_1 p_1 = u_2 p_2 = f_1$, there is a morphism $h: X \to P$ such that $m_1 h = p_1$ and $m_2 h = p_2$. Thus m_1 and m_2 are bimorphisms, i.e. isomorphisms by hypothesis. Let $v = m_2 m_1^{-1}$. Then $u_2 v = u_1$, and $vp_1 = m_2 m_1^{-1} p_1 = m_2 h = p_2$. ✳

Now we investigate the existence of a particular triangular decomposition of a morphism.

PROPOSITION 19.2. *Let $\mathscr{C}$ be a category with kernels. Assume that $f: X \to Y$ is a morphism in $\mathscr{C}$ and that f has an image. Then the canonical morphism $p: X \to$ $\to \mathrm{Im}(f)$ is an epimorphism i.e. f has a triangular decomposition. In particular, if $\mathscr{C}$ is a category with kernels and images then every morphism of $\mathscr{C}$ has a triangular decomposition.*

Proof. Let $(h, g): \mathrm{Im}(f) \to Z$ be a pair of morphisms such that $hp = gp$. Then p factors through $j: \mathrm{Ker}(h, g) \to \mathrm{Im}(f)$. Since j is a monomorphism, then by the definition of image we check that f factors through uj, or, equivalently, that j is an isomorphism, i.e. $g = h$. ✳

A triangular decomposition $X \xrightarrow{p_1} X_1 \xrightarrow{u_1} Y$ of a morphism f is called *minimal* if for every other triangular decomposition $X \xrightarrow{p_2} X_2 \xrightarrow{u_2} Y$ of f there is a morphism

$h: X_1 \to X_2$ such that $u_2h = u_1$ and $hp_1 = p_2$. It is easy to see that the morphism h is a bimorphism and that any two minimal triangular decompositions are isomorphic.

COROLLARY 19.3. *In a category with kernels and images every morphism has a minimal triangular decomposition. Moreover, in a complete wellpowered category, every morphism has a minimal triangular decomposition. If in addition the category is balanced, then every morphism has a unique triangular decomposition.*

Proof. We claim that the decomposition $X \xrightarrow{p} \mathrm{Im}(f) \xrightarrow{u} Y$ is a minimal triangular decomposition of f. Indeed, if

$$X \xrightarrow{p'} Z \xrightarrow{u'} Y$$

is another triangular decomposition of f, then we must have $\mathrm{Im}(f) \subseteq Z$, i.e. there is a morphism $h: \mathrm{Im}(f) \to Z$ such that $u'h = u$. But then $f = u'p' = up = u'hp$, which gives $p' = hp$. The remaining assertions follow from Propositions 18.2 and 19.1 and the above results. ✳

Of course, the above results can be dualized for coimages and triangular decompositions.

COROLLARY 19.4. *In a category with cokernels and coimages every morphism has a maximal triangular decomposition. Moreover, in a cocomplete wellcopowered category every morphism has a maximal triangular decomposition. If in addition the category is balanced, then every morphism has a unique triangular decomposition.* ✳

The notion of triangular decomposition is related to the existence of limits and colimits. This is illustrated by the following results.

THEOREM 19.5. (Pupier [1], Theorem 1.6.2). *Let $\mathscr{C}$ be a wellpowered and wellcopowered category with products and coproducts. The following are equivalent:*

1) *$\mathscr{C}$ is complete.*

2) *$\mathscr{C}$ is cocomplete.*

3) *Every morphism of $\mathscr{C}$ has a triangular decomposition.*

Proof. The implications 1) ⇒ 3) and 2) ⇒ 3) are given in Corollaries 19.3 and 19.4.

3) ⇒ 2). By Theorem 9.2 it will suffice to check that $\mathscr{C}$ is a category with fibered coproducts. Consider the coangle $X_1 \xleftarrow{f_1} Y \xrightarrow{f_2} X_2$, let $S = X_1 \coprod X_2$ and let $u_j: X_j \to S$, $j = 1, 2$, be structural injections. Denote by $\{q_i, S_i\}_{i \in \mathscr{I}}$ the set of all quotient objects of S defined such that $q_iu_1f_1 = q_iu_2f_2$. Let us put $Q = \coprod S_i$ and let $p_i: Q \to S_i$ be structural projections and $q: S \to Q$ the unique morphism such that $p_iq = q_i$ for all $i \in \mathscr{I}$. Since $p_iqu_1f_1 = p_iqu_2f_2$ for all $i \in \mathscr{I}$ we see that $qu_1f_1 = qu_2f_2$. Furthermore, let $q = up$ be a triangular decomposition of q. Obviously $pu_1f_1 = pu_2f_2$ since u is a monomorphism.

We shall prove that the diagram

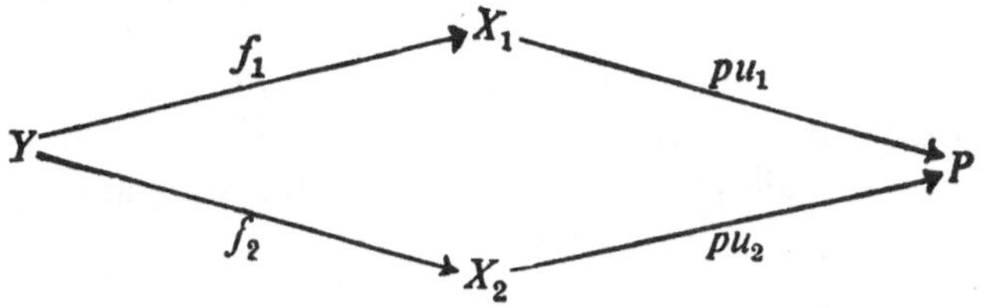

is cocartesian, where P is the codomain of p. To this end, let $g_j: X_j \to X, j = 1, 2$, be such that $g_1 f_1 = g_2 f_2$ and let $g: S \to X$ be the unique morphism such that $gu_j = g_j$, $j = 1, 2$. Thus, $gu_1 f_1 = g_1 f_1 = g_2 f_2$. Hence, if $u'p'$ is the triangular decomposition of g, then $p'u_1 f_1 = p'u_2 f_2$. So if P' is the codomain of p', then there is an index $i \in \mathscr{I}$ such that the epimorphisms p' and q_1 define the same quotient object of S. Consequently, we have that the diagram

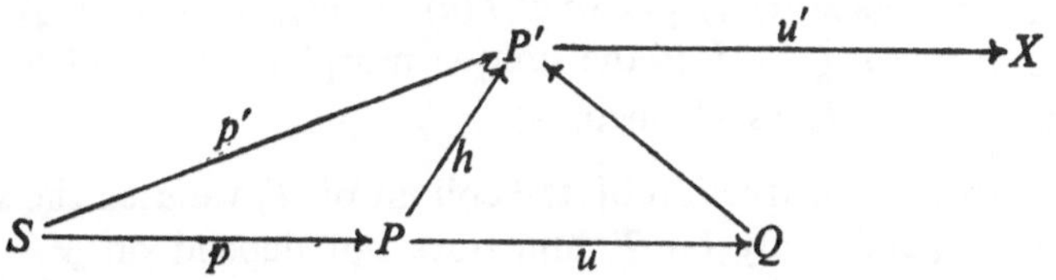

is commutative.

Furthermore, $u'hpu_j = u'p'u_j = gu_j = g_j$, $j = 1, 2$. The proof will be complete once we prove that $u'h$ is the unique morphism, which composed with pu_j, gives g_j, $j = 1, 2$. Indeed, if $g': P \to X$ is another morphism such that $g'pu_j = g_j$, $j = 1, 2$, then $u'hpu_j = g_j = g'pu_j$, $j = 1, 2$, i.e. $u'hp = g'p$ and finally $u'h = g'$. The implication 3) $\Rightarrow$ 1) is obtained by duality. ✳

It is clear that every small category is right bounded, so that a category with bounded colimits is a cocomplete category. Under certain conditions, the converse is also true.

THEOREM 19.6. *Let $\mathscr{C}$ be a cocomplete category, such that for each $X \in \mathscr{C}$ every class of quotient objects of Y has a lower bound. Then $\mathscr{C}$ is a category with bounded colimits.*

Proof. Let $\mathscr{I}$ be a right bounded category, $\mathscr{J}$ a right bounding subcategory and $T: \mathscr{I} \to \mathscr{C}$ a functor. In what follows, we give the construction of the colimit of this functor. Let $X = \coprod_{j \in \mathscr{J}} T(j)$ and $t_j: T(j) \to X$ be the structural morphisms. Consider the class $\mathscr{A}$ of all quotient objects (p, P) of X defined by the condition that for each $i \in \mathscr{I}$, for each $u: i \to j$ and for each $v: i \to k$, with $j, k \in \mathscr{J}$, we have $pt_j T(u) = pt_k T(v)$. Let $(q, Q) = \inf \{(p, P)\}_{(p, P) \in \mathscr{A}}$. We prove that $(q, Q) \in \mathscr{A}$, and consequently that $\mathscr{A}$ has a minimum element with respect to the ordering of quotient objects. Indeed, let $i \in \mathscr{I}$, $u: i \to j$, $v: i \to k$, $j, k \in \mathscr{J}$, and $(m, M) = \operatorname{Coker}(t_j T(u), t_k T(v))$. It is obvious that (m, M) is a quotient object of X and $(m, M) \leqslant (p, P)$ for every $(p, P) \in \mathscr{A}$. Thus, $(m, M) \leqslant (q, Q)$ and therefore $qt_j T(u) = qt_k T(v)$. Hence $(q, Q) \in \mathscr{A}$, as required.

Now for each $i \in \mathscr{I}$, let $f_i: T(i) \to Q$ be the morphism $qt_jT(u)$, where $u: i \to j$ is a morphism of $\mathscr{I}$ and $j \in \mathscr{J}$. By the above considerations we see that f_i does not depend on the choice of u or j. Obviously, the morphisms $\{f_i\}_{i \in \mathscr{I}}$ define a morphism f from T to Q. We claim that (f, Q) is the colimit of T. In order to prove this, let (g, Y) be a cocone over T. Then there is a morphism $\bar{g}: X \to Y$ such that $\bar{g}t_j = g_j$ for all $j \in \mathscr{J}$. Now consider the diagram

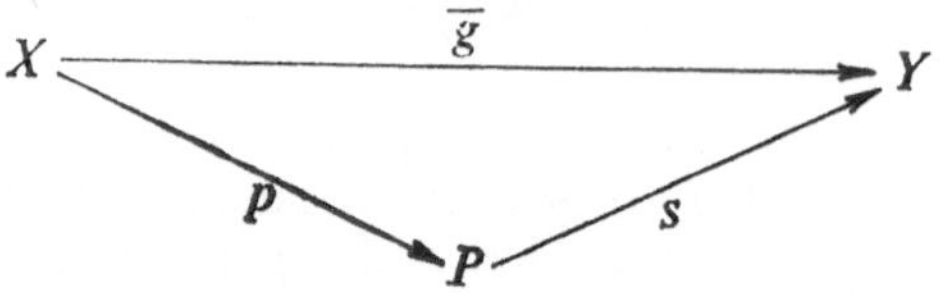

that gives the triangular factorization of $\bar{g}$ (see Exercise 19.10). It is easy to check that $(p, P) \in \mathscr{A}$, and thus there is a morphism $m: Q \to P$ such that $mq = p$. Thus for every $i \in \mathscr{I}$, we have $(sm)f_i = smqt_jT(u) = spt_jT(u) = \bar{g}t_jT(u) = g_jT(u) = g_i$. Finally, we see that $sm: Q \to Y$ is the unique morphism for which $(sm)f = g$, i.e. (f, Q) is the colimit of T, as claimed. ✳

Note 19.7. In the construction of the colimit of T, we used the subcategory $\mathscr{J}$, but this colimit depends only on T, and does not depend on $\mathscr{J}$.

COROLLARY 19.8. *Every cocomplete wellcopowered category is a category with bounded colimits.*

The proof follows from Corollary 19.4 and Theorem 19.6. ✳

Exercises

19.1. Let $f: X \to Y$ be a morphism in a balanced category and assume that f has an image. Prove: If $X \xrightarrow{p} X' \xrightarrow{v} Y$ is a triangular decomposition of f, then $X' = \mathrm{Im}(f)$ and $v = u_f$.

19.2. We say that the morphism $f: X \to Y$ has an *epimorphic image* if f has an image and the canonical morphism $p: X \to \mathrm{Im}(f)$ is an epimorphism. Show that the morphism f has an epimorphic image if and only if it has an image and a triangular decomposition.

A category in which every morphism has an epimorphic image is called a *category with epimorphic images*. Dually, w· have the notions of a *monomorphic coimage* and of a *category with monomorphic coimages*.

19.3. Let $\mathscr{C}$ be a balanced category with epimorphic images. Show that if $f: X \to Y$, $g: Y \to Z$ and X' is a subobject of X, then $g(f(X')) = gf(X')$.

19.4. Assume that the morphism $f\colon X \to Y$ in a category with fibered products has an image. Show that the canonical monomorphism $u\colon \mathrm{Im}(f) \to Y$ is a bimorphism if and only if f is an epimorphism.

19.5. Assume that $f\colon X \to Y$ is a morphism in a category with images and fibered products. Prove that the canonical morphism $p\colon X \to \mathrm{Im}(f)$ is an epimorphism. Thus, in such a category every morphism has a minimal triangular decomposition.

19.6. An epimorphism $f\colon X \to Y$ is called an *external epimorphism* if $\mathrm{Im}(f) = Y$ and $u_f = 1_Y$. Prove that retractions, strict epimorphisms and normal epimorphisms are examples of external epimorphisms and, moreover, if $f = up$ is a unique triangular decomposition of f, then p is an extremal epimorphism. Consider the category $\mathscr{C}$ whose objects are X_1, X_2, X_3, X_4, X_5; the morphisms of $\mathscr{C}$ are the identity morphisms and the morphisms pictured in the following diagram:

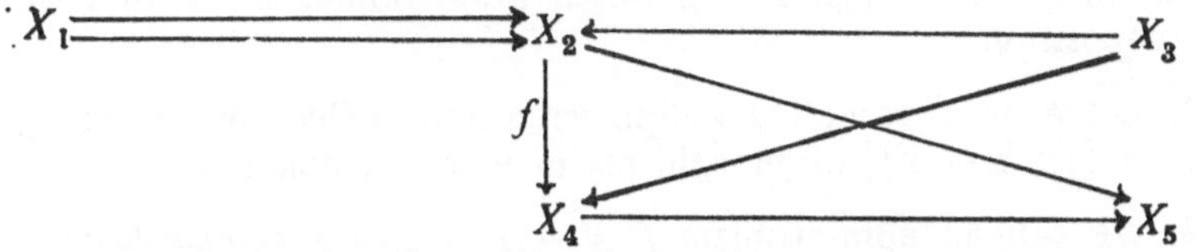

Assume that all occurring diagrams are commutative. Show that f is an extremal, but not a strict, epimorphism. Prove that an extremal bimorphism is an isomorphism and, moreover, that a category $\mathscr{C}$ is balanced if and only if every epimorphism of $\mathscr{C}$ is extremal.

19.7. Show that if the morphism $f\colon X \to Y$ has an epimorphic image, then p is an extremal epimorphism. Show that if moreover, $X \xrightarrow{p'} X' \xrightarrow{u'} Y$ is a triangular decomposition of f such that p' is an extremal epimorphism then it is isomorphic to the decomposition $X \xrightarrow{p} \mathrm{Im}(f) \xrightarrow{u} Y$, i.e. this is the unique triangular decomposition of f such that p is extremal.

19.8. Assume that the morphism $f\colon X \to Y$ has an epimorphic image and a monomorphic coimage. Show that there is a unique isomorphism $\bar{f}\colon \mathrm{Coim}(f) \to$ $\to \mathrm{Im}(f)$ such that the diagram

$$\begin{array}{ccc} X & \xrightarrow{\quad f \quad} & Y \\ {\scriptstyle q}\downarrow & & \uparrow{\scriptstyle u} \\ \mathrm{Coim}(f) & \xrightarrow{\quad \bar{f} \quad} & \mathrm{Im}(f) \end{array}$$

is commutative.

19.9. Let $f: X \to Y$ be a morphism of a finitely complete and cocomplete category $\mathscr{C}$. Prove that there is a commutative diagram:

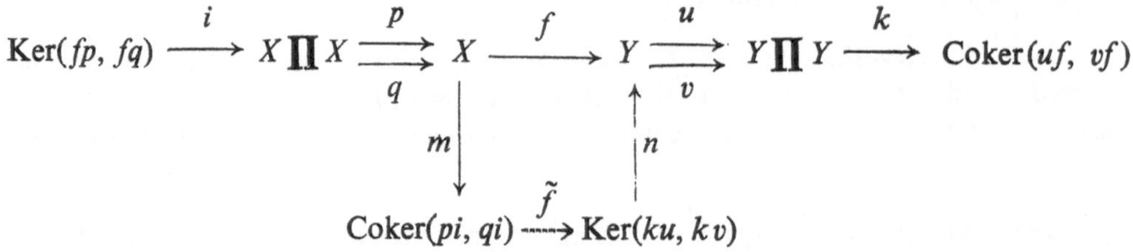

i
p
q
f
u
v
k
Ker(fp, fq)
X∐X
X
Y
Y∐Y
Coker(uf, vf)
m
n
f̃
Coker(pi, qi)
Ker(ku, kv)

where p, q are structural projections and u, v are structural injections. If $\tilde{f}$ is an isomorphism for every f, then $\mathscr{C}$ is called *perfect*. Prove that: the categories $\mathscr{S}et$, $\mathscr{G}r$ and $\mathscr{A}b$ are perfect; every perfect category is balanced, and every epimorphism (monomorphism) of a perfect category is effective.

19.10. Let $f: X \to Y$ be a morphism in a category with cokernels. Assume that every family of quotient objects of X has a lower bound. Show that f has a triangular decomposition.

19.11. Let $\mathscr{C}$ be a complete wellpowered and wellcopowered category. Prove that each set $\{f_i: X \to Y\}_i$ of morphisms of $\mathscr{C}$ has a cokernel.

19.12. We call an epimorphism $f: X \to Y$ a *strong epimorphism* if for every commutative diagram with u a monomorphism:

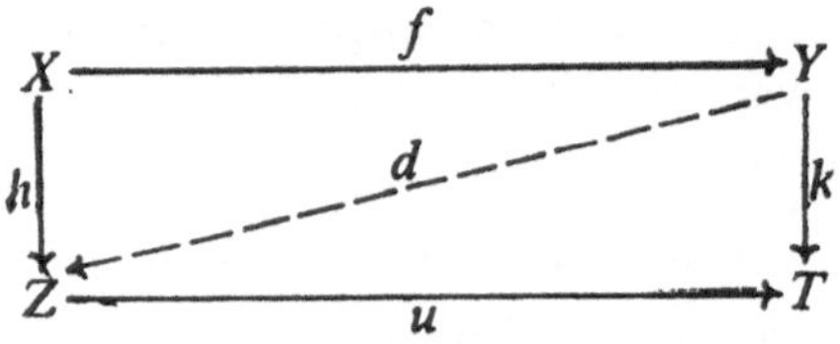

there exists a (necessarily unique) morphism $d: Y \to Z$ such that $ud = k$. (Then we obtain also that $df = h$.) Prove:

a) Every strict epimorphism is a strong epimorphism.

b) The class of all strong epimorphisms of a category is a replete multiplicative system.

c) If fg is a strong epimorphism, then f is a strong epimorphism.

d) Let $F, F': \mathscr{I} \to \mathscr{C}$ be two functors and let $u: F \to F'$ be a functorial morphism. Assume that for each i, $u_i: F_i \to F'_i$ is a strong epimorphism and that $\varinjlim F$, $\varinjlim F'$ are defined. Then the morphism $\varinjlim u$ is a strong epimorphism.

e) If f is both a strong epimorphism and a monomorphism, then it is an isomorphism.

f) Every strong epimorphism is an extremal epimorphism. Conversely, in a category with fibered products every strong epimorphism is an extremal epimorphism.

Dually, we have the notion of *strong monomorphism*. Also, such notions as *strong subobject* and *strong quotient object* are defined in an obvious way.

g) A cointersection of strong quotient objects is again a strong quotient object.

19.13. A *left (right) strict factorization* of a morphism f in a category $\mathscr{C}$ is a factorization $f = up$, where p is a strict epimorphism (u is a strict monomorphism) and were $ph = pk$ whenever $fh = fk$ ($hu = ku$ whenever $hf = kf$).

Prove:

a) Two left (right) strict factorizations of a morphism are canonically isomorphic.

b) If f has a left strict factorization $f = up$, then u is an isomorphism if and only if f is a strict epimorphism. Moreover, p is an isomorphism if and only if f is a monomorphism.

c) If $f: X \to Y$ has a left strict factorization $f = up$, then p defines the smallest strict quotient object of X through which f factors. Conversely, if there is a smallest strict quotient object $X \xrightarrow{p} X'$ of X through which f factors and $\mathscr{C}$ is a category with cokernels, then f has a left strict factorization $f = up$.

d) If $\mathscr{C}$ is a category with kernel pairs and cokernels, then every morphism has a left strict factorization.

e) If $\mathscr{C}$ is a category with cokernels and cointersections and every object has only one set of strict quotient objects, then every morphism of $\mathscr{C}$ has a left strict factorization.

19.14. A *left (right) strict decomposition* of a morphism f is a left (right) strict factorization $f = up$ such that u is a monomorphism (p is an epimorphism). Assume that every morphism of $\mathscr{C}$ has a left strict factorization. Show that the following are equivalent:

1) Strict epimorphisms are closed under composition.

2) Every left strict factorization is a left strict decomposition.

19.15. Show that every morphism in a perfect category has both a left and a right strict decomposition, and these decompositions are isomorphic.

19.16. A *left (right) strong decomposition* of a morphism f is a triangular decomposition $f = up$ where p is a strong epimorphism (u is a strong monomorphism). Prove:

a) Two left (right) strong decompositions of a morphism are canonically isomorphic.

b) If every morphism of $\mathscr{C}$ has a left strong decomposition, then every extremal epimorphism is a strong epimorphism.

c) If every epimorphism of $\mathscr{C}$ is a strong epimorphism, then every monomorphism is a strong monomorphism.

d) Let $\mathscr{C}$ be a category with left strict factorizations and cointersections. Given a morphism $f: X \to Y$ define for each ordinal α a factorization $f = u_\alpha p_\alpha$ in the following way:

(i) Let $p_0 = 1_X$ and $u_0 = f$.

(ii) Suppose u_α and p_α are defined and let $u_\alpha = up$ be a left strict factorization of u_α. Then $u_{\alpha+1} = u$ and $p_{\alpha+1} = pp_\alpha$.

(iii) Suppose α is a limit ordinal and let p_α be the cointersection of all p_β with $\beta < \alpha$. If p_α is stationary for $\alpha \geqslant \lambda$ (which will be the case if X has only one set of strong quotient objects), then $f = u_\alpha p_\alpha$ is a left strong decomposition of f.

19.17. Show that a category $\mathscr{C}$ is a category with left strong decompositions if either one of the following two assertions holds:

(i) $\mathscr{C}$ admits intersections, kernels and fibered products and is wellpowered.

(ii) $\mathscr{C}$ admits cointersections and cokernels, and each object has only one set of strong quotient objects.

19.18. A *coequalizer (equalizer) decomposition* of a morphism f is a triangular decomposition $f = up$ such that p is a coequalizer (u is an equalizer). Prove:

a) Every coequalizer decomposition is unique.

b) If every morphism of $\mathscr{C}$ has a coequalizer decomposition, then the class of all coequalizers of $\mathscr{C}$ is a multiplicative system.

c) Assume that $\mathscr{C}$ is a category with kernel pairs and that every kernel pair has a coequalizer. If the class of coequalizers is a multiplicative system, then every morphism of $\mathscr{C}$ has a coequalizer decomposition.

d) Let $\mathscr{C}$ be a category with fibered products. If the coequalizers of $\mathscr{C}$ are universal, then they form a multiplicative system.

19.19. Let $\mathscr{C}$ be a category and Σ a multiplicative system of morphisms of $\mathscr{C}$ such that $\mathscr{C}(\Sigma^{-1})$ is defined and the canonical functor is exact. Prove:

a) If $\mathscr{C}$ is perfect then so is $\mathscr{C}(\Sigma^{-1})$.

b) If every epimorphism (monomorphism) in $\mathscr{C}$ is universal (couniversal), then the same is true for $\mathscr{C}(\Sigma^{-1})$.

c) If every morphism of $\mathscr{C}$ has a coequalizer decomposition, then the same is true for $\mathscr{C}(\Sigma^{-1})$.

19.20. Let $F: \mathscr{I} \to \mathscr{C}$ be a functor, (h, X) a colimit of F and (g, Y) a cocone over F. Assume that the unique morphism $f: X \to Y$ such that $fh = g$ has a unique triangular decomposition $f = up$, and that also for every $i \in \mathscr{I}$, $g_i = u_i p_i$ is a unique triangular decomposition. Then show that the subobject of Y defined by u is the upper bound of the class of subobjects defined by u_i for $i \in \mathscr{I}$. Show that if, moreover, every morphism of $\mathscr{C}$ has a unique triangular decomposition and $\mathscr{C}$ is a category with coproducts (finite coproducts) then every set (finite set) of subobjects of an object has an upper bound.

1.20. Relative triangular decomposition of morphisms

In this section we define two relative triangular decompositions of a morphism, and attempt to show what these triangular decompositions imply about adjoints of certain functors.

References: Baron [2], Freyd [1], Isbell [8], Kennison [1], [2], [3], D. Popescu [1], Schubert [1].

Let $F: \mathscr{C} \to \mathscr{C}'$ be a functor, $X \in \mathscr{C}$ and $X' \in \mathscr{C}'$. We say that a morphism $f: X' \to F(X)$ is an *epimorphism relative to* F, or simply an *F-epimorphism*, if for every pair (u, v) where $u, v: X \to Y$ are morphisms of $\mathscr{C}$, the condition $F(u)f = F(v)f$ implies $u = v$. Dually, we have the notion of an *F-monomorphism*. In this way we will speak of the notions *F-quotient object* and *F-subobject* of an object X' of $\mathscr{C}'$. An F-quotient object of X' is a pair (f, X), where $X \in \mathscr{C}$ and $f: X' \to F(X)$ is an F-epimorphism.

We could define concepts such as lower bound and upper bound of a family of F-quotient objects (F-subobjects) of an object of $\mathscr{C}'$. For instance, we say that a family $\{f_i: X \to F(X_i)\}_i$ of F-quotient objects of X' has an *F-lower bound* if there is an F-quotient object $f: X' \to F(X)$ of X', and if for each index i a morphism $g_i: X \to X_i$ of $\mathscr{C}$ is defined such that $F(g_i)f = f_i$, and furthermore whenever $h: X \to F(Y)$ is another F-quotient object of X such that $F(Y) \subseteq F(X_i)$ for all i, then $F(Y) \subseteq F(X)$.

Lemma 20.1. *Let $F: \mathscr{C} \to \mathscr{C}'$ be a left exact functor. Assume that $\mathscr{C}$ is finitely complete, and $f: X' \to F(X)$ is a morphism. The following are equivalent:*

1) *f is an F-epimorphism.*

2) *If $h: Y \to X$ is a monomorphism such that f can be factored through $F(h): F(Y) \to F(X)$, then h is a bimorphism.*

Proof. 1) $\Rightarrow$ 2). Let $u, v: X \to Z$ be a pair of morphisms such that $uh = vh$ and let $p: X' \to F(Y)$ be a morphism such that $F(h)p = f$. Then $F(u)F(h)p = F(v)F(h)p$. Since f is an F-epimorphism, we get that $u = v$.

2) $\Rightarrow$ 1). Let $u, v: X \to Z$ be such that $F(u)f = F(v)f$. Consider the following cartesian square in $\mathscr{C}$:

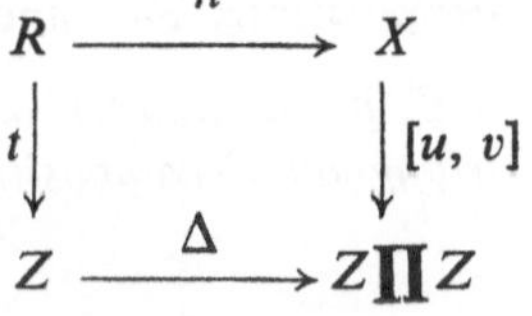

where Δ is the diagonal morphism. Furthermore, consider the cartesian square in $\mathscr{C}'$

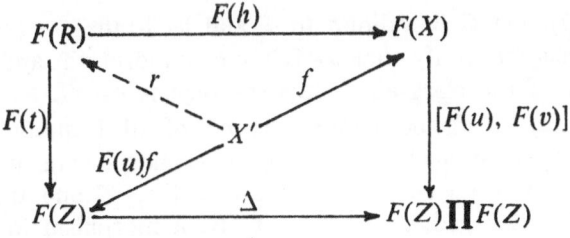

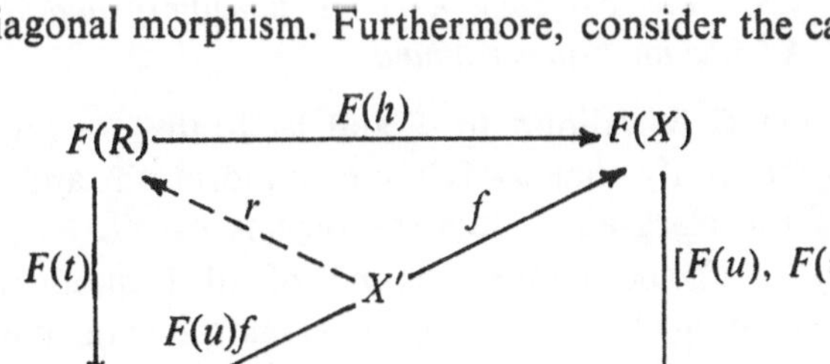

where r is canonically defined, such that all the subdiagrams are commutative. But then we get that h is a bimorphism. Hence the equality $[u, v]h = \Delta t$ implies that $u = v$. Thus f is an F-epimorphism. ✳

We say that the morphism $f\colon X' \to F(X)$ of $\mathscr{C}'$ has an *F-triangular decomposition* if there exists a monomorphism $j\colon Y \to X$ of $\mathscr{C}$ such that $f = F(f)p$, where $p\colon X' \to F(Y)$. A morphism f may have several F-triangular decompositions. Nevertheless, in some categories a morphism has in some sense a unique F-triangular decomposition. Two F-triangular decompositions $X' \xrightarrow{p_1} F(Y_1) \xrightarrow{F(j_1)} F(X)$. $X' \xrightarrow{p_2} F(Y_2) \xrightarrow{F(j_2)} F(X)$ of a morphism $f\colon X' \to F(X)$ are *F-isomorphic* if there is an isomorphism $v\colon Y_1 \to Y_2$ such that $F(v)p_1 = p_2$ and $j_2 v = j_1$.

PROPOSITION 20.2. *Let $F\colon \mathscr{C} \to \mathscr{C}'$ be a functor that commutes with limits. Assume that $\mathscr{C}$ is finitely complete and that every family of subobjects of an object of $\mathscr{C}$ has an intersection. Then every morphism $f\colon X' \to F(X)$ in $\mathscr{C}'$ has an F-triangular decomposition. Moreover, if $\mathscr{C}$ is balanced, then every morphism $f\colon X' \to F(X)$ has a unique triangular decomposition.*

Proof. Let $\mathscr{M}$ be the family of all subobjects $j\colon Y \to X$ of X such that f can be factored through $F(j)$ and let $j\colon P \to X$ be the intersection of this family of subobjects. It is easy to see that $(P, j) \in \mathscr{M}$ (since F commutes with colimits) and that the unique morphism $p\colon X' \to F(P)$ such that $F(j)p = f$ is an F-epimorphism. Hence $X' \xrightarrow{p} F(P) \xrightarrow{F(j)} F(X)$ is an F-triangular decomposition of f. The last part of the proposition follows from Lemma 20.1. ✳

COROLLARY 20.3. *Let $F\colon \mathscr{C} \to \mathscr{C}'$ be a functor which commutes with limits. Assume that $\mathscr{C}$ is a wellpowered complete category. Then every morphism $f\colon X \to F(X)$ has an F-triangular decomposition. Moreover, if $\mathscr{C}$ is balanced, then every morphism $f\colon X' \to F(X)$ has a unique triangular decomposition.*

The proof follows from Propositions 17.1 and 20.2. ✳

LEMMA 20.4. *Let $F\colon \mathscr{C} \to \mathscr{C}'$ be a functor and assume that every morphism $f\colon X' \to F(X)$ in $\mathscr{C}'$ has an F-triangular decomposition. Then the following are equivalent:*

1) *F has an adjoint.*

2) *For every $X' \in \mathscr{C}'$, the category X'/F is nonempty and the family of all F-quotient objects of X' has an F-lower bound.*

Proof. 1) ⇒ 2). Let G be adjoint to F and let $u\colon \mathrm{Id}\mathscr{C}' \to FG$ be an arrow of adjunction. It is easy to verify that u_X is an F-epimorphism and that $(u_{X'}, G(X'))$ is a lower bound of the class all F-quotient objects of X'.

2) ⇒ 1). Let $(u_X, G(X))$ be a lower bound of all F-quotient objects of X'. It is clear that $G(X')$ is uniquely defined (up to isomorphism). We shall prove that the assignment $X' \rightsquigarrow G(X')$ defines a functor $G\colon \mathscr{C}' \to \mathscr{C}$ and that this functor G is adjoint to F. To this end, let $f\colon X' \to Y'$ be a morphism in $\mathscr{C}'$. Consider the

diagram

$$\begin{array}{ccc} X' & \xrightarrow{u_{X'}} & FG(X') \\ f\downarrow & & \\ Y' & \xrightarrow{u_{Y'}} & FG(Y') \end{array}$$

and let $X' \xrightarrow{p} F(X) \xrightarrow{F(u)} FG(Y')$ be an F-triangular decomposition of $u_{Y'}f$. Now, since p is an F-epimorphism, there is a (necessarily unique) morphism $h\colon G(X') \to X$ such that $F(h)u_{X'} = p$. But then $F(uh)u_{X'} = u_{Y'}f$. Put $uh = G(f)$. It is clear that $G(f)$ is uniquely determined (since $u_{X'}$ is an F-epimorphism), so that the assignments $X' \rightsquigarrow G(X')$ and $f \rightsquigarrow G(f)$ define a functor $G\colon \mathscr{C}' \to \mathscr{C}$ which is adjoint to F. The remaining details are left for the reader. ⁜

COROLLARY 20.5. *Let $F\colon \mathscr{C} \to \mathscr{C}'$ be a functor that commutes with limits. Assume that $\mathscr{C}$ is complete and wellpowered, the family of all F-quotient objects of any object X' of $\mathscr{C}'$ is a set, and the category X'/F is nonempty. Then F has an adjoint.*

Proof. By Proposition 20.2, every morphism $f\colon X' \to F(X)$ in $\mathscr{C}'$ has an F-triangular decomposition. Let $\{f_i\colon X' \to F(X_i)\}_{i\in\mathscr{I}}$ be the set of all F-quotient objects of X' and let $f = [f_i]_i\colon X' \to \prod_i F(X_i) = F(\prod_i X_i)$. If $X' \xrightarrow{p} F(Y) \xrightarrow{F(u)} F(\prod_i X_i)$ is an F-triangular decomposition of f, then (p, Y) is a lower bound of the family of all F-quotient objects of X'. Now the proof follows from Lemma 20.4. ⁜

A subcategory $\mathscr{C}'$ of a category $\mathscr{C}$ is called *replete in* $\mathscr{C}$, or simply *replete*, if $X \in \mathscr{C}'$ and Y is equivalent to X imply $Y \in \mathscr{C}'$.

COROLLARY 20.6. (Freyd [1]). *Let $\mathscr{C}$ be a complete wellpowered and wellcopowered category that has a final object. Let $\mathscr{C}'$ be a full replete subcategory of $\mathscr{C}$ such that $\mathscr{C}'$ is closed under products and subobjects. Then $\mathscr{C}'$ is a reflective subcategory of $\mathscr{C}$.*

Proof. We use Lemma 20.4. By Theorem 9.1 we get that $\mathscr{C}'$ is complete, and the inclusion functor $I\colon \mathscr{C}' \to \mathscr{C}$ commutes with limits. It is obvious that $\mathscr{C}'$ is wellpowered.

Now let $f\colon Y \to I(X')$ be a morphism in $\mathscr{C}$ and let $X \xrightarrow{p} Y \xrightarrow{u} I(X')$ be a triangular decomposition of f. Since $\mathscr{C}'$ is closed under subobjects and replete, we get that $Y = I(X_1')$ and $u = I(u')$. Hence $p\colon X \to I(X_1')$ is an I-epimorphism and an epimorphism of $\mathscr{C}$. Furthermore, let $\{p_i\colon X \to I(X_1')\}_i$ be the set of all quotient objects of X that lie in $\mathscr{C}'$ and let $p = [p_i]\colon X \to \prod_i I(X_i') = I(\prod_i X_i')$. Also, let $X \xrightarrow{q} I(Y') \xrightarrow{I(r)} I(\prod_i X_i')$ be an I-triangular decomposition of p. It is easy to see that $(q, I(Y'))$ is a lower bound of all I-quotient objects of X. Now the proof follows from Lemma 20.4. ⁜

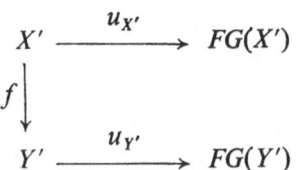

By an *epireflective (monoreflective)* subcategory of a category $\mathscr{C}$, we mean a reflective subcategory $\mathscr{C}'$ such that for each $X \in \mathscr{C}$, the arrow of reflection $X \to \overline{X}$ is an epimorphism (a monomorphism). It is easy to see that a reflective subcategory $\mathscr{C}'$ which is obtained from Corollary 20.6 is an epireflective subcategory. This is closely related to the fact that $\mathscr{C}'$ is closed under subobjects. In the following discussion, we consider a sort of triangular decomposition which permits us to generalize Corollary 20.6.

A *replete (*or *gorged!) multiplicative system* of a category $\mathscr{C}$ is a multiplicative system Σ that contains all isomorphisms of $\mathscr{C}$.

Let $\mathscr{I}$ and $\mathscr{P}$ be replete multiplicative systems of a category $\mathscr{C}$. We say that a morphism f in $\mathscr{C}$ has an $(\mathscr{I}, \mathscr{P})$-triangular decomposition if $f = f_1 f_0$, where $f_1 \in \mathscr{I}$ and $f_0 \in \mathscr{P}$. Moreover, this factorization is unique up to an equivalence in the sense that if $f = gh$ and $g \in \mathscr{I}$, $h \in \mathscr{P}$, then there exists an isomorphism u such that $uf_0 = h$ and $gu = f_1$.

A *right bicategory structure* on $\mathscr{C}$ is a pair $(\mathscr{I}, \mathscr{P})$ of replete multiplicative systems of morphisms in $\mathscr{C}$ such that:

B_1) Every morphism in $\mathscr{C}$ has a unique $(\mathscr{I}, \mathscr{P})$-triangular decomposition.

B_2) Every element of $\mathscr{P}$ is an epimorphism.

We obtain the concept of *left bicategory structure* if B_2) is replaced by the following condition:

B_3) Every element of $\mathscr{I}$ is a monomorphism.

Finally, by a *bicategory structure on* $\mathscr{C}$ we mean a pair $(\mathscr{I}, \mathscr{P})$ which is simultaneously a right and a left bicategory structure.

Let $\mathscr{A}$ be a class of morphisms in a category $\mathscr{C}$. An *$\mathscr{A}$-subobject* of an object X of $\mathscr{C}$ is a pair (X', f), where $f \in \mathscr{A}$. Two $\mathscr{A}$-subobjects (X', f) and (X_1', f_1) are equal if and only if there exists an isomorphism $u: X' \to X_1'$ such that $f_1 u = f$. The notion of an *$\mathscr{A}$-quotient object* is defined by duality. The notions of an $\mathscr{A}$-wellpowered object, an $\mathscr{A}$-wellpowered category, and also their duals are defined in the obvious manner. Also the notions of lower bound of a family of $\mathscr{A}$-subobjects, etc. have the obvious meanings.

LEMMA 20.7. *Let $\mathscr{C}$ be a category and let $(\mathscr{I}, \mathscr{P})$ be a right bicategory structure on $\mathscr{C}$. The following are equivalent for a full replete subcategory $\mathscr{C}'$ of $\mathscr{C}$:*

1) *$\mathscr{C}'$ is an epireflective subcategory of $\mathscr{C}$ and all arrows of reflection belong to $\mathscr{P}$.*

2) *For each $X \in \mathscr{C}$ the category $X/\mathscr{C}'$ is nonempty, the family of all $\mathscr{P}$-quotient objects (f, Y) of X with $Y \in \mathscr{C}'$ has a lower bound, and any $\mathscr{I}$-subobject of an object of $\mathscr{C}'$ also belongs to $\mathscr{C}'$.*

Proof. 1) $\Rightarrow$ 2). If $u_X: X \to X'$ is an arrow of reflection, then it is quite clear that this $\mathscr{P}$-quotient object of X is the lower bound of all $\mathscr{P}$-quotient objects (f, Y) of X with $Y \in \mathscr{C}'$.

Furthermore, let $X \in \mathscr{C}'$ and let $f: Y \to X$ be a morphism in $\mathscr{I}$. We must prove that $Y \in \mathscr{C}'$. To do this, consider the commutative diagram

$$\begin{array}{ccc} Y & \xrightarrow{u_Y} & Y' \\ {\scriptstyle f}\downarrow & & \downarrow{\scriptstyle f'} \\ X & \xrightarrow{u_X} & X' \end{array} .$$

Let $f' = f_1' f_0'$ be an $(\mathscr{I}, \mathscr{P})$-decomposition of f'. Then $f_1' f_0' u_Y = u_X f \in \mathscr{I}$, so that $f_0' u_Y$ is necessarily an isomorphism. Hence u_Y is also an isomorphism. Therefore Y is an object of $\mathscr{C}'$.

2) $\Rightarrow$ 1). Let $X \in \mathscr{C}$ and let $f: X \to Y$ be a morphism such that $Y \in \mathscr{C}'$. Then $f = f_1 f_0$ where $f_1: Z \to Y$ and $f_1 \in \mathscr{I}$. But then $Z \in \mathscr{C}'$, so that the family of all P-quotient objects (h, Z) with $Z \in \mathscr{C}'$ is nonempty. Let (u_X, X') be a lower bound of this family of quotient objects. We claim that u_X is an arrow of reflection. To see this, let $g: X \to Y$ be a morphism in $\mathscr{C}$ such that $Y \in \mathscr{C}'$. But then g can be decomposed as $X \xrightarrow{g_0} Y' \xrightarrow{g_1} Y$ with $g_0 \in \mathscr{P}$ and $g_1 \in \mathscr{I}$. Now, by hypothesis, $Y' \in \mathscr{C}'$, so that there is a morphism $p: X' \to Y'$ such that $p u_X = g_0$. Finally, it is clear that $g_1 p: X' \to Y$ is the unique morphism such that $g_1 p u_X = g$, in other words, u_X is an arrow of reflection. ∗

THEOREM 20.8. *Let $\mathscr{C}$ be a category with products and $(\mathscr{I}, \mathscr{P})$ a right bicategory structure on $\mathscr{C}$ such that $\mathscr{C}$ is $\mathscr{P}$-wellcopowered. The following are equivalent for a full replete subcategory $\mathscr{C}'$ of $\mathscr{C}$:*

1) *$\mathscr{C}$ is reflective and all arrows of reflection belong to $\mathscr{P}$.*

2) *$\mathscr{C}'$ is closed under products, for every $X \in \mathscr{C}$ the category $X/\mathscr{C}'$ is nonempty, and any $\mathscr{I}$-subobject of an object of $\mathscr{C}'$ also belongs to $\mathscr{C}'$.*

Proof. The implication 1) $\Rightarrow$ 2) follows from Lemma 20.7 and Proposition 13.12.

2) $\Rightarrow$ 1). As in the proof of Lemma 20.7, we see that for each $X \in \mathscr{C}$ there exist $\mathscr{P}$-quotient objects (p, X') where $X' \in \mathscr{C}'$. Let $\{p_i, X_i\}_{i \in \mathscr{I}}$ be the set of all $\mathscr{P}$-quotient objects of X such that $X_i \in \mathscr{C}'$ for all $i \in \mathscr{I}$. Also let $q = [p_i]: X \to \to \coprod_i X_i$ and let $q = q_1 q_0$ be an $(\mathscr{I}, \mathscr{P})$-decomposition of q with $q_0: X \to X'$. By hypothesis we see that $X' \in \mathscr{C}'$. We leave it for the reader to show that (q_0, X') is a lower bound of the family $\{p_i, X_i\}_{i \in \mathscr{I}}$. Now the proof follows from Lemma 20.7. ∗

COROLLARY 20.9. *Let $\mathscr{C}$ be a wellcopowered category. Assume that $\mathscr{C}$ is a category with products, and that every morphism in $\mathscr{C}$ has a unique triangular decomposition. Then a full replete subcategory $\mathscr{C}'$ of $\mathscr{C}$ is epireflective if and only if $\mathscr{C}'$ is closed under subobjects and products; furthermore, for every $X \in \mathscr{C}$ the category $X/\mathscr{C}'$ is nonempty.* ∗

COROLLARY 20.9 enables us to determine all full replete subcategories of $\mathscr{S}et$ that are reflective.

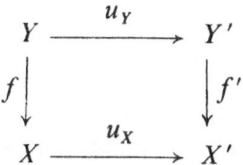

Let $\mathscr{C}$ be a full, replete and reflective subcategory of $\mathscr{S}et$. Two cases are possible:

a) Each object of $\mathscr{C}$ contains no more than one element (i.e. each object of $\mathscr{C}$ is empty or it is isomorphic to a set with one element).

b) There are objects of $\mathscr{C}$ that contain more than one element. Then we claim that every arrow of reflection is a surjection. Indeed, let X be a nonempty set such that $u_X: X \to X'$, the corresponding arrow of reflection is not a surjection. Further, let Y be an object of $\mathscr{C}$ such that Y contains at least two elements, and let $f: X \to Y$ be a map. It is clear that we can find two distinct maps $g, g': X' \to Y$ such that $gu_X = g'u_X = f$, which is a contradiction.

Hence it follows from Corollary 20.9 that $\mathscr{C}$ is necessarily equal to $\mathscr{S}et$. Furthermore, if X is a set with exactly one element, then the arrow of reflection u_X is clearly an isomorphism, so that u_X is an isomorphism for every $X \in \mathscr{S}et$ (since every set is a coproduct of sets which consist of one element). We conclude with the following corollary.

COROLLARY 20.10. *Let $\mathscr{C}$ be a full, replete and reflective subcategory of $\mathscr{S}et$. Then the functor of reflection $F: \mathscr{S}et \to \mathscr{C}$ is an equivalence of categories, or $\mathscr{C}$ is equivalent to the category arrow (that is, every object of $\mathscr{C}$ contains no more than one element).* ✳

Exercises

20.1. Let $F: \mathscr{C} \to \mathscr{C}'$ be a functor. We say that a morphism $f: X' \to F(X)$ *generates* X if each monomorphism $u: X_1 \to X$ such that f may be factored through $F(u): F(X_1) \to F(X)$ must necessarily be an isomorphism. Show that the following hold:

a) If (f, X) is an initial object of X'/F, then $F: X' \to F(X)$ is an F-epimorphism and generates X.

b) Assume that $\mathscr{C}$ is a category with kernels and F is left exact. Then a morphism $f: X' \to F(X)$ which generates X is an F-epimorphism. Conversely, if $\mathscr{C}$ is a balanced category then an F-epimorphism $f: X' \to F(X)$ generates X.

20.2. Let $F: \mathscr{C} \to \mathscr{C}'$ be a functor that commutes with limits. Assume that $\mathscr{C}$ is wellpowered and complete. Let X' be an object of $\mathscr{C}'$ such that the category X'/F is nonempty and is a left bounded category. Prove that the class of all F-quotient objects of X has an F-lower bound and that this lower bound is an initial object of X'/F.

20.3. a) Let $\mathscr{C}'$ be a subcategory of $\mathscr{C}$. Prove that there is a smallest full replete subcategory of $\mathscr{C}$ that contains $\mathscr{C}'$.

b) Assume that $\mathscr{C}$ is a category as in Corollary 20.6 and that $\mathscr{C}'$ is a full subcategory of $\mathscr{C}$ closed under products. Denote by $\mathscr{C}''$ the smallest full replete subcategory of $\mathscr{C}$ that contains all subobjects of every object of $\mathscr{C}'$. Show that $\mathscr{C}''$ is a reflective subcategory of $\mathscr{C}$.

20.4. Let $\mathscr{C}'$ be a subcategory of a category $\mathscr{C}$. Assume that $\mathscr{C}$ is a category with products. Prove that there is a smallest full subcategory $\mathscr{C}''$ of $\mathscr{C}$ that is closed under products and such that $\mathscr{C}''$ contains $\mathscr{C}'$. (Hint: There are two ways to construct $\mathscr{C}''$. On the one hand, $\mathscr{C}''$ is the intersection of all subcategories $\mathscr{C}$ of $\mathscr{C}'$ that are full and closed under products. On the other hand, we may use Exercise 7.8 and Proposition 5.5 to define $\mathscr{C}''$.)

20.5. Assume that $\mathscr{C}$ is a category as in Corollary 20.6. Prove that every full subcategory of $\mathscr{C}$ may be embedded in a smallest reflective replete subcategory of $\mathscr{C}$ that is closed under subobjects.

20.6. Use Corollary 20.6 to prove that the following hold:

a) $\mathscr{H}d$ is a reflective subcategory of $\mathscr{T}op$.

b) $\mathscr{A}b$ is a reflective subcategory of $\mathscr{G}r$.

c) The arrow category is equivalent to a full reflective subcategory of $\mathscr{S}et$.

d) $\mathscr{S}et$ is a full reflective subcategory of $\mathscr{T}op$.

20.7. Let $\mathscr{C}$ be a wellpowered complete category and let $F: \mathscr{C} \to \mathscr{C}'$ be a representative functor that commutes with limits. Prove that F has an adjoint.

20.8. Let $(\mathscr{I}, \mathscr{P})$ be a right bicategory structure on $\mathscr{C}'$. Prove:

1) $f \in \mathscr{I}$ if and only if f_0 is an isomorphism and $f \in \mathscr{P}$ if and only if f_1 is an isomorphism. Thus $\mathscr{I} \cap \mathscr{P}$ is precisely the class of all isomorphisms of $\mathscr{C}$.

2) $fg \in \mathscr{I}$ implies $g \in \mathscr{I}$.

3) If $gh \in \mathscr{P}$ and h is an epimorphism then $g \in \mathscr{P}$.

4) $\mathscr{I} = \{g | g = fh$ and $h \in \mathscr{P}$ implies h is an isomorphism$\}$. Thus $\mathscr{I}$ is uniquely determined by $\mathscr{P}$.

5) $\mathscr{P} = \{f | f = gh$, h epimorphism and $g \in \mathscr{I}$ implies g is an epimorphism$\}$. Thus $\mathscr{P}$ is uniquely determined by $\mathscr{I}$.

6) If $\{f_i\}_i$ is a set of morphisms in $\mathscr{I}$ (in $\mathscr{P}$), then $\coprod_i f_i$ ($\prod_i f_i$) — if it exists! — belongs to $\mathscr{I}$ (to $\mathscr{P}$).

20.9. Let $\mathscr{C}$ be a category with products, $(\mathscr{I}, \mathscr{P})$ be a right bicategory structure on $\mathscr{C}$ such that $\mathscr{C}$ is $\mathscr{P}$-wellcopowered, $\mathscr{C}'$ be a full subcategory of $\mathscr{C}$ that is closed under products, and $\mathscr{C}_1'$ be the full subcategory of all $\mathscr{I}$-subobjects of members of $\mathscr{C}'$. Assume, further, that $X/\mathscr{C}'$ is nonempty for all $X \in \mathscr{C}$. Prove that $\mathscr{C}_1'$ is the smallest reflective subcategory of $\mathscr{C}$ that contains $\mathscr{C}'$ and every arrow of reflection belongs to $\mathscr{P}$.

20.10. Let $\mathscr{C}$ be a cocomplete category. Let $\mathscr{P}$ be a replete multiplicative system of epimorphisms in $\mathscr{C}$ for which $\mathscr{C}$ is $\mathscr{P}$-wellcopowered. Prove that there exists a unique replete multiplicative system $\mathscr{I}$ such that $(\mathscr{I}, \mathscr{P})$ is a right bicategory structure on $\mathscr{C}$, if for every set $\{f_i\}_i$ of elements of $\mathscr{P}$ we have that $\coprod_i f_i \in \mathscr{P}$.

20.11. Prove that a full replete subcategory $\mathscr{C}$ of $\mathscr{T}op$ is coreflective if and only if the following three conditions hold:

(i) $\mathscr{C}$ contains nonempty spaces.

(ii) If $X \in \mathscr{C}$ then every quotient space (in the topological sense) of X also belongs to $\mathscr{C}$.

(iii) $\mathscr{C}$ is closed under products.

Show that, moreover, for every $Y \in \mathscr{T}\!op$, the coarrow of coreflection $Y' \to Y$ is a bijection.

20.12. Let $\mathscr{C}$ be a reflective full replete subcategory of $\mathscr{T}\!op$. Prove that:

a) The arrows of reflection are bijective maps if and only if $\mathscr{C}$ is closed under products, closed under subobjects and contains all indiscrete spaces.

b) The arrows of reflection are surjective maps if and only if $\mathscr{C}$ is closed under products and subobjects.

c) Assume that every space of $\mathscr{C}$ is Hausdorff. Then every arrow of reflection $u_X: X \to X'$ is an embedding (i.e. $u_X(X)$ is dense in X') if and only if $\mathscr{C}$ is closed under products and an object Y belongs to $\mathscr{C}$ whenever Y is a closed subset with the relative topology of some $X \in \mathscr{C}$.

20.13. Prove that if $X \xrightarrow{q} Y \xrightarrow{u} Z$ is a unique triangular decomposition of a morphism, then $u: Y \to Z$ is an extremal subobject of Z. Let $\mathscr{C}$ be a complete wellpowered and wellcopowered category. Show that a full subcategory $\mathscr{C}'$ of $\mathscr{C}$ is an epireflective subcategory of $\mathscr{C}$ if and only if it is closed under products and extremal subobjects in $\mathscr{C}$.

20.14. Let $\mathscr{C}$ be a category as in Exercise 20.13 and let $\mathscr{A}$ be a subcategory of $\mathscr{C}$. Let $\mathscr{C}'$ denote the full subcategory of $\mathscr{C}$ whose objects are the extremal subobjects in $\mathscr{C}$ of products of objects of $\mathscr{A}$. Furthermore, denote by $\mathscr{C}''$ the subcategory of $\mathscr{C}'$ whose objects are the extremal subobjects of $\mathscr{C}'$ of products of objects of $\mathscr{A}$. Prove that:

1) $\mathscr{C}'$ is the smallest epireflective subcategory of $\mathscr{C}$ that contains $\mathscr{A}$.

2) If $\mathscr{D}$ is any reflective subcategory of $\mathscr{C}$ that contains $\mathscr{A}$, then $\mathscr{D}$ also contains $\mathscr{C}''$. Thus a necessary condition for $\mathscr{A}$ to be reflective is that $\mathscr{A} = \mathscr{C}''$.

3) If $\mathscr{C}'$ is wellcopowered, then $\mathscr{C}''$ is the smallest reflective subcategory of $\mathscr{C}$ that contains $\mathscr{A}$.

Chapter 2

Completion of categories

2.1. Proper functors

Let $F, G: \mathscr{C} \to \mathscr{S}et$ be functors. We know that the class $[F, G]$ of all functorial morphisms from F into G is not always a set, so that we can not speak in general about the category of all functors from $\mathscr{C}$ into $\mathscr{S}et$. But instead of all functors from $\mathscr{C}$ into $\mathscr{S}et$, we can consider only so-called proper functors, which were defined by Isbell [3]. In this way we can define a category $\{\mathscr{C}, \mathscr{S}et\}$, the category of all proper functors from $\mathscr{C}$ into $\mathscr{S}et$. This category contains $\mathscr{C}$ as a full subcategory and it coincides with the category $[\mathscr{C}, \mathscr{S}et]$ in the case when $\mathscr{C}$ is a small category.

References: Bénabou [1], Freyd [1], Isbell [3], Lambeck [1], Gabriel—Ulmer [1].

Let $F: \mathscr{C} \to \mathscr{S}et$ be a functor. Denote by $\mathscr{P}_F$ the pointed category associated with F. Recall that the objects of $\mathscr{P}_F$ are pairs (X, x) where $X \in \mathscr{C}$ and $x \in F(X$. and a morphism $f: (X, x) \to (Y, y)$ of $\mathscr{P}_F$ is by definition a morphism $f: X \to Y$ of $\mathscr{C}$ such that $F(f)(x) = y$.

THEOREM 1.1. *Let $F: \mathscr{C} \to \mathscr{S}et$ be a functor. The following are equivalent:*

1) *The category $\mathscr{P}_F$ is left bounded.*

2) *There is a set $\{X_i\}_{i \in \mathscr{I}}$ of objects of $\mathscr{C}$ and a functorial epimorphism*

$$\coprod_{i \in \mathscr{I}} H^{X_i} \xrightarrow{u} F.$$

3) *There is a set $\{X_i\}_{i \in \mathscr{I}}$ of objects of $\mathscr{C}$ such that for each $X \in \mathscr{C}$* we have hat

$$F(X) = \bigcup_{\substack{i \in \mathscr{I} \\ f \in \mathscr{C}(X_i, X)}} F(f)(F(X_i)).$$

Proof. 1) $\Rightarrow$ 2). By the definition of a left bounded category, there is a set $\{(X_i, x_i)\}_{i \in \mathscr{I}}$ of objects of $\mathscr{P}_F$ such that for each object (X, x) of $\mathscr{P}_F$ one has an index $i \in \mathscr{I}$ and a morphism $f: X_i \to X$ such that $F(f)(x_i) = x$. Now consider the morphism $\coprod_i H^{X_i} \xrightarrow{u = \langle u^i \rangle_i} F$, where u^i is the functorial morphism associated with x_i. By the dual of Proposition 10.5, to prove that u is a functorial epimorphism, it will suffice to show that u is an argumentwise epimorphism, i.e. for each

$X \in \mathscr{C}$, the map u_X is a surjection. To show that let $x \in F(X)$; then $x = F(f)(x_i)$ where $i \in \mathscr{I}$ and $f \in \mathscr{C}(X_i, X) = H^{X_i}(X)$, hence $u_X(f) = u_X^i(f) = F(f)(x_i) = x$.

The other implications are left for the reader. ✳

A functor $F: \mathscr{C} \to \mathscr{S}et$ as in the previous theorem is called a *proper* functor. The set $\{X_i\}_{i \in \mathscr{I}}$ of objects of $\mathscr{C}$ stated in Theorem 1.1 is called a *dominant set* for F, or a set that *dominates* F. A set $\{(X_i, x_i)\}_i$ of objects of $\mathscr{P}_F$ is a *system of generators* for F if for each $(X, x) \in \mathscr{P}_F$ there is an index i, such that $\mathscr{P}_F((X_i, x_i)), (X, x) \neq \emptyset$. If α is a cardinal number, the functor F will be called *α-generated* if there exists a system of generators $\{(X_i, x_i)\}_{i \in \mathscr{I}}$ for F such that $\mathrm{Card}(\mathscr{I}) \leqslant \alpha$. In particular, a functor F is said to be *finitely generated* if it is α-generated where α is a finite cardinal.

Example 1.2. For each $X \in \mathscr{C}$ the functor H^X is proper. Moreover the set $\{X\}$ dominates H^X and $(X, 1_X)$ is a system of generators for H^X.

Example 1.3. If $\mathscr{C}$ is a small category, then every functor $F: \mathscr{C} \to \mathscr{S}et$ is proper.

LEMMA 1.4. *Let $F, G: \mathscr{C} \to \mathscr{S}et$ be functors, let $\{X_i\}_{i \in \mathscr{I}}$ be a set of objects of $\mathscr{C}$ which dominates F, and let u, v: $F \to G$ be functorial morphisms. The following are equivalent:*

a) $u = v$.

b) $u_{X_i} = v_{X_i}$ *for each* $i \in \mathscr{I}$.

Proof. b) $\Rightarrow$ a). Let $X \in \mathscr{C}$ and $x \in F(X)$. Since $\{X_i\}_{i \in \mathscr{I}}$ dominates F, there are an $i \in \mathscr{I}$, $f: X_i \to X$ and $x_i \in F(X_i)$ such that $F(f)(x_i) = x$. Thus we get $u_X(x) = u_X F(f)(x_i) = G(f)\, u_{X_i}(x_i)$. Similarly, $v_X(x) = G(f)\, v_{X_i}(x_i)$ and since $u_{X_i}(x_i) = v_{X_i}(x_i)$ it follows that $u_X(x) = v_X(x)$, i.e. that $u = v$. The implication a)$\Rightarrow$b) is trivial. ✳

Let $F, F': \mathscr{C} \to \mathscr{S}et$ be functors. In the same way as in Ch. 1, Example 3.18, we may show that the class $[F, F']$ is not generally a set. Nevertheless, the following assertion is true.

PROPOSITION 1.5. *Let $F, F': \mathscr{C} \to \mathscr{S}et$ be functors, such that F is proper. Then $[F, F']$ is a set.*

Proof. Let $\{X_i\}_{i \in \mathscr{I}}$ be a set of objects of $\mathscr{C}$ which dominates F. Let us consider the set $M = \coprod_{i \in \mathscr{I}} \mathscr{S}et(F(X_i), F'(X_i))$ and the map $f: [F, F'] \to M$ defined by the assignment $f(u) = \{u_{X_i}\}_{i \in \mathscr{I}}$, $u \in [F, F']$. By Lemma 1.4, f is an injection. Hence $[F, F']$ is a set, as claimed. ✳

For each category $\mathscr{C}$, denote by $\{\mathscr{C}, \mathscr{S}et\}$ the category whose objects are all proper functors from $\mathscr{C}$ into $\mathscr{S}et$ and whose morphisms are the functorial morphisms. If $\mathscr{C}$ is a small category, then $\{\mathscr{C}, \mathscr{S}et\}$ is isomorphic to $[\mathscr{C}, \mathscr{S}et]$. If F and G are objects of $\{\mathscr{C}, \mathscr{S}et\}$, for the sake of simplicity we shall write $[F, G]$ instead of $\{\mathscr{C}, \mathscr{S}et\}(F, G)$. This notation is quite adequate since F and G are functors (see Ch. 1, Section 1.3).

PROPOSITION 1.6. *For every category $\mathscr{C}$, the category $\{\mathscr{C}, \mathscr{S}et\}$ is cocomplete. The colimits in $\{\mathscr{C}, \mathscr{S}et\}$ are formed argumentwise.*

Proof. Let $F\colon \mathscr{Y}\{\mathscr{C},\mathscr{S}et\}$ be a small functor, i.e. a diagram of functors. By Proposition 10.1, Ch. 1, this diagram has a colimit. We shall prove that this colimit is an object of $\{\mathscr{C},\mathscr{S}et\}$. Since a colimit of a diagram of functors is a quotient functor of a coproduct of functors, it suffices by Theorem 1.1 to show that, if $\{T_i\}_{i\in\mathscr{I}}$ is a set of proper functors, then $\coprod_{i\in\mathscr{I}}T_i$ is also a proper functor (this last coproduct is a coproduct of functors). But if $\{M_i\}_{i\in I}$ is a set of objects of $\mathscr{C}$ that dominates T_i, then $\bigcup_{i\in\mathscr{I}} M_i$ is a set of objects of $\mathscr{C}$ that dominates $\coprod_{i\in\mathscr{I}}T_i$. ⁎

As a consequence of the dual of Proposition 10.5 of Ch. 1, we have the following result.

COROLLARY 1.7. *A morphism $u\colon T\to T'$ of $\{\mathscr{C},\mathscr{S}et\}$ is an epimorphism if and only if it is an argumentwise epimorphism, i.e. if and only if for each $X\in\mathscr{C}$, the map $u_X\colon T(X)\to T'(X)$ is a surjection.* ⁎

Using Corollary 1.7 and the proof of Proposition 17.16 of Ch. 1, we obtain the following result.

COROLLARY 1.8. *Let T be an object of $\{\mathscr{C},\mathscr{S}et\}$. Then every class of quotient objects of T has a lower bound and an upper bound. Moreover, every morphism of $\{\mathscr{C},\mathscr{S}et\}$ has a unique triangular decomposition (see also Exercise 1.17).* ⁎

COROLLARY 1.9. *For every category $\mathscr{C}$, the category $\{\mathscr{C},\mathscr{S}et\}$ has bounded colimits.*

The proof follows from Theorem 19.6, Ch. 1, Proposition 1.6, and Corollary 1.8. ⁎

Note that bounded colimits in $\{\mathscr{C},\mathscr{S}et\}$ may be formed argumentwise in a manner similar to colimits of small functors. ⁎

The existence of limits in the category $\{\mathscr{C},\mathscr{S}et\}$ is a rather difficult problem and is still incompletely solved. From Exercise 1.9 it may be inferred that generally $\{\mathscr{C},\mathscr{S}et\}$ is not a complete category. Also Exercise 1.9 shows that a subfunctor of a proper functor need not be proper. These examples, as well as the example given in Exercise 1.1, indicate the difficulty of studies concerning the limits in $\{\mathscr{C},\mathscr{S}et\}$.

Now let $\mathscr{C}$ be a category and let $D\colon\mathscr{C}^0\to\mathscr{C}$ be the duality functor. Also let $F\colon\mathscr{C}\to\mathscr{S}et$ be a cofunctor. We say that F is *proper* if the functor FD is proper. This means that a set $\{X_i\}_{i\in\mathscr{I}}$ of objects of $\mathscr{C}$ exists such that for any $X\in\mathscr{C}$ and an $x\in F(X)$, there are an $i\in\mathscr{I}$, a morphism $f\colon X\to X_i$ and an $x_i\in F(X_i)$ such that $F(f)(x_i)=x$. Let $\{\mathscr{C}^0,\mathscr{S}et\}$ denote the category of all proper cofunctors from $\mathscr{C}$ to $\mathscr{S}et$. As above, the morphisms are all functorial morphisms. Since $H_X\colon\mathscr{C}\to\mathscr{S}et$ is a proper cofunctor, the assignments $X\rightsquigarrow H_X$ and $f\rightsquigarrow H(f)$ give a functor $H^{\mathscr{C}}\colon\mathscr{C}\to\{\mathscr{C}^0,\mathscr{S}et\}$ called the *natural*, or *canonical*, *functor*. (We shall write simply H, if there is no danger of confusion.)

By Corollary 3.22, Ch. 1, H is full and faithful, and by Theorem 10.4, Ch. 1, H commutes with limits. By Exercise 1.14, we see that H need not commute with colimits. When $F\colon\mathscr{C}\to\mathscr{C}'$ is a full and faithful functor we say that F defines $\mathscr{C}'$ as an *extension* of $\mathscr{C}$. Therefore H defines $\{\mathscr{C}^0,\mathscr{S}et\}$ as an extension of $\mathscr{C}$.

Note 1.10. An object of $\{\mathscr{C}^0,\mathscr{S}et\}$ is called a *presheaf* over $\mathscr{C}$, whereas a proper functor $T\colon\mathscr{C}\to\mathscr{S}et$ is called a *copresheaf*. This terminology is imposed by the applications of these functors in Algebraic Geometry (see Grothendieck—Verdier [1]).

Sometimes $\{\mathscr{C}^0, \mathscr{S}et\}$ is called *the category of presheaves* over $\mathscr{C}$ and $\{\mathscr{C}, \mathscr{S}et\}$, *the category of copresheaves* over $\mathscr{C}$.

There is another extension of $\mathscr{C}$ defined as follows. Let $H^0_{\mathscr{C}}: \mathscr{C} \to \{\mathscr{C}, \mathscr{S}et\}$ (or simply H^0) denote the functor defined by $H^0(X) = H^X$. According to Corollary 3.22 and Proposition 12.3, Ch. 1, we see that H^0 is full and faithful and commutes with colimits. Again let $H'_{\mathscr{C}}: \mathscr{C} \to \{\mathscr{C}, \mathscr{S}et\}^0$ (or simply H') denote the composition of H^0 with the duality functor. Clearly, H' is full and faithful and commutes with colimits, i.e. H' defines $\{\mathscr{C}, \mathscr{S}et\}^0$ as an extension of $\mathscr{C}$. By Corollary 1.9 it follows that $\{\mathscr{C}, \mathscr{S}et\}^0$ is a category with bounded limits, but generally this is not a category with colimits. Moreover the functor H' does not commute with limits.

Exercises

1.1. Let $\mathscr{I}$ be a discrete category. A functor $F: \mathscr{I} \to \mathscr{S}et$ is proper if and only if the class X of all objects of $\mathscr{I}$ such that $F(X) \neq \emptyset$ is a set. Show that if $\mathscr{I}$ is not a set then the category $\{\mathscr{I}, \mathscr{S}et\}$ does not have a final object, even though it does have an initial object. Also show that $\{\mathscr{I}, \mathscr{S}et\}$ is a complete category.

1.2. Let $\mathscr{C}$ be a full subcategory of $\mathscr{S}et$. Show that the inclusion functor $I: \mathscr{C} \to \mathscr{S}et$ is proper.

1.3. If a functor $F: \mathscr{C} \to \mathscr{S}et$ has an adjoint, prove that F is proper.

1.4. Prove that every proper functor $F: \mathscr{C} \to \mathscr{S}et$ has a limit. Prove that $\mathscr{S}et$ is a category with bounded limits.

1.5. Let $F, G: \mathscr{C} \to \mathscr{S}et$ be functors. Show that if F is a proper functor and $p: F \to G$ is a functorial epimorphism, then G is also proper.

1.6. Prove that a functor $F: \mathscr{C} \to \mathscr{S}et$ is 1-generated if and only if there exists a functorial epimorphism $p: H^X \to F$.

1.7. Let $F, F': \mathscr{C} \to \mathscr{S}et$ be functors, and let $\{(X_i, x_i)\}_{i \in \mathscr{I}}$ be a set of generators of F. Show that two functorial morphisms $u, v: F \to F'_i$ are equal if and only if $u_{X_i}(x_i) = v_{X_i}(x_i)$ for any $i \in \mathscr{I}$. Show that, moreover, a functorial morphism $w: F' \to F$ is a functorial epimorphism if and only if $x_i \in w_{X_i}(F'(X_i))$ for any $i \in \mathscr{I}$.

1.8. Let $X \in \mathscr{C}$. Show that the argument functor $A^X: \{\mathscr{C}, \mathscr{S}et\} \to \mathscr{S}et$ is proper. (Hint: This functor is in fact representable.)

1.9. Let $\mathscr{A}$ be the class of all cardinal numbers α such that $\alpha \geqslant 1$ and let $\{M_\alpha\}_{\alpha \in \mathscr{A}}$ be a class of sets such that for any α, $\mathrm{Card}(M_\alpha) = \alpha$, and $\alpha \leqslant \alpha'$ implies $M_\alpha \subseteq M_{\alpha'}$. Let $\mathscr{C}_1$ be the subcategory of $\mathscr{S}et$ whose objects are the sets $\{M_\alpha\}_{\substack{\alpha \in \mathscr{A} \\ \alpha \geqslant 2}}$ and whose morphisms are the canonical inclusions (i.e. $\mathscr{C}_1$ is a preordered category). Denote by $\mathscr{C}$ the subcategory of $\mathscr{S}et$ whose objects are $\{M_\alpha\}_{\alpha \in \mathscr{A}}$. Assume that $\mathscr{C}(M_\alpha, M_1) = \emptyset$ and $\mathscr{C}(M_1, M_\alpha) = \mathscr{S}et(M_1, M_\alpha)$ for $\alpha \geqslant 2$. Moreover, let $\mathscr{C}(M_\alpha, M_{\alpha'}) =$

$= \mathcal{S}et(M_\alpha, M_{\alpha'})$ if $\alpha, \alpha' \neq 1$. Finally let $\mathcal{C}_1 \xrightarrow{I_1} \mathcal{C} \xrightarrow{I} \mathcal{S}et$ be the inclusion functors. Show that:

a) I is a proper functor.

b) I_1 is not proper.

c) The functor $T: \mathcal{C} \to \mathcal{S}et$ defined by $T(M_1) = \emptyset$ and $T(M_\alpha) = I(M_\alpha)$ for $\alpha \neq 1$ is not proper.

d) T is a subfunctor of I.

1.10. Let I be the functor defined as in Exercise 1.9. Prove that the product $I \coprod I$ does not exist in $\{\mathcal{C}, \mathcal{S}et\}$. (Hint: Note first that the product of functors, i.e. the functor defined argumentwise is not proper. Now define the functors $L^\alpha: \mathcal{C} \to \mathcal{S}et$, $\alpha \geqslant 1$, by $L^\alpha(M_\beta) = \emptyset$ if $\alpha > \beta$ and $L^\alpha(M_\beta) = M_1$ if $\alpha \leqslant \beta$. If the product $I \coprod I$ would exist in $\{\mathcal{C}, \mathcal{S}et\}$ then we should have $(I \coprod I)(M_\alpha) = M_\alpha \coprod M_\alpha$ for every $\alpha \geqslant 1$. The contradiction arises if we make use of the morphisms which result from the universal property of the product by considering the distinct inclusions of L^α in I.)

1.11. Let $F: \mathcal{C} \to \mathcal{C}'$ be a representative functor and let $G: \mathcal{C}' \to \mathcal{S}et$ be a functor. Show that if GF is proper, then G is proper.

1.12. Let $F: \mathcal{C} \to \mathcal{D}$ be a functor and let $\mathcal{B}$ be the full subcategory of $\{\mathcal{D}^0, \mathcal{S}et\}$ whose objects are all cofunctors $G: \mathcal{D} \to \mathcal{S}et$ such that GF is proper. Show that the canonical functor $F_*: \mathcal{B} \to \{\mathcal{C}^0, \mathcal{S}et\}$ defined by $F_*(G) = GF$, commutes with the bounded colimits of $\mathcal{B}$, which exist in $\{\mathcal{D}^0, \mathcal{S}et\}$.

1.13. Let $\mathcal{I}$ be a preordered set. Show that the canonical functor $H: \mathcal{I} \to [\mathcal{I}^0, \mathcal{S}et]$ does not commute with coproducts.

1.14. Let $\mathcal{C}$ be the ordered category of cardinal numbers (i.e. the objects are cardinal numbers α, and for any two cardinal numbers α, β we have a morphism from α into β if and only if $\alpha \leqslant \beta$). Let $\{a, b\}$ be a set of two elements and $F: \mathcal{C} \to \mathcal{S}et$ the constant functor associated with $\{a, b\}$. For every cardinal number α denote by $L^\alpha: \mathcal{C} \to \mathcal{S}et$ the functor defined as follows:

$$L^\alpha(\beta) = \{a, b\} \text{ if } \beta \leqslant \alpha \text{ and } L^\alpha(\beta) = \{c\}, \text{ if } \beta > \alpha$$

where $\{c\}$ is a set of one element. Furthermore $L^\alpha(\beta \leqslant \gamma) = id$ if $\gamma \leqslant \alpha$, $L^\alpha(\beta \leqslant \gamma) = id$ if $\beta > \alpha$ and $L^\alpha(\beta \leqslant \gamma) =$ the unique morphism $\{a, b\} \to \{c\}$, if $\beta \leqslant \alpha$ and $\gamma > \alpha$. Prove that $\{L^\alpha\}$ and F are proper functors, and L^α is a quotient object of F for every α. Hence the quotient objects of F do not form a set.

1.15. Show that a monomorphism in $\{\mathcal{C}, \mathcal{S}et\}$ is an argumentwise monomorphism. Show that if the following diagram of $\{\mathcal{C}, \mathcal{S}et\}$

$$\begin{array}{ccc} F & \longrightarrow & G \\ {\scriptstyle u'}\downarrow & & \downarrow{\scriptstyle u} \\ F' & \longrightarrow & G' \end{array}$$

is cartesian and u is an epimorphism, then u' is also an epimorphism.

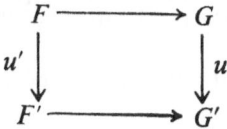

1.16. Let X be an object of $\mathscr{C}$ and let $\{f_i, X_i\}_i$ be a family of objects of $X/\mathscr{C}$. By the *left ideal over X generated by the family* $\{f_i, X_i\}_i$ we mean the smallest left ideal over X that contains all objects $\{f_i, X_i\}_i$. This left ideal over X will be denoted by $\langle f_i \rangle_i$. Show that:

a) An object (f, X) of $X/\mathscr{C}$ belongs to $\langle f_i \rangle_i$ if and only if there is an index i such that f can be factored through f_i.

b) A subfunctor of H^X is proper if and only if the left ideal over X associated with it is generated by a set of objects of $X/\mathscr{C}$.

c) If $\mathscr{C}$ is a category with fibered coproducts, then the intersection of a finite set of proper subfunctors of H^X is also a proper subfunctor. Moreover, if every copencil of $\mathscr{C}$ has a fibered coproduct, then the intersection of every set of proper subfunctors of H^X is likewise a proper functor.

1.17. Prove that $\{\mathscr{C}, \mathscr{S}et\}$ is a category with unions, cointersections and counions. Also every morphism of $\{\mathscr{C}, \mathscr{S}et\}$ has an image, a coimage and a unique triangular decomposition. Moreover $\{\mathscr{C}, \mathscr{S}et\}$ is a balanced category.

1.18. Show that every epimorphism of $\{\mathscr{C}, \mathscr{S}et\}$ is a strong and a strict epimorphism. (Hint: Use Theorem 3.19, Ch. 1.)

1.19. Let $\mathscr{C}$ be a category and $\mathscr{M}$ a family (not necessarily a set) of objects of $\mathscr{C}$. We say that $\mathscr{M}$ is a *family of separators* of $\mathscr{C}$ if for each object X of $\mathscr{C}$, there is a subset $\mathscr{M}'$ of $\mathscr{M}$ such that for every pair (f, g) of distinct morphisms with the same domain X and the same codomain, there exists a $Y \in \mathscr{M}'$ and a morphism $h: Y \to X$ such that $fh \neq gh$. The family $\mathscr{M}$ is called a *family of generators (family of quasigenerators)* if for each $X \in \mathscr{C}$ there is a subset $\mathscr{M}'$ of $\mathscr{M}$ such that for every proper subobject $f: X' \to X$ of X (for every monomorphism $f: X' \to X$ which is not a bimorphism) there is an object Y of $\mathscr{M}'$ and a morphism $g: Y \to X$ which is not factored through f.

The family $\mathscr{M}$ is a *family of strong separators* (a *family of strong generators)* if:

i) $\mathscr{M}$ is a family of separators (generators).

ii) A morphism $f: X \to Y$ of $\mathscr{C}$ is an isomorphism if and only if for each object Z of $\mathscr{M}$, the map $H^Z(f): H^Z(X) \to X^Z(Y)$ is a bijection.

If we assume that $\mathscr{M}$ is a set, we obtain again the notions: *set of separators, set of generators (quasigenerators), set of strong generators* and *set of strong separators.*

Let $\mathscr{M}$ be a family of objects of a category $\mathscr{C}$. Assume that the coproduct of every subset of objects of $\mathscr{M}$ is defined.

a) Prove that $\mathscr{M}$ is a family of separators if and only if for each $X \in \mathscr{C}$ there exists subset $\{U_i\}_i$ of objects of $\mathscr{M}$ and an epimorphism $\coprod_i U_i \to X$.

b) Prove that if $\mathscr{C}$ is a category with triangular decomposition and $\mathscr{M}$ is a family of generators, then there exist a subset $\{U_i\}_i$ of $\mathscr{M}$ and an epimorphism $\coprod_i U_i \to X$. If in addition $\mathscr{C}$ is a balanced category, the converse also holds.

c) Consider the following:

1) For each object X of $\mathscr{C}$ there is a subset $\{U_i\}_i$ of $\mathscr{M}$ and a strong epimorphism: $\coprod_i U_i \to X$.

2) A morphism $f: X \to Y$ such that every morphism $g: Z \to Y$, $Z \in \mathscr{M}$ may be factored through f is necessarily an epimorphism.

3) A morphism $f: X \to Y$ such that $H^Z(f)$ is a bijection for every object Z of $\mathcal{M}$ is an isomorphism.

4) A monomorphism $F: X \to Y$ such that $H^Z(f)$ is a bijection for all $Z \in \mathcal{M}$ is an isomorphism.

Show that we always have the implications 1) $\Rightarrow$ 2), 1) $\Rightarrow$ 3) $\Rightarrow$ 4). If $\mathcal{M}$ is a set, or if we suppose that $\mathcal{M}$ is a family of separators, then 2) $\Rightarrow$ 1). If $\mathcal{M}$ is a set and every morphism of $\mathscr{C}$ has a left strong decomposition, then all the four assertions are equivalent and thus $\mathcal{M}$ is a set of strong generators.

d) Show that for every category $\mathscr{C}$, the objects $\{H^X\}_{X \in \mathscr{C}}$ of $\{\mathscr{C}^0, \mathscr{S}et\}$ give a family of separators, a family of generators, and a family of strong generators of this category.

1.20. An object X of a category $\mathscr{D}$ is called *O-presentable* if the functor $H^X: \mathscr{D} \to \mathscr{S}et$ commutes with colimits. Show that for any object X of a category $\mathscr{C}$, the object $X^0(X) = H^X$ of $\{\mathscr{C}, \mathscr{S}et\}$ is O-presentable. Moreover, an object F of $\{\mathscr{C}^0, \mathscr{S}et\}$ is O-presentable if and only if there exist an object X of $\mathscr{C}$ and a right invertible epimorphism $p: H^X \to F$. (Hint: Use Ch. 1, Theorem 10.6 and Proposition 9.7.)

2.2. The extension theorem

In this section we give a slight generalization of the so-called "Kan Extension Theorem". Some properties of this extension are given.

References: Freyd [1], Gabriel—Ulmer [1], Isbell [3], [8], Kan [1], Lambek [1], Ulmer [1], Grothendieck—Verdier [1].

Let $\mathscr{C}$ be a category and let $H: \mathscr{C} \to \{\mathscr{C}^0, \mathscr{S}et\}$ be the canonical functor. With each object T of $\{\mathscr{C}^0, \mathscr{S}et\}$, we have associated the pointed category $\mathscr{P}_T$. Now let $s_T: \mathscr{P}_T \to \mathscr{C}$ denote the *underlying functor associated with T*, defined as follows. If $(X, x) \in \mathscr{P}_T$, then $s_T(X, x) = X$; the action of s_T on morphisms is obvious.

As a direct consequence of Theorem 10.6, Ch. 1, we have the following result.

PROPOSITION 2.1. *Let T be a presheaf over a category $\mathscr{C}$. For each $(X, x) \in \mathscr{P}_T$, let $\varphi^{(X, x)}: H_X = Hs_T(X, x) \to T$ denote the functorial morphism u^x associated with x. Then the morphisms*

$$\{\varphi^{(X, x)}\}_{(X, x) \in \mathscr{P}_T}$$

define a functorial morphism $\varphi: Hs_T \to T$, and the pair (φ, T) is a colimit of the cofunctor Hs_T. ✳

The above proposition can obviously be dualized for the category $\{\mathscr{C}, \mathscr{S}et\}^0$ as follows.

PROPOSITION 2.2. *Let T be a proper functor over $\mathscr{C}$, and let $s_T: \mathscr{P}_T \to \mathscr{C}$ be the underlying functor. Then the object T^0 of $\{\mathscr{C}, \mathscr{S}et\}$ is canonically a limit of the functor $H's_T$.*

The details are left for the reader. ✳

The following result plays an important role in the theory of completions of categories, as well as in some related problems.

THEOREM 2.3. (Extension Theorem). *Let $T:\mathscr{C}\to\mathscr{D}$ be a functor, and $\mathscr{D}$ be a category with bounded colimits. Then there is a unique functor $\bar{T}:\{\mathscr{C}^0,\mathscr{S}et\}\to\mathscr{D}$ such that:*

a) $\bar{T}H\simeq T$.

b) *$\bar{T}$ commutes with bounded colimits.*

Proof. Let F be an object of $\{\mathscr{C}^0,\mathscr{S}et\}$. Then by Proposition 2.1, we have that $F=\varinjlim(Hs_F)$, where $s_F:\mathscr{P}_F\to\mathscr{C}$ is the underlying functor. Now we define $\bar{T}(F)=\varinjlim(Ts_F)$. We see that this colimit can be calculated in $\mathscr{D}$, since $\mathscr{P}_F$ is a right bounded category (see Section 2.1 and Exercise 9.5, Ch. 1). If $F=H_X$, with $X\in\mathscr{C}$, then $(X,1_X)$ is a final object of $\mathscr{P}_{H_X}$, so that $H_X=\varinjlim(Hs_{H_X})$. Hence $\bar{T}(H_X)=T(X)$.

Let $u\colon F\to G$ be a functorial morphism and let $u_*:\mathscr{P}_F\to\mathscr{P}_G$ denote the functor defined by the assignments $(X,x)\rightsquigarrow(X,u_X(x))$ and $f\rightsquigarrow f$.

By Exercise 10.3, Ch. 1, it follows that u_* induces a unique morphism $\bar{T}(u)\colon\bar{T}(F)\to\to\bar{T}(G)$. If $v\colon G\to R$ is another functorial morphism, then we have that $(vu)_*=v_*u_*$, that is, $\bar{T}(vu)=\bar{T}(v)\bar{T}(u)$. So the construction of $\bar{T}$ is finished and the proof of a) is complete. We make use of the following lemma in proving b).

LEMMA 2.4. *Let $\mathscr{I}$ be a category and let $K:\mathscr{I}\to\{\mathscr{C}^0,\mathscr{S}et\}$ be a functor such that $\varinjlim K$ is an $\{\mathscr{C}^0,\mathscr{S}et\}$ and that this colimit is calculated argumentwise. Then the pair $(\bar{T}u,\ \bar{T}(F))$ is a colimit of $\bar{T}K$.*

Proof. First we note that $(\bar{T}u,\ \bar{T}(F))$ is a cocone over $\bar{T}K$. Now let (v,D) be another cocone over $\bar{T}K$. For each $X\in\mathscr{C}$ and each $i\in\mathscr{I}$, let $v_X^i\colon K_i(X)\to\mathscr{D}(T(X),D)$, denote the map that assigns to $x\in K_i(X)$ the composition of morphisms of $\mathscr{D}$:

$$T(X)=\bar{T}(H_X)\xrightarrow{\bar{T}(u^x)}\bar{T}(K_i)=(\bar{T}K)_i\xrightarrow{v_i}D$$

i.e. $v_X^i(x)=v_i\bar{T}(u^x)$. (Here $u^x\colon H_X\to K_i$ is as always the functorial morphism associated with x.) We claim that the maps $\{v_X^i\}_{i\in\mathscr{I}}$ define a functorial morphism $v_X\colon A^XK\to\mathscr{D}(T(X),D)$, A^X being the argument functor associated with X. In order to show that, let $\alpha\colon i\to j$ be a morphism of $\mathscr{I}$; we must prove that the diagram

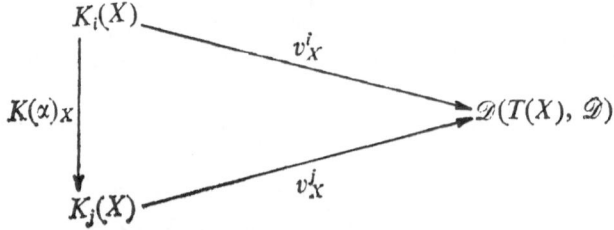

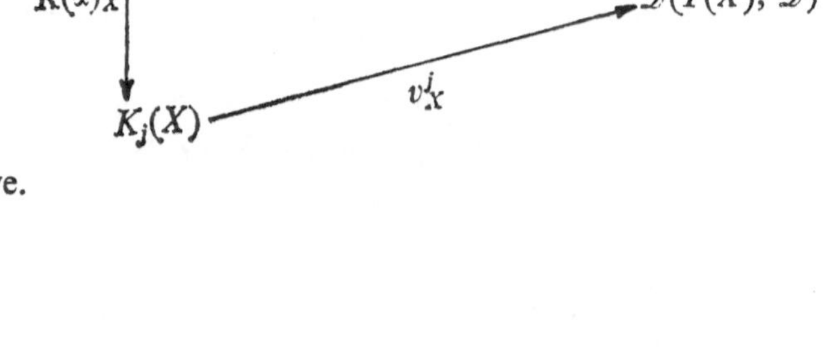

is commutative.

Indeed, if $x \in K_i(X)$, consider the following commutative diagram of $\mathscr{D}$:

$$\begin{array}{ccccc} & & (\overline{T}K)_i & & \\ & \overset{\overline{T}(u^x)}{\nearrow} & & \overset{v_i}{\searrow} & \\ T(X)=\overline{T}(H_X) & & \Big\downarrow{\scriptstyle \overline{T}K(\alpha)} & & D \\ & \underset{\overline{T}(u^{K(\alpha)_X(x)})}{\searrow} & & \underset{v_j}{\nearrow} & \\ & & (\overline{T}K)_j & & \end{array}$$

and so, $v^i_X(x) = v_i\overline{T}(u^x) = v_j\overline{T}K(\alpha)\overline{T}(u^x) = v_j\overline{T}(u^{K(\alpha)_X(x)}) = v^j_X K(\alpha)_X(x)$. Since $(A^X u, A^X F = F(X))$ is a colimit of $A^X K$, there is a unique map $\overline{v}_X\colon F(X) \to \mathscr{D}(T(X), D) = H_D T(X)$, so that $\overline{v}_X A^X u = v_X$. Now we prove that the maps $\{\overline{v}_X\}_{X\in\mathscr{C}}$ define a functorial morphism $\overline{v}\colon F \to H_D T$. (Note that $H_D T$ is not generally a proper cofunctor.) First we prove that the morphisms $\{v^i_X\}_{X\in\mathscr{C}}$ define a functorial morphism $v^i\colon K_i \to H_D T$. For this, let $f\colon Y \to X$ be a morphism of $\mathscr{C}$; we must prove that the diagram

$$\begin{array}{ccc} K_i(X) & \xrightarrow{\;v^i_X\;} & \mathscr{D}(T(X), D) = H_D T(X) \\ {\scriptstyle K_i(f)}\Big\downarrow & & \Big\downarrow{\scriptstyle H_D(T(f))} \\ K_i(Y) & \xrightarrow{\;v^i_Y\;} & \mathscr{D}(T(Y), D) = H_D T(Y) \end{array} \qquad (1)$$

is commutative.

But this follows, since if $x \in K_i(X)$, we have that $v^i_Y K_i(f)(x) = v_i\overline{T}(u^{K_i(f)(x)}) = v_i\overline{T}(u^x H(f)) = v_i\overline{T}(u^x)\overline{T}(H(f)) = v^i_X(x)T(f) = H_D(T(f))\,v^i_X(x)$. (Here $u^{K_i(f)(x)}\colon H_Y \to K_i$ is the functorial morphism associated with $K_i(f)(x)$ and we made use of the fact that the diagram

$$\begin{array}{ccc} H_Y & & \\ & \overset{u^{K_i(f)(x)}}{\searrow} & \\ {\scriptstyle H(f)}\Big\downarrow & & K_i \\ & \underset{u^x}{\nearrow} & \\ H_X & & \end{array}$$

is commutative in $\{\mathscr{C}^0, \mathscr{S}et\}$. Furthermore we must check that $\overline{v}\colon F \to H_D T$ defined above is in fact a functorial morphism. For this purpose let $f\colon Y \to X$ be a morphism of $\mathscr{C}$; we get the following diagram in $\mathscr{S}et$:

$$\begin{array}{ccc} F(X) & \xrightarrow{\;\overline{v}_X\;} & H_D T(X) \\ {\scriptstyle F(f)}\Big\downarrow & & \Big\downarrow{\scriptstyle H_D T(f)} \\ F(Y) & \xrightarrow{\;\overline{v}_Y\;} & H_D T(Y) \end{array} \qquad (2)$$

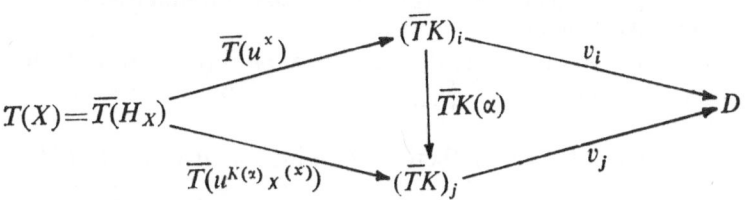

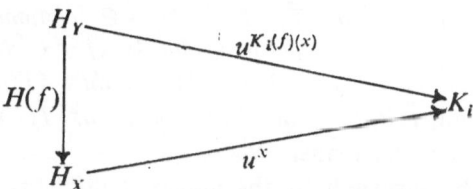

Thus for each $i \in \mathscr{I}$, we have that $H_D T(f)\bar{v}_X(u_i)_X = H_D T(f) v^i_X$. We have also that $\bar{v}_Y F(f)(u_i)_X = \bar{v}_Y(u_i)_Y K_i(f) = v^i_Y K_i(f)$, so that the commutativity of (2) follows from the commutativity of (1).

For each object (X, x) of $\mathscr{P}_F$, we define $t_{(X,x)}$ to be $\bar{v}_X(x)$. From the last diagram we deduce that the morphisms $\{t_{(X,x)}\}_{(X,x)\in\mathscr{P}_F}$ define a functorial morphism $t\colon Ts_F \to D$.

Now, since $\bar{T}(F) = \varinjlim (Ts_F)$ there is a unique morphism $\bar{t}\colon \bar{T}(F) \to D$ such that $\bar{t}w = t$, where $w\colon Ts_F \to \bar{T}(F)$ is the structural morphism. Therefore, for each object (X, x) of $\mathscr{P}_F$ we have that $\bar{t}w_{(X,x)} = t_{(X,x)}$. Since $\bar{T}(H_X) = T(X)$ for every $X \in \mathscr{C}$, we can make use of Proposition 2.1 and may consider that $w_{(X,x)} = \bar{T}(u^x)$. Hence we have $\bar{t}\bar{T}(u^x) = t_{(X,x)}$, for every object (X, x) of $\mathscr{P}_F$.

If $Z \in \mathscr{C}$, $i \in \mathscr{I}$ and $z \in K_i(Z)$, then the following composition of morphisms

$$H_Z \xrightarrow{u^z} K_i \xrightarrow{u_i} F$$

defines a morphism $H_Z \to F$ (actually equal to $u^{(u_i)_Z(z)}$). Thus as above we have $\bar{t}\bar{T}(u^{(u_i)_Z(z)}) = t_{(Z,(u_i)_Z(z))} = \bar{t}\bar{T}(u_i)\bar{T}(u^z) = v^i\,\bar{T}(u^z)$. Thus $\bar{t}\bar{T}(u_i) = v^i$ since $K_i = \varinjlim (Ts_{K_i})$. Consequently we have shown that $\bar{t}\bar{T}u = v$. Now, making use of the above construction we note that $\bar{t}\colon \bar{T}(F) \to D$ is the unique morphism such that $\bar{t}\bar{T}u = v$. In conclusion $(\bar{T}u, \bar{T}(F))$ is a colimit of $\bar{T}K$. The functor $\bar{T}$ is called the *extension* of T. ✳

Let $F\colon \mathscr{I} \to \mathscr{C}$ be a functor, X an object of $\mathscr{C}$ and $u\colon F \to X$ a functorial morphism. We say that (F, u) is an *epicocone* (*monococone*) if for each $i \in \mathscr{I}$, the morphism $u_i\colon F_i \to X$ is an epimorphism (a monomorphism). In an obvious manner we define the notions of *epicone* and *monocone*.

COROLLARY 2.5. *The functor* $\bar{T}\colon \{\mathscr{C}^0, \mathscr{S}et\} \to \mathscr{D}$ *commutes with epicocones. Moreover if* $\{p_i, P_i\}$ *is a class of quotient objects of* $F \in \{\mathscr{C}^0, \mathscr{S}et\}$ *and* (p, P) *is the upper bound of this class of quotient objects, then* $(\bar{T}(p), \bar{T}(P))$ *is the upper bound of the class* $\{\bar{T}(p_i), \bar{T}(P_i)\}_i$ *of quotient objects of* $\bar{T}(F)$.

The proof is left for the reader. ✳

The reader may be surprised by the apparent difficulty of the proof that $\bar{T}$ commutes with bounded colimits. One could ask why we did not prove this result by simply constructing a coadjoint to $\bar{T}$. The answer is that one can not generally be sure of the existence of such a coadjoint to $\bar{T}$ (as will soon be seen). Obviously, if there existed a coadjoint $S\colon \mathscr{D} \to \{\mathscr{C}^0, \mathscr{S}et\}$ to $\bar{T}$, then we would have had that

$$\mathscr{D}(T(X), D) \simeq \mathscr{D}(\bar{T}(H_X), D) \simeq [H_X, S(D)] \simeq S(D)(X)$$

or every $X \in \mathscr{C}$. Hence, the cofunctor $H_D T$ would be proper for every $D \in \mathscr{D}$. But T was arbitrarily chosen and hence we do not have in general that $H_D T$ is proper for every $D \in \mathscr{D}$. This is shown in the following Note.

Note 2.6. Let $T: \mathscr{C} \to \mathscr{D}$ be a functor. The assignment $F \leadsto FT$ does not always define a functor $\{\mathscr{D}^0, \mathscr{S}et\} \to \{\mathscr{C}^0, \mathscr{S}et\}$. Indeed, in Exercise 1.9 the composite functor II_1 is not proper although I is proper. Thus let us consider the diagram of categories and functors

$$\begin{array}{ccc} \mathscr{C} & \xrightarrow{H^{\mathscr{C}}} & \{\mathscr{C}^0, \mathscr{S}et\} \\ T\big\downarrow & & \big\uparrow T_* \\ \mathscr{D} & \xrightarrow{H^{\mathscr{D}}} & \{\mathscr{D}^0, \mathscr{S}et\} \end{array}$$

(where T_* is the restriction functor).

If the functor $L = \overline{HT}$ has a coadjoint, S, then for every $X \in \mathscr{C}$ and every $D \in \{\mathscr{D}^0, \mathscr{S}et\}$ we have that

$$D(T(X)) = DT(X) \simeq [H_{(T,X)}, D] \simeq [L(H_X), D] \simeq (H_X, S(D)] \simeq S(D)(X)$$

all occurring isomorphisms being functorial. But then it is clear that $S \simeq T_*$, which is a contradiction. Hence generally the functor $\overline{T}$ does not always have a coadjoint.

However one can obtain the following positive result.

COROLLARY 2.7. *Let $\mathscr{D}$ be a category with bounded colimits and let $T: \mathscr{C} \to \mathscr{D}$ be a functor. Assume that for each object D of $\mathscr{D}$, the cofunctor $H_D T: \mathscr{C} \to \mathscr{S}et$ is proper. Then the extension functor $\overline{T}: \{\mathscr{C}^0, \mathscr{S}et\} \to \mathscr{D}$ has a coadjoint. Moreover, the functor $S^T: \mathscr{D} \to \{\mathscr{C}^0, \mathscr{S}et\}$, defined by $S^T(D) = H_D T$, is a coadjoint to $\overline{T}$.*

Proof. Let X be an object of $\mathscr{C}$. Then we have the following functorial isomorphisms:

$$\mathscr{D}(\overline{T}(H_X), D) \simeq \mathscr{D}(T(X), D) = H_D T(X) \simeq [H_X, S(D)].$$

Now, if $F \in \{\mathscr{C}^0, \mathscr{S}et\}$, then $F \simeq \varinjlim (Hs_F)$ (see Proposition 2.1) so that we have

$$\mathscr{D}(\overline{T}(F), D) \simeq \mathscr{D}(\overline{T} \varinjlim (Hs_F)), D) \simeq \mathscr{D}(\varinjlim \overline{T}(Hs_F), D) \simeq \varprojlim \mathscr{D}(Ts_F, D) \simeq$$

$$\simeq \varprojlim (H_D(Ts_F)) \simeq \varprojlim [Hs_F, S^T(D)] \simeq [\varinjlim (Hs_F), S^T(D)] \simeq [F, S^T(D)].$$

We leave for the reader to check that all these isomorphisms are functorial. ✳

COROLLARY 2.8. (Kan [1]). *Let $\mathscr{C}$ be a small category. Then for every cocomplete category $\mathscr{D}$, and for every functor $T: \mathscr{C} \to \mathscr{D}$, a unique functor $\overline{T}: [\mathscr{C}^0, \mathscr{S}et] \to \mathscr{D}$ may be found, such that*

a) *$\overline{T}H \simeq T$, and*

b) *$\overline{T}$ commutes with colimits and has a coadjoint.*

Moreover, every functor $L: [\mathscr{C}^0, \mathscr{S}et] \to \mathscr{D}$ that commutes with colimits has a coadjoint.

Proof. If F is an object of $[\mathscr{C}^0, \mathscr{S}et]$, then the category $\mathscr{P}_F$ is small, so that by the construction of the extension functor (given in the proof of Theorem 2.3), we may deduce the existence of the functor $\overline{T}$, such that a) holds. By Corollary 2.7 we see that b) is also valid. The last part of the corollary is obvious. ✳

The following result will be useful later.

PROPOSITION 2.9. *Let $T: \mathscr{C} \to \mathscr{C}'$ be a functor. Then there is a unique functor $T^*: \{\mathscr{C}^0, \mathscr{S}et\} \to \{\mathscr{C}'^0, \mathscr{S}et\}$ such that:*

a) *The following diagram of categories and functors*

$$\begin{array}{ccc} \mathscr{C} & \xrightarrow{H^{\mathscr{C}}} & \{\mathscr{C}^0, \mathscr{S}et\} \\ {\scriptstyle T}\downarrow & & \downarrow{\scriptstyle T^*} \\ \mathscr{C}' & \xrightarrow{H^{\mathscr{C}'}} & \{\mathscr{C}'^0, \mathscr{S}et\} \end{array}$$

is commutative, and

b) *T^* commutes with bounded colimits.*

Proof. It suffices to define T^* to be $\overline{H^{\mathscr{C}'}T}$. ✳

COROLLARY 2.10. *Let $\mathscr{C}$ be a small category and let $T: \mathscr{C} \to \mathscr{C}'$ be a functor. There is a unique functor $T^*: [\mathscr{C}^0, \mathscr{S}et] \to \{\mathscr{C}'^0, \mathscr{S}et]$ such that:*

a) *The following diagram of categories and functors*

$$\begin{array}{ccc} \mathscr{C} & \xrightarrow{H^{\mathscr{C}}} & [\mathscr{C}^0, \mathscr{S}et] \\ {\scriptstyle T}\downarrow & & \downarrow{\scriptstyle T^*} \\ \mathscr{C}' & \xrightarrow{H^{\mathscr{C}'}} & \{\mathscr{C}'^0, \mathscr{S}et\} \end{array}$$

is commutative, and

b) *T^* has a coadjoint.*

Moreover, the restriction functor $[T, \mathscr{S}et]$ is a coadjoint to T^* (this coadjoint is denoted by T_*).

Proof. a) follows by the above proposition. Now, according to Corollary 2.7, T^* has as coadjoint the functor $S: \{\mathscr{C}'^0, \mathscr{S}et\} \to [\mathscr{C}^0, \mathscr{S}et]$ defined by $S(F) = H_F H^{\mathscr{C}'}T$ for every copresheaf $F: \mathscr{C} \to \mathscr{S}et$. Thus for each $X \in \mathscr{C}$ we have that

$$S(F)(X) = H_F H^{\mathscr{C}'}T(X) = [H_{T(X)}, F] \simeq F(T(X)) = FT(X) \simeq T_*(F)(X).$$

Hence, by an easy computation we have $S \simeq T_*$. ✳

T^* is called the *functor induced by T*. It is clear that the induced functor is closely related to the extension functor: however there is a slight difference between them.

COROLLARY 2.11. *For every category $\mathscr{C}$ and every object X of $\mathscr{C}$, the argument functor $A^X: \{\mathscr{C}^0, \mathscr{S}et\} \to \mathscr{S}et$ has an adjoint.*

Proof. Let $\{0\}$ be the point category, and $T\colon \{0\} \to \mathscr{C}$ the functor that assigns to the unique object of $\{0\}$ the object X of $\mathscr{C}$. Thus $[\{0\}^0, \mathscr{S}et] \simeq \mathscr{S}et$ and the functor $T^*\colon \mathscr{S}et \to \{\mathscr{C}^0, \mathscr{S}et\}$ has a coadjoint. We leave it for the reader to check that this coadjoint is in fact the argument functor A^X. ⁎

Note 2.12. Let $\mathscr{D}$ be a category with bounded colimits and let $T\colon \mathscr{C} \to \mathscr{D}$ be a functor. It is easy to check that the extension functor $\overline{T}\colon \{\mathscr{C}^0, \mathscr{S}et\} \to \mathscr{D}$ has a coadjoint if and only if for each $D \in \mathscr{D}$, the functor $H_D T$ is proper. In this case, T is called a *properly defined functor*. A subcategory $\mathscr{C}'$ of $\mathscr{C}$ is called *properly defined subcategory* if the inclusion functor $\mathscr{C}' \to \mathscr{C}$ is properly defined.

Exercises

2.1. Prove that every functor $T\colon \mathscr{S}et \to \mathscr{S}et$ that commutes with colimits is representable. Dually, every functor $F\colon \mathscr{S}et \to \mathscr{S}et$ that commutes with limits is representable.

2.2. Let $\mathscr{C}$ be the ordered category of all cardinal numbers (see Exercise 1.15) and let $\overline{\mathscr{C}}$ be the category that is obtained from $\mathscr{C}$ by adding a final object ω. Let $\{0 \to 1\}$ be the arrow category and $F\colon \mathscr{C} \to \{0 \to 1\}$ the functor defined by $F(\alpha) = 0$ for any cardinal number α, and $F(\omega) = 1$. Prove that F commutes with bounded colimits of $\overline{\mathscr{C}}$ but does not preserve the upper bounds of quotient objects.

2.3. Let $T\colon \{\mathscr{C}, \mathscr{S}et\} \to \mathscr{D}$ be a functor that commutes with colimits and let $\{q^i, Q_i\}_i$ be a class of quotient objects of an object F of $\{\mathscr{C}^0, \mathscr{S}et\}$. Assume that $(q, Q) = \sup_i (q^i, Q_i)$. Show that $(T(q), T(Q))$ is the upper bound of the class $\{T(q^i), T(Q_i)\}$ of quotient objects of $T(Q)$, but $\{T(q), T(Q)\}$ is not the upper bound of the whole class of all quotient objects of $T(Q)$, but only in the class of all quotient objects (p, P) of $T(Q)$ having the following property: for any object X of $\mathscr{C}$ and every $x, x' \in F(X)$, such that there is an $i \in \mathscr{I}$ with $q^i_X(x) = q^i_X(x')$, we have $pT(u^x) = pT(u^{x'})$. Making use of this exercise, prove the second assertion of Corollary 2.5.

2.4. Let $\mathscr{C}$ be a left bounded category and let $F\colon \{\mathscr{C}, \mathscr{S}et\}^0 \to \mathscr{S}et$ be a functor that commutes with bounded limits. Show that there is an object X in $\{\mathscr{C}, \mathscr{S}et\}^0$ such that for any $Y \in \mathscr{C}$ we have $F(Y) = \bigcup_{f \in \mathscr{C}(X, Y)} F(f) F(X))$.

2.5. Show that for every category $\mathscr{C}$, the functor $H^*\colon \{\mathscr{C}^0, \mathscr{S}et\} \to \{\{\mathscr{C}^0, \mathscr{S}et\}^0$ $\mathscr{S}et\}$ induced by the canonical functor $H\colon \mathscr{C} \to \{\mathscr{C}^0, \mathscr{S}et\}$ is full and faithful and has a coadjoint. Also show that the functor $H'^*\colon \{\mathscr{C}, \mathscr{S}et\}^0 \to \{\{\mathscr{C}, \mathscr{S}et\}^0, \mathscr{S}et\}^0$ induced by the canonical functor $H'\colon \mathscr{C} \to \{\mathscr{C}, \mathscr{S}et\}^0$ is full and faithful and has an adjoint.

2.6. With hypothesis and notation as in Corollary 2.10, prove that the restriction functor T_* also has a coadjoint (Hint: Let $F \in [\mathscr{C}^0, \mathscr{S}et]$, $X' \in \mathscr{C}'$ and let $r_{X'}\colon T/X' \to \mathscr{C}$

be the underlying functor. Denote $T^{+}(F)(X')$ by $\underleftarrow{\lim}(Fr_{X'})$. If $f: X' \to Y'$ is a morphism of $\mathscr{C}'$, then f defines a functor f^*: $F/X' \to F/Y'$ and this functor defines in a canonical way a map $T^{+}(F)(f)$: $T^{+}(F)(X') \to T^{+}(F)(Y')$. Thus we obtain a functor T^{+}: $[\mathscr{C}^0, \mathscr{S}et] \to \{\mathscr{C}'^0, \mathscr{S}et\}$ a coadjoint of T_*.) (Compare with Exercise 11.14, Ch. 1.)

2.7. With the notation and hypothesis as in Corollary 2.10, assume that $\mathscr{C}$ is finitely complete and T is left exact. Show that the functor T^* is left exact.

2.8. With the notation and the hypothesis as in Exercise 2.6, show that the following are equivalent:

1) The functor T is full and faithful.
2) The functor T^* is full and faithful.
3) The functor T^+ is full and faithful.

2.9. Let $F: \mathscr{C} \to \mathscr{C}'$ and $G: \mathscr{C} \to \mathscr{C}''$ be functors. A pair $(u^F, E_F(G))$, where $E_F(G): \mathscr{C}' \to \mathscr{C}''$ is a functor and u^F: $G \to E_F(G)F$, is called a *Kan extension of G relative to F*, if for every functor $T: \mathscr{C}' \to \mathscr{C}''$ the map

$$[E_F(G), T] \to [G, TF], \quad v \rightsquigarrow (vF)u^F$$

is a bijection. In an obvious fashion we may define the *Kan coextension of a functor G relative to a functor F*. Show that:

a) Two Kan extensions of G relative to F are canonically isomorphic.

b) Let $F': \mathscr{C}' \to \mathscr{D}$ be a functor. Assume that $E_{F'F}(G)$ and $E_{F'}(E_F(G))$ are defined. Then $E_{F'F}(G) \simeq E_{F'}(E_F(G))$.

c) Let $F: \mathscr{C} \to \mathscr{C}'$ be a functor and X an object of $\mathscr{C}$. Then the functor $H^X: \mathscr{C} \to \mathscr{S}et$ has a Kan extension relative to F. Moreover, $E_F(H^{(X)}) = H^{F(X)}$, and u^F: $H^X \to H^{F(X)}F$ is defined by: $u^F_Y(f) = F(f)$ where $Y \in \mathscr{C}$ and $f \in \mathscr{C}(X, Y)$.

d) Let $G: \mathscr{C} \to \mathscr{D}$ be a properly defined functor. Then the functor $S^G: \mathscr{D} \to \{\mathscr{C}^0, \mathscr{S}et\}$ is a Kan extension of the canonical functor $H: \mathscr{C} \to \{\mathscr{C}^0, \mathscr{S}et\}$ relative to G.

2.10. We say that an object X is a *retract* of an object Y if there are morphisms $X \xrightarrow{u} Y \xrightarrow{p} X$ such that $pu = 1_X$. We say that $\mathscr{C}$ is a *category with retracts* if all idempotents of $\mathscr{C}$ split. Show that:

a) The idempotent $f: X \to X$ of $\mathscr{C}$ splits if and only if the double arrow $(f, 1_X): X \to X$ has a kernel, or equivalently, if and only if the double arrow $(f, 1_X)$ has a cokernel.

b) The category $\{\mathscr{C}^0, \mathscr{S}et\}$ is a category with retracts. Moreover, if f: $F \to F$ is an idempotent of $\{\mathscr{C}^0, \mathscr{S}et\}$, (F_1, u) is a kernel of $(f, 1_F)$, (q, F_2) is a cokernel of $(f, 1_F)$ and v: $F_2 \to F$ is a morphism such that $vq = f$, then the morphisms u: $F_1 \to F$ and v: $F_2 \to F$ define F as a coproduct of F_1 with F_2.

2.11. Let $\mathscr{C}$ be a category. Denote by $K_0(\mathscr{C})$ the full and replete subcategory of $\{\mathscr{C}^0, \mathscr{S}et\}$ whose objects are retracts of representable presheaves. It is clear that the canonical functor $H: \mathscr{C} \to \{\mathscr{C}^0, \mathscr{S}et\}$ defines a functor $H_0: \mathscr{C} \to K_0(\mathscr{C})$. The pair $(H_0\ K_0(\mathscr{C}))$ is called the *0-cocompletion* of $\mathscr{C}$. Show that:

a) $K_0(\mathscr{C})$ is category with retracts.

b) If $F: \mathscr{C} \to \mathscr{C}'$ is a functor, and $\mathscr{C}'$ is a category with retracts, then there is a unique functor $F_0: K_0(\mathscr{C}) \to \mathscr{C}'$ such that $F_0 H_0 = F$. (Hint: Define $F_0(H_X) = F(X)$ for all $X \in \mathscr{C}$. If $R \in K_0(\mathscr{C})$ then there are morphisms $R \xrightarrow{u} H_X \xrightarrow{p} R$ such that $pu = 1_R$ and we can define the coexact diagram

$$H_X \underset{i}{\overset{up}{\rightrightarrows}} H_X \xrightarrow{p} R.$$

Define $F_0(R) = \text{Coker } (F(1_X), F(up))$.)

Moreover if F is full and faithful then F_0 is too.

c) Let $\mathscr{C}$ and $\mathscr{D}$ be categories such that $\{\mathscr{C}^0, \mathscr{S}et\}$ and $\{\mathscr{D}^0, \mathscr{S}et\}$ are equivalent. Then $K_0(\mathscr{C})$ is equivalent to $K_0(\mathscr{D})$.

d) If $\mathscr{C}$ is a category with retracts, then the functor $H_0: \mathscr{C} \to K_0(\mathscr{C})$ is an equivalence of categories.

e) Denote by $K^0(\mathscr{C})$ the full and replete subcategory of $\{\mathscr{C}, \mathscr{S}et\}$ generated by all retracts of representable functors. Prove that $K^0(\mathscr{C})$ is equivalent to $(K_0(\mathscr{C}))^0$. The pair $(H^0, K^0(\mathscr{C}))$, where H^0 is defined by the canonical functor $H^0: \mathscr{C} \to \{\mathscr{C}, \mathscr{S}et\}$ is called a *0-completion* of $\mathscr{C}$.

2.12. An infinite cardinal number α is called *regular* if for every set $\{\alpha_i\}_{i \in \mathscr{I}}$ of cardinal numbers such that $\alpha_i < \alpha$ for all i, and where $\text{Card}(\mathscr{I}) < \alpha$, we get that $\Sigma_i \alpha_i < \alpha$. Let α be a regular cardinal number. We say that a category $\mathscr{I}$ is *α-small* if $\text{Card}(\text{Ob}(\mathscr{I})) < \alpha$ and $\text{Card}(\mathscr{I}(i,j)) < \alpha$ for all $i, j \in \mathscr{I}$. A category $\mathscr{I}$ is called *α-right bounded* if it has an α-small right bounding subcategory. We shall say that a category $\mathscr{C}$ is *α-cocomplete* (has *α-bounded colimits*) if every functor $F: \mathscr{I} \to \mathscr{C}$ with $\mathscr{I}$ α-small ($\mathscr{I}$ α-right bounded) has a colimit. In an obvious way we define the notions of *α-left bounded category*, *α-complete category* and *category with α-bounded limits*.

Let $\mathscr{C}$ be a category and α a regular cardinal number. Let $K_\alpha(\mathscr{C})$ (resp. $\bar{K}_\alpha(\mathscr{C})$) denote the full and replete subcategory of $\{\mathscr{C}^0, \mathscr{S}et\}$ defined as follows. First we consider the full and replete subcategory $\mathscr{C}'$ of $\{\mathscr{C}^0, \mathscr{S}et\}$ generated by all *α-coproducts* of representable functors (i.e. the coproducts indexed by α-small sets). Furthermore let $K_\alpha(\mathscr{C})$ (resp. $\bar{K}_\alpha(\mathscr{C})$) be the full and replete subcategory of $\{\mathscr{C}^0, \mathscr{S}et\}$ generated by cokernels of all double arrows of $\mathscr{C}'$ (generated by all quotient objects in $\{\mathscr{C}^0, \mathscr{S}et\}$ of all objects of $\mathscr{C}'$). It is clear that the functor $H: \mathscr{C} \to \{\mathscr{C}^0, \mathscr{S}et\}$ determines a functor $H_\alpha: \mathscr{C} \to K_\alpha(\mathscr{C})$ (resp. $\bar{H}_\alpha: \mathscr{C} \to \bar{K}_\alpha(\mathscr{C})$). The pair $(H_\alpha, K_\alpha(\mathscr{C}))$ (resp. $(\bar{H}_\alpha, \bar{K}_\alpha(\mathscr{C}))$) is called the *$\alpha$-cocompletion* (resp. the *α-bounding cocompletion*) of $\mathscr{C}$.

a) Show that $K_\alpha(\mathscr{C})$ (resp. $\bar{K}_\alpha(\mathscr{C})$) is an α-cocomplete category (resp. a category with α-bounded colimits). The colimits in these categories are formed argumentwise.

b) If $\mathscr{C}$ is a discrete category then $K_\alpha(\mathscr{C})$ and $\bar{K}_\alpha(\mathscr{C})$ are identical. Show that generally $K_\alpha(\mathscr{C})$ and $\bar{K}_\alpha(\mathscr{C})$ are not identical.

c) Let $F: \mathscr{C} \to \mathscr{C}'$ be a functor so that $\mathscr{C}'$ is an α-complete category (a category with α-bounded colimits). There is a unique functor F_α: $K_\alpha(\mathscr{C}) \to \mathscr{C}'$ (resp.

$\bar{F}_\alpha: \bar{K}_\alpha(\mathscr{C}) \to \mathscr{C}'$ so that $F_\alpha H_\alpha = F$ (resp. $\bar{F}_\alpha \bar{H}_\alpha = F$). Moreover F_α (resp. $\bar{F}_\alpha$) commutes with α-limits (resp. with α-bounded colimits). With some reasonable conditions F_α (resp. $\bar{F}_\alpha$) has a coadjoint. What are those conditions?

The *α-completion* (*α-bounding completion*) of a category $\mathscr{C}$ is defined in an obvious way using the category $\{\mathscr{C}, \mathscr{S}et\}$. It is denoted by $K^\alpha(\mathscr{C})$ (resp. $\bar{K}^\alpha(\mathscr{C})$). Generally $(K^\alpha(\mathscr{C}))^0$ and $K_\alpha(\mathscr{C})$ (resp. $(\bar{K}^{-\alpha}(\mathscr{C}))^0$ and $\bar{K}_\alpha(\mathscr{C})$) are not equivalent.

2.13. Let $\mathscr{C}$ be a category and let $K_\infty(\mathscr{C})$ be the full and replete subcategory of $\{\mathscr{C}^0, \mathscr{S}et\}$ generated by all objects which are colimits of small functors of representable presheaves, and let $H_\infty: \mathscr{C} \to K_\infty(\mathscr{C})$ be the canonical functor.

c) Show that $K_\infty(\mathscr{C})$ is a cocomplete category and, moreover, that if $F: \mathscr{C} \to \mathscr{C}'$ is a functor and $\mathscr{C}'$ is a complete category, then there is a unique functor $F_\infty: K_\infty(\mathscr{C}) \to \mathscr{C}'$ such that $F_\infty H_\infty = F$. (Compare with Exercise 9.9, Ch. 1.)

b) Show that the following are equivalent:

i) The inclusion functor $K_\infty(\mathscr{C}) \to \{\mathscr{C}^0, \mathscr{S}et\}$ has an adjoint.

ii) The inclusion functor $K_\infty(\mathscr{C}) \to \{\mathscr{C}^0, \mathscr{S}et\}$ is an equivalence of categories.

iii) $K_\infty(\mathscr{C})$ is a category with bounded colimits.

These conditions hold if $\mathscr{C}$ is a small or a discrete category.

2.3. Dense functors

Dense or left adequate functors were defined by Isbell [2]. This notion plays an important role in the study of the completion of categories. In this section we define dense functors and study some of their properties. A characterization of the category $\{\mathscr{C}^0, \mathscr{S}et\}$ is obtained.

References: Gabriel—Ulmer [1], Isbell [3], [8], [9], Lambek [1], Pereigis [1], Ulmer [1], [2].

Let $T: \mathscr{C} \to \mathscr{D}$ be a functor and D an object of $\mathscr{D}$. Recall that by T/D we mean (see Section 1.11) the category whose objects are pairs (X,f), where X is an object of $\mathscr{C}$ and $f: T(X) \to D$ is a morphism in $\mathscr{D}$. A morphism $g: (X,f) \to (X',f')$ of T/D is a morphism $g: X \to X'$ such that $f'T(g) = f$. Now let $s_D: T/D \to \mathscr{C}$ denote the underlying functor, i.e. the functor that assigns to each object (X,f) of T/D the object X of $\mathscr{C}$; the action of s_D on morphisms is obvious. Furthermore, for each object (X,f) of T/D let $\varphi_{(X,f)} = f: T(X) \to D$. It is easy to see that the morphisms $\{\varphi_{(X,f)}\}_{(X,f) \in T/D}$ define $\varphi: Ts_D \to D$, i.e. (φ, D) is a cocone over Ts_D, called the *canonical cocone associated with T and D.*

THEOREM 3.1. *Let $T: \mathscr{C} \to \mathscr{D}$ be a functor. The following are equivalent for each object D of $\mathscr{D}$:*

1) *The canonical cocone associated with T and D is a colimit of Ts_D.*

2) *For each object D' of $\mathscr{D}$, the assignment $g \rightsquigarrow H(g)T$ defines a bijection between $\mathscr{D}(D, D')$ and $[H_D T, H_{D'} T]$.*

Proof. 2) $\Rightarrow$ 1). Let (φ, D) be the canonical cocone associated with T and D. We want to show that (φ, D) is a colimit of Ts_D. For this purpose let (Ψ, D') be another cocone over Ts_D and let $\bar{\Psi}: \mathscr{D}(T(X), D) \to \mathscr{D}(T(X), D')$ denote the map that assigns to each morphism $f: T(X) \to D$, the morphism $\Psi_{(X,f)}: T(X) \to D'$ (the component of Ψ associated with the object (X, f) of T/D). It can be shown that since (Ψ, D') is a cocone over Ts_D, the morphisms $\{\bar{\Psi}_X\}_{X \in \mathscr{C}}$ define a functorial morphism $\bar{\Psi}: H_D T \to H_{D'} T$. By hypothesis, there is a unique morphism $g: D \to D'$ such that $\bar{\Psi} = H(g)T$. In particular, we get for any $X \in \mathscr{C}$ and any $f: F(X) \to D$, that $\Psi_{(X,f)} = \bar{\Psi}_X(f) = (H(g)T)_X(f) = gf = g\varphi_{(X,f)}$. Hence $g\varphi = \Psi$. If $g': D \to D'$ is another morphism of $\mathscr{D}$ such that $\Psi = g'\varphi$ then for every $X \in \mathscr{C}$ and every $f \in \mathscr{D}(T(X), D)$ we have that $\bar{\Psi}_X(f) = \Psi_{(X,f)} = g'\varphi_{(X,f)} = g'f = (H(g)T)_X(f)$. Consequently $\bar{\Psi} = H(g)\,T = H(g')T$, and thus $g = g'$ by hypothesis.

1) $\Rightarrow$ 2). Assume that (φ, D) is a colimit of Ts_D. We shall prove that the map $g \rightsquigarrow H(g)T$ is a bijection. Let $t: H_D T \to H_{D'} T$ be a functorial morphism. For each $X \in \mathscr{C}$ and each $f \in \mathscr{D}(T(X), D) = (H_D T)(X)$, consider the morphism $\Psi_{(X,f)} = t_X(f): T(X) \to D'$. Since t is a functorial morphism, the morphisms $\{\Psi_{(X,f)}\}_{(X,f) \in T/D}$ define a functorial morphism $\Psi: Ts_D \to D'$. Then, by hypothesis, there is a unique morphism $g: D \to D'$ such that $g\varphi = \Psi$. Now for every $X \in \mathscr{C}$ and every $f \in \mathscr{D}(T(X), D)$ we have that $gf = g\varphi_{(X,f)} = \Psi_{(X,f)} = t_X(f)$, or equivalently $H(g)T = t$. Finally, if $g': D \to D'$ is another morphism such that $H(g')T = t$, then for every $X \in \mathscr{C}$ and every $f \in \mathscr{D}(T(X), D)$ we have that $g'f = t_X(f) = gf = g\varphi_{(X,f)} = g'\varphi_{(X,f)}$. It follows from the definition of colimits that $g = g'$. The proof is now complete. ✳

A functor $T: \mathscr{C} \to \mathscr{D}$ satisfying the (equivalent) conditions of the above theorem is called *left adequate*, or *dense*, *for D*. If T is dense for every object D of $\mathscr{D}$, then it is called a *left adequate*, or *dense*, *functor*. A subcategory $\mathscr{C}'$ of a category $\mathscr{C}$ is called a *dense subcategory* if the inclusion functor $I: \mathscr{C}' \to \mathscr{C}$ is dense. A family $\{X_i\}$ of objects of $\mathscr{C}$ is called a family of *dense generators* if the full subcategory generated by them is dense.

Dually, a functor $T: \mathscr{C} \to \mathscr{D}$ is called *right adequate* or *codense* if the canonical functor $T^0: \mathscr{C}^0 \to \mathscr{D}^0$ is dense. The notion of *codense subcategory* is obtained in the obvious way.

As a consequence of Proposition 2.1, we have the following corollaries:

COROLLARY 3.2. *For every category $\mathscr{C}$, the canonical functor $H: \mathscr{C} \to \{\mathscr{C}^0, \mathscr{S}et\}$ is dense.* ✳

COROLLARY 3.3. *Let $T: \mathscr{C} \to \mathscr{D}$ be a properly defined functor. The following are equivalent:*

1) *The functor T is dense.*

2) *The functor $S^T: \mathscr{D} \to \{\mathscr{C}^0, \mathscr{S}et\}$ defined by $S^T(D) = H_D T$ is full and faithful.* ✳

We now give a characterization of the category $\{\mathscr{C}^0, \mathscr{S}et\}$.

THEOREM 3.4. *The following are equivalent for a category $\mathscr{D}$:*

1) *There is a category $\mathscr{C}$ such that $\mathscr{D}$ is equivalent to $\{\mathscr{C}^0, \mathscr{S}et\}$.*

2) *The following hold:*

a) *$\mathscr{D}$ is a category with bounded colimits.*

b) *$\mathscr{D}$ contains a full subcategory $\mathscr{C}$ such that the objects of $\mathscr{C}$ give a family of strong separators, and every object of $\mathscr{C}$ is 0-presentable in $\mathscr{D}$.*

c) *For every $D \in \mathscr{D}$, the category $\mathscr{C}/D$ is left bounded.*

3) *$\mathscr{D}$ is a category with bounded colimits and contains a full properly dense subcategory $\mathscr{C}$, whose objects are 0-presentable in $\mathscr{D}$.*

Proof. The implications 1) $\Rightarrow$2) and 1) $\Rightarrow$ 3) are obtained from Corollary 1.8, Exercises 3.9 and 1.17, and Theorem 1.1.

2) $\Rightarrow$ 1). Let $T: \mathscr{C} \to \mathscr{D}$ be the inclusion functor. It is easy to see that T is a properly defined functor.

By Theorem 2.3 and Corollary 2.7, the extension functor $\bar{T}: \{\mathscr{C}^0, \mathscr{S}et\} \to \mathscr{D}$ has a coadjoint $S = S^T$. Now we show that $\bar{T}$ and S define an equivalence of categories.

First, we show that the functor S commutes with bounded colimits. For this purpose let $F: \mathscr{I} \to \mathscr{D}$ be a functor such that $\mathscr{I}$ is right bounded; then $S(\varinjlim F)$: $\mathscr{C}^0 \to \mathscr{S}et$ is the presheaf defined as follows:

$$S(\varinjlim F)(X^0) = \mathscr{D}(X, \varinjlim F).$$

Now, since X is *0*-presentable for every $X \in \mathscr{C}$,

$$\mathscr{D}(X, \varinjlim F) \simeq \varinjlim H^X F = (\varinjlim SF)(X^0).$$

Since the last isomorphisms hold for every $X \in \mathscr{C}$, and since the bounded colimits of $\{\mathscr{C}^0, \mathscr{S}et\}$ are defined argumentwise, we see that

$$S(\varinjlim F) = \varinjlim(SF),$$

i.e. S commutes with bounded colimits.

Furthermore, let u: Id $\{\mathscr{C}^0, \mathscr{S}et\} \to S\bar{T}$ be an arrow of adjunction and v a coarrow quasi-inverse to u. We shall prove that u is a functorial isomorphism. First note that for each $X \in \mathscr{C}$, $u_{H_{X^0}}: H_{X^0} \to S\bar{T}(H_{X^0})$ is an isomorphism since T is full and faithful, $ST \simeq H$ and $\bar{T}H \simeq T$. If $F \in \{\mathscr{C}^0, \mathscr{S}et\}$, then by Proposition 2.1, F is the colimit of the functor Hs_F and $\bar{T}(F) = \varinjlim (Ts_F)$. But then $S\bar{T}(F) = S(\varinjlim Ts_F) \simeq \varinjlim(STs_F) \simeq \varinjlim(Hs_F) \simeq F$. It is easy to see that this last isomorphism is isomorphic to u_F. Hence u is a functorial isomorphism.

The proof of the implication 2) $\Rightarrow$ 1) will be complete if v is also a functorial isomorphism (see Theorem 4.1, Ch. 1). We note that for each $X \in \mathscr{C}$, the morphism $v_X\colon \overline{T}S(X) \to X$ is an isomorphism. If D is an object of $\mathscr{D}$, then we can define the commutative diagram

$$\begin{array}{ccc} \coprod_i \overline{T}S(X_i) & \xrightarrow{\overline{T}S(p)} & \overline{T}S(D) \\ {\scriptstyle v_{\coprod_i X_i}}\big\downarrow \wr & & \big\downarrow {\scriptstyle v_D} \\ \coprod_i X_i & \xrightarrow{\quad p \quad} & D \end{array} ;$$

$X_i \in \mathscr{C}$ for all i, where p is an epimorphism, and $v_{\coprod_i X_i}$ is an isomorphism since S commutes with coproducts. Then we construct in $\{\mathscr{C}^0, \mathscr{S}et\}$ the diagram

$$\begin{array}{ccc} \coprod_i S\overline{T}S(X_i) & \xrightarrow{S\overline{T}S(p)} & S\overline{T}S(D) \\ {\scriptstyle Sv_{\coprod_i X_i}}\big\downarrow \wr & & {\scriptstyle u_{S(D)}}\big\uparrow \; \big\downarrow {\scriptstyle Sv_D} \\ \coprod_i S(X_i) & \xrightarrow{\quad S(p) \quad} & S(D) \end{array} .$$

By the conditions (Ad) of Section 1.11, $u_{S(D)}$ is right inverse to Sv_D and by the last diagram we see that $u_{S(D)}$ is necessarily an epimorphism, since $S(p)$ and $S\overline{T}S(p)$ are epimorphisms. Hence $u_{S(D)}$ is an isomorphism, so that $Sv_D = S(v_D)$ is also an isomorphism. But this means that for each $X \in \mathscr{C}$, the morphism $S(v_D)_X$: $S\overline{T}S(D)(X) = \mathscr{D}(X, \overline{T}S(D)) \to \mathscr{D}(X, D) = S(D)(X)$ is an isomorphism, so that v_D is an isomorphism by hypothesis. Therefore the functorial coarrow v is a functorial isomorphism.

3) $\Rightarrow$ 1). As in the proof of implication 2) $\Rightarrow$ 1) we check that the functor $S\colon \mathscr{D} \to \{\mathscr{C}^0, \mathscr{S}et\}$ commutes with bounded colimits, which will imply that the arrow of adjunction $u\colon \mathrm{Id}\{\mathscr{C}^0, \mathscr{S}et\} \to S\overline{T}$ is an isomorphism. Now from Corollary 3.3 it follows that the functor $S^T = S\colon \mathscr{D} \to \{\mathscr{C}^0, \mathscr{S}et\}$ is full and faithful. Thus by Theorem 13.10, Ch. 1, it follows that v, the coarrow quasi-inverse to u, is also a functorial isomorphism. Again we use Theorem 4.1, Ch. 1. ✳

COROLLARY 3.5. (Gabriel—Ulmer [1]). *The following are equivalent for a category* $\mathscr{D}$:

1) *There is a small category* $\mathscr{C}$ *such that* $\mathscr{D}$ *is equivalent to* $[\mathscr{C}^0, \mathscr{S}et]$.
2) $\mathscr{D}$ *is cocomplete and has a set of 0-presentable strict separators.*
3) $\mathscr{D}$ *contains a small dense and full subcategory whose objects are 0-presentable in* $\mathscr{D}$. ✳

COROLLARY 3.6. (Bunge [1], Gabriel (unpublished)). *The following are equivalent for a category* $\mathscr{C}$:

1) *There is a small category* $\mathscr{I}$ *such that* $\mathscr{C}$ *is equivalent to* $[\mathscr{I}^0, \mathscr{S}et]$.

2) $\mathscr{C}$ *contains a set* $\{X_i\}_{i \in \mathscr{I}}$ *of objects, such that:*

a) $\mathscr{C}$ *is a category with kernel pairs and coequalizers.*

b) *If* $\{Y_j\}_j$ *is a set of objects of* $\mathscr{C}$ *such that for each* j *there is an* $i \in \mathscr{I}$ *so that* $Y_j \simeq X_i$ *then the coproduct* $\coprod_j Y_j$ *is defined in* $\mathscr{C}$.

c) *A morphism* $p: X \to Y$ *is a coequalizer in* $\mathscr{C}$ *if and only if for every* $i \in \mathscr{I}$, $H^{X_i}(p)$ *is a surjection.*

d) *The double arrow* $(f, g): X \to Y$ *of* $\mathscr{C}$ *is the kernel pair of the morphism* $p: Y \to Z$ *if and only if* $(H^{X_i}(f), H^{X_i}(g))$ *is a kernel pair of* $H^{X_i}(p)$ *for all* $i \in \mathscr{I}$.

e) *For every* $i \in \mathscr{I}$ *the functor* $H^{X_i}: \mathscr{C} \to \mathscr{S}et$ *commutes with coproducts.*

Proof. By Corollary 3.5, it suffices to show that 2) ⇒ 1). We shall show that condition 2) of Corollary 3.5 is verified. The proof is divided into several steps.

i) A morphism $f: X \to Y$ of $\mathscr{C}$ is an isomorphism if and only if for each $i \in \mathscr{I}$, $H^{X_i}(f)$ is an isomorphism.

Assume for each $i \in \mathscr{I}$ that $H^{X_i}(f)$ is an isomorphism. Consider the cartesian square

$$\begin{array}{ccc} Z & \xrightarrow{u} & X \\ {\scriptstyle v}\downarrow & & \downarrow{\scriptstyle f} \\ X & \xrightarrow{f} & Y \end{array}$$

By c), f is a coequalizer of (u, v) so that $H^{X_i}(f)$ is an equalizer of $(H^{X_i}(u), H^{X_i}(v))$; hence $H^{X_i}(u)$ and $H^{X_i}(v)$ are isomorphisms for all $i \in \mathscr{I}$. Now if t is a common right inverse of u and v we see that $H^{X_i}(t)$ is also an isomorphism, i.e. t is a coequalizer according to c). Therefore u and v are isomorphisms so that f is a monomorphism and finally an isomorphism (since f is a coequalizer by c)).

ii) The objects $\{X_i\}_{i \in \mathscr{I}}$ give a set of separators of $\mathscr{C}$.

To see that let $u, v: X \to Y$ be morphisms such that $H^{X_i}(u) = H^{X_i}(v)$ for all $i \in \mathscr{I}$. We claim that $u = v$. Indeed, if $(f, g): Z \to X$ is a kernel pair of u, then by d), $(H^{X_i}(f), H^{X_i}(g))$ is a kernel pair of $H^{X_i}(v)$ and thus (f, g) is a kernel pair of v, so that $uf = vf$ i.e. $u = v$ since f is an epimorphism.

iii) $\mathscr{C}$ is a cocomplete category. Since $\mathscr{C}$ is a category with cokernels, then by Theorem 9.2, Ch. 1, it suffices to show that $\mathscr{C}$ is a category with coproducts.

Consider an object X of $\mathscr{C}$ and let $\mathscr{F}$ denote the set of all pairs (Y, f), where $Y \in \{X_i\}_{i \in \mathscr{I}}$ and $f: Y \to X$ is a morphism of $\mathscr{C}$. Now let $\bar{X}$ denote $\coprod_{(Y, f) \in \mathscr{F}} Y$, and let $p_X: \bar{X} \to X$ be the unique morphism such that $p_X u_{(Y, f)} = f$ for all $(Y, f) \in \mathscr{F}$, where $u_{(Y, f)}$ is the structural morphism. According to its definition, for each $i \in \mathscr{I}$, $H^{X_i}(p_X)$ is a surjection. Hence, by c), p_X is a coequalizer.

Now let $\{Y_j\}_j$ be a set of objects of $\mathscr{C}$. We can define for each j the following coequalizer diagram:

$$Z_j \underset{v_j}{\overset{u_j}{\rightrightarrows}} \bar{Y}_j \xrightarrow{p_{Y_j}} Y_j,$$

where (u_j, v_j) is a kernel pair of p_{Y_j}. Let f_j denote $u_j p_{Z_j}$ and g_j denote $v_j p_{Z_j}$. Then

$$\bar{Z}_j \underset{g_j}{\overset{f_j}{\rightrightarrows}} \bar{Y}_j \xrightarrow{p_{Y_j}} Y_j$$

is also a coequalizer diagram.

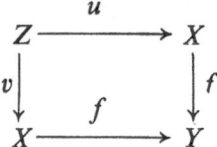

According to b), we can define the coproducts $\coprod_j \bar{Z}_j$ and $\coprod_j \bar{Y}$. Let f denote $\coprod f_j$ and g denote $\coprod g_j$. We leave it for the reader to check that the object Coker (f, g) becomes (in a canonical way) a coproduct of the family $\{Y_j\}_j$ of objects.

iv) The reader may check that for each $i \in \mathscr{I}$ the functor H^{X_i} is right exact.

Finally, by i) and ii) we see that the objects $\{X_i\}_{i \in \mathscr{I}}$ give a set of strict generators for $\mathscr{C}$. By iii) we set that $\mathscr{C}$ is cocomplete and by e) and iv) we see that the objects $\{X_i\}_{i \in \mathscr{I}}$ are 0-presentable. Hence the proof follows from Corollary 3.5, as claimed. ✳

Exercises

3.1. Let $T: \mathscr{C} \to \mathscr{D}$ be a dense functor. Assume that T carries distinct objects into distinct objects. Show that the subcategory ImT of $\mathscr{D}$ is dense.

3.2. Let $T: \mathscr{C} \to \mathscr{D}$ be a functor and S a coadjoint of T. Show that the following are equivalent:

1) S is full and faithful.
2) T is dense.

Moreover, if $K: \mathscr{C}' \to \mathscr{C}$ is a dense functor, the following are equivalent:

1) S is full and faithful.
2) TK is dense.

3.3. Let $T: \mathscr{C} \to \mathscr{D}$ be a dense functor, and $F: \mathscr{D} \to \mathscr{E}$ a functor such that F commutes with the limits of all the functors Ts_D, $D \in \mathscr{D}$. Prove that F is dense for an object E of $\mathscr{E}$, whereas FT is dense for $\mathscr{E}$ and, moreover, that if F is dense then FT is dense.

Note. If $F: \mathscr{C} \to \mathscr{D}$ and $G: \mathscr{D} \to \mathscr{E}$ are dense functors, then the functor GF is not generally dense. Indeed, we have the following counterexample due to Isbell [2]. Let $\mathscr{E}$ be the category with three objects, W, X, Y, and with the nine sets $\mathscr{E}(A, B)$ defined as follows: $\mathscr{E}(W, W)$ is a group with two elements; $\mathscr{E}(X, Y)$ contains two elements; $\mathscr{E}(X, W)$, $\mathscr{E}(Y, W)$ and $\mathscr{E}(Y, X)$ are empty; the others are one element sets. The composition of morphisms is defined in an obvious manner. The full subcategory of $\mathscr{E}$ generated by W and X is dense; in it the full subcategory whose only object is W is dense, but this subcategory is not dense in $\mathscr{E}$.

3.4. Prove that the canonical functor $H: \mathscr{C} \to \{\mathscr{C}^0, \mathscr{S}et\}$ is not codense.

3.5. Prove that $\mathscr{S}etf$ is a dense subcategory of $\mathscr{S}et$.

3.6. A subcategory $\mathscr{C}'$ of a category $\mathscr{C}$ is called an *adequate subcategory* if it is both dense and codense.

Let $\mathscr{C}$ be an adequate subcategory of $\mathscr{C}'$, and $\mathscr{C}'$ an adequate subcategory of $\mathscr{C}''$. Show that $\mathscr{C}$ is an adequate subcategory of $\mathscr{C}''$.

3.7. Assume that $\mathscr{C}$ is a dense subcategory of $\mathscr{C}'$, and let $F, F': \mathscr{C}' \to \mathscr{C}''$ be functors that commute with colimits. Show that if $I: \mathscr{C} \to \mathscr{C}'$ is the inclusion functor and $FI \simeq F'I$, then $F \simeq F'$.

3.8. Let $F: \mathscr{C} \to \mathscr{D}$ be a properly dense functor. Show that the following are equivalent:

a) $\mathscr{D}$ is a category with bounded colimits.

b) The induced functor $S^F: \mathscr{D} \to \{\mathscr{C}^0, \mathscr{S}et\}$ has an adjoint.

3.9. Show that for each object F of $\{\mathscr{C}^0, \mathscr{S}et\}$ the functor $H_F H: \mathscr{C} \to \mathscr{S}et$ is canonically isomorphic to the functor FD, where $D: \mathscr{C} \to \mathscr{C}^0$ is the duality functor.

3.10. Prove that the following are equivalent for a category $\mathscr{D}$:

1) $\mathscr{D}$ is equivalent to a full reflective subcategory of the category $[\mathscr{C}^0, \mathscr{S}et]$, where $\mathscr{C}$ is a small category.

2) $\mathscr{D}$ contains a full and dense small subcategory.

A category $\mathscr{D}$ as in Exercise 3.10 is called a *retract*.

3.11. Let $\mathscr{C}$ be a small full and dense subcategory of $\mathscr{D}$. Assume that $Y = \coprod_{X \in \mathscr{C}} X$ is an object of $\mathscr{D}$ and that for all $X, X' \in \mathscr{C}$, the set $\mathscr{C}(X, X')$ is non-empty. Show that the full subcategory of $\mathscr{D}$ generated by Y is a dense subcategory of $\mathscr{D}$.

3.12. A set $\{X_i\}_{i \in \mathscr{I}}$ of objects of a category $\mathscr{C}$ is called a *set of strict generators* of $\mathscr{C}$ if and only if:

1) For each $X \in \mathscr{C}$, the coproduct $\coprod_{i \in \mathscr{I},\, f \in \mathscr{C}(X_i, X)} Y_{(i,f)}$ exists, where $Y_{(i,f)} = X_i$, for all $i \in \mathscr{I}$ and for all $f \in \mathscr{C}(X_i, X)$, and

2) The canonical morphism $p: \coprod_{i \in \mathscr{I},\, f \in \mathscr{C}(X_i, X)} Y_{(i,f)} \to X$ defined such that $pu_{(i,f)} = f$ (where the $u_{(i,f)}$ are structural injections) is a strict epimorphism.

Let $\mathscr{C}$ be a category with fibered products and universal coproducts and let $\mathscr{C}'$ be a full and small subcategory of $\mathscr{C}$. Prove that $\mathscr{C}'$ is a dense subcategory of $\mathscr{C}$ if and only if the objects of $\mathscr{C}'$ give a set of strict generators of $\mathscr{C}$.

3.13. Let $T = (u, v, F)$ be a cotriple in a category $\mathscr{C}$. Let $\mathscr{C}'$ denote the full subcategory of $\mathscr{C}$ generated by all objects $F(X)$, when X runs through all objects of $\mathscr{C}$. Prove that $\mathscr{C}'$ is a dense subcategory of $\mathscr{C}$ if and only if for every object X of $\mathscr{C}$, the diagram

$$F^2(X) \underset{(uF)_X}{\overset{Fu_X}{\rightrightarrows}} F(X) \xrightarrow{u_X} F(X)$$

is coexact.

3.14. Show that $I \coprod I$, where I is the unit interval of the real line, is a dense cogenerator of the category $\mathscr{C}om$ of compact topological spaces. (See Isbell [2] and Gabriel—Ulmer [1].)

3.15. Let F be a presheaf over $\mathscr{C}$. We define the functor $F^*: \mathscr{C} \to \mathscr{S}et$ as follows. For each $X \in \mathscr{C}$, $F^*(X) = [F, H_X]$, and for each morphism f in $\mathscr{C}$, $F^*(f) = [F, H(f)]$. The quasifunctor $(F^*)^*$ is denoted by F^{**}. There is a canonical functorial morphism $F \xrightarrow{\alpha} F^{**}$, called the *evaluation functorial morphism*, defined as follows. If $x \in F(X)$, then $\alpha_X(x) \in [F^*, H^X] = F^{**}(X)$; and for each $Y \in \mathscr{C}$, $u \in F^*(Y) = [F, H_Y]$, we have that $[\alpha_X(x)]_Y(u) = u_X(x)$. We call F a *reflexive presheaf* if F^* is proper and the evaluation functor morphism is a functorial isomorphism.

Let $\mathscr{C}'$ be a full dense and properly defined subcategory of $\mathscr{C}$. Show that $\mathscr{C}'$ is properly codense if and only if for each $X \in \mathscr{C}$ the cofunctor $H_X I$ (where $I: \mathscr{C}' \to \mathscr{C}$ is the inclusion functor) is reflexive.

2.4. Σ-sheaves

This section deals with the following problem. Let Σ be a class of morphisms in $\{\mathscr{C}^0, \mathscr{S}et\}$ and let $\tilde{\mathscr{C}}_\Sigma$ be the full subcategory of all Σ-sheaves. Under what conditions does the inclusion functor $\tilde{\mathscr{C}}_\Sigma \to \{\mathscr{C}^0, \mathscr{S}et\}$ have an adjoint? The main result is Theorem 4.13 that contains as special cases many results in this field. Example 4.18 shows that this problem does not always have a solution.

References: Gabriel—Ulmer [1], Gabriel—Zisman [1], D. Popescu [3], [5].

Let Σ be a class of morphisms in $\{\mathscr{C}^0, \mathscr{S}et\}$. An object F of $\{\mathscr{C}^0, \mathscr{S}et\}$ is called a *Σ-sheaf* (resp. *Σ-separated presheaf*) if for every $s \in \Sigma$, $s: A \to B$, the map $[s, F]: [B, F] \to [A, F]$ induced by s is a bijection (resp. an injection). Using the terminology of Section 1.15, the object F is a Σ-sheaf if and only if it is left closed for Σ. When there is no danger of confusion, we shall simply call F a sheaf (resp. separated presheaf). The full subcategory of $\{\mathscr{C}^0, \mathscr{S}et\}$ whose objects are all Σ-sheaves (resp. Σ-separated presheaves) will be denoted by $\tilde{\mathscr{C}}_\Sigma$ (resp. $\overline{\mathscr{C}}_\Sigma$).

Let $\mathscr{D}$ be a nonempty full subcategory of $\{\mathscr{C}^0, \mathscr{S}et\}$. Let $\mathscr{C}sp(\mathscr{D})$ denote the full subcategory of $\{\mathscr{C}^0, \mathscr{S}et\}$ that is formed by all the objects T satisfying the following condition: for each $X \in \mathscr{C}$ and $x, x' \in T(X)$ with $x \neq x'$ there is an $F \in \mathscr{D}$ and $f: T \to F$ such that $f_X(x) \neq f_X(x')$. $\mathscr{C}sp(\mathscr{D})$ is called the *coseparated subcategory associated* with $\mathscr{D}$. An object of $\mathscr{C}sp(\mathscr{D})$ will be called a *$\mathscr{D}$-coseparated presheaf*. The study of the category of Σ-sheaves is interconnected with the study of the category $\mathscr{C}sp(\tilde{\mathscr{C}}_\Sigma)$, which contains $\tilde{\mathscr{C}}_\Sigma$ and is contained in $\overline{\mathscr{C}}_\Sigma$.

We say that a functor $T: \mathscr{C} \to \mathscr{C}'$ *generates* (*cogenerates*) $\mathscr{C}'$ if the objects $\{T(X)\}_{X \in \mathscr{C}}$ give a class of generators (cogenerators) of $\mathscr{C}'$. A subcategory $\mathscr{C}$ of a category $\mathscr{C}'$, generates (cogenerates) $\mathscr{C}$ if the inclusion functor generates (cogenerates) $\mathscr{C}'$.

Note 4.1. i) $\mathscr{C}sp(\mathscr{D})$ is closed under subobjects, that is, for every monomorphism $u: T' \to T$ of $\{\mathscr{C}^0, \mathscr{S}et\}$, with $T \in \mathscr{C}sp(\mathscr{D})$ we have that $T' \in \mathscr{C}sp(\mathscr{D})$.

ii) It is clear that $\mathscr{D} \subseteq \mathscr{C}sp(\mathscr{D})$ cogenerates $\mathscr{C}sp(\mathscr{D})$. Precisely, if $\mathscr{D}$ is small, then $\mathscr{C}sp(\mathscr{D})$ has a set of cogenerators.

iii) Let $T: \mathscr{C} \to \mathscr{S}et$ be a presheaf. The category $T/\mathscr{C}sp(\mathscr{D})$ may be empty. In fact, if $\mathscr{C} = \{0\}$ then $[\{0\}^0, \mathscr{S}et] = \mathscr{S}et$, and if $\mathscr{D}$ is the full subcategory of

$\mathscr{S}et$ consisting of a single object, the empty set, then $\mathscr{C}sh(\mathscr{D}) = \mathscr{D}$ and $T/\mathscr{D}$ is empty even though T is not the empty set.

PROPOSITION 4.2. *Let Σ be a class of morphisms of $\{\mathscr{C}^0, \mathscr{S}et\}$. Then, for any presheaf $T: \mathscr{C} \to \mathscr{S}et$, the categories $T/\tilde{\mathscr{C}}_\Sigma$, $T/\mathscr{C}sh(\mathscr{C})$ and $T/\overline{\mathscr{C}}_\Sigma$ are nonempty.*

Proof. Let $\overline{T}$ denote the presheaf on $\mathscr{C}$ defined as follows:

$$\overline{T}(X) = \begin{cases} \{*\} & \text{if} \quad T(X) \neq \emptyset, \\ \emptyset & \text{if} \quad T(X) = \emptyset. \end{cases}$$

It is clear that $\overline{T}$ is a Σ-sheaf for every class Σ of morphisms of $\{\mathscr{C}^0, \mathscr{S}et\}$, and there is a canonical epimorphism $T \to \overline{T}$. ✳

PROPOSITION 4.3. *Assume that for each object T of $\{\mathscr{C}^0, \mathscr{S}et\}$ the category $T/\mathscr{C}sh(\mathscr{D})$ is nonempty. Then $\mathscr{C}sh(\mathscr{D})$ is a reflective subcategory of $\{\mathscr{C}^0, \mathscr{S}et\}$.*

Proof. We use Lemma 20.7, Ch. 1. By Note 4.1, i), it suffices to check that the lower bound $(t, \overline{T})$ of all quotient objects of T that belong to $\mathscr{C}sh(\mathscr{D})$ (the lower bound is calculated in $\{\mathscr{C}^0, \mathscr{S}et\}$), also belongs to $\mathscr{C}sh(\mathscr{D})$. For this, let $X \in \mathscr{C}$, and x, $x' \in \overline{T}(X)$, with $x \neq x'$. By the construction of $\overline{T}$ (see the proof of Proposition 17.16, Ch. 1) there is a quotient object (p, P) of T, with $P \in \mathscr{C}sh(\mathscr{D})$ such that $q_X(x) \neq q_X(x')$ (where $q: \overline{T} \to P$ is canonically defined). Then there are an object $F \in \mathscr{C}sh(\mathscr{D})$ and a morphism $f: P \to F$ such that $f_X q_X(x) \neq f_X q_X(x')$. Hence $\overline{T} \in \mathscr{C}sh(\mathscr{D})$. ✳

Note 4.4. i) Since $(t, \overline{T})$ is an initial object of $T/\mathscr{C}sh(\mathscr{D})$, Proposition 4.3 follows easily from Theorem 11.8, Ch. 1.

ii) If T is not proper, the previous result still holds in a more general setting. Therefore a cofunctor $F: \mathscr{C} \to \mathscr{S}et$ (not necessarily proper) is said to be a *$\mathscr{D}$-coseparated object* if for each $X \in \mathscr{C}$, and for each $x, x' \in F(X)$, with $x \neq x'$, there is a $D \in \mathscr{D}$ and a functorial morphism $f: F \to D$ such that $f_X(x) \neq f_X(x')$. With these changes, Proposition 4.3 becomes:

"Let $T: \mathscr{C} \to \mathscr{S}et$ be a cofunctor such that there is a functorial morphism $f: T \to F$, where F is $\mathscr{D}$-coseparated. Then there is a functorial epimorphism $p: T \to \overline{T}$, with $\overline{T}$ being $\mathscr{D}$-coseparated such that for any functorial morphism $f: T \to F$ with F being $\mathscr{D}$-coseparated, there is a unique functorial morphism $\overline{f}: \overline{T} \to F$ such that $\overline{f}p = f$."

Observe that the $\mathscr{D}$-coseparated objects need not form a category.

COROLLARY 4.5. *For each class Σ of morphisms of $\{\mathscr{C}°, \mathscr{S}et\}$, the inclusion functor $\mathscr{C}sh(\overline{\mathscr{C}}_\Sigma) \to \{\mathscr{C}°, \mathscr{S}et\})$ (resp. $\mathscr{C}sh(\tilde{\mathscr{C}}_\Sigma) \to \{\mathscr{C}°, \mathscr{S}et\})$ has an adjoint.*

The proof follows directly from Propositions 4.2 and 4.3. ✳

PROPOSITION 4.6. *Let Σ be a class of morphisms from $\{\mathscr{C}^0, \mathscr{S}et\}$. Then we have that:*

1) $\mathscr{C}sh(\overline{\mathscr{C}}_\Sigma) = \overline{\overline{\mathscr{C}}}_\Sigma$.

2) $\mathscr{C}sh(\tilde{\mathscr{C}}_\Sigma) \subseteq \overline{\mathscr{C}}_\Sigma$.

3) *The inclusion functor* $\overline{\mathscr{C}}_\Sigma \to \{\mathscr{C}^0, \mathscr{S}et\}$ *has an adjoint, i. e.* $\overline{\mathscr{C}}_\Sigma$ *is a full reflective subcategory of* $\{\mathscr{C}^0, \mathscr{S}et\}$.

4) $\mathscr{C}sh(\tilde{\mathscr{C}}_\Sigma)$ *and* $\overline{\mathscr{C}}_\Sigma$ *are categories with bounded colimits. The monomorphisms of these categories are argumentwise monomorphisms.*

Proof. In order to obtain 1) it will suffice to show that if $T \in \mathscr{C}sh(\overline{\mathscr{C}}_\Sigma)$ and if $s \in \Sigma$, then $H_T(s)$ is a monomorphism (see Note 4.1, ii)). For that, let $s: A \to B$, and $f, g: B \to T$ be such that $fs = gs$. Then for every morphism $t: T \to F$, with $F \in \overline{\mathscr{C}}_\Sigma$ we have that $tfs = tgs$. But then $tf = tg$ since F is Σ-separated. Thus for each $X \in \mathscr{C}$ and $x \in B(X)$ we get $t_X f_X(x) = t_X g_X(x)$. Since $T \in \mathscr{C}sh(\overline{\mathscr{C}}_\Sigma)$ we see that $f_X(x) = g_X(x)$, i.e. $f = g$. Now since $\tilde{\mathscr{C}}_\Sigma \subseteq \overline{\mathscr{C}}_\Sigma$ we conclude that $\mathscr{C}sh(\tilde{\mathscr{C}}_\Sigma) \subseteq \mathscr{C}sh(\overline{\mathscr{C}}_\Sigma)$, and from this we get 2).

3) follows from 1) and Corollary 4.5; 4) follows from Proposition 4.3. ✳

Generally, the inclusion in 2) is not an equality, as may be seen from Exercise 4.5.

Let us denote by L_Σ(resp. $\overline{L}_\Sigma$) an adjoint to the inclusion functor $I_\Sigma: \mathscr{C}sh(\tilde{\mathscr{C}}_\Sigma) \to \{\mathscr{C}^0, \mathscr{S}et\}$ (resp. $\overline{I}_\Sigma: \overline{\mathscr{C}}_\Sigma \to \{\mathscr{C}^0, \mathscr{S}et\}$). However, when no confusion is possible, we shall denote these functors by L and $\overline{L}$.

Let Σ be a class of morphisms of $\{\mathscr{C}^0, \mathscr{S}et\}$ and $L(\Sigma) = \{L(s)\}_{s \in \Sigma}$. It is easy to see that any Σ-sheaf is an $L(\Sigma)$-sheaf. However, in general not every $L(\Sigma)$-sheaf is a Σ-sheaf (see Exercise 4.5).

If Σ is a class of morphisms of $\{\mathscr{C}^0, \mathscr{S}et\}$, let $\tilde{\Sigma}$ (resp. $\overline{\Sigma}$) denote the class of all morphisms t such that $[t, F]$ is a bijection (an injection) for every Σ-sheaf (Σ-separated presheaf). It is clear that $\Sigma \subseteq \overline{\Sigma}$ and $\Sigma \subseteq \tilde{\Sigma}$

COROLLARY 4.7. *For every class* Σ *of morphisms of* $\{\mathscr{C}^0, \mathscr{S}et\}$ *the categories* $\tilde{\mathscr{C}}_\Sigma$ *and* $\tilde{\mathscr{C}}_{\tilde{\Sigma}}$ *are identical.*

The proof is left for the reader. ✳

LEMMA 4.8. *Let* Σ *be a class of morphisms of* $\{\mathscr{C}^0, \mathscr{S}et\}$. *Then:*

a) *A morphism* t *of* $\{\mathscr{C}^0, \mathscr{S}et\}$ *belongs to* $\tilde{\Sigma}$ *if and only if* $L(t)$ *belongs to* $\tilde{\Sigma}$.

b) *A morphism* t *of* $\{\mathscr{C}^0, \mathscr{S}et\}$ *belongs to* $\overline{\Sigma}$ *if and only if* $\overline{L}(t)$ *is an epimorphism of* $\overline{\mathscr{C}}_\Sigma$.

c) *For each* $t \in \tilde{\Sigma}$, *the morphism* $L(t)$ *is a bimorphism of* $\mathscr{C}sh(\mathscr{C})$.

Moreover, for every $t \in \tilde{\Sigma}$, $L(t)$ *is an argumentwise monomorphism*, i.e. *a monomorphism of* $\{\mathscr{C}^0, \mathscr{S}et\}$.

Proof. a) Consider the following commutative diagram of $\{\mathscr{C}^0, \mathscr{S}et\}$:

$$\begin{array}{ccc} F & \xrightarrow{\quad t \quad} & G \\ {\scriptstyle \varphi_F}\downarrow & & \downarrow{\scriptstyle \varphi_G} \\ L(F) & \xrightarrow{\quad L(t) \quad} & L(G) \end{array} \qquad (3)$$

where φ_F and φ_G are arrows of reflection. Assume that $t \in \tilde{\Sigma}$. Let $u, v\colon L(G) \to R$ be such that $uL(t) = vL(t)$ and R is a Σ-sheaf. Then $uL(t)\varphi_F = vL(t)\varphi_F = u\varphi_G t = v\varphi_G t$, i.e. $u\varphi_G = v\varphi_G$, since $t \in \tilde{\Sigma}$. But then $u = v$ since φ_G is an arrow of reflection. Also let $u\colon L(F) \to R$ be such that $R \in \tilde{\mathscr{C}}_\Sigma$. By hypothesis there is a unique morphism $u'\colon G \to R$ such that $u't = u\varphi_F$. Since $R \in \mathscr{C}\mathit{sh}(\tilde{\mathscr{C}}_\Sigma)$, there is a unique morphism $\bar{u}\colon L(G) \to R$ with $\bar{u}\varphi_G = u'$. But then $\bar{u}L(t)\varphi_F = \bar{u}\varphi_G t = u't = u\varphi_F$, i.e. $\bar{u}L(t) = u$. Hence $L(t)$ belongs to $\tilde{\Sigma}$.

Conversely, assume that $L(t) \in \tilde{\Sigma}$. Let $u, v\colon G \to R$ be such that $ut = vt$ and $R \in \tilde{\mathscr{C}}_\Sigma$. There are morphisms $\bar{u}, \bar{v}\colon L(G) \to R$ with $\bar{u}\varphi_G = u$ and $\bar{v}\varphi_G = v$. Hence $\bar{u}L(t)\varphi_F = \bar{u}\varphi_G t = ut = vt = \bar{v}\varphi_G t = \bar{v}L(t)\varphi_F$ so that $\bar{u}L(t) = \bar{v}L(t)$. Therefore $\bar{u} = \bar{v}$, and necessarily $u = v$. Furthermore, let $u\colon F \to R$ be such that $R \in \tilde{\mathscr{C}}_\Sigma$. There is a unique morphism $u'\colon L(F) \to R$ with $u'\varphi_F = u$. Since $L(t) \in \tilde{\Sigma}$, there is a unique morphism $u''\colon L(G) \to R$ with $u''L(t) = u'$. Then $u''\varphi_G t = u''L(t)\varphi_F = u'\varphi_F = u$. Hence $t \in \tilde{\Sigma}$.

b) Consider the following commutative diagram of $\{\mathscr{C}^0, \mathscr{S}\mathit{et}\}$:

$$\begin{array}{ccc} F & \xrightarrow{\quad t \quad} & G \\ {\scriptstyle \psi_F}\downarrow & & \downarrow{\scriptstyle \psi_G} \\ \bar{L}(F) & \xrightarrow{\quad \bar{L}(t) \quad} & \bar{L}(G) \end{array} \qquad (4)$$

where ψ_F and ψ_G are arrows of reflection. Assume that $t \in \bar{\Sigma}$ and let $u, v\colon \bar{L}(G) \to R$ be such that $u\bar{L}(t) = v\bar{L}(t)$, and $R \in \bar{\mathscr{C}}_\Sigma$. Then $u\bar{L}(t)\,\psi_F = v\bar{L}(t)\,\psi_F = u\psi_G t = v\psi_G t$, i.e. $u\psi_G = v\psi_G$. Hence $u = v$, since ψ_G is an arrow of reflection. Hence $\bar{L}(t)$ is an epimorphism.

Conversely, assume that $\bar{L}(t)$ is an epimorphism in $\bar{\mathscr{C}}_\Sigma$, and let $u, v\colon G \to R$ be such that $ut = vt$ and $R \in \bar{\mathscr{C}}_\Sigma$. There are morphisms $\bar{u}, \bar{v}\colon \bar{L}(G) \to R$ with $\bar{u}\psi_G = u$, $\bar{v}\psi_G = v$. Then $\bar{u}\bar{L}(t)\psi_F = \bar{u}\psi_G t = ut = vt = \bar{v}\psi_G t = \bar{v}\bar{L}(t)\,\psi_F$, i.e. $\bar{u}\bar{L}(t) = \bar{v}\bar{L}(t)$. But then $\bar{u} = \bar{v}$, since $\bar{L}(t)$ is an epimorphism. Therefore $u = v$, i.e. $t \in \bar{\Sigma}$.

c) Consider diagram (3) where $t \in \tilde{\Sigma}$. If $X \in \mathscr{C}$, and $x, x' \in L(F)(X)$ with $x \neq x'$; there exist $R \in \tilde{\mathscr{C}}_\Sigma$ and an $f\colon L(F) \to R$ such that $f_X(x) \neq f_X(x')$. By a) there is a unique morphism $\bar{f}\colon L(G) \to R$ such that $\bar{f}L(t) = f$. Hence $L(t)$ is a functorial monomorphism. Furthermore, let $u, v\colon L(G) \to K$ be such that $uL(t) = vL(t)$ and $K \in \mathscr{C}\mathit{sh}(\tilde{\mathscr{C}}_\Sigma)$. Assume that, to the contrary, $u \neq v$. Then there are an $X \in \mathscr{C}$ and an $x \in L(G)(X)$ such that $u_X(x) \neq v_X(x)$. Now let $R \in \tilde{\mathscr{C}}_\Sigma$, and $f\colon K \to R$ such that $f_X(u_X(x)) \neq f_X(v_X(x))$. It is clear that $fuL(t) = fvL(t)$, so that by a) $fu = fv$, since $L(t) \in \tilde{\Sigma}$. But then $f_X u_X(x) = f_X v_X(x)$, which is a contradiction. Hence $L(t)$ is an epimorphism of $\mathscr{C}\mathit{sh}(\tilde{\mathscr{C}}_\Sigma)$. ✻

COROLLARY 4.9. *Let Σ be a class of morphisms of $\{\mathscr{C}^0, \mathscr{S}\mathit{et}\}$. Denote by Σ^1(resp. $\bar{\Sigma}^1$) the class of all morphisms f which are made invertible by L(resp. $\bar{L}$). Then:*

a) $\Sigma^1 \subseteq \tilde{\Sigma}$ (*resp.* $\bar{\Sigma}^1 \subseteq \bar{\Sigma}$).

b) *$\tilde{\Sigma}$ (resp. $\bar{\Sigma}$) is a left calculable and saturated system of morphisms of* $\{\mathscr{C}^0, \mathscr{S}et\}$.

Proof. a) Assume that $L(t)$ is invertible. Then for each $F \in \tilde{\mathscr{C}}_\Sigma$ the map $[L(t), F]$ is a bijection so that $L(t) \in \tilde{\Sigma}$. But then $t \in \tilde{\Sigma}$, by Lemma 4.8. a. Hence $\Sigma^1 \subseteq \tilde{\Sigma}$. Likewise, from Lemma 4.8.b it follows that $\bar{\Sigma}^1 \subseteq \bar{\Sigma}$.

b) It is clear that $\tilde{\Sigma}$ is a multiplicative system. Now consider the cocartesian square of $\{\mathscr{C}^0, \mathscr{S}et\}$

$$\begin{array}{ccc} F & \xrightarrow{\ t\ } & G \\ {\scriptstyle f}\downarrow & & \downarrow{\scriptstyle f'} \\ K & \xrightarrow{\ t'\ } & K\coprod_F G \end{array}$$

where $t \in \tilde{\Sigma}$.

Since L has a coadjoint, the following square of $\mathscr{C}o\!\not\!p(\tilde{\mathscr{C}}_\Sigma)$ is again cartesian

$$\begin{array}{ccc} L(f) & \xrightarrow{\ L(t)\ } & L(G) \\ {\scriptstyle L(f)}\downarrow & & \downarrow{\scriptstyle L(f')} \\ L(K) & \xrightarrow{\ L(t')\ } & L(K\coprod_F G) \end{array}$$

and $L(t) \in \tilde{\Sigma}$ by Lemma 4.8. a. Now we claim that $L(t') \in \tilde{\Sigma}$. To see this let $R \in \tilde{\mathscr{C}}_\Sigma$, and u, v: $L(K\coprod_F G) \to R$ be such that $uL(t) = vL(t')$. Then $uL(t')L(f) = vL(t')L(f) = uL(f')L(t) = vL(f')L(t)$ so that $uL(f') = vL(f')$. Hence $u = v$, since the square is cocartesian. Furthermore, let u: $L(K) \to R$ with $R \in \tilde{\mathscr{C}}_\Sigma$. Then there is a unique morphism u': $L(G) \to R$ such that $u'L(t) = uL(f)$. Hence there is a unique morphism $\bar{u}$: $L(K\coprod_F G) \to R$ such that $\bar{u}L(t') = u$. Therefore, $L(t') \in \tilde{\Sigma}$ and thus $t' \in \tilde{\Sigma}$ by Lemma 4.8. a.

Now consider the diagram

$$\begin{array}{ccccc} F & \xrightarrow{\ t\ } & G & \overset{u}{\underset{v}{\rightrightarrows}} & K \\ {\scriptstyle \varphi_F}\downarrow & & \downarrow{\scriptstyle \varphi_G} & & \downarrow{\scriptstyle \varphi_K} \\ L(F) & \xrightarrow{\ L(t)\ } & L(G) & \overset{L(u)}{\underset{L(v)}{\rightrightarrows}} & L(K) \end{array}$$

where $ut = vt$ and $t \in \tilde{\Sigma}$.

A few computations prove that $L(u) = L(v)$ (since $L(t)$ is a bimorphism of $\mathscr{C}o\!\not\!p(\tilde{\mathscr{C}}_\Sigma)$, by Lemma 4.8. c) so that $\varphi_K u = \varphi_K v$. But it is clear that $\varphi_K \in \Sigma^1 \subseteq \tilde{\Sigma}$. Hence $\tilde{\Sigma}$ is a left calculable system.

For proving that $\tilde{\Sigma}$ is saturated, consider the commutative diagram of $\{\mathscr{C}^0, \mathscr{S}et\}$

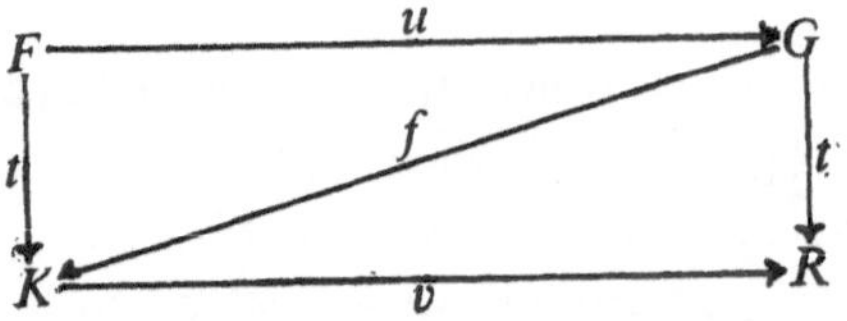

where $t, t' \in \tilde{\Sigma}$ (see Exercise 14.6 a, Ch. 1). Then $f \in \tilde{\Sigma}$. Indeed, if $P \in \tilde{\mathscr{C}}_\Sigma$ and $p, q\colon K \to P$ are such that $pf = qf$, then $pt = qt$ that is $p = q$. Also, if $h\colon G \to P$ with $P \in \tilde{\mathscr{C}}_\Sigma$, there is a unique morphism $h'\colon R \to P$ so that $h't' = h$. Then $h'vf = h't' = h$. Therefore $f \in \tilde{\Sigma}$, i.e. $\tilde{\Sigma}$ is saturated. In the same way we obtain that $\bar{\Sigma}$ is saturated. ✳

Now we again consider the problem of the existence of an adjoint to the inclusion functor $U\colon \tilde{\mathscr{C}}_\Sigma \to \{\mathscr{C}^0, \mathscr{S}et\}$. We get the following result.

THEOREM 4.10. *Let Σ be a class of morphisms of $\{\mathscr{C}^0, \mathscr{S}et\}$. The following are equivalent:*

1) $\mathscr{C}osp(\tilde{\mathscr{C}}_\Sigma) = \tilde{\mathscr{C}}_\Sigma$.

2) $\tilde{\Sigma} = \Sigma^1$.

3) $\Sigma \subseteq \Sigma^1$.

4) *There is an adjoint $\tilde{L}$ to U and the arrow of reflection $\alpha_F\colon F \to \tilde{L}(F)$ is an epimorphism of $\{\mathscr{C}^0, \mathscr{S}et\}$ for every F.*

Proof. 1) $\Rightarrow$ 2). By Corollary 4.9. a, $\Sigma^1 \subseteq \tilde{\Sigma}$, and we need only to check that $\tilde{\Sigma} \subseteq \Sigma^1$. For that consider diagram (3) where $t \in \tilde{\Sigma}$. Since $L(t) \in \tilde{\Sigma}$ (see Lemma 4.8.a) there is a unique morphism $k\colon L(G) \to L(F)$ such that $kL(t) = 1_{L(F)}$. But then $L(t)kL(t) = L(t)$ so that $L(t)k = 1_{L(G)}$. Hence $t \in \Sigma^1$.

3) $\Rightarrow$ 1). Let $R \in \mathscr{C}osp(\tilde{\mathscr{C}}_\Sigma)$ and $t \in \Sigma$. We shall prove that $[t, R]$ is a bijection. Again consider diagram (3). Let $u, v\colon G \to R$ be such that $ut = vt$. There are $\bar{u}, \bar{v}$: $L(G) \to R$ with $\bar{u}\varphi_G = \bar{v}\varphi_G = v$. Then $\bar{u}L(t) = \bar{v}L(t)$, i.e. $\bar{u} = \bar{v}$ and necessarily $u = v$. Now let $u\colon F \to R$ be a morphism, k an inverse of $L(t)$ and $u'\colon L(F) \to R$ with $u'\varphi_F = u$. Then $u'k\,\varphi_G\colon G \to R$ is the unique morphism such that $u'k\,\varphi_G t = u'kL(t)\varphi_F = u'\varphi_F = u$.

4) $\Rightarrow$ 1). Since $\tilde{\mathscr{C}}_\Sigma \subseteq \mathscr{C}osp\,(\tilde{\mathscr{C}}_\Sigma)$, for each $F \in \{\mathscr{C}^0, \mathscr{S}et\}$ there is a unique morphism $\gamma_F\colon L(F) \to \tilde{L}(F)$ such that $\gamma_F\varphi_F = \alpha$. By hypothesis, γ_F is a functorial epimorphism. Actually we claim that it is also a functorial monomorphism. Indeed, let $X \in \mathscr{C}$ and $x, x' \in L(F)(X)$ be such that $x \neq x'$. There are a $K \in \tilde{\mathscr{C}}_\Sigma$ and $f\colon L(F) \to K$ such that $f_X(x) \neq f_X(x')$. Since α_F is an arrow of reflection there is an $f'\colon \tilde{L}(F) \to K$ such that $f'\gamma_F\varphi_F = f'\alpha_F = f\varphi_F$. But then $f'\gamma_F = f$, so that γ_F is necessarily a monomorphism. Therefore γ_F is in fact an isomorphism of $\{\mathscr{C}^0, \mathscr{S}et\}$. The details and the remaining implications are left for the reader. ✳

Now we give an application of Theorem 4.10.

COROLLARY 4.11. *Let Σ be a class of epimorphisms of $\{\mathscr{C}^0, \mathscr{S}et\}$. Then $\mathscr{C}osp(\tilde{\mathscr{C}}_\Sigma) = \tilde{\mathscr{C}}_\Sigma$.*

Proof. We prove that $\Sigma \subseteq \Sigma^1$. If $s\colon F \to G$ is an element of Σ then $\varphi_G s$ is also an epimorphism (as always $\varphi_G\colon G \to L(G)$ is an arrow of reflection), since by Proposition 4.3, φ_G is an epimorphism. Then $L(s)\varphi_F$ is also an epimorphism so that $L(s)$ is an epimorphism, actually in $\{\mathscr{C}^0, \mathscr{S}et\}$. Moreover, by Lemma 4.8. c, $L(s)$ is a functorial monomorphism, so that $L(s)$ is an isomorphism. ✳

Exercise 4.9 shows that the converse of Corollary 4.11 is not true, i.e. there is a class Σ of morphisms of $\mathscr{C}$ such that $\tilde{\mathscr{C}}_\Sigma = \mathscr{C}sp(\tilde{\mathscr{C}}_\Sigma)$ where the morphisms of Σ are not epimorphisms.

Proposition 4.12. *Let Σ be a class of morphisms of $\{\mathscr{C}^0, \mathscr{S}et\}$. The following are equivalent:*

1) *The inclusion functor $U\colon \tilde{\mathscr{C}}_\Sigma \to \{\mathscr{C}^0, \mathscr{S}et\}$ has an adjoint.*

2) *For each $F \in \{\mathscr{C}^0, \mathscr{S}et\}$, the category $F/\tilde{\Sigma}$ has a final object.*

Proof. 1)$\Rightarrow$2). Recall that $F/\tilde{\Sigma}$ is the full subcategory of $F/\{\mathscr{C}^0, \mathscr{S}et\}$, whose objects are (t, G) with $t \in \tilde{\Sigma}$. Now if $\tilde{L}$ is an adjoint to U, and $\alpha_F\colon F \to \tilde{L}(F)$ is an arrow of reflection, then $\alpha_F \in \tilde{\Sigma}$ and $(\alpha_F, \tilde{L}(F))$ is a final object of $F/\tilde{\Sigma}$.

2) $\Rightarrow$ 1). Let $(\alpha_F, \tilde{F})$ be a final object of $F/\tilde{\Sigma}$. First we prove that $\tilde{F} \in \tilde{\mathscr{C}}_\Sigma$. For that let $t\colon A \to B$ be an element of $\tilde{\Sigma}$ and let $v\colon A \to \tilde{F}$ be a morphism. Then we have the commutative diagram

$$\begin{array}{ccccc} & & A & \xrightarrow{\ t\ } & B \\ & & {\scriptstyle v}\downarrow & & \downarrow{\scriptstyle v'} \\ F & \xrightarrow{\ \alpha_F\ } & \tilde{F} & \xrightarrow{\ t'\ } & B' \end{array}$$

where $t' \in \tilde{\Sigma}$, since $\tilde{\Sigma}$ is left calculable (see Corollary 4.9). But then $(t'\alpha_F, B') \in F/\tilde{\Sigma}$, so that there is a unique morphism $k\colon B' \to \tilde{F}$ with $kt'\alpha_F = \alpha_F$. Now since $1_{\tilde{F}}\alpha_F = \alpha_F$ we see that $kt' = 1_{\tilde{F}}$. Hence $kv'\colon B \to \tilde{F}$ is the unique morphism such that $kv't = kt'v = v$. Now, using Corollary 4.7, we get that $\tilde{F} \in \tilde{\mathscr{C}}_\Sigma$.

Furthermore, $(\alpha_F, \tilde{F})$ is an initial object of the category $F/\tilde{\mathscr{C}}_\Sigma$ since $\alpha_F \in \tilde{\Sigma}$. ✳

In the sequel we shall give sufficient conditions for the category $F/\tilde{\Sigma}$ to have a final object, or equivalently, the inclusion functor $U\colon \tilde{\mathscr{C}}_\Sigma \to \{\mathscr{C}^0, \mathscr{S}et\}$ to have an adjoint.

Let Σ be a class of morphisms of $\{\mathscr{C}^0, \mathscr{S}et\}$. For the sake of simplicity we let A_s denote the domain and B_s denote the codomain of $s \in \Sigma$. Also for every presheaf F we let $\Sigma(F)$ denote $\{s \mid s \in \Sigma \text{ and } [A_s, F] \neq \emptyset\}$. For each $s \in \Sigma(F)$ and each $h \in [A_s, F]$ we consider the following cocartesian square of $\{\mathscr{C}^0, \mathscr{S}et\}$

$$\begin{array}{ccc} A_s & \xrightarrow{\ s\ } & B_s \\ {\scriptstyle h}\downarrow & & \downarrow{\scriptstyle h'} \\ F & \xrightarrow{\ t'_{(s,h)}\ } & F'_{(s,h)} \end{array} \qquad (5)$$

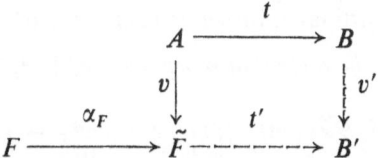

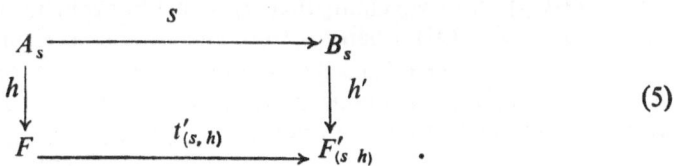

Then, applying L, we obtain the following cocartesian square of $\mathscr{C}op(\tilde{\mathscr{C}}_\Sigma)$

$$\begin{array}{ccc} L(A_s) & \xrightarrow{L(s)} & L(B_s) \\ {\scriptstyle L(h)}\downarrow & & \downarrow{\scriptstyle L(h')} \\ L(F) & \xrightarrow{L(t'_{(s,h)}) = t_{(s,h)}} & L(F'_{(s,h)}) = F_{(s,h)} \end{array} \quad . \tag{6}$$

By Corollary 4.9. b, $t'_{(s,h)} \in \tilde{\Sigma}$ so that by Lemma 4.8 $t_{(s,h)}$ is an epimorphism of $\mathscr{C}op(\tilde{\mathscr{C}}_\Sigma)$. Now denote by $\mathscr{I}_F$ the full subcategory of $L(F)/\mathscr{C}op(\tilde{\mathscr{C}}_\Sigma)$ whose objects are

$$\{t_{(s,h)}, F_{(s,h)})\}_{\substack{s\in\Sigma(F)\\ h\in[A_s, F]}}$$

and let $K_F: \mathscr{I}_F \to L(F)/\mathscr{C}op(\tilde{\mathscr{C}}_\Sigma)$ denote the inclusion functor. These notations will be used later.

We need the following definition. An ordinal number θ is called *inaccessible* if it is a limit ordinal and for every set $\{\alpha_i\}_{i\in\mathscr{I}}$ of ordinal numbers such that $\alpha_i < \theta$ for all i and also such that $\mathrm{Card}(\mathscr{I}) < \mathrm{Card}(\theta)$, there is an ordinal number $\alpha < \theta$ such that $\alpha_i \leqslant \alpha$ for all i. It is easy to see that for every set $\{\alpha_i\}_i$ of ordinal numbers there is an inaccessible ordinal number θ such that $\alpha_i < \theta$ for all i.

THEOREM 4.13. *Let Σ be a class of morphisms of $\{\mathscr{C}^0, \mathscr{S}et\}$ such that the following conditions are satisfied:*

(∗) *For each $F \in \{\mathscr{C}^0, \mathscr{S}et\}$, the functor $K_F: \mathscr{I}_F \to L(F)/\mathscr{C}op(\tilde{\mathscr{C}}_\Sigma)$ has a colimit (in $L(F)/\mathscr{C}op(\tilde{\mathscr{C}}_\Sigma)$).*

(∗∗) *There is a set $\mathscr{D}$ of objects of $\mathscr{C}$ such that $\mathscr{D}$ dominates A_s for all $s \in \Sigma$.*

(∗∗∗) *There is a cardinal β such that $\mathrm{Card}(A_s(X)) \leqslant \beta$ for every $X \in \mathscr{D}$ and every $s \in \Sigma$.*

Then the inclusion functor $U: \tilde{\mathscr{C}}_\Sigma \to \{\mathscr{C}^0, \mathscr{S}et\}$ has an adjoint.

Proof. Let θ be an inaccessible ordinal number such that $\mathrm{Card}(\theta) > \beta\mathrm{Card}(\mathscr{O}b(\mathscr{D}))$. For each ordinal $\delta \leqslant \theta$ we shall define a pair $(\alpha_F^\delta, F^\delta)$, where $F^\delta \in \mathscr{C}op(\tilde{\mathscr{C}}_\Sigma)$ and $\alpha_F^\delta: L(F) \to F^\delta$ is a bimorphism belonging to $\tilde{\Sigma}$.

a) Let F^1 be the colimit of the functor K_F (the colimit is defined in $L(F)/\mathscr{C}op(\tilde{\mathscr{C}}_\Sigma)$). Let $u^1: K_F \to F^1$ be the structural morphism. Then for every $s, s' \in \Sigma(F)$ and every $h \in [A_s, F]$, and $h' \in [A_{s'}, F]$, we have $u^1_{(s,h)} t_{(s,h)} = u^1_{(s',h')} t_{(s',h')} = \alpha_F^1$; in this way we have defined the morphism α_F^1. Since the colimit is calculated in $\mathscr{C}op(\tilde{\mathscr{C}}_\Sigma)$ then $F^1 \in \mathscr{C}op(\tilde{\mathscr{C}}_\Sigma)$. Also we claim that α_F^1 is a bimorphism and $\alpha_F^1 \in \tilde{\Sigma}$. Indeed, let $X \in \mathscr{C}$ and $x, x' \in L(F)(X)$ where $x \neq x'$. Then there is a morphism $f: L(F) \to R$ with $R \in \tilde{\mathscr{C}}_\Sigma$ and such that $f_X(x) \neq f_X(x')$. Now since $t_{(s,h)} \in \tilde{\Sigma}$ for every $s \in \Sigma(F)$ and every $h \in [A_s, F]$, there is a unique morphism $f_{(s,h)}: F_{(s,h)} \to R$ such that $f_{(s,h)} t_{(s,h)} = f$. It is easy to check that the morphisms $(f_{(s,h)})_{\substack{s\in\Sigma(F)\\ h\in[A_s, F]}}$ give a functorial morphism

$f: K_{F'} \to R$ so that there is a unique morphism $f^1: F^1 \to R$ with $f^1\alpha_F^1 = f$. Hence α_F^1 is a functorial monomorphism (and a bimorphism in $\mathscr{C}o\not{p}(\mathscr{C}_\Sigma)$) belonging to $\tilde{\Sigma}$.

b) Now if δ is an ordinal number such that $(\alpha_F^\delta, F^\delta)$ has been defined, then $F^{\delta+1} = (F^\delta)^1$ and $\alpha_F^{\delta+1} = (\alpha_{F^\delta})^1\alpha_F^\delta$ where $(F^\delta)^1$ and $(\alpha_{F^\delta})^1$ have the same meaning as F^1 and α_F^1. (We observe that $L(F^\delta) \equiv F^\delta$, if $\delta \geqslant 1$.)

c) Furthermore, if δ is a limit ordinal number and $(\alpha_F^\gamma, F^\gamma)$ has been defined for each $\gamma < \delta$, then F^δ will be defined as the colimit of the direct system of $L(F)/\mathscr{C}o\not{p}(\tilde{\mathscr{C}}_\Sigma)$: $(F^\gamma, \alpha_F^{\gamma\gamma'})_{\gamma<\delta}$. Here $\alpha_F^{\gamma\gamma'}: F^\gamma \to F^{\gamma'}$, $\gamma \leqslant \gamma'$, is the unique morphism such that $\alpha_F^{\gamma\gamma'}\alpha_F^\gamma = \alpha_F^{\gamma'}$. Also, $\alpha_F^\delta: L(F) \to F^\delta$ will be defined as the composition $u^\gamma\alpha_F^\gamma$, where $\gamma < \delta$ and $u^\gamma: F^\gamma \to F^\delta$ is the structural injection. We leave it for the reader to show that α_F^δ is a bimorphism of $\tilde{\Sigma}$, and that u^γ is a functorial monomorphism. (In fact u^γ is a bimorphism and belongs to $\tilde{\Sigma}$.)

Now we claim that $F^\theta \in \tilde{\mathscr{C}}_\Sigma$. Let $s \in \Sigma$ and let $h: A_s \to F^\theta$ be a morphism in $\{\mathscr{C}^0, \mathscr{S}et\}$. Also let $(X_i, x_i)_{i\in\mathscr{I}}$ be a set of generators for A_s when $X_i \in \mathscr{D}$ and $x_i \in A_s(X_i)$ for all $i \in \mathscr{I}$. Since F^θ is the colimit of the direct system $(F^\gamma, \alpha_F^{\gamma\gamma'})_{\gamma<\theta}$ and the structural morphisms $u^\gamma: F^\gamma \to F^\theta$ are functorial monomorphisms, we get that for each $i \in \mathscr{I}$, there is an ordinal number $\gamma_i < \theta$ and a $y_i \in F^{\gamma_i}(X_i)$ such that $u_{X_i}^{\gamma_i}(y_i) = h_{X_i}(x_i)$. Now since $\mathrm{Card}(\mathscr{I}) \leqslant \mathrm{Card}(\mathscr{D})\beta < \mathrm{Card}(\theta)$, we can choose an ordinal number $\gamma < \theta$, such that $\gamma_i \leqslant \gamma$ for all $i \in \mathscr{I}$. But then it is clear that $\mathrm{Im}(h) \subseteq \mathrm{Im}(u^*)$ so that there is a morphism $h': A_s \to F^\gamma$ with $u^\gamma h' = h$. Furthermore, consider the commutative diagram

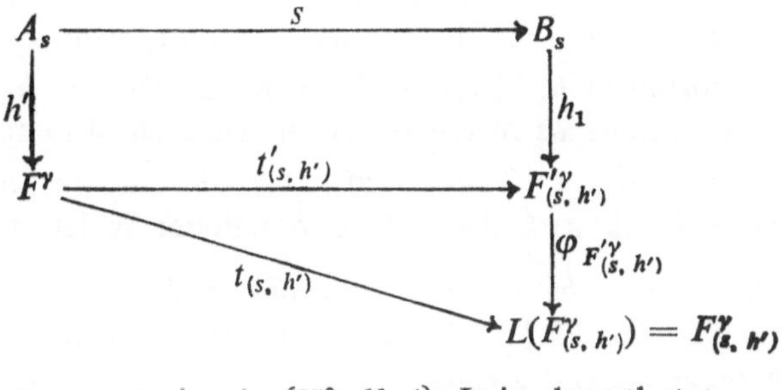

where the square is cocartesian in $\{\mathscr{C}^0, \mathscr{S}et\}$. It is clear that $t_{(s,h')}$ is a bimorphism of $\tilde{\Sigma}$, so that the object $(t_{(s,h')}, F^\gamma_{(s,h')})$ belongs to the category $\mathscr{I}_{F^\gamma}$. Now let $v_{(s,h')}: F^\gamma_{(s,h')} \to F^{\gamma+1}$ be the structural injection (see the definition of $F^{\gamma+1}$ given in b)). Then

$$u^{\gamma+1}v_{(s,h')}\varphi_{F'^\gamma_{(s,h')}}h_1 s = u^{\gamma+1}v_{(s,h')}\varphi_{F'^\gamma_{(s,h')}}t'_{(s,h')}h'$$
$$= u^{\gamma+1}v_{(s,h')}t_{(s,h')}h' = u^{\gamma+1}\alpha_F^{\gamma,\gamma+1}h' = u^\gamma h' = h.$$

This last relation proves that $F^\theta \in \tilde{\mathscr{C}}_\Sigma$.

Finally, let α_F be the composition $F \xrightarrow{\varphi_F} L(F) \xrightarrow{\alpha_F^\theta} F^\theta = \tilde{F}$. The reader should check that $(\alpha_F, \tilde{F})$ is an initial object of the category $F/\tilde{\mathscr{C}}_\Sigma$, so that by Theorem 11.8, Ch. 1, the inclusion functor $U: \tilde{\mathscr{C}}_\Sigma \to \{\mathscr{C}^0, \mathscr{S}et\}$ has an adjoint. ✻

COROLLARY 4.14. *Let Σ be a class of morphisms of $\{\mathscr{C}^0, \mathscr{S}et\}$ such that the conditions (**) and (***) of Theorem 4.13 are satisfied. Assume that for each $F \in \{\mathscr{C}^0, \mathscr{S}et\}$ the class $\Sigma(F) = \{s \in \Sigma | [A_s, F] \neq \emptyset\}$ is a set. Then the inclusion functor $U: \tilde{\mathscr{C}}_\Sigma \to \{\mathscr{C}^0, \mathscr{S}et\}$ has an adjoint.*

Proof. Using the notation from the proof of Theorem 4.13, it is easy to see that the category $\mathscr{I}_F$ is small for every $F \in \{\mathscr{C}^0, \mathscr{S}et\}$. Hence condition (*) of Theorem 4.13 is also satisfied. ✳

COROLLARY 4.15. *For every set Σ of morphisms of $\{\mathscr{C}^0, \mathscr{S}et\}$ the inclusion functor $U: \tilde{\mathscr{C}}_\Sigma \to \{\mathscr{C}^0, \mathscr{S}et\}$ has an adjoint.* ✳

Another interesting case when the inclusion functor $\tilde{\mathscr{C}}_\Sigma \to \{\mathscr{C}^0, \mathscr{S}et\}$ has an adjoint is given in the following lemma.

LEMMA 4.16. *Let Σ be a class of morphisms of $\{\mathscr{C}^0, \mathscr{S}et\}$. Assume that condition (*) of Theorem 4.13 is satisfied. Also assume that the objects $\{L(A_s)\}_{s \in \Sigma}$ satisfy conditions (**) and (***) of the same theorem. Then the inclusion functor $\tilde{\mathscr{C}}_\Sigma \to \{\mathscr{C}^0, \mathscr{S}et\}$ has an adjoint.*

Proof. Choose an inaccessible ordinal number θ such that $\text{Card}(\theta) > \beta\text{Card}(\mathscr{O}b(\mathscr{D}))$ where $\mathscr{D}$ is a common dominating set of objects of $\mathscr{C}$, for all $L(A_s)$ and $s \in \Sigma$, and β is defined such that $\text{Card}L(A_s)(\mathscr{D})) \leqslant \beta$ for all $s \in \Sigma$ and $D \in \mathscr{D}$. As in the proof of Theorem 4.13 we can define the pair $(\alpha^\theta, F^\theta)$. Now we must prove that $F^\theta \in \tilde{\mathscr{C}}_\Sigma$. For that let $s \in \Sigma$ and $h: A_s \to F^\theta$ be a functorial morphism. Since $F^\theta \in \mathscr{C}sh(\tilde{\mathscr{C}}_\Sigma)$, there is a unique morphism $\bar{h}: L(A_s) \to F^\theta$ such that $\bar{h}\varphi_{A_s} = h$. Furthermore, we may use the same procedure as in the proof of Theorem 4.13 to see that there is a unique morphism $u: L(B_s) \to F^\theta$ such that $uL(s) = \bar{h}$. But then $u\varphi_{B_s}s = uL(s)\varphi_{A_s} = \bar{h}\varphi_{A_s} = h$. Hence F^θ is a Σ-sheaf. The remainder is left for the reader. ✳

THEOREM 4.17. *Let $\mathscr{C}$ be a small category and Σ a class of morphisms of $\mathscr{C}^0, \mathscr{S}et]$ such that the class $\{B_s\}_{s \in \Sigma}$ has a set of distinct elements. Then the inclusion functor $\tilde{\mathscr{C}}_\Sigma \to [\mathscr{C}^0, \mathscr{S}et]$ has an adjoint.*

Proof. Let $s \in \Sigma$ and let $h: A_s \to F$ be a morphism of $[\mathscr{C}^0, \mathscr{S}et]$. Consider diagrams (5) and (6). By Proposition 1.5, $F'_{(s,h)}$ is a quotient object of $F \coprod B_s$. Also by Proposition 4.3, $L(F'_{(s,h)}) = F_{(s,h)}$ is a quotient object of $F'_{(s,h)}$. Now since $[\mathscr{C}^0, \mathscr{S}et]$ is locally cosmall (see Exercise 16.6, Ch. 1) it follows that the category $\mathscr{I}_F$ defined in Theorem 4.13 is equivalent to a small category. Hence condition (*) of Theorem 4.13 is satisfied. Furthermore, by Lemma 4.8.c we see that $L(A_s)$ is a subobject of $L(B_s)$, so that conditions (**) and (***) of Theorem 4.13 are clearly satisfied for the objects $\{L(A_s)\}_{s \in \Sigma}$. Hence Theorem 4.17 follows from Lemma 4.16. ✳

It is natural to ask if for each class Σ of morphisms of $\{\mathscr{C}^0, \mathscr{S}et\}$ the inclusion functor $\tilde{\mathscr{C}}_\Sigma \to \{\mathscr{C}^0, \mathscr{S}et\}$ has an adjoint. This is not always possible as the following example shows.

Example 4.18. Let Λ be the class of all cardinal numbers. For each $\alpha \in \Lambda$, choose a set M_α such that $\mathrm{Card}(M_\alpha) = \alpha$, and $M_\alpha \subseteq M_{\alpha'}$ when $\alpha \leqslant \alpha'$. Let $\mathscr{C}^0$ denote the subcategory of $\mathscr{S}et$ whose objects are $\{M_\alpha\}_{\alpha \in \Lambda}$ and where morphisms are defined as follows:

a) For each $\alpha \in \Lambda$, $\alpha \geqslant 1$ we have $\mathscr{C}^0(M_1, M_\alpha) = \mathscr{S}et(M_1, M_\alpha)$ and $\mathscr{C}^0(M_\alpha, M_1) = \mathscr{S}et(M_\alpha, M_1)$.

b) For each α and β, with $\alpha > \beta > 1$, we have that $\mathscr{C}^0(M_\beta, M_\alpha) =$ the inclusion $M_\beta \subseteq M_\alpha$, denoted $\bar{s}^\alpha_\beta$, and $\mathscr{C}^0(M_\alpha, M_\beta) = \emptyset$. Also we have $\mathscr{C}^0(M_\alpha, M_\alpha) = \{1_{M_\alpha}\}$ for every α.

Let $U^0: \mathscr{C}^0 \to \mathscr{S}et$ be the inclusion functor and $D: \mathscr{C} \to \mathscr{C}^0$ the duality cofunctor. Let U denote $U^0 D$. It is easy to see that U is a proper cofunctor generated by M_1. Let s^α_β denote the dual of $\bar{s}^\alpha_\beta$. Consider in the category $\{\mathscr{C}^0, \mathscr{S}et\}$ the class $\Sigma = \{H(s^\alpha_\beta)\}_{1<\beta<\alpha}$ where as always $H: \mathscr{C} \to \{\mathscr{C}^0, \mathscr{S}et\}$ is the canonical functor. We assert that the inclusion functor $\mathscr{C}_\Sigma \to \{\mathscr{C}^0, \mathscr{S}et\}$ does not have an adjoint.

Assume, to the contrary, that this functor does have an adjoint. Then the cofunctor U will be associated through an adjunction with a Σ-sheaf $\ddot{U}$; let $u: U \to \ddot{U}$ be the canonical morphism. Now consider the functors $\{L^\alpha: \mathscr{C}^0 \to \mathscr{S}et\}_{\alpha>1}$ defined as follows:

$$L^\alpha(M_\beta) = \begin{cases} M_\beta & \text{if } \beta > 1 \\ M_1 & \text{if } \beta = 1. \end{cases}$$

Before defining L^α for morphisms in $\mathscr{C}^0$, we make some observations. For each $\alpha \in \Lambda$, fix an element x_α of M_α and let $\bar{f}^\alpha_\beta: M_\beta \to M_\alpha$ denote the map defined by

i) $\bar{f}^\alpha_\beta = \bar{s}^\alpha_\beta$ if $\beta < \alpha$;

ii) $\bar{f}^\alpha_\beta(x) = \begin{cases} x & \text{if } x \in M_\alpha, \\ x_\alpha & \text{if } x \notin M_\alpha; \end{cases}$ and if $\alpha < \beta$

iii) $\bar{f}^\alpha_\alpha = 1_M$.

Then, if $g: M_\beta \to M_\gamma$ is a morphism in $\mathscr{C}^0$, define L^α by $L^\alpha(g) = 1_{M_\alpha}$ for $\beta, \gamma > 1$ and $L^\alpha(g) = \bar{f}^\alpha_\gamma g$ for $\beta = 1$ and $\gamma > 1$. Since M_1 has only one element, for every $\gamma \geqslant 1$, there is a unique morphism $t_\gamma: M_\gamma \to M$ so that $L^\alpha(t_\gamma) = t_\alpha$ if $\gamma > 1$ and $L^\alpha(t_1) = L^\alpha(1_{M_1}) = 1_{M_1}$.

It is easy to check that each L^α is dominated by M_1 and L^α is a Σ-sheaf for each α. Also for every $\alpha > 1$, the maps $\{\bar{f}^\alpha_\beta\}_\beta$ define a functorial morphism $f^\alpha: U \to L^\alpha$. By the adjunction property it follows that there is a unique morphism $\tilde{f^\alpha}: \ddot{U} \to L^\alpha$ such that $\tilde{f^\alpha} u = f^\alpha$. In particular, we get that $\tilde{f}^\alpha_{M_\alpha} u_{M_\alpha} = f^\alpha_{M_\alpha} = \bar{f}^\alpha_\alpha = 1_{M_\alpha}$, so that $\tilde{f}^\alpha_{M_\alpha}$ is a surjection. Hence for $\alpha > 2$, $\mathrm{Card}(\ddot{U}(M_2)) = \mathrm{Card}(\ddot{U}(M_\alpha)) \geqslant \mathrm{Card}(L^\alpha(M_\alpha)) = \mathrm{Card}(M_\alpha) = \alpha$. The first equality results from the fact that $\ddot{U}$ is a Σ-sheaf. In conclusion, it follows that in the category of sets there is a set of maximal cardinality, which is a contradiction.

Exercises

4.1. Let Σ be a class of morphisms of $\{\mathscr{C}^0, \mathscr{S}et\}$ and $s: F \to G$, where $s \in \Sigma$. Show that if G is Σ-separated and F is a Σ-sheaf, then s is an isomorphism, and moreover, that $\tilde{\mathscr{C}}_\Sigma = \{\mathscr{C}^0, \mathscr{S}et\}$ if and only if every element of Σ is an isomorphism.

4.2. Let $\mathscr{A}$ be the class of all nonempty full subcategories of $\{\mathscr{C}^0, \mathscr{S}et\}$. Show that the assignment $\mathscr{D} \rightsquigarrow \mathscr{C}sp(\mathscr{D})$ gives a "closure operator" on $\mathscr{A}$, i.e.:

i) $\mathscr{D} \subseteq \mathscr{C}sp(\mathscr{D})$.

ii) $\mathscr{C}sp(\mathscr{C}sp(\mathscr{D})) = \mathscr{C}sp(\mathscr{D})$.

iii) $\mathscr{C}sp(\mathscr{D} \cup \mathscr{D}') = \mathscr{C}sp(\mathscr{D}) \cup \mathscr{C}sp(\mathscr{D}')$.

4.3. Let Σ be a class of morphisms of $\{\mathscr{C}^0, \mathscr{S}et\}$ and let $\psi_F: F \to \bar{L}_2$ be an arrow of reflection, and $\Sigma_1 = \{\psi_F\}_{F \in \{\mathscr{C}^0, \mathscr{S}et\}}$. Prove that a presheaf G is a Σ-separated presheaf if and only if it is a Σ_1-sheaf. (Hint: Observe that ψ_F is an epimorphism and has a retraction whereas F is a Σ_1-sheaf.)

4.4. Let Σ be a class of morphisms of $\{\mathscr{C}^0, \mathscr{S}et\}$ and $F: \mathscr{I} \to \tilde{\mathscr{C}}_\Sigma$ a functor such that $\varprojlim F$ exists in $\{\mathscr{C}^0, \mathscr{S}et\}$. Prove that $\varprojlim F$ is also a Σ-sheaf. Hence $\tilde{\mathscr{C}}_\Sigma$ is complete whereas $\mathscr{C}$ is small.

4.5. Let $s: \{1, 2\} \to \{1\}$ be the morphism of $\mathscr{S}et = \{\{0 \to 1\}^0, \mathscr{S}et\}$ and $\Sigma = \{s\}$. Show that $\{0 \to 1\}_\Sigma = \mathscr{S}et$ while $\mathscr{C}sp(\widetilde{\{0 \to 1\}}_\Sigma) \simeq \widetilde{\{0 \to 1\}}_{\bar{\Sigma}}$. Also show that $L_\Sigma(s)$ is an isomorphism, so that $\widetilde{\{0 \to 1\}}_\Sigma = \mathscr{S}et$.

4.6. Let $F: \mathscr{I} \to \mathscr{C}$ be a functor. By a *coseparator cone* over F we mean a cone (X, u) over F such that for every double arrow $(f, g): Y \to X$, the condition $uf = ug$ implies $f = g$. Dually, we have the notion of *separator cocone*.

a) Let $\mathscr{D}$ be a full (and nonempty) subcategory of $\{\mathscr{C}^0, \mathscr{S}et\}$. Prove that an object F belongs to $\mathscr{C}sp(\mathscr{D})$ if and only if there is a functor $T: \mathscr{I} \to \mathscr{D}$ such that F is the vertex of a coseparator cone over T.

b) Let $\mathscr{C}$ be a category and $\mathscr{D}$ a full subcategory of $\mathscr{C}$. We shall say that an object X of $\mathscr{C}$ is a *$\mathscr{D}$-coseparated object* if there is a functor $F: \mathscr{I} \to \mathscr{D}$ and a functorial morphism $u: X \to F$ such that (X, u) is a coseparator cone. Denote by $\mathscr{C}sp(\mathscr{D})$ the full subcategory of $\mathscr{C}$ whose objects are all $\mathscr{D}$-coseparated objects. Prove that $\mathscr{C}sp(\mathscr{D})$ is closed under subobjects and limits. Moreover, if $\mathscr{C}$ is a complete and co-locally small category, and every morphism of $\mathscr{C}$ has a triangular decomposition, and furthermore, for every $X \in \mathscr{C}$, the category $X/\mathscr{C}sp(\mathscr{D})$ is not empty, then $\mathscr{C}sp(\mathscr{D})$ is a reflective subcategory.

c) Let $\mathscr{C}$ be a small category and $\mathscr{D}$ a nonempty and full subcategory of $[\mathscr{C}^0, \mathscr{S}et]$. Show that an object F of $[\mathscr{C}^0, \mathscr{S}et]$ belongs to $\mathscr{C}sp(\mathscr{D})$ if and only if there is a monomorphism to F into a product of objects of $\mathscr{D}$. Moreover $\mathscr{C}sp(\mathscr{D})$ is the smallest complete subcategory of $[\mathscr{C}^0, \mathscr{S}et]$ that contains $\mathscr{D}$.

2.5. Topologies and sheaves

The notions of topology and sheaves which we discuss here are a generalization of the notions of Grothendieck topology and sheaves defined over a topological space. These notions can be applied to algebraic categories, the completion of categories, and elsewhere.

References: Gabriel—Ulmer [1], Kennison [4], Lambek [1], D. Popescu [5].

A *topology* on a category $\mathscr{C}$ is a class J whose elements are triples (F, ψ, X) where F is a right bounded functor in $\mathscr{C}$ and (ψ, X) is a cocone over F.

Example 5.1. The *canonical topology* K of a category $\mathscr{C}$ is the topology whose elements are (F, ψ, X) where F is a small functor in $\mathscr{C}$ and (ψ, X) is a colimit of F.

Let J be a topology on a category $\mathscr{C}$; a functor $T: \mathscr{C} \to \mathscr{C}'$ is called a $(\mathscr{C}, J)$-*cosheaf with values in* $\mathscr{C}'$ if for every $(F, \psi, X) \in K$ we have that $(T\psi,(TX))$ is a colimit of TF. Dually, a cofunctor $T: \mathscr{C} \to \mathscr{C}'$ is a called $(\mathscr{C}, J)$-*sheaf with values in* $\mathscr{C}'$, if for every $(F, \psi, X) \in J$ we have that $(T(X), T\psi)$ is a limit of TF. The functor (cofunctor) $T: \mathscr{C} \to \mathscr{C}'$ is called a $(\mathscr{C}, J)$-*separated functor (a* $(\mathscr{C}, J)$-*separated cofunctor) with values in* $\mathscr{C}'$ if for every $(F, \psi, X) \in J$, we have that $(T\psi, T(X))$ (resp. $(T(X), T\psi)$) is an epicocone (a monocone) over TF. If $\mathscr{C}' = \mathscr{S}et$ we call these a $(\mathscr{C}, J)$-*sheaf*, $(\mathscr{C}, J)$-*cosheaf*, $(\mathscr{C}, J)$-*separated presheaf* and $(\mathscr{C}, J)$-*separated copresheaf*.

The following result is a direct consequence of Theorems 6.10 and 6.11 in Ch. 1.

PROPOSITION 5.2. *Let $T: \mathscr{C} \to \mathscr{C}'$ be a functor (a cofunctor) and J a topology on $\mathscr{C}$. The following are equivalent:*

a) T is a $(\mathscr{C}, J)$-cosheaf (a $(\mathscr{C}, J)$-sheaf) over $\mathscr{C}'$.

b) $H_{X'}T$ (resp. $H^{X'}T$) is a $(\mathscr{C}, J)$-sheaf for every object X' of $\mathscr{C}'$.

An analogous result holds for $(\mathscr{C}, J)$-separated functors and cofunctors.

Let J be a topology on $\mathscr{C}$. We shall denote by $\tilde{\mathscr{C}}_J$ (resp. $\bar{\mathscr{C}}_J$) the full subcategory of $\{\mathscr{C}^0, \mathscr{S}et\}$ whose objects are the $(\mathscr{C}, J)$-sheaves ($(\mathscr{C}, J)$-separated presheaves). In particular, we denote the full subcategory of $\{\mathscr{C}^0, \mathscr{S}et\}$ whose objects are the sheaves (separated presheaves) in the canonical topology by $\tilde{\mathscr{C}}$ (resp. $\bar{\mathscr{C}}$).

PROPOSITION 5.3. *Let $\mathscr{D}$ be a full subcategory of $\{\mathscr{C}^0, \mathscr{S}et\}$ so that for each $X \in \mathscr{C}$, $X(F) = H_X \in \mathscr{D}$. The following are equivalent:*

a) *The canonical functor $K: \mathscr{C} \to \mathscr{D}(K(X) = H_X)$ is a $(\mathscr{C}, J)$-cosheaf with values in $\mathscr{D}$.*

b) $\mathscr{D} \subseteq \tilde{\mathscr{C}}_J$.

Proof. By Proposition 5.2, K is a $(\mathscr{C}, J)$-cosheaf with values in $\mathscr{D}$ if and only if for each $F \in \mathscr{D}$ the cofunctor $H_F K$ is a $(\mathscr{C}, J)$-sheaf. But by Theorem 3.19, Ch. 1, for each $X \in \mathscr{C}$ we have the functorial isomorphism

$$(H_F K)(X) = [K(X), F] \simeq F(X),$$

and this completes the proof since the colimits in $\{\mathscr{C}^0, \mathscr{S}et\}$ are defined argumentwise. $\divideontimes$

LEMMA 5.4. *For each $X \in \mathscr{C}$ the presheaf $H_X = H(X)$ belongs to $\tilde{\mathscr{C}}$. Moreover, the assignment $X \rightsquigarrow H_X$ gives a full and faithful functor $\tilde{H}: \mathscr{C} \to \tilde{\mathscr{C}}$.* $\divideontimes$

The functor $\tilde{H}: \mathscr{C} \to \tilde{\mathscr{C}}$ is called the *natural extension* of $\mathscr{C}$.

THEOREM 5.5. (Lambek [1]). *For every category $\mathscr{C}$, the following are true:*

a) *The natural extension $\tilde{H}: \mathscr{C} \to \tilde{\mathscr{C}}$ is a dense functor that commutes with limits and colimits.*

b) *If $T: \mathscr{C} \to \mathscr{D}$ is a full and faithful dense and properly defined functor that commutes with limits and colimits, then there is a unique full and faithful functors $\tilde{S}^T: \mathscr{D} \to \tilde{\mathscr{C}}$ such that $\tilde{S}^T T \simeq \tilde{H}$.*

(We shall say that $\tilde{\mathscr{C}}$ is the largest left-adequate extension of $\mathscr{C}$ such that $\tilde{H}$ commutes with limits and colimits.)

Proof. The proof of part a) is routine and is left for the reader. To get b), observe that the functor $S^T: \mathscr{D} \to \{\mathscr{C}^0, \mathscr{S}et\}$ defined in Corollary 3.3 has values in $\tilde{\mathscr{C}}$, i.e. S^T determines the functor $\tilde{S}^T$. $\divideontimes$

In Lambek's paper, the category $\tilde{\mathscr{C}}$ is denoted by $[\mathscr{C}^0, \mathscr{S}et]_{\text{inf}}$.

It is clear that for each $X \in \mathscr{C}$ the presheaf H_X belongs to $\overline{\mathscr{C}}$ so that the functor $H: \mathscr{C} \to \{\mathscr{C}^0, \mathscr{S}et\}$ defines a functor $\overline{H}: \mathscr{C} \to \overline{\mathscr{C}}$. The following result is obvious.

COROLLARY 5.6. *The embedding $\overline{H}: \mathscr{C} \to \overline{\mathscr{C}}$ is the largest dense extension of $\mathscr{C}$ that commutes with limits and epicocones. Moreover, $\overline{H}$ is a dense functor that preserves monocones.* $\divideontimes$

Let J be a topology on $\mathscr{C}$. We shall prove that the category $\tilde{\mathscr{C}}_J$ of $(\mathscr{C}, J)$-sheaves coincides with the category $\tilde{\mathscr{C}}_\Sigma$ of Σ-sheaves, where Σ is a suitable class of morphisms of $\{\mathscr{C}^0, \mathscr{S}et\}$.

THEOREM 5.7. *Let $J = \{F_r, \Psi^r, X_r\}_{r \in R}$ be a topology on a category $\mathscr{C}$. For each $r \in R$ fix a colimit (t^r, Y_r) of HF_r in $\{\mathscr{C}^0, \mathscr{S}et\}$ and let $s_r: Y_r \to H_{X_r}$ denote the unique morphism such that $s_r t^r = H\Psi^r$. Also let $\Sigma(J)$ or Σ denote $\{s_r\}_{r \in R}$. Then a presheaf F is a $(\mathscr{C}, J)$-sheaf (a $(\mathscr{C}, J)$-separated presheaf) if and only if it is a Σ-sheaf (a Σ-separated presheaf).*

Proof. Let F be an object of $\{\mathscr{C}^0, \mathscr{S}et\}$. For each $X \in \mathscr{C}$, define the map $\varphi_X: [H_X, F] \to F(X)$ by $\varphi_X(u) = u_X(1_X)$. By Theorem 3.19, Ch. 1, it is known that φ_X is a bijection, and the morphisms $\{\varphi_X\}_{X \in \mathscr{C}}$ define a functorial morphism (actually a functorial isomorphism) $\varphi: H_F H \to F$ where $H: \mathscr{C} \to \{\mathscr{C}^0, \mathscr{S}et\}$ is the canonical embedding. Now let $r \in R$; since H_F commutes with colimits, we have that $(H_F(Y_r), H_F t^r)$ is a limit of $H_F H F_r$. Hence, since $H_F H \simeq F$, we see that F is a $(\mathscr{C}, J)$-sheaf if and only if, for each $r \in R$, $H_F(s_r) = [s_r, F]$ is a bijection.

Now we give a more explicit proof, essentially a detailed version of the above proof. Assume that F is a $(\mathscr{C}, J)$-sheaf. We must prove that the morphism

$$[s_r, F]: [H(X_r), F] \to [Y_r, F] \tag{7}$$

is an isomorphism, for each $r \in R$. For that let $u, u' \colon H(X_r) \to F$ be functorial morphisms such that $us_r = u's_r$. Since $H(X_r) = H_{X_r}$, we may assume that $u = u^x$ and $u' = u^{x'}$, where $x, x' \in F(X_r)$. Furthermore, for each object i of $\mathcal{I}_r$ (the domain of F_r), let x_i denote $F(\Psi^r_i)(x)$ and x'_i denote $F(\Psi^r_i)(x')$. If $k \colon i \to j$ is a morphism of $\mathcal{I}_r$, then obviously $F(F_r(k))(x_j) = x_i$ and $F(F_r(k))(x'_j) = x'_i$. Hence the morphisms $\{u^{x_i}\}_{i \in \mathcal{I}_r}$ (resp. $\{u^{x'_i}\}_{i \in \mathcal{I}_r}$) define a morphism $\bar{u} \colon HF_r \to F$ (resp. $\bar{u}' \colon HF_r \to F$). Since $x_i = F(\Psi^r_i)(x)$, then $u^{x_i} = uH(\Psi^r_i)$, i.e. $\bar{u} = uH\Psi^r$. Analogously, $\bar{u}' = u'H\Psi^r$. But then $\bar{u} = uH\Psi^r = us_rt^r = u's_rt^r = u'H\Psi^r = \bar{u}'$. Hence $u^{x_i} = u^{x'_i}$, i.e. $x_i = x'_i$ for all $i \in \mathcal{I}_r$. But this implies that $x = x'$, since $(F(X_r), F\Psi^r)$ is a limit of FF_r. Therefore $u = u'$, or equivalently the map $[s_r, F]$ is injective.

Now let $u \colon Y_r \to F$ be a functorial morphism. For each $i \in \mathcal{I}_r$, choose an element $x_i \in FF_r(i)$ such that $ut^r_i = u^{x_i}$. If $k \colon i \to j$ is a morphism of $\mathcal{I}_r$, then the equality $ut^r_jH(F_r(k)) = ut^r_i$ implies that $F(F_r(k))(x_j) = x_i$. Hence the elements $(x_i)_{i \in \mathcal{I}_r}$ form a coherent family i.e. they define an element of $\varprojlim FF_r$. Thus by the assumption that $(F(X_r), F\Psi^r)$ is a limit of FF_r, there is a unique element $x \in F(X_r)$ such that $F(\Psi^r_i)(x) = x_i$ for all $i \in \mathcal{I}_r$. Therefore, the morphism $u^x \colon H(X_r) \to$ $\to F$ is the unique morphism such that $u^x_{s_r} = u$. In conclusion the map $[s_r, F]$ is bijective.

Conversely, let F be a Σ-sheaf; then for each $r \in R$, the map (7) is a bijection. Then for each object i of $\mathcal{I}_r$ we have the commutative diagram

$$\begin{array}{ccccc}
[H(X_r), F] & \xrightarrow[\sim]{[s_r, F]} & [Y_r, F] & \xrightarrow{[t^r_i, F]} & [H(F_r)(i)), F] \\
\wr\Big\downarrow \varphi & & \Big\| & & \wr\Big\downarrow \varphi^i \\
F(X_r) & \longrightarrow & [Y_r, F] & \longrightarrow & F(F_r(i))
\end{array}$$

where the composition of maps that occurs on the bottom row is in fact $F(\Psi^r_i)$. Now if $x, x' \in F(X_r)$ are such that $F(\Psi^r_i)(x) = F(\Psi^r_i)(x') = x_i$ for all $i \in \mathcal{I}_r$, then $\varphi^i(x_i) = u^{x_i}$, and the morphisms $\{u^{x'_i}\}_{i \in \mathcal{I}_r}$ define a functorial morphism $u \colon HF_r \to F$ such that $u^xH\Psi^r = u^{x'}H\Psi^r = u$. Hence $u^xs_r = u^{x'}s_r$, i.e. $u^x = u^{x'}$ or equivalently $x = x'$.

Finally, let $x_i \in F_r(i)$, $i \in \mathcal{I}_r$, be such that $F_r(k)(x_i) = x_j$ for each morphism $k \colon i \to j$ of $\mathcal{I}_r$. Then the morphisms $\{u^{x_i}\}_i$ give a functorial morphism $u \colon HF_r \to F$. Thus there is a unique morphism $\bar{u} \colon Y_r \to F$ such that $\bar{u}t^r = u$. By hypothesis $\bar{u} = u's_r$ and $u' = u^x$, $x \in F(X_r)$. An easy computation shows that $F(\Psi^r_i)(x) = x_i$ for all $i \in \mathcal{I}_r$. Hence $(F(X_r), F(\Psi^r))$ is a limit of FF_r, i.e. any Σ-sheaf is a $(\mathscr{C}, j)$-sheaf. Similarly, it may be shown that the object F from $\{\mathscr{C}^0, \mathscr{S}et\}$ is $(\mathscr{C}, J)$-separated if and only if it is Σ-separated. $\divideontimes$

Theorem 5.7 shows that the study of categories of $(\mathscr{C}, J)$-sheaves is equivalent to the study of categories of Σ-sheaves for a suitable class Σ of morphisms of $\{\mathscr{C}^0, \mathscr{S}et\}$. Using this we obtain the following result.

THEOREM 5.8. (D. Popescu [5]). *Let J be a topology on the small category $\mathscr{C}$. Then the inclusion functor $\tilde{\mathscr{C}}_J \to [\mathscr{C}^0, \mathscr{S}et]$ has an adjoint.*

Proof. By Theorem 5.7, $\tilde{\mathscr{C}}_J = \tilde{\mathscr{C}}_{\Sigma(J)}$. Now we observe that the class $\Sigma(J)$ satisfies the conditions of Theorem 4.17, since the codomain of every $s \in \Sigma(J)$ is an object of the form H_X with $X \in \mathscr{C}$ and $\mathscr{C}$ is small. ✳

COROLLARY 5.9. (Kennison [1]). *For every small category $\mathscr{C}$, the inclusion functor $\tilde{\mathscr{C}} \to [\mathscr{C}^0, \mathscr{S}et]$ has an adjoint. In particular, $\tilde{\mathscr{C}}$ is a complete and cocomplete category.*

The proof follows from Theorem 5.8 and Theorem 12.4, Ch. 1. ✳

Terminology. Let J be a topology on a small category $\mathscr{C}$ such that the inclusion functor $\tilde{\mathscr{C}}_J \xrightarrow{U} [\mathscr{C}^0, \mathscr{S}et]$ has an adjoint, that we denote by A. Then for each object F of $[\mathscr{C}^0, \mathscr{S}et]$ the object $UA(F)$ is sometimes denoted by $\tilde{F}_J$, or simply by $\tilde{F}$ and is called the *J-sheaf associated with F*. The arrow of reflection $F \to \tilde{F}$ is also called the *natural arrow* from the presheaf F to its associated sheaf.

Exercises

5.1. Let X be a topological space and let $\mathscr{D}(X)$ be the set of open sets of X ordered by inclusion. Also let $(I_k, T_k)_k$ be the set of all pairs, where I_k is a full subcategory of $\mathscr{D}(X)$ closed under intersections (i.e. if $U_1, U_2 \in I_k$, then $U_1 \cap U_2 \in I_k$), and where $T_k: I_k \to \mathscr{D}(X)$ is the inclusion functor. Obviously, for each k, $\varinjlim T_k = \bigcup_{U \in I_k} U = U_k$. Consider in $\mathscr{D}(X)$ the topology $I = \{(T_k, \Psi_k, U_k)\}_k$ where the $\Psi_k: T_k \to U_k$ are structural inclusions. Prove that a presheaf F on $\mathscr{D}(X)$ is a $(D(X), J)$-sheaf if and only if it is a sheaf over the topological space X (see Section 1.4.)

5.2. With the notation and hypothesis as in the previous exercise, prove that $\widetilde{\mathscr{D}(X)}_J$ has a set of cogenerators and $H(U_k) = \varinjlim (HT_k)$ for every k. (Here $H: \mathscr{D}(X) \to [\mathscr{D}(X)^0, \mathscr{S}et] = \mathscr{P}(X)$ is the canonical functor and the colimit is calculated in $\widetilde{\mathscr{D}(X)}_J$.)

5.3. Prove that the composition of functors

$$\mathscr{C} \xrightarrow[D]{} \mathscr{C}^0 \xrightarrow[\widetilde{H^{\mathscr{C}^0}}]{} \widetilde{\mathscr{C}^0} \xrightarrow[D']{} (\widetilde{\mathscr{C}^0})^0$$

where D and D' are the duality functors, is the largest dense embedding that preserves the limits.

5.4. If $\mathscr{C}$ is a discrete (not necessarily small) category show that the inclusion functor $\tilde{\mathscr{C}} \to \{\mathscr{C}^0, \mathscr{S}et\}$ has an adjoint.

2.6. Some adjoint theorems

An important problem in the theory of categories is that of determining under what conditions a functor has an adjoint or a coadjoint. The first results in this field were obtained by Freyd [1]. Freyd's results were expanded by Benaboù [1], Isbell [8], and Lambek [1]. The results in this area are conventionally called "The Adjoint Theorems". In this section we present the main part of these adjoint theorems.

References: Benaboù [1], Freyd [1], Isbell [5], [8]. Lambek [1], MacDonald [1].

LEMMA 6.1. *Let $\mathscr{C}$ be a cocomplete category, $T: \mathscr{C} \to \mathscr{S}et$ a proper co-functor that commutes with fibered coproducts, $\mathscr{P}_T$ the pointed category associated with T, and $s_T: \mathscr{P}_T \to \mathscr{C}$ the forgetful functor. Then the functor s_T has a colimit in $\mathscr{C}$.*

Proof. Let $\{(X_i, x_i)\}_{i \in \mathscr{I}}$ be a set of generators of T. Let $\mathscr{A}$ denote the full subcategory of $\mathscr{P}_T$ whose objects are $\{(X_i, x_i)\}_{i \in \mathscr{I}}$. Since $\mathscr{A}$ is small there is a colimit (Ψ, Y) of the functor s_T, where $U: \mathscr{A} \to \mathscr{P}_T$ is the inclusion functor. We shall extend Ψ to a functorial morphism $\overline{\Psi}: s_T \to Y$ such that $(\overline{\Psi}, Y)$ becomes canonically a colimit of s_T. It suffices to check that $\mathscr{A}$ is a final subcategory of $\mathscr{P}_T$. For this, consider the following coangle in $\mathscr{P}_T$:

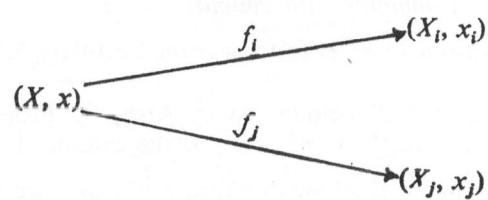

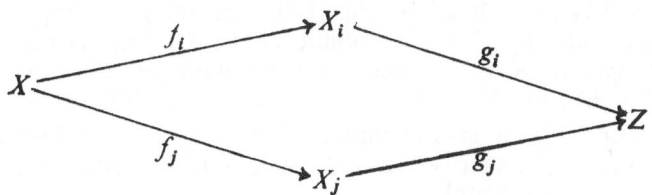

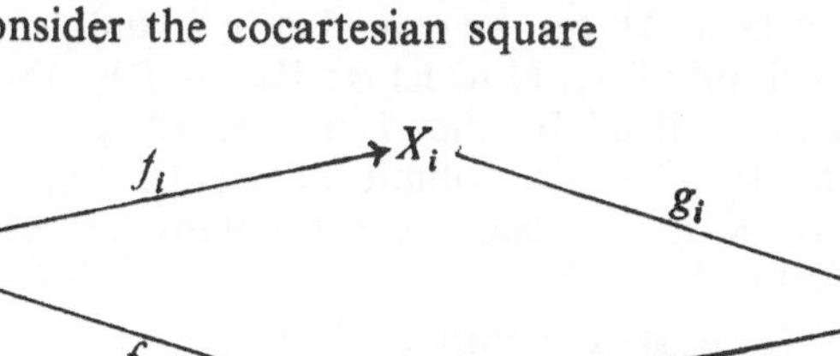

Furthermore, consider the cocartesian square

in $\mathscr{C}$ and note that we have the cartesian square

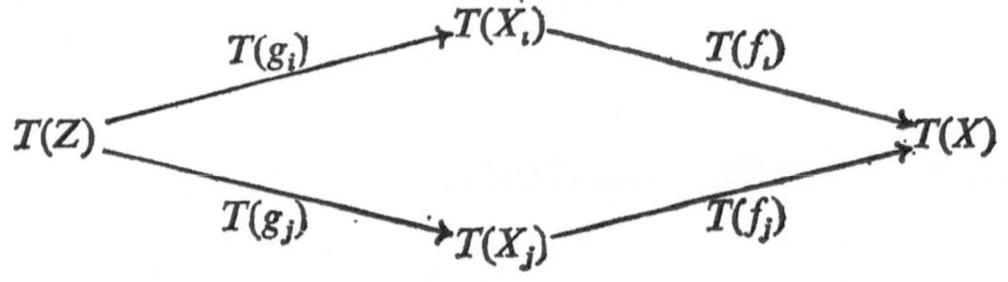

in $\mathscr{S}et$ since T commutes with fibered coproducts. Now since $T(f_i)(x_i)=T(f_j)(x_j)=x$, it follows that there is a unique element $z \in T(Z)$ such that $T(g_i)(z) = x_i$ and $T(g_j)(z) = x_j$. Since T is proper there is a $k \in \mathscr{I}$ and an $h: Z \to X_k$ such that $T(h)(x_k) = = z$. Hence we have the commutative square

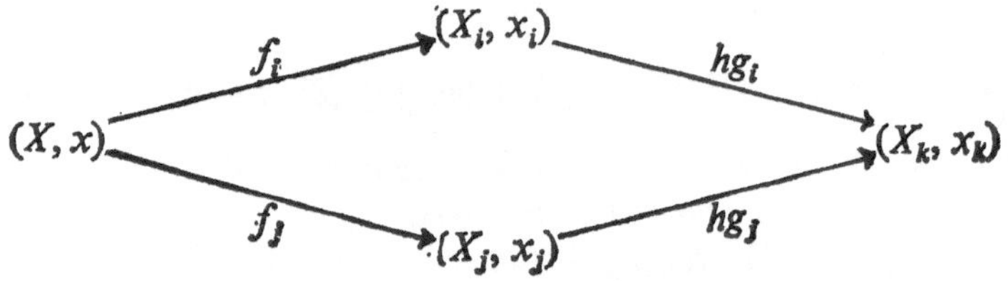

in $\mathscr{P}_T$, i.e. $\mathscr{A}$ is a final subcategory of $\mathscr{P}_T$, so that the proof follows from Theorem 6.12, Ch. 1. ✳

Recall that by $\tilde{\mathscr{C}}$ we understand (see Section 2.5) the full subcategory of $\{\mathscr{C}^0, \mathscr{S}et\}$ whose objects are the sheaves in the canonical topology on $\mathscr{C}$.

THEOREM 6.2. *Let $\mathscr{C}$ be a cocomplete category and let $T: \mathscr{C} \to \mathscr{S}et$ be a cofunctor. The following are equivalent:*

1) *T is representable.*

2) *T is proper and commutes with colimits.*

Proof. The implication 1) ⇒ 2) follows from Corollary 3.27, Proposition 12.3, Ch. 1, and Theorem 1.1.

2) ⇒ 1). It is clear that T belongs to $\tilde{\mathscr{C}}$. Also, by Proposition 2.1 we have that $T = \varinjlim (Hs_T)$ where $H: \mathscr{C} \to \{\mathscr{C}^0, \mathscr{S}et\}$ is the canonical embedding. Furthermore, the category $\mathscr{P}_T$ has a final small subcategory $\mathscr{A}$ (see the proof of Lemma 6.1). Let $U: \mathscr{A} \to \mathscr{P}_T$ be the inclusion functor. Consider a colimit (Ψ, Y) of $s_T U$. By Lemma 6.1 the morphism Ψ can be extended to a unique morphism $\overline{\Psi}: s_T \to Y$ such that $(\overline{\Psi}, Y)$ is a colimit of s_T. Now let $\varphi: Hs_T \to T$ be the structural morphism (see Proposition 2.1) and let φ' be the restriction of φ to $Hs_T U$. Since $\mathscr{A}$ is a final subcategory, then (φ', T) is a colimit of $Hs_T U$. Furthermore, we see that $(s_T U, \Psi, Y) \in K$ (where K is the canonical topology on $\mathscr{C}$) so that the unique morphism $u: T \to H(Y) = H_Y$ defined such that $u\varphi' = H\Psi$ belongs to $\Sigma(K)$ (see Theorem 5.7). But then u is an isomorphism of $\tilde{\mathscr{C}}$ since T and H_Y are $\Sigma(K)$-sheaves (see Theorem 5.7 and Exercise 4.1). Hence u is an isomorphism of $\{\mathscr{C}^0, \mathscr{S}et\}$ or equivalently T is representable. ✳

COROLLARY 6.3. (Adjoint Functor Theorem). *Let $\mathscr{C}$ be a cocomplete category. A functor $F:\mathscr{C} \to \mathscr{C}'$ has a coadjoint if and only if it commutes with colimits and if, for each $X' \in \mathscr{C}'$ the cofunctor $H_{X'}: F:\mathscr{C} \to \mathscr{S}et$ is proper.*

The proof is a direct consequence of Theorem 11.9, Ch. 1, and of Theorem 6.2. ✳

COROLLARY 6.4. (Bénabou [1]). *If $\mathscr{C}$ is a complete category, then a functor $F:\mathscr{C} \to \mathscr{S}et$ is representable if and only if it commutes with limits and is proper.* ✳

COROLLARY 6.5. *Let $\mathscr{C}$ be a complete category. A functor $F:\mathscr{C} \to \mathscr{C}'$ has an adjoint if and only if it commutes with limits and if for each $X' \in \mathscr{C}'$, the functor $H^{X'}F:\mathscr{C} \to \mathscr{S}et$ is proper.* ✳

We already know that the categories having limits often also have some colimits. Thus it is natural to seek those conditions under which the existence of limits will imply the existence of some colimits. Let us suppose that the category $\mathscr{C}$ has the following property:

⊕Any proper functor $T:\mathscr{C} \to \mathscr{S}et$ that commutes with limits is representable.

The question now becomes: What can we say about the limits and colimits of $\mathscr{C}$?

A partial answer to this question is given by the following proposition.

PROPOSITION 6.6. *Let $\mathscr{C}$ be a complete category that satisfies condition ⊕: If $F:\mathscr{I} \to \mathscr{C}$ is a functor and $\mathscr{I}$ is right bounded then the following are equivalent:*

1) *F has a colimit.*

2) *The functor $[F, ?]:\mathscr{C} \to \mathscr{S}et$ defined by the assignment $X \rightsquigarrow [F, X]$ (see Section 1.6) is proper.*

Proof. Since $\mathscr{I}$ is right bounded, it follows from Exercise 1.12 that $[F, X]$ is a set for all $X \in \mathscr{C}$. Now the theorem is an immediate consequence of Theorem 6.2. ✳

COROLLARY 6.7. *If $\mathscr{C}$ is a complete category, then any functor $F:\mathscr{I} \to \mathscr{C}$ with $\mathscr{I}$ right bounded and such that $[F, ?]:\mathscr{C} \to \mathscr{S}et$ is proper, has a colimit.* ✳

COROLLARY 6.8. *Let $\mathscr{C}$ be a complete category. The following are equivalent:*

a) *$\mathscr{C}$ is cocomplete (has bounded colimits).*

b) *For each functor $F:\mathscr{I} \to \mathscr{C}$ with $\mathscr{I}$ small ($\mathscr{I}$ right bounded) the functor $[F, ?]:\mathscr{C} \to \mathscr{S}et$ is proper.* ✳

We have shown (see Section 2.5) that if the topology J on $\mathscr{C}$ satisfies certain conditions, then the inclusion functor $\tilde{\mathscr{C}}_J \to [\mathscr{C}^0, \mathscr{S}et\}$ has an adjoint, and thus, $\tilde{\mathscr{C}}_J$ has bounded colimits. Thus, if $\mathscr{C}$ is small, then the above-mentioned conditions are satisfied and consequently $\tilde{\mathscr{C}}_J$ has bounded colimits. We cannot say the same in the case when $\mathscr{C}$ is not small. Nonetheless we have the following result.

THEOREM 6.9. *A functor $\tilde{T}:(\tilde{\mathscr{C}})^0 \to \mathscr{S}et$ that commutes with bounded limits is representable.*

Proof. Let $\tilde{H}:\mathscr{C} \to \tilde{\mathscr{C}}$ be the natural extension of $\mathscr{C}$ (i.e. $\tilde{H}(X) = H_X$) and let $\tilde{H}^0:\mathscr{C} \to \tilde{\mathscr{C}}^0$ be the composition of the duality functor $D:\tilde{\mathscr{C}} \to \tilde{\mathscr{C}}^0$ with $\tilde{H}$. By Corollary 5.6, $\tilde{H}^0$ is dense and commutes with limits and colimits. Consequently, $T\tilde{H}^0$

commutes with bounded limits. Suppose, for the moment, that $T\tilde{H}^0$ is also proper. Then we may consider $T\tilde{H}^0$ as an object of $\tilde{\mathscr{C}}$. Taking $T' = \tilde{\mathscr{C}}^0(T\tilde{H}^0, ?) = [T\tilde{H}^0, ?]: \mathscr{C}^0 \to \mathscr{S}et$, we have that for each $X \in \mathscr{C}$, there is the following functorial isomorphism:

$$T'(\tilde{H}^0(X)) = [T\tilde{H}^0, \tilde{H}^0(X)] \simeq T\tilde{H}^0(X).$$

This implies that $T \simeq T'$ since $\tilde{H}^0$ is dense and T and T' commute with bounded limits. Hence T is representable. In order to finish the proof it will suffice to show that $T\tilde{H}^0$ is proper. This is established by the following lemma. ✱

LEMMA 6.10. *Let $\mathscr{C}$ be a full subcategory of $\mathscr{D}$ such that the inclusion functor $U: \mathscr{C} \to \mathscr{D}$ is proper. If $F: \mathscr{D} \to \mathscr{S}et$ is a proper cofunctor, then FU is also proper.*

Proof. Consider the following diagram of categories and functors

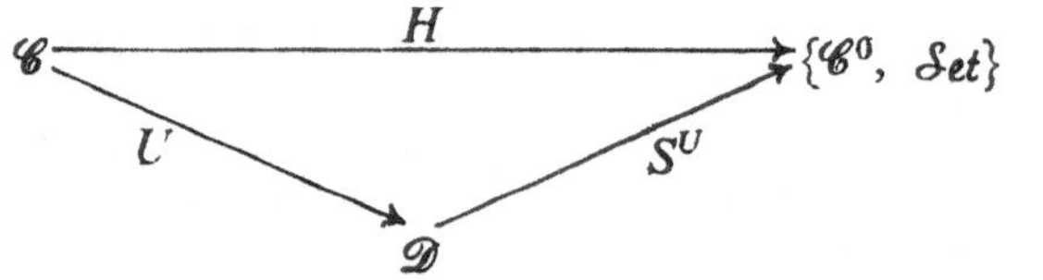

where S^U is defined as in Corollary 3.3. Now let $\{D_i\}_{i \in \mathscr{I}}$ be a set of objects of $\mathscr{D}$ that dominates F, and let $T_i = S^U(D_i) = H_{D_i}U$. For each $i \in \mathscr{I}$ consider a dominating set F_i of objects from $\mathscr{C}$, for T_i. We assert that $\bigcup_{i \in \mathscr{I}} \mathscr{F}_i = \mathscr{F}$ is a dominating set for FU. Indeed, let $X \in \mathscr{C}$ and $x \in FU(X)$; then there is an $i \in \mathscr{I}$, $f: U(X) \to D_i$ and an $x_i \in F(D_i)$ such that $F(f)(x_i) = x$. Consider the element $f \in T_i(X) = \mathscr{D}(U(X), D_i)$. By hypothesis there are a $Y \in F_i$ and a $g: X \to Y$ such that $T_i(g)(h) = f$, where $h \in T_i(Y)$. Now, since $H = S^U U$, there is a unique morphism $g': U(X) \to U(Y)$ such that $S^U(g') = Hg$. But then $T_i(g) = S^U(D_i)(g')$, so that $T_i(g)(h) = hg' = f$. Hence $F(g')(F(h)(x_i)) = F(f)(x_i) = x$, i.e. FU is dominated by $\mathscr{F}$. ✱

Consider the following property on a category $\mathscr{C}$:

(++) Any functor $R: \mathscr{C} \to \mathscr{S}et$ that commutes with limits is proper.

Categories with property (++) are interesting since it is clear that if they have limits, then a functor $F: \mathscr{C} \to \mathscr{D}$ has an adjoint if and only if F commutes with limits. In the following theorem we give sufficient conditions for a category to have the property (++).

THEOREM 6.11. *Let $\mathscr{C}$ be a complete category and suppose that $\mathscr{C}$ satisfies one of the following conditions:*

Case 1. $\mathscr{C}$ has a small codense subcategory.

Case 2. $\mathscr{C}$ is a locally small category and has a set of coseparators.

Then every functor $T: \mathscr{C} \to \mathscr{S}et$ that commutes with limits is proper, and thus representable.

Proof. Let $\mathscr{C}'$ be a small subcategory of $\mathscr{C}$, and let $U\colon \mathscr{C}' \to \mathscr{C}$ be the inclusion functor. Also let $T\colon \mathscr{C} \to \mathscr{S}et$ be a functor that commutes with limits. Consider the category $\mathscr{P} = \mathscr{P}_{TU}$, the pointed category associated with TU and let $s\colon \mathscr{P} \to \mathscr{C}'$ be the underlying functor. It is clear that $\mathscr{P}$ is a small category, since $\mathscr{C}'$ is small. Assume that the objects of $\mathscr{C}'$ give a set of separators of $\mathscr{C}$.

Now if (X, φ) is a limit of Us, then we shall prove that in Case 1 the set $\{X\}$ dominates T, and in Case 2 we shall show that the set of subobjects of X dominates T. Let Y be an arbitrary object of $\mathscr{C}$, and let $F_Y = H^Y U\colon \mathscr{C}' \to \mathscr{S}et$. Also let $\mathscr{P}'$ be the pointed category associated with F_Y, $s'\colon P' \to \mathscr{C}'$ the underlying functor, and (M, γ) a limit of Us'. Since $\mathscr{C}'$ coseparates $\mathscr{C}$, it is easy to check that $m\colon Y \to M$ is a monomorphism, where m is the unique morphism with the property that: if for each $(A, f) \in \mathscr{P}'$ (where $f \in F_Y(A) = \mathscr{C}(Y, U(A))$, then $\gamma_{(A,f)} m = f$. In both Case 1 and Case 2, $\mathscr{C}'$ coseparates $\mathscr{C}$. The distinction between the two cases consists in the fact that in Case 1 the monomorphism m is an isomorphism. Suppose, for the moment, that having been given an element $y \in T(Y)$ we found a morphism $t\colon X \to M$ and $x \in T(X)$ such that $T(t)(x) = T(m)(y)$. In Case 1 since m is an isomorphism, we have thus shown that $\{X\}$ dominates T. In Case 2, consider in $\mathscr{C}$ the cartesian square

$$\begin{array}{ccc} Y & \xrightarrow{\quad m \quad} & M \\ {\scriptstyle t'}\uparrow & & \uparrow{\scriptstyle t} \\ D & \xrightarrow{\quad m' \quad} & X \end{array} \quad .$$

Obviously, the morphism m' will be a monomorphism. Since T commutes with limits we get in $\mathscr{S}et$ the cartesian square

$$\begin{array}{ccc} T(Y) & \xrightarrow{\quad T(m) \quad} & T(M) \\ {\scriptstyle T(t')}\uparrow & & \uparrow{\scriptstyle T(t)} \\ T(D) & \xrightarrow{\quad T(m') \quad} & T(X) \end{array} \quad .$$

Also, since $T(t)(x) = T(m)(y)$ it follows that there is an element $d \in T(D)$ such that $T(t')(d) = y$ and $T(m')(d) = x$. Consequently, since D is a subobject of X we have shown that the subobjects of X dominate T.

Now, in order to finish the proof we shall show that for each $y \in T(Y)$ there is a morphism $t\colon X \to M$ and an element $x \in T(X)$ such that $T(t)(x) = T(m)(y)$. Let us define a functorial morphism $\Psi\colon X \to Us'$, as follows. Let $(A, f) \in \mathscr{P}'$; here $f \in F_Y(A) = \mathscr{C}(Y, U(A))$. Consider the element $y_f = T(f)(y) \in T(U(A))$ and let $\Psi_{(A,f)} = \varphi_{(A, y_f)}\colon X \to U(A) = Us'(A, f) = Us(A, y_f)$. We leave it for the reader to check that the morphisms $\{\Psi_{(A,f)}\}_{(A,f) \in \mathscr{P}'}$ define the desired functorial morphism $\Psi\colon X \to Us'$. But then there is a unique morphism $t\colon X \to M$ such that $\gamma t = \Psi$. By Exercise 6.7 it follows that an element $x \in T(X)$ exists, such that, for every $A \in \mathscr{C}'$, and every $a \in T(U(A))$ we have that $T(\Psi_{(A,a)})(x) = a$. In particular, we get

$T(\varphi_{(A,y_f)})(x) = y_f$ for every $A \in \mathscr{C}'$ and every $f \in \mathscr{C}(Y, U(A))$. Thus we have that $T(\Psi_{(A,f)})(x) = y_f = T(f)(y)$, and this implies that $T(\gamma_{(A,f)})(T(t)(x) = T(\Psi_{(A,f)})(x) = T(f)(y)$, for every $(A,f) \in \mathscr{P}'$. Since T commutes with limits, we have that $(T(M), T\gamma)$ is a limit of TUs'. Finally, $T(\gamma_{(A,f)})T(t)(x) = T(f)(y) = T(\gamma_{(A,f)})T(m)(y)$ for every $(A,f) \in \mathscr{P}'$. But then $T(t)(x) = T(m)(y)$. ✳

COROLLARY 6.12. *If $\mathscr{C}$ is a category that satisfies the conditions of Theorem 6.11, then a functor $F: \mathscr{C} \to \mathscr{D}$ has an adjoint if and only if it commutes with limits.* ✳

Examples of such categories are: $\mathscr{S}et$, $\mathscr{T}op$, $\mathscr{A}b$ and the category of sheaves of sets over a topological space (see Exercise 4.4, Ch. 1, and Exercise 6.4).

PROPOSITION 6.13. *Let $\mathscr{C}$ be a complete category having the property* $(++)$. *Then $\mathscr{C}$ is a category with bounded colimits.*

Proof. Note that the canonical functor $H: \mathscr{C} \to \{\mathscr{C}^0, \mathscr{S}et\}$ commutes with limits. It follows from Corollary 6.12 that H has an adjoint. ✳

COROLLARY 6.14. *A complete category $\mathscr{C}$ has bounded colimits in either of the following situations:*

1) *$\mathscr{C}$ contains a small codense subcategory.*

2) *$\mathscr{C}$ is locally small and has a set of coseparators.* ✳

Exercises

6.1. Using Corollary 6.3, show that if $\mathscr{C}$ is a small category, then for every topology J on $\mathscr{C}$, the inclusion functor $\tilde{\mathscr{C}}_J \to [\mathscr{C}^0, \mathscr{S}et]$ has an adjoint. (Prove this without using the results of Section 2.5.)

6.2. Prove: If $\mathscr{C}$ is a small category and J a topology on $\mathscr{C}$, then for each $X \in \mathscr{C}$ the argument functor $A^X: \tilde{\mathscr{C}}_J \to \mathscr{S}et$ has an adjoint.

6.3. Let $\mathscr{A}$ and $\mathscr{B}$ be preordered sets and $f: \mathscr{A} \to \mathscr{B}$ a nondecreasing map. Assume that $\mathscr{A}$ is complete and f commutes with limits. Give a construction of the adjoint of f that exists according to Corollary 6.3.

6.4. Let X be a topological space and let $F(X)$ be the category of sheaves of sets on X. Prove that $\mathscr{F}(X)$ has the property $(++)$.

6.5. Let $\mathbf{N}$ be the set of natural numbers with the usual order relation defined on it. Show that the constant functor $F: \mathbf{N} \to \mathscr{S}et$ defined by the assignment $n \rightsquigarrow \{*\}$ is not representable although it commutes with limits.

6.6. Let M be a preordered set. Show that the following are equivalent:

a) M is complete.

b) Any nondecreasing map $f: M \to X$, where X is a preordered set, commutes with limits if and only if it has an adjoint.

c) Any nondecreasing map $f\colon M \to X$ where X is a preordered set commutes with colimits if and only if it has a coadjoint.

6.7. Let $\mathscr{C}$ be a category with property $(++)$. Show that the following are equivalent:

a) $\mathscr{C}$ is complete.

b) A functor $F\colon \mathscr{C} \to \mathscr{S}et$ is representable if and only if it commutes with limits.

6.8. Let $G\colon \mathscr{C} \to \mathscr{S}et$ be a functor. We say that $\mathscr{C}$ has *minimal G-factorizations* if for each $(X, x) \in \mathscr{P}_G$ there is a subobject $f\colon X' \to X$ of X in $\mathscr{C}$, minimal with respect to the property that $f\colon (X', x') \to (X, x)$, for some $x' \in G(X')$, and furthermore it is required that if $g, h\colon (X', x') \to (Y, y)$, then $g = h$. Show that if $\mathscr{C}$ is a wellpowered and complete category and $G\colon \mathscr{C} \to \mathscr{S}et$ commutes with limits, then $\mathscr{C}$ has a minimal G-factorization.

6.9. Let $\mathscr{C}$ be a wellpowered, copowered, and complete category, and let $\mathscr{C}'$ be a full subcategory of $\mathscr{C}$ containing a copy of $\coprod_i X_i$ for every set $\{X_i\}_i$ of its objects. Let $I\colon \mathscr{C}' \to \mathscr{C}$ be the inclusion functor. Suppose that if F is a subfunctor of $H^X I$, where $X \in \mathscr{C}$, such that for every $X' \in \mathscr{C}'$ and $f \in F(X') \subseteq \mathscr{C}(X, X')$, then f has a triangular decomposition $X \xrightarrow{p} Y' \to X'$ with $Y' \in \mathscr{C}'$ and $p \in F(Y')$. Show that F is a representable functor, i.e. $F \simeq H^Z$, for some $Z \in \mathscr{C}$ and the inclusion $F \to H^X I$ is induced by an epimorphism $p\colon X \to Z$.

6.10. Let $\mathscr{C}$ be a category with products, coproducts, and triangular decompositions. Let $\mathscr{C}'$ be a full subcategory of $\mathscr{C}$, $U\colon \mathscr{C}' \to \mathscr{C}$ the inclusion functor, and $F\colon \mathscr{C}' \to \mathscr{S}et$ a reflexive presheaf. Show that there is an object X of $\mathscr{C}$ such that $F \simeq H_X U$.

2.7. A generalization of the extension theorem

The Extension Theorem given in Section 2.2 will now be extended to some full subcategories of $\{\mathscr{C}^0, \mathscr{S}et\}$. The notation used is the same as in Sections 2.4 and 2.5.

References: Freyd [2], Gabriel—Ulmer [1], D. Popescu [5].

LEMMA 7.1. *Let $F\colon \mathscr{C} \to \mathscr{C}'$ be a functor, G a full and faithful coadjoint to F, and $u\colon Id\,\mathscr{C} \to GF$ an arrow of adjunction. If $T\colon \mathscr{C} \to \mathscr{D}$ is a functor, then there is a functor $T'\colon \mathscr{C} \to \mathscr{D}$ such that $T'F \simeq F$ if and only if $T(u_X)$ is an isomorphism for all $X \in \mathscr{C}$. Moreover the functor T' is uniquely determined by T, and T commutes with colimits if and only if T commutes with colimits.*

The proof is left for the reader. ⁎

THEOREM 7.2. *Let Σ be a class of morphisms of $\{\mathscr{C}^0, \mathscr{S}et\}$, and $T: \{\mathscr{C}^0, \mathscr{S}et\} \to \mathscr{D}$ a functor that commutes with colimits and such that $T(s)$ is an epimorphism for every $s \in \Sigma$. Then there is a unique functor $\bar{T}: \mathscr{C}_\Sigma \to \mathscr{D}$ such that:*

a) *$\bar{T}\bar{L} \simeq T$ (where $\bar{L} = \bar{L}_\Sigma: \{\mathscr{C}^0, \mathscr{S}et\} \to \mathscr{C}_\Sigma$ is the adjoint to the inclusion functor $\bar{I} = \bar{I}_\Sigma: \mathscr{C}_\Sigma \to \{\mathscr{C}^0, \mathscr{S}et\}$).*

b) *$\bar{T}$ commutes with colimits.*

Proof. We shall make use of Lemma 7.1, i.e. we shall show that $T(\Psi_F)$ is an isomorphism for every presheaf F. (Here, as in Sections 2.3 and 2.4, $\Psi_F: F \to \bar{L}(F)$ is an arrow of reflection. Let $(p^i, P_i)_{i \in \mathscr{I}} = \mathscr{A}$ be the class of all quotient objects of F such that $T(p^i)$ is an isomorphism, and let (p, P) be a lower bound of this class of quotient objects. Now, since the lower bound is in fact a colimit in $\{\mathscr{C}^0, \mathscr{S}et\}$ (see Proposition 17.16, Ch. 1) and T commutes with colimits, we get that $T(p)$ is an isomorphism. Now we claim that $P \in \bar{\mathscr{C}}_\Sigma$. Let $s \in \Sigma$, $s: A \to B$, and $u, v: B \to P$ be a pair of morphisms such that $us = vs$. Also, let (q, Q) be a cokernel of (u, v). Since T commutes with colimits, $(T(q), T(Q))$ is a cokernel of $(T(u), T(v))$. Hence $T(q)$ is an isomorphism, since $T(s)$ is an epimorphism, i.e. $(qp, Q) \in A$ so that there is an epimorphism $h: Q \to P$ with $hqp = p$. But then q is an isomorphism, that is $u = v$. Therefore $P \in \mathscr{C}_\Sigma$, and hence there is a unique morphism $\bar{p}: \bar{L}(F) \to P$ such that $\bar{p}\Psi_F = p$. Thus, we have that $T(p) = T(\bar{p})T(\Psi_F)$ is an isomorphism, and this implies that $T(\Psi_F)$ is a left invertible epimorphism, i.e. an isomorphism. Finally, $\bar{T} = T\bar{I}$ and it is easy to see that $\bar{T}$ commutes with colimits (see Lemma 7.1). ✳

Now let Σ be a class of morphisms of $\{\mathscr{C}^0, \mathscr{S}et\}$. Consider diagram (4) in Section 2.4. For each $F \in \{\mathscr{C}^0, \mathscr{S}et\}$ denote by $\mathscr{I}'_F$ the full subcategory of $F/\{\mathscr{C}^0, \mathscr{S}et\}$ whose objects are $(t'_{(s,h)}, F'_{(s,h)})_{s \in \Sigma(F),\, h \in [A_s, F]}$ and by K'_F the inclusion functor. Consider the following condition:

(*′) For each $F \in \{\mathscr{C}^0, \mathscr{S}et\}$ the functor K'_F has a colimit.

It is easy to see that condition (*′) implies condition (*) of Theorem 4.13.

THEOREM 7.3. *Let Σ be a class of morphisms from $\{\mathscr{C}^0, \mathscr{S}et\}$ such that condition (*′) and conditions (**) and (***) of Theorem 4.13 are satisfied. Assume that $T: \{\mathscr{C}^0, \mathscr{S}et\} \to \mathscr{D}$ is a functor that commutes with colimits and makes invertible all elements of Σ. Then there is a unique functor $\tilde{T}: \mathscr{C}_\Sigma \to \mathscr{D}$ such that:*

a) *$\tilde{T}\tilde{L} \simeq \tilde{T}$, where $\tilde{L}$ is the adjoint to the inclusion functor $\tilde{\mathscr{C}}^0_\Sigma \to \{\mathscr{C}^0, \mathscr{S}et\}$.*

b) *$\tilde{T}$ commutes with colimits.*

Proof. We again use Lemma 7.1. If $\alpha_F: F \to \tilde{L}(F)$ is an arrow of reflection, we shall show that $T(\alpha_F)$ is an isomorphism. The proof will be divided into several steps.

i) Since the hypothesis of Theorem 7.2 is satisfied for every $F \in \{\mathscr{C}^0, \mathscr{S}et\}$, we can define a quotient object $\{u_0, F_0\}$ such that this quotient object is the lower bound of all quotient objects (q, R) of F where q is such that $T(q)$ is an isomorphism. Obviously $T(u_0)$ is such an isomorphism and F_0 is a Σ-separated presheaf.

An object F such that u_0 is a isomorphism will be called *T-separated.* It is clear that F_0 is T-separated; F_0 is called the *T-separated presheaf associated with F.*

ii) Let (u_1', F_1') be a colimit of the functor K_{F_0}' in $F_0/\{\mathscr{C}^0, \mathscr{S}et\}$. See diagram (4). Since $T(s)$ is an isomorphism and T commutes with colimits we get that $T(t'_{(s,h)})$ is an isomorphism for every $s \in \Sigma$ and for every $h \in [A_s, F_0]$, hence $T(u_1')$ is also an isomorphism.

Furthermore, let (p, F_1) be the lower bound of all quotient objects (h, Q) of F_1' such that $T(h)$ is an isomorphism in D. Let u_1 denote pu_1'. Then u_1 is a functorial monomorphism. Indeed, if

$$F_0 \xrightarrow{\;q\;} \mathrm{Im}(u_1) \xrightarrow{\;v\;} F_1$$

is the triangular decomposition of u_1, then $T(q)$ is a left invertible epimorphism, so that it is invertible. Hence q is an isomorphism, by the definition of F_0 (see step *i*)). In conclusion, $v = u_1$, i.e. u_1 is a monomorphism in $\{\mathscr{C}^0, \mathscr{S}et\}$. It is clear that $T(u_1)$ is an isomorphism.

iii) Choose an inaccessible ordinal number θ as in the proof of Theorem 4.13. For each ordinal number $\delta \leqslant \theta$, define a pair (u_δ, F_δ), where $u_\delta: F_0 \to F_\delta$ is a functorial monomorphism, $T(u_\delta)$ is an isomorphism and F_δ is T-separated.

—(u_1, F_1) was defined in step ii). In the same way we define $(u_{\delta+1}, F_{\delta+1})$ when (u_δ, F_δ) has been defined.

— If δ is a limit ordinal and (u_γ, F_γ) has been defined for every $\gamma < \delta$, then consider F_δ' the colimit of the direct system $(F_\gamma, u_{\gamma\gamma'})_{\gamma<\delta}$, and let $u_\delta': F_0 \to F_\delta'$ be the canonical morphism. Also let (p, F_δ) be the T-separated presheaf associated with F_δ' (see i)) and $u_\delta = pu_\delta'$. As in step ii) we get that u_δ is a functorial monomorphism. Also, the morphism $u_{\gamma\delta} = pu_{\gamma\delta}'$ (where $u_{\gamma\delta}': F_\gamma \to F_\delta'$ is the structural morphism), is a monomorphism since F_γ is T-separated for all $\gamma < \delta$. Since T commutes with colimits, $T(u_\delta)$ is an isomorphism.

Now using the same procedure as in the proof of Theorem 4.13 we see that F_θ is a Σ-sheaf.

iv) Consider the commutative diagram

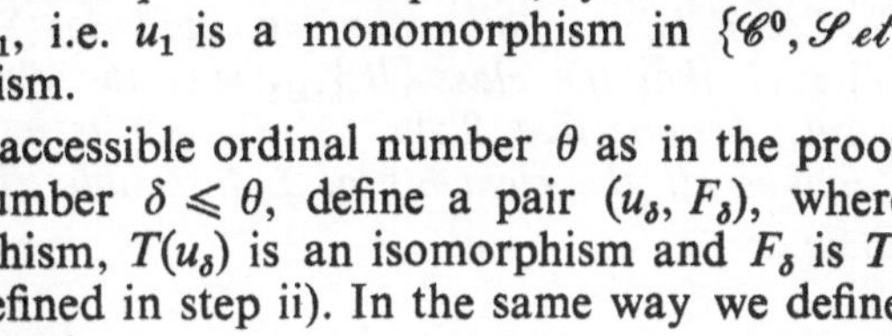

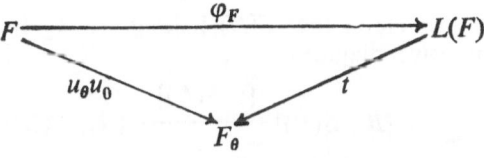

where t is defined, since $F_\theta \in \tilde{\mathscr{C}}_\Sigma \subseteq \mathscr{C}o\mathscr{p}(\tilde{\mathscr{C}}_\Sigma)$ and φ_F is an arrow of reflection. Hence $T(\varphi_F)$ is a left invertible epimorphism (see Proposition 4.3), i.e. an isomorphism.

v) Assume that $F \in \mathscr{C}o\mathscr{p}(\tilde{\mathscr{C}}_\Sigma)$. Again consider the cocartesian diagram (4) in Section 2.4. Since T commutes with colimits and $T(s)$ is an isomorphism, we obtain that $T(t'_{(s,h)})$ is also an isomorphism. Hence if (α^1, F^1) is defined as in the proof of Theorem 4.13, $T(\alpha^1)$ will be an isomorphism. Furthermore, since T commutes with colimits, then we see that $T(\alpha_F)$ is an isomorphism, where $\alpha_F: F \to \tilde{L}(F)$ is an arrow of reflection.

vi) Finally let F be an arbitrary object of $\{\mathscr{C}^0, \mathscr{S}et\}$; then by Theorem 4.13 the arrow of reflection $\alpha_F: F \to L(F)$ can be factored as

$$F \xrightarrow{\varphi_F} L(F) \xrightarrow{\alpha_{L(F)}} \tilde{L}(F).$$

Now by step iv) it follows that $T(\varphi_F)$ is an isomorphism and by step v) it follows that $T(\alpha_{L(F)})$ is an isomorphism. Hence $T(\alpha_F)$ is an isomorphism. ✳

Note 7.4. In the proof of Theorem 7.3 we did not use condition (*′) but a weaker condition, namely:

(*″). For every T-separated presheaf F, the functor K'_F has a colimit.

This condition will be used in the following theorem.

THEOREM 7.5. *Let $\mathscr{C}$ be a small category, and let Σ be a class of morphisms of $[\mathscr{C}^0, \mathscr{S}et]$ such that the class $\{B_s\}_{s\in\Sigma}$ (i.e. the class of codomains of all morphisms s) is a set. Assume that $T: [\mathscr{C}^0, \mathscr{S}et] \to \mathscr{D}$ is a functor that commutes with colimits and makes all the elements of Σ invertible. Then there is a functor $\tilde{T}: \tilde{\mathscr{C}}_\Sigma \to \mathscr{D}$ such that:*

a) $\tilde{T}\tilde{L} \simeq T$.

b) *$\tilde{T}$ commutes with colimits and has a coadjoint.*

Moreover, every functor $R: \mathscr{C} \to \mathscr{D}$ that commutes with colimits has a coadjoint.

Proof. First we show that condition (*″) is satisfied. Indeed, let F be a T-separated presheaf. Consider diagram (4) of Section 2.4. If $F \xrightarrow{q} \mathrm{Im}(t'_{(s,h)}) \xrightarrow{\bullet} F'_{(s,h)}$ is the triangular decomposition of $t'_{(s,h)}$, then $T(q)$ is an isomorphism since $T(t'_{(s,h)})$ is an isomorphism. Hence q is an isomorphism, because F is T-separated. Thus $t'_{(s,h)} = v$. Now condition (*″) follows, so that $[\mathscr{C}^0, \mathscr{S}et]$ is locally small and locally cosmall (see the proof of Theorem 4.17). Now the existence of $\tilde{T}$ follows from Theorems 4.17 and 7.3.

Furthermore let $T' = TH$ where $H: \mathscr{C} \to [\mathscr{C}^0, \mathscr{S}et]$ is the canonical embedding. We let $S: \mathscr{D} \to [\mathscr{C}^0, \mathscr{S}et]$ denote the functor $S(D) = H_D T'$. By Exercise 2, S is a coadjoint of T. Now we claim that $S(D)$ belongs to $\tilde{\mathscr{C}}_\Sigma$ for every $D \in \mathscr{D}$. Indeed, consider the commutative diagram

$$\begin{array}{ccc} [B_s, S(D)] & \xrightarrow{[s,\, S(D)]} & [A_s, S(D)] \\ \wr\Big| & & \Big|\wr \\ \mathscr{D}(T(B_s), D) & \xrightarrow{H_D(T(s))} & \mathscr{D}(T(A_s), D), \end{array}$$

where $s: A_s \to B_s$ belongs to Σ and the vertical maps are isomorphisms of adjunction. Hence $[s, S(D)]$ is also a bijection, so that $S(D) \in \tilde{\mathscr{C}}_\Sigma$. In conclusion S defines a functor $\tilde{S}: \mathscr{D} \to \tilde{\mathscr{C}}_\Sigma$ such that $\tilde{S}$ is a coadjoint to $\tilde{T}$, and $U\tilde{S} = S$, where $U: \tilde{\mathscr{C}}_\Sigma \to [\mathscr{C}^0, \mathscr{S}et]$ is the canonical inclusion. ✳

Note 7.6. If in Theorems 7.2 and 7.3 the functor TH is properly defined, then we may say that the functors $\overline{T}$ and $\tilde{T}$ each have a coadjoint.

COROLLARY 7.7. *Let $\mathscr{C}$ be a small category, J a topology on $\mathscr{C}$ and $T:\mathscr{C}\to\mathscr{D}$ a $(\mathscr{C}, J)$-cosheaf with values in $\mathscr{D}$. We assume that $\mathscr{D}$ is a cocomplete category. Then there is a unique functor $\tilde{T}:\tilde{\mathscr{C}}_J\to\mathscr{D}$ such that:*

a) $\tilde{T}\tilde{H}_J \simeq T$.

b) *$\tilde{T}$ commutes with colimits and has a coadjoint.*

Moreover, every functor $R:\tilde{\mathscr{C}}_J\to\mathscr{D}$ that commutes with colimits has a coadjoint.

Proof. Let $J=(F_r,\ \Psi^f,\ X_r)_{r\in R}$ be the given topology, (t^r, R) a colimit of HF_r and $s_r\colon P_r\to H(X_r)$ the unique morphism such that $s_r t^r = H\Psi^r$, $r\in R$. Let $\Sigma(J)$ denote $\{s_r\}_{r\in R}$. By Theorem 5.7, the presheaf F belongs to $\tilde{\mathscr{C}}_J$ if and only if it is a $\Sigma(J)$-sheaf.

Now let $\bar{T}\colon[\mathscr{C}^0, \mathscr{S}et]\to\mathscr{D}$ be the extension functor that was defined in Theorem 2.3 and Corollary 2.8. We claim that $\bar{T}$ makes the elements of $\Sigma(J)$ invertible. Indeed, since $\bar{T}$ commutes with colimits, then $(\bar{T}(t^r),\ \bar{T}(P_r))$ is a colimit of $\bar{T}HF_r$. On the other hand, we see that $\bar{T}HF_r = TF_r$ and by hypothesis $(T(\Psi^r),\ T(X_r))$ is a colimit of TF_r. Hence $\bar{T}(s_r)$ is necessarily an isomorphism. Finally, the proof is concluded by using Theorem 7.5. ✳

COROLLARY 7.8. *Let $\mathscr{C}$ be a small category, $\mathscr{D}$ a cocomplete category and $T:\mathscr{C}\to\mathscr{D}$ a cosheaf in the canonical topology on $\mathscr{C}$, with values in $\mathscr{D}$. Then there is a functor $\tilde{T}:\tilde{\mathscr{C}}\to\mathscr{D}$ such that:*

a) $\tilde{T}\tilde{H} = T$.

b) *$\tilde{T}$ commutes with colimits and has a coadjoint.*

Moreover every functor $R:\tilde{\mathscr{C}}\to\mathscr{D}$ that commutes with colimits has a coadjoint. ✳

Exercises

7.1. Let Σ be a class of epimorphisms of $\{\mathscr{C}^0, \mathscr{S}et\}$ that satisfies the condition (*′). Assume that $T\colon\{\mathscr{C}^0, \mathscr{S}et\}\to\mathscr{D}$ is a functor that commutes with colimits and makes all elements of Σ invertible. Show that then there is a unique functor $\tilde{T}\colon\mathscr{C}_\Sigma\to\mathscr{D}$ such that:

a) $\tilde{T}\tilde{L}\simeq T$.

b) T commutes with colimits.

(Hint: Use Theorem 4.4 and the steps i), ii) of the proof of Theorem 7.3. Observe that u_1 is an epimorphism.)

7.2. Let Σ be a class of morphisms from $\{\mathscr{C}^0, \mathscr{S}et\}$ and let $T\colon\{\mathscr{C}^0, \mathscr{S}et\}\to\mathscr{S}et$ be a functor that commutes with colimits and makes all the elements of Σ invertible. Let $\mathscr{D}$ be the full subcategory of $\{\mathscr{C}^0, \mathscr{S}et\}$ whose objects are T-separated presheaves. Prove that the inclusion functor $\mathscr{D}\to\{\mathscr{C}^0, \mathscr{S}et\}$ has an adjoint.

7.3. Let (X, t) be a topological space. Show that for each sheaf on X, there is a unique functor $\bar{F}\colon F(X) \to \mathcal{S}et^0$ that commutes with colimits and such that the diagram

$$\begin{array}{ccc} t & \longrightarrow & \tilde{t} = E(X) \\ {\scriptstyle F}\downarrow & & \downarrow{\scriptstyle \bar{F}} \\ \mathcal{S}et & \xrightarrow{\;D\;} & \mathcal{S}et^0 \end{array}$$

is commutative, where D is the duality functor.

7.4. Let α be a regular cardinal number. An object of $\{\mathscr{C}^0, \mathcal{S}et\}$ is called an *α-continuous presheaf* if it is the colimit of an α-filtered system of representable presheaves. Let $\mathrm{Cont}_\alpha(\mathscr{C})$ denote the full subcategory of $\{\mathscr{C}^0, \mathcal{S}et\}$ whose objects are α- continuous presheaves. It is clear that the canonical functor $H\colon \mathscr{C} \to \{\mathscr{C}^0, \mathcal{S}et\}$ defines a full and faithful functor $C_\alpha\colon \mathscr{C} \to \mathrm{Cont}_\alpha(\mathscr{C})$.

a) Show that $\mathrm{Cont}_\alpha(\mathscr{C})$ is a category with α-filtered colimits and these are formed argumentwise.

b) Prove that α-filtered colimits in $\{\mathscr{C}^0, \mathcal{S}et\}$ commute with α-limits and, moreover, that if $\mathscr{C}$ is a small category, then $\mathrm{Cont}_\alpha(\mathscr{C})$ is an α-complete category.

c) Let $\mathscr{D}$ be a category with α-filtered colimits and $F\colon \mathscr{C} \to \mathscr{D}$ a functor. Show that there is a unique functor $F_\alpha\colon \mathrm{Cont}_\alpha(\mathscr{C}) \to \mathscr{D}$ such that $F_\alpha C_\alpha \simeq F$, and F_α commutes with α-filtered colimits, and, moreover, if $T\colon \mathrm{Cont}_\alpha(\mathscr{C}) \to \mathscr{D}$ is a functor, then $(TC_\alpha)_\alpha \simeq T$ if and only if T commutes with α-filtered colimits.

7.5. Let $\mathscr{C}$ be a category with α-bounded colimits. Show that then the category $\mathrm{Cont}_\alpha(\mathscr{C})$ is cocomplete.

7.6. Let $\mathscr{C}$ be a small and α-cocomplete category. Prove that $\mathrm{Cont}_\alpha(\mathscr{C})$ is a coreflective subcategory of $[\mathscr{C}^0, \mathcal{S}et]$.

7.7. Let $\mathscr{C}$ and $\mathscr{C}'$ be categories. Let $\mathrm{Ad}(\mathscr{C}, \mathscr{C}')$ denote the (generally illegitimate) category whose objects are pairs (T, S), where $T\colon \mathscr{C} \to \mathscr{C}'$ is a functor and S is a coadjoint to T. The morphisms are pairs $(t, s)\colon (T, S) \to (T', S')$ where $t\colon T \to T'$, $s\colon S' \to S$ are functorial morphisms such that the diagrams

$$\begin{array}{ccc} \mathrm{Id}\,\mathscr{C} & \xrightarrow{\;u\;} & ST \\ {\scriptstyle u'}\downarrow & & \downarrow{\scriptstyle St} \\ S'T' & \xrightarrow{\;sT'\;} & ST' \end{array} \qquad \begin{array}{ccc} TS' & \xrightarrow{\;Ts\;} & TS \\ {\scriptstyle tS'}\downarrow & & \downarrow{\scriptstyle v} \\ T'S' & \xrightarrow{\;v'\;} & \mathrm{Id}\mathscr{C}' \end{array}$$

are commutative, where u, u' and v, v' are corresponding arrows and coarrows of adjunction.

Using the hypothesis and notation of Corollary 7.7, show that the assignment

$$T \rightsquigarrow (\tilde{T}, \tilde{S})$$

defines in a canonical way an equivalence between the category of all $(\mathscr{C}, J)$-cosheaves with values in $\mathscr{D}$ and $\mathrm{Ad}(\tilde{\mathscr{C}}_J, \mathscr{D})$.

Generally, $\mathrm{Ad}(\mathscr{C}, \mathscr{C}')$ is called the *category of adjoint pairs* from $\mathscr{C}$ into $\mathscr{C}'$.

2.8. Completion of categories

Some aspects of the completion of categories are presented. These are related to the so-called Dedekind—MacNeille completion of an ordered set.

References: Birkhoff [2], Gabriel—Ulmer [1], Isbell [8], [9], Kennison [5], Lambeck [1], Ulmer [1].

THEOREM 8.1. *Let $\mathscr{C}$ be a small category and let $\tilde{H}: \mathscr{C} \to \tilde{\mathscr{C}}$ be the natural extension. Then the following hold:*

1) *$\tilde{H}$ is full, faithful, dense and commutes with limits, colimits and monocones.*

2) *If $T: \mathscr{C} \to \mathscr{D}$ is a full, faithful and dense functor that commutes with limits and colimits, then there is a unique full and faithful functor $\tilde{S}^T: \mathscr{D} \to \tilde{\mathscr{C}}$ such that $\tilde{S}^T T = \tilde{H}$.*

3) *If $\mathscr{C}'$ is a cocomplete category and $F: \mathscr{C} \to \mathscr{C}'$ is a functor that commutes with colimits, then there is a unique functor $\tilde{F}: \tilde{\mathscr{C}} \to \mathscr{C}'$ such that $\tilde{F}$ commutes with colimits and has a coadjoint.*

The proof follows from Theorem 5.5 and Corollary 7.8. ✳

COROLLARY 8.2. *Let $\mathscr{M}$ be an ordered set. Then there is an ordered set $\mathscr{C}(\mathscr{M})$ and an injective nondecreasing map $h_{\mathscr{M}}: \mathscr{M} \to \mathscr{C}(\mathscr{M})$ such that:*

1) *h_M commutes with limits and colimits.*

2) *$\mathscr{C}(\mathscr{M})$ is a complete and cocomplete lattice.*

3) *If $f: \mathscr{M} \to \mathscr{N}$ is a nondecreasing map, and $\mathscr{N}$ is a cocomplete lattice such that f commutes with colimits, then there is a unique map $\bar{f}: \mathscr{C}(\mathscr{M}) \to \mathscr{N}$ that commutes with colimits and $\bar{f} h_{\mathscr{M}} = f$.*

Proof. By Exercise 12.11, Ch. 1, if $F: \mathscr{M} \to \mathscr{S}et$ is a functor that commutes with limits, then for every $m \in \mathscr{M}$, $F(m)$ is a set that contains at most one element (this happens since F commutes with products and $m \coprod m = m$ for every $m \in \mathscr{M}$). We consider the category arrow $\{0 \to 1\}$ and let $K: \mathscr{S}et \to \{0 \to 1\}$ be the functor defined by $K(\emptyset) = 0$ and $K(X) = 1$ if $X \neq \emptyset$. It is clear that K is an adjoint to the functor $V: \{0 \to 1\} \to \mathscr{S}et$ (where V is defined by $V(0) = \emptyset$ and $V(1) = \{1\}$) so that V commutes with limits. From this we may conclude that for each functor $F: \mathscr{M}^0 \to \mathscr{S}et$ that commutes with limits, there is a functor (in fact a nondecreasing map) $F_1: \mathscr{M}^0 \to \{0 \to 1\}$ that also commutes with limits and is such that $VF_1 \simeq F$. Hence, for each $m \in \mathscr{M}$, $F_1(m) = 1$ if $F(m) \neq \emptyset$, and $F_1(m) = \emptyset$ if $F(m) = \emptyset$. Thus the assignment $F \rightsquigarrow F_1$ defines an equivalence between $\tilde{\mathscr{M}}$ and the category

$\mathscr{C}(\mathscr{M})$ the full subcategory of $[\mathscr{M}^0, \{0 \to 1\}]$ generated by all the objects that commute with limits. (We note that $\mathscr{C}(\mathscr{M})$ is in fact the ordered set of all nondecreasing maps $f: \mathscr{M}^0 \to \{0 \to 1\}$ such that $f \leqslant g$ if and only if $f(m) \leqslant g(m)$ for all $m \in \mathscr{M}$). Now we define $h_{\mathscr{M}}$ to be KH. The theorem now follows from Theorem 8.1. ✳

From Theorem 8.1 it follows that $\tilde{\mathscr{C}}$ is an interesting extension of $\mathscr{C}$. However, the problem of the construction of colimits in $\tilde{\mathscr{C}}$ is complicated (see Corollary 7.8). On the other hand, it is obvious that the smaller dense extension of $\mathscr{C}$ considered previously preserves the greater number of the properties of $\mathscr{C}$. For this reason we shall consider now only extensions of $\mathscr{C}$ for which the colimits can be easily constructed. For this, some further definitions are necessary.

We say that a functor $F: \mathscr{C} \to \mathscr{C}'$ is *predense* if for each object X' of $\mathscr{C}'$ there is a small functor I in $\mathscr{C}$ such that X is the colimit of FI. A subcategory $\mathscr{C}$ of $\mathscr{C}'$ is called *predense*, if the inclusion functor is predense. Dually, we have the notions of *precodense functor* and *precodense subcategory*.

LEMMA 8.3. *Let $\mathscr{C}$ be a full subcategory of a complete category $\mathscr{D}$. Let $\mathscr{H}$ denote the class of all full and replete subcategories $\mathscr{C}'$ of $\mathscr{D}$ that contain $\mathscr{C}$ and are close under limits. Then, relative to inclusion, $\mathscr{H}$ is an ordered class and every class of elements of $\mathscr{H}$ has a lower bound and an upper bound.*

Proof. If $\{\mathscr{C}_i'\}_{i \in \mathscr{I}}$ is a class of elements of $\mathscr{H}$, then $\bigcap_{i \in \mathscr{I}} \mathscr{C}_I' \in \mathscr{H}$ and obviously $\bigcap_i \mathscr{C}_i' = \inf_i(\mathscr{C}_i')$. Now since $\mathscr{D} \in \mathscr{H}$, it follows that every class of elements of $\mathscr{H}$ has an upper bound. ✳

We let $\mathscr{C}^1$ denote $\bigcap \mathscr{C}'$ where $\mathscr{C}'$ runs through $\mathscr{H}$; $\mathscr{C}^1$ is called the *completion* of $\mathscr{C}$ in $\mathscr{D}$.

PROPOSITION 8.4. *Let $\mathscr{C}$ be a small full subcategory of a complete category $\mathscr{D}$. Then the inclusion functor $V: \mathscr{C} \to \mathscr{C}^1$ commutes with colimits and is precodense.*

Proof. Let $\mathscr{C}' \in \mathscr{H}$. Let $\mathscr{E}$ denote the full subcategory of $\mathscr{C}'$ generated by all the objects $X \in \mathscr{C}'$, such that H_X', the restriction of H_X to $\mathscr{C}$ carries colimits to limits (i.e. if F is a functor in $\mathscr{C}$ and $\varinjlim F$ is in $\mathscr{C}$, then $H_X'(\varinjlim F) = \varprojlim H_X' F$). Obviously, $\mathscr{C} \subseteq \mathscr{E}$ and the inclusion functor $\mathscr{C} \to \mathscr{E}$ commutes with colimits (see Theorem 6.11, Ch. 1). Moreover, we claim that $\mathscr{E} \in \mathscr{H}$. Indeed, let $F: \mathscr{I} \to \mathscr{E}$ be a functor having a limit in $\mathscr{C}'$. Also let $G: \mathscr{J} \to \mathscr{C}$ be a functor having a colimit in $\mathscr{C}$. Then we have the following functorial isomorphisms:

$$H'_{\varprojlim F}(\varinjlim G) = \mathscr{E}(\varinjlim G, \varprojlim F) = \varprojlim_{i \in \mathscr{I}} \mathscr{E}(\varinjlim G, F_i).$$

Now since for each $i \in \mathscr{I}$, $F_i \in \mathscr{E}$, we have that $\mathscr{E}(\varinjlim G, F_i) = \varprojlim_{j \in \mathscr{J}} \mathscr{E}(G_j, F_i)$.

Hence

$$\varinjlim_{i \in \mathscr{I}} \mathscr{E}(\varinjlim G, F_i) = \varprojlim_{i \in \mathscr{I}} (\varprojlim_{j \in \mathscr{J}} \mathscr{E}(G_j, F_i)) = \varprojlim_{j \in \mathscr{J}} (\varprojlim_{i \in \mathscr{I}} \mathscr{E}(G_j, F_i) =$$

$$= \varprojlim_{j \in \mathscr{J}} \mathscr{E}(G_j, \varprojlim_i F_i) = (\varprojlim H'_{\varprojlim F})G.$$

In conclusion we see that $\varprojlim F \in \mathcal{E}$, so that $\mathcal{E} \in \mathcal{H}$. Hence the inclusion functor $V: \mathcal{C} \to \mathcal{C}^1$ commutes with colimits.

Finally we assert that the functor V is precodense. For that let $\mathcal{F}$ be the full subcategory of $\mathcal{D}$ defined as follows: an object X of $\mathcal{D}$ belongs to $\mathcal{F}$ if and only if there are a set $\{X_i\}_i$ of objects of $\mathcal{C}$ and a monomorphism $X \to \coprod_i X_i$. It is clear that $\mathcal{F}$ is a replete full and complete subcategory of $\mathcal{D}$ that contains $\mathcal{C}$, i.e. $\mathcal{F} \in \mathcal{H}$. Therefore the inclusion functor $V: \mathcal{C} \to \mathcal{C}^1$ coseparates $\mathcal{C}^1$ and is precodense. ✳

THEOREM 8.5. *Let $\mathcal{D}$ be a complete category such that every class of subobjects of an object of $\mathcal{D}$ has an intersection. Let $\mathcal{C}$ be a small subcategory of $\mathcal{D}$. The following hold:*

1) *$\mathcal{C}^1$, the completion of $\mathcal{C}$ in $\mathcal{D}$, is a complete and cocomplete subcategory.*

2) *The inclusion functor $V: \mathcal{C} \to \mathcal{C}^1$ commutes with colimits and is precodense.*

3) *If $\mathcal{D}$ is locally small then a functor $F: \mathcal{C}^1 \to \mathcal{E}$ has an adjoint if and only if it commutes with limits. In particular, the inclusion functor $\mathcal{C}^1 \to \mathcal{D}$ has an adjoint.*

Proof. 1) follows from the definition of $\mathcal{C}^1$ and Theorem 17.11, Ch. 1.

2) follows from Theorem 8.4.

3) follows from Theorem 8.4 and Theorem 6.11. ✳

COROLLARY 8.6. *For every small category $\mathcal{C}$, there is a full embedding $K: \mathcal{C} \to \mathcal{C}^1$ such that:*

1) *K is a dense, predense, and precodense functor that commutes with limits and colimits.*

2) *$\mathcal{C}^1$ is a complete and cocomplete category.*

Proof. Let $\tilde{H}: \mathcal{C} \to \tilde{\mathcal{C}}$ be the natural extension of $\mathcal{C}'$ and let $\mathcal{D} = \mathrm{Im}(\tilde{H})$. Since $\tilde{H}$ is an embedding we get that $\tilde{H}$ defines an isomorphism between $\mathcal{C}$ and $\mathcal{D}$. Let $\mathcal{C}^1 = \mathcal{D}^1$ and $K = V\tilde{H}$ where $V: \mathcal{D} \to \mathcal{D}^1$ is the inclusion. The proof follows from Theorems 8.1 and 8.5. ✳

Example 8.7. Let $\mathcal{M}$ be an ordered set. Let $\mathcal{M}'$ denote the full subcategory of $[\mathcal{M}^0, \mathcal{S}et]$ whose objects are (small) products of representable functors. By Theorem 9.4, Ch. 1, we get that $\mathcal{M}'$ is a full subcategory of $\mathcal{M}$ and $\mathcal{M}'$ is a complete category. Let $\mathcal{D}(\mathcal{M})$ be a skeleton of $\mathcal{M}'$ and let $\alpha_{\mathcal{M}}: \mathcal{M} \to \mathcal{D}(\mathcal{M})$ be the functor determined by the natural functor $H: \mathcal{M} \to [\mathcal{M}^0, \mathcal{S}et]$ and the canonical functor $T: \mathcal{M}' \to \mathcal{D}(\mathcal{M})$.

THEOREM 8.8. (Dedekind—McNeille). *Let $\mathcal{M}$ be an ordered set and $\mathcal{D}(\mathcal{M})$ and $\alpha_{\mathcal{M}}$ be as defined above. Then the following hold:*

1) *$\mathcal{D}(\mathcal{M})$ is a complete and cocomplete ordered set.*

2) *The functor $\alpha_{\mathcal{M}}$ is a predense and precodense embedding that commutes with limits and colimits.*

Proof. Let $\mathcal{M}_1$ be the image of H (since H is an embedding then $\mathcal{M}_1$ is isomorphic to $\mathcal{M}$). Then $\mathcal{M}^1$ as defined above is in fact the same as $\mathcal{M}_1^1$ which was defined in Corollary 8.6. Hence $\mathcal{D}(\mathcal{M})$ is complete and cocomplete. Now if F is a representable cofunctor from $\mathcal{M}$ into $\mathcal{S}et$, then for each $m \in \mathcal{M}$, $F(m)$ contains at most one element. Hence, every product of representable functors has the same property, so that $\mathcal{M}'$ is a preordered category. Furthermore, for every representable functor F, we have that $F\coprod F \simeq F$ so that $\mathscr{D}(\mathscr{M})$ is necessarily a small category. Now Theorem 8.8 follows from Corollary 8.6.

The ordered set $\mathscr{D}(\mathscr{M})$ is called the *Dedekind—McNeille completion* of $\mathscr{M}$. ✳

It is easy to see that $\mathscr{D}(\mathscr{M})$ is a full subcategory of the category $\mathscr{C}(\mathscr{M})$ which was defined in Corollary 8.2. Moreover the inclusion functor $\mathscr{D}(\mathscr{M}) \to \mathscr{C}(\mathscr{M})$ has an adjoint.

Let $\mathscr{D}$ be a category, let $\mathscr{C}$ be a small and full subcategory of $\mathscr{D}$ and let $U: \mathscr{C} \to \mathscr{D}$ be the inclusion functor. We say that $(U, \mathscr{D})$ is a *Lambek extension* of $\mathscr{C}$ if U is predense and precodense. Theorem 8.8 shows that every ordered set has a complete and cocomplete Lambek extension such that the inclusion functor is dense and codense. It is also easy to check that the category $\mathscr{S}et$ is a complete and cocomplete Lambek extension of the category $\mathscr{S}etf$, and $\mathscr{S}etf$ is a dense subcategory of $\mathscr{S}et$.

Now it is natural to ask whether every small category has a complete and cocomplete Lambek extension. The following example shows that the answer is no.

Example 8.9 (Isbell). Let $\mathbf{Z}_4$ be the cyclic group with four elements (in fact the elements of $\mathbf{Z}_4$ are the cosets of the integers modulo 4), which is viewed as a category with a single object denoted A (see Section 1.2). We shall show that no Lambek extension of the small category $\mathbf{Z}_4$ has finite limits. For that, let $U: \mathbf{Z}_4 \to \mathscr{C}$ be a full faithful functor such that $(U, \mathscr{C})$ is a Lambek extension of $\mathbf{Z}_4$. For the sake of simplicity we shall identify $\mathbf{Z}_4$ with the image of U. Hence we write $U(A) = A$ and $U(f) = f$ for each morphism f of $\mathbf{Z}_4$.

LEMMA 8.10. *Any object of $\mathscr{C}$ that is not initial or final will be both a coproduct and a product of copies of A.*

Proof. We say that a functor $F: \mathscr{I} \to \mathbf{Z}_4$ is *consistent* if for every pair $d, d': i \to i'$ of morphisms of $\mathscr{I}$ we have that $F(d) = F(d')$. If F is an inconsistent functor in $\mathbf{Z}_4$ and (X, u) is a limit of $UF(= F)$ in $\mathscr{C}$, then X is an initial object of $\mathscr{C}$. Indeed, since F is inconsistent, there are morphisms $d, d': i \to i'$, in $\mathscr{I}$ such that $F(d) \neq F(d')$. However, $F(d)u_i = F(d')u_i = u_i$. This implies that $\mathscr{C}(A, X) = \varnothing$ (otherwise, if a morphism $f: A \to X$ existed we would have that $F(d)u_i f = F(d')u_i f$, and consequently $F(d) = F(d')$, since $\mathbf{Z}_4$ is a groupoid, which is a contradiction). Hence the only functor G in $\mathbf{Z}_4$ such that X is a limit of G is the empty functor, so that X is an initial object of $\mathscr{C}$. Dually, if X is a colimit of an inconsistent functor in $\mathbf{Z}_4$, then X is a final object.

Now let $F: \mathscr{I} \to \mathbf{Z}_4$ be a nonempty consistent functor and let (X, u) be a limit of F, where X is not a final or an initial object in $\mathscr{C}$. We shall prove that X is a product of copies of A. Since X is not an initial object, there is a morphism $f: A \to X$ in $\mathscr{C}$. Initially, assume that $\mathscr{I}$ is a connected category. Then we claim that for each $i \in \mathscr{I}$ the structural component $u_i: X \to F_i = A$ is an isomorphism. Indeed, since $u_i f$ is an isomorphism, then u_i must be an epimorphism, so that to prove that u_i is an isomorphism, it will suffice to check that it is a monomorphism. For that, let $(g, h): Y \to X$ be a pair of morphisms such that $u_i g = u_i h$. We shall show that $u_j g = u_j h$ for every object j of $\mathscr{I}$. Indeed, since $\mathscr{I}$ is connected the diagram

$$i \xrightarrow{d_1} i_1 \xleftarrow{d_2} i_2 \longrightarrow i_3 \longleftarrow \cdots \xleftarrow{d_{n-1}} i_{n-1} \xrightarrow{d_n} i_n = j$$

exists in $\mathscr{I}$.

Then

$$u_j g = F(d_n)u_{i_{n-1}}g = F(d_n)F(d_{n-1})^{-1}u_{i_{n-2}}g = \ldots = F(d_n)F(d_{n-1})^{-1}$$
$$F(d_{n-2})\ldots F(d_2)^{-1}F(d_1)u_i g = F(d_n)F(d_{n-1})^{-1}\ldots F(d_2)^{-1}F(d_1)u_i h$$
$$= \ldots = F(d_n)u_{i_{n-1}}h = u_j h,$$

since $u_i h = u_i g$. Therefore $h = g$, since (X, u) is a limit of F, so that u_i is a right invertible monomorphism, i.e. an isomorphism.

Furthermore, if $\mathscr{I}$ is not connected, then it is a coproduct of a class $\{\mathscr{I}_\alpha\}_\alpha$ of connected categories. For each α, let $F_\alpha: \mathscr{I}_\alpha \to \mathbf{Z}_4$ denote the restriction of F to $\mathscr{I}_\alpha$, and let i_α be a fixed object of $\mathscr{I}_\alpha$. Let A_α denote F_{i_α}, and p_α denote u_{i_α}, the component of the structural morphism $u: X \to F$. We show that X together with the morphisms $\{p_\alpha\}_\alpha$ defines a product of the class $\{A_\alpha\}_\alpha$ of objects of $\mathscr{C}$. For that, assume that for each α a morphism $g_\alpha: Y \to A_\alpha$ is given. Let $v^\alpha: Y \to F_\alpha$ denote the functorial morphism defined as follows: $v^\alpha_{i_\alpha} = g_\alpha$; if j is another object of $\mathscr{I}_\alpha$, then there exists a diagram:

$$i_\alpha \xrightarrow{d_1} i_1 \xleftarrow{d_2} i_2 \xrightarrow{d_3} \ldots \xleftarrow{d_{n-1}} i_{n-1} \xrightarrow{d_n} j,$$

and we define $v^\alpha_j = F(d_n)F(d_{n-1})^{-1}\ldots F(d_2)^{-1}F(d_1)g_\alpha$. Since p_α is an epimorphism, an easy computation proves that the definition of v^α_i does not depend on the diagram considered, and the morphism $\{v^\alpha_j\}_{j\in\mathscr{I}_\alpha}$ determines the required functorial morphism v^α. Finally, it is easy to see that the functorial morphisms $\{v^\alpha\}_\alpha$ define a functorial morphism $v: Y \to F$. Hence a unique morphism $g: Y \to X$ exists such that $ug = v$. In particular, $p_\alpha g = g_\alpha$, i.e. $\{X, p_\alpha\}_\alpha$ gives a product of the class $\{A_\alpha\}_\alpha$. ✳

LEMMA 8.11. *Let $(U, \mathscr{C})$ be a Lambek extension of $\mathbf{Z}_4$. Then $\mathscr{C}$ has finite coproducts but no finite limits.*

Proof. Assume, to the contrary, that $\mathscr{C}$ has finite limits and finite coproducts. By Lemma 8.10, the object $A^{(2)} = A \coprod A$ (which is not initial or final in $\mathscr{C}$) is of the form $A^{\mathscr{I}}$, where $\mathscr{I}$ is a suitable set, and let $p_i: A^{(2)} = A^{\mathscr{I}} \to A$ be the structural projections. Furthermore, let $\pi_1, \pi_2: A^2 \to A$ be the structural projections, and let $t: A^2 \to A^{\mathscr{I}}$ be the uniquely defined morphism of $\mathscr{C}$ such that $p_{i_1} t = \pi_1$, where i_1 is a fixed element of $\mathscr{I}$ and $p_i t = \pi_2$ for all $i \neq i_1$. Analogously, there is a unique morphism $p: A^{\mathscr{I}} \to A^2$ such that $\pi_1 p = p_{i_1}$ and $\pi_2 p = p_{i_2}$ where i_2 is another element of $\mathscr{I}$ distinct from i_1. It is clear that $pt = 1_{A^2}$ so that p is an epimorphism in $\mathscr{C}$. But then the map $H_A(p): \mathscr{C}(A^2, A) \to \mathscr{C}(A^{\mathscr{I}}, A)$ is an injection. But it is clear that $\mathscr{C}(A^{\mathscr{I}}, A)$ is equivalent to $\mathscr{C}(A^{(2)}, A)$ so that $\mathrm{Card}(\mathscr{C}(A^{\mathscr{I}}, A)) = 4^2$. Consequently $\mathrm{Card}(\mathscr{C}(A^2, A)) \leqslant 4^2$. Now, if we use Lemma 8.10, we get that A^2 is of the form $A^{(M)}$, where M is a set and $\mathrm{Card}(M) \geqslant 2$, so that $\mathrm{Card}(\mathscr{C}(A^2, A)) = 4^{\mathrm{Card}(M)} \leqslant 4^2$. Hence $\mathrm{Card}(M) = 2$ and thus A^2 and $A^{(2)}$ are isomorphic objects of $\mathscr{C}$.

For the sake of simplicity we assume that the elements of $\mathbf{Z}_4$ are $g^0 = 1_A$, g, g^2, g^3 and that the law of composition is denoted multiplicatively.

Eight of the morphisms $A^2 \to A$ are the translations of the projections, that is they are of the form $g^i\pi_1$ or $g^j\pi_2$, $0 \leqslant i, j \leqslant 3$. The other eight morphisms $A^2 \to A$

are of the type $g^k m$, $g^l n$, for $0 \leqslant k, l \leqslant 3$ and m, n are morphisms $A^2 \to A$ that are distinct from the first eight. Let $f: \mathscr{C}(A^2, A) \to \mathscr{C}(A, A)$ be the map defined by the diagonal morphism $\Delta: A \to A^2$. This map is a surjection since $f(g^i h) = g^i f(h)$ for every i, $0 \leqslant i \leqslant 3$, and $\mathbf{Z}_4$ is a cyclic group. Hence f defines canonically an equivalence relation on $\mathscr{C}(A^2, A)$ with four equivalence classes. It may be immediately noticed that every equivalence class contains four elements which are of the type $g^i\pi_1$, $g^j\pi_2$, $g^k m$, $g^l n$ and moreover we get $m\Delta = f(m) = f(n) = n\Delta$.

Let us consider the automorphisms $v = \pi_2 \coprod \pi_1$ and $t = g\pi_1 \coprod \pi_2$ of A^2. Then for every $x, y \in \mathbf{Z}_4$ we have that $v[x, y] = [y, x] t[x, y] = [gx, y]$, and also that $mv\Delta = m\Delta = n\Delta = nv\Delta$.

It follows that mv coincides with m or n because otherwise we would have that $mv = g^i\pi_1$ (or $mv = g^j\pi_2$) and consequently, $m = mvv = g^i\pi_1 v = g^i\pi_2$ (or $m = g^j\pi_1$), which is an impossibility. However, we have that $mt\Delta = g^i m\Delta = gn^j\Delta$ since $\mathbf{Z}_4$ is a finite group and thus mt coincides with $g^i m$ or $g^j n$ (otherwise, we would have that $mt = g^k\pi_1$ or $mt = g^l\pi_2$) and consequently, $m = mt^4 = g^k\pi_1 t^3 = g^{k+1}\pi_1 t^2 = g^{k+2}\pi_1 t = g^{k+3}\pi_1$ (because $\pi_1 t = g\pi_1$, according to the definition of t, or analogously $m = g^{l+3}\pi_2$ which is a contradiction). Analogously, it may be shown that nv coincides with m or n and that nt is of the form $g^i m$ or $g^j n$. Obviously we get the following four cases:

1) $mv = m$, $mt = g^i m$.

In this case the commutativity of $\mathbf{Z}_4$ implies that $gm[x, y] = m[x, y]g = m[gx, gy] = mv[gy, gx] = mvt[y, gx] = mvtvt[gx, y] = mvtvt[x, y]$ and from this we get that $gm = mvtvt$, with x, y arbitrarily chosen elements of $\mathbf{Z}_4$ and $A^{(2)}$ a coproduct of two copies of A. If $gm = mtvt = mvtvt = g^i mvt = g^{2i} m$, then we have $g = g^{2i}$, with m being an epimorphism of $\mathscr{C}$, since $m\Delta$ is an isomorphism. To conclude we get $1 \equiv 2i \pmod 4$, which is nonsense.

2) $mv = n$, $mt = g^j n$.

If $mt = g^j n$, then the morphism nt is necessarily of the form $g^i m$. By computation similar to the above, we get that $gm = mvtvt = ntvt = g^i mvt = g^i nt = g^{2i} m$ and therefore $g = g^{2i}$, which is a contradiction, since $g^4 = 1_A$.

3) $mv = m$, $mt = g^j n$.

We have the identities: $nv = n$, $nt = g^i m$ and from these relations it results that $gm = mvtvt = mtvt = g^j nvt = g^j nvt = g^j nt = g^{i+j} m$. To conclude we get $g = g^{i+j}$ and therefore $1 \equiv i + j \pmod 4$.

On the other hand, we have: $m = mt^4 = g^j nt^3 = g^{i+j} mt^2 = g^{i+2j} nt = g^{2(i+j)}$ and this implies that $1_A = g^0 = g^{2(i+j)}$ and therefore $0 \equiv 2(i + j) \pmod 4$. Furthermore since $i + j \equiv 1 \pmod 4$ we see that $0 \equiv 2 \pmod 4$, which is an impossibility.

4) $mv = n$ and $mt = g^i m$.

Analogously with case 3), it results that $nv = m$ (the morphism nt being of the form $g^j n$) and $1 \equiv i + j \pmod 4$. Obviously, we have now that $m[g^r, g^s] = mt^r[g^0, g^s] = mt^r v[g^s, g^0] = mt^r vt^s[g^0, g^0] = g^{ri} mvt^s[1_A, 1_A] = g^{ri} nt^s[1_A, 1_A] = g^{ri+sj} m[1_A, 1_A]$ and $n[g^r, g^s] = g^{rj+si} n[1_A, 1_A]$.

We assert that $i \not\equiv 0 \pmod 4$. Indeed, otherwise we should have that $n[g^r, g^s] = g^r n\Delta = n\Delta\pi_1[g^r, g^s]$, since $i \equiv 0 \pmod 4$ would imply that $j \equiv 1 \pmod 4$. Consequently $n = n\Delta\pi_1$ since A^2 is a coproduct of copies of A.

Taking $n\Delta = g^l$ we would get that $n = g^l\pi_1$, which is nonsense. To conclude we have obtained that $i \not\equiv 0 \pmod 4$. Similarly $j \not\equiv 0 \pmod 4$ and therefore the only solutions of the above congruences are:

a) $i \equiv 3 \pmod 4$, $j \equiv 2 \pmod 4$,

b) $j \equiv 3 \pmod 4$, $i \equiv 2 \pmod 4$.

In case a) let us denote $p = (m\Delta)^{-1}m\colon A^2 \to A$ and consider in $\mathscr{S}et$ the kernel of the double arrow $(H^A(\pi_1), H^A(p))\colon \mathscr{C}(A, A^2) \to \mathscr{C}(A, A)$. Obviously, $H^A(p)([g^r, g^s]) = (m\Delta)^{-1}m[g^r, g^s] = (m[1_A, 1_A])^{-1}g^{3r+2s}m\Delta = g^{3r+2s}$ and this implies that $H^A(\pi_1)([g^r, g^s]) = H^A(p)([g^r, g^s])$ if and only if $3r + 2s \equiv r \pmod 4$ or equivalently $2(r+s) \equiv 0 \pmod 4$. This congruence has in $\mathbf{Z}_4$ the following eight solutions: (0, 0), (1, 1), (2, 2), (3, 3), (0, 2), (2, 0), (1, 3) and (3, 1).

We assert that $\operatorname{Ker}(\pi_1, p)$ does not exist in $\mathscr{C}$. Indeed, if (K, u) were a kernel of (π_1, p) in $\mathscr{C}$, then we should have that $\operatorname{Card}(\mathscr{C}(A, K)) = 8$ which is absurd (because, by Lemma 8.10, $\operatorname{Card}(\mathscr{C}(A, K))$ would be zero or a power of four).

In case b) the proof reduces to that of case a) by taking $p = (n\Delta)^{-1}n$. ⁎

Note 8.12. Let $\mathscr{C}$ be a small category and let $\mathscr{C}_1$ be the full and replete subcategory of $[\mathscr{C}^0, \mathscr{S}et]$ whose objects are limits of representable functors. Obviously $\mathscr{C}_1$ is a Lambek extension of $\mathscr{C}$ and $\mathscr{C}$ need not be codense in $\mathscr{C}_1$. From Example 8.9 we see that $\mathscr{C}_1$ is not a complete category, although it always has products, by its definition (see Exercise 8.2).

Relative to the category $\mathscr{C}_1$ considered in Note 8.12 we have the following result.

Proposition 8.13. *Let $\mathscr{C}$ be a small category. The following are equivalent:*

a) *$\mathscr{C}$ has a Lambek extension that is complete and cocomplete, and $\mathscr{C}$ is dense in it.*

b) *The category $\mathscr{C}_1$ is complete and cocomplete.*

c) *$\mathscr{C}_1$ is a category with kernels.*

d) *$\mathscr{C}_1$ is a complete category.*

e) *$\mathscr{C}_1$ is a cocomplete category.*

f) *The inclusion functor $\mathscr{C}_1 \to [\mathscr{C}^0, \mathscr{S}et]$ has an adjoint.*

Proof. The implications f) ⇒ a) ⇒ b) ⇒ c) are obvious.

c) ⇒ d). By Exercise 8. 2, $\mathscr{C}_1$ is a category with products so that by Theorem 9.1, Ch. 1, we see that $\mathscr{C}_1$ is complete.

d) ⇒ e). Use Theorem 17.11, Ch. 1.

e) ⇒ f). Use Corollary 2.8. ⁎

Proposition 8.14. *Let $\mathscr{C}$ be a small category and $\mathscr{D}$ a full subcategory of $[\mathscr{C}^0, \mathscr{S}et]$. For each ordinal α let $\mathscr{D}_\alpha$ denote the full and replete subcategory of $[\mathscr{C}^0, \mathscr{S}et]$ defined as follows:*

$\mathscr{D}_0 = \mathscr{D}$.

$\mathscr{D}_{\alpha+1}$ is the full and replete subcategory of $[\mathscr{C}^0, \mathscr{S}et]$ whose objects are limits of objects of $\mathscr{D}_\alpha$.

If α is a limit ordinal, then

$\mathscr{D}_\alpha = \bigcup_{\beta<\alpha}\mathscr{D}_\beta$.

Then $\mathscr{D}^1$, the completion of $\mathscr{D}$ in $[\mathscr{C}^0, \mathscr{S}et]$, is equal to $\bigcup_{\alpha\geqslant 1}\mathscr{D}_\alpha$.

Proof. Since $\mathscr{D} \subseteq \mathscr{D}^1$ and $\mathscr{D}^1$ is a full and replete subcategory of $[\mathscr{C}^0, \mathscr{S}et]$ which is closed under limits, we see that $\mathscr{D}_1 \subseteq \mathscr{D}^1$. But then $\mathscr{D}_\alpha \subseteq \mathscr{D}^1$ for every α, so that $\bigcup_{\alpha \geq 1} \mathscr{D}_\alpha \subseteq \mathscr{D}^1$.

In order to finish the proof it will suffice to show that $\bigcup_{\alpha \geq 1} \mathscr{D}_\alpha$ is a full and replete subcategory of $[\mathscr{C}^0, \mathscr{S}et]$, closed under limits (see the definition of $\mathscr{D}^1$). Since $\bigcup_{\alpha \geq 1} \mathscr{D}_\alpha$ is obviously full and replete, we have only to show that it is closed under limits.

Let $F: \mathscr{I} \to \bigcup_{\alpha \geq 1} \mathscr{D}_\alpha$ be a small functor. Then, for each $i \in \mathscr{I}$, there is an ordinal number α_i such that $F_i \in \mathscr{D}_{\alpha_i}$. Since $\mathscr{I}$ is a small category, there is an ordinal α such that $F_i \in \mathscr{D}_\alpha$ for every i, i.e. $\underrightarrow{\lim}\, F$, the limit of F in $[\mathscr{C}^0, \mathscr{S}et]$ belongs to $\mathscr{D}_{\alpha+1}$. Hence $\bigcup_{\alpha \geq 1} \mathscr{D}_\alpha$ is closed under limits in $[\mathscr{C}^0, \mathscr{S}et]$. ✳

Exercises

8.1. Let $\mathscr{M}$ be an ordered set. Show that there is no topology J on $\mathscr{M}$ such that $\mathscr{M}_J$ is equivalent to the Dedekind—McNeille completion.

8.2. Let $\mathscr{C}$ be a small category. Show that the category $\mathscr{C}_1$ defined in Note 8.12 is a category with products.

8.3. Prove that the double arrow category $\{\mathbf{0} \rightrightarrows \mathbf{1}\}$ has no complete and cocomplete Lambek extension $(U, \mathscr{C})$ such that $\{\mathbf{0} \rightrightarrows \mathbf{1}\}$ is codense in $\mathscr{C}$. Show that the category $\{\mathbf{0} \rightrightarrows \mathbf{1}\}_1$ defined as in Note 8.12 is the full subcategory of $[\{\mathbf{0} \rightrightarrows \mathbf{1}\}^0, \mathscr{S}et]$, whose objects are:

— the initial and final objects of $[\{\mathbf{0} \rightrightarrows \mathbf{1}\}^0, \mathscr{S}et]$;
— the objects of the form $H_1^{\mathscr{I}}$, $\mathscr{I}$ being a set;
— the objects of the form $H_1^{\mathscr{I}} \coprod H_0$, $\mathscr{I}$ being a set.

Note that the coproduct $H_0^{(3)}$ does not exist in $\{\mathbf{0} \rightrightarrows \mathbf{1}\}_1$.

8.4. Let $\mathscr{G}$ be a groupoid, $\mathscr{I}$ a connected category and $F: \mathscr{I} \to \mathscr{G}$ a functor. Assume that (X, u) is a cone over F. Show that for every functor $T: \mathscr{G} \to \mathscr{S}et$, the cone $(T(X), Tu)$ is a limit of TF.

8.5. Let $\mathscr{G}$ be a group and $F: \mathscr{I} \to \mathscr{G}$ a functor. Assume that $\mathscr{I}$ is a connected category. Prove that F has a limit in $\mathscr{G}$ if and only if there exists a cone over F. If $\mathscr{G}$ has more than one element and $\mathscr{I}$ is disconnected, then F has no limit in $\mathscr{G}$. In particular, every functor $T: \mathscr{G} \to \mathscr{S}et$ commutes with limits.

8.6. Show that if $\mathscr{G}$ is a group then the canonical functor $H: \mathscr{G} \to [\mathscr{G}^0, \mathscr{S}et]$ commutes with limits and colimits.

8.7. Show that Lemma 8.10 still holds if $\mathbf{Z}_4$ is replaced by an arbitrary group $\mathscr{G}$.

8.8. Let $\mathscr{A}$ be a small full subcategory of a complete and cocomplete category $\mathscr{C}$. Prove: If $\mathscr{C}$ admits a bicategory structure $(\mathscr{I}, \mathscr{P})$, then there is a full subcategory $\mathscr{N}$ of $\mathscr{C}$ such that:

i) $\mathscr{A} \subseteq \mathscr{N}$ and $\mathscr{N}$ is a subcategory closed under limits and colimits.

ii) There are no proper intermediate complete or cocomplete categories between $\mathscr{A}$ and $\mathscr{N}$ (i.e. if $\mathscr{A} \subseteq \mathscr{B} \subseteq \mathscr{N}$ and $\mathscr{B}$ complete or cocomplete, then $\mathscr{B} = \mathscr{N}$).

iii) $\mathscr{A}$ is closed under any small limits and any small colimits that might exist in $\mathscr{A}$.

iv) The inclusion functor $\mathscr{N} \to \mathscr{C}$ has an adjoint.

Moreover, if $\mathscr{N}$ is a category that contains a small and full subcategory $\mathscr{A}$, such that i) and ii) are satisfied, then $\mathscr{N}$ is a wellpowered and copowered category.

2.9. Grothendieck topologies

The Grothendieck topologies are a particular, but important, type of topology on a category. In this section we define the Grothendieck topologies on a small category, and study the categories of sheaves associated with them. Recall that the Grothendieck topologies were defined by Grothendieck in [1] and are closely related to some problems of Algebraic Geometry. They were first investigated by Giraud [1], M. Artin [1] and Grothendieck—Verdier [1].

References: M. Artin [1], Gabriel [1], Giraud [1], Grothendieck—Verdier [1], N. Popescu [2], [3], Roos [2], Schubert [1], Stenström [1].

Let $\mathscr{C}$ be a small category. We say that a *Grothendieck pretopology* is defined on $\mathscr{C}$ if for each $X \in \mathscr{C}$ a set $\tau(X)$ of right ideals over X is given so that:

GT 1) For each $X \in \mathscr{C}$, the right ideal $\mathscr{C}/X$ generated by 1_X belongs to $\tau(X)$.

GT 2) If $f\colon Y \to X$ is a morphism of $\mathscr{C}$, and $\mathscr{A} \in \tau(X)$, then $f^*(\mathscr{A})$, the inverse image of $\mathscr{A}$ under f, belongs to $\tau(Y)$.

Let $\mathscr{A}$ be a right ideal over X. Let $U_{\mathscr{A}}\colon \mathscr{A} \to \mathscr{C}$ denote the underlying functor. If $(Y,f) \in \mathscr{A}$, then there is a morphism $U^{\mathscr{A}}_{(Y,f)} = f\colon U_{\mathscr{A}}(Y,f) = Y \to X$. The morphisms $\{u^{\mathscr{A}}_{(Y,f)}\}_{(Y,f)\in\mathscr{A}}$ define a functorial morphism $u^{\mathscr{A}}\colon U_{\mathscr{A}} \to X$. In this way with each right ideal $\mathscr{A}$ over X we associate in a canonical manner a triple $(U_{\mathscr{A}}, u^{\mathscr{A}}, X)$. Hence a Grothendieck pretopology on a category $\mathscr{C}$ may be viewed as a topology in the sense of Section 2.5, also denoted by τ. In this way we speak of the category $\tilde{\mathscr{C}}_\tau$ (resp. $\mathscr{C}_\tau$), of the full subcategory of $[\mathscr{C}^0, \mathscr{S}et]$ whose objects are $(\mathscr{C}, \tau)$-sheaves ($(\mathscr{C}, \tau)$-*separated presheaves*) or simply τ-*sheaves* (τ-*separated presheaves*) over $\mathscr{C}$. By Theorem 5.8 we see that the inclusion functor $I\colon \tilde{\mathscr{C}}_\tau \to [\mathscr{C}^0, \mathscr{S}et]$ has an adjoint A (we shall write I_τ and A_τ if there is any danger of confusion).

In some special circumstances the functor A is left exact, i.e. commutes with finite limits. To show that, we develop some terminology concerning Grothendieck pretopologies.

We say that a Grothendieck pretopology τ on $\mathscr{C}$ is *cofiltered* if for each $X \in \mathscr{C}$, the set $\tau(X)$ is cofiltered (relative to inclusion of ideals).

A *Grothendieck topology* on a category $\mathscr{C}$ is a Grothendieck pretopology τ which satisfies the following condition:

GT 3) If $X \in \mathscr{C}$, and $\mathscr{A}$ and $\mathscr{A}'$ are right ideals over X, such that $\mathscr{A} \in \tau(X)$, and furthermore, if for each $(Y,f) \in \mathscr{A}$ we have that $f^*(\mathscr{A}') \in \tau(Y)$, then $\mathscr{A}' \in \tau(X)$.

Denote by $\tau_0(\mathscr{C})$, $\tau'(\mathscr{C})$ and $\tau(\mathscr{C})$ respectively the set of Grothendieck pretopologies, cofiltered Grothendieck pretopologies and Grothendieck topologies on a category $\mathscr{C}$. By Exercise 9.1 we have the inclusions $\tau(\mathscr{C}) \subseteq \tau'(\mathscr{C}) \subseteq \tau_0(\mathscr{C})$.

Let $\tau, \tau' \in \tau_0(\mathscr{C})$. We write $\tau \leqslant \tau'$ if, for each $X \in \mathscr{C}$, we have that $\tau(X) \subseteq \tau'(X)$. In this case we say that τ' is *finer* than τ or that τ is included in τ'. In this way $\tau_0(\mathscr{C})$, $\tau'(\mathscr{C})$ and $\tau(\mathscr{C})$ are ordered sets (see Exercise 9.2).

PROPOSITION 9.1. *Let τ be a Grothendieck pretopology on $\mathscr{C}$. There is a smallest Grothendieck topology $\bar{\tau}$ that contains τ.*

Proof. For every ordinal α, denote by τ_α the element of $\tau_0(\mathscr{C})$ defined as follows:

$$\tau_0 = \tau;$$

If τ_α has been defined, then $\tau_{\alpha+1}$ is defined as follows: a right ideal $\mathscr{A}$ over X belongs to $\tau_{\alpha+1}(X)$ if and only if there is an element $\mathscr{A}' \in \tau_\alpha(X)$ such that, for every $(Y, F) \in \mathscr{A}'$, $f^*(\mathscr{A}) \in \tau_\alpha(Y)$;

If α is a limit ordinal then $\tau_\alpha = \bigcup_{\beta<\alpha} \tau_\beta$.

Let $\bar{\tau}$ denote $\bigcup_{\alpha \geqslant 0} \tau_\alpha$. We claim that $\bar{\tau}$ is a Grothendieck topology. We first show that τ_1 is a Grothendieck pretopology. Indeed, if $\mathscr{A} \in \tau_1(X)$ then there is an $\mathscr{A}' \in \tau_0(X) = \tau(X)$ such that $f^*(\mathscr{A}) \in \tau(Y)$ for all $(Y, f) \in \mathscr{A}'$. Let $g: Z \to X$ be a morphism of $\mathscr{C}$. Then $g^*(\mathscr{A}') \in \tau(Z)$. Furthermore, if $(H, h) \in g^*(\mathscr{A}')$, then $(H, gh) \in \mathscr{A}'$, so that $(gh)^*(\mathscr{A}) \in \tau(H)$, by hypothesis. But it is clear that $(gh)^*(\mathscr{A}) = h^*(g^*(\mathscr{A}))$. Hence we get that $g^*(\mathscr{A}) \in \tau_1(Z)$, i.e. τ_1 is a Grothendieck pretopology. Now it is easy to show that τ_α is a Grothendieck pretopology for every α, and finally that $\bar{\tau}$ is a Grothendieck topology. ✳

$\bar{\tau}$ is called the *Grothendieck topology associated with the Grothendieck pretopology* τ.

Now let τ be a Grothendieck pretopology on a small category $\mathscr{C}$. In what follows we give another construction of the sheaf associated with a presheaf on $\mathscr{C}$, one that is even simpler than that of Theorem 4.13. This construction is suggested by the work of Grothendieck and Verdier [1], and it throws some light on the categories of sheaves considered here.

Let $\mathscr{A}$ be right ideal over X; recall that $\mathscr{A}^c$ denotes the subfunctor of H_X associated with $\mathscr{A}$ (see Section 1.16); by $s_X^{\mathscr{A}}$, or simply s, we denote the canonical inclusion $\mathscr{A}^c \to H_X$. In this way we may consider the ordered set $\{\mathscr{A}^c\}_{\mathscr{A} \in \tau(X)}$ of subcofunctors of H_X.

Let T be an object of $[\mathscr{C}^0, \mathscr{S}et]$, and θ an inaccessible ordinal such that

$$\text{Card } \theta > (\max_{(X,Y) \in \mathscr{C}} \text{Card}(\mathscr{C}(X, Y) \times \text{Card}(\mathscr{O}b(\mathscr{C}))$$

(here the symbol "$\times$" denotes the product of cardinal numbers). For each ordinal number $\alpha \leqslant \theta$ we define a pair (u^α, T_α) where T_α is a presheaf over $\mathscr{C}$, and $u^\alpha: T \to T_\alpha$ is a functorial morphism.

Let $T_0 = T$ and $u^0 = 1_T$. If T_α has been defined then we define $T_{\alpha+1}$ by

$$T_{\alpha+1}(X) = \varinjlim_{\mathscr{A} \in \tau(X)} [\mathscr{A}^c, T]. \tag{4}$$

(Observe that the colimit is over the ordered set $\tau(X)$ of right ideals of X.)

Since $\mathscr{C}/X \in \tau(X)$, and $(\mathscr{C}/X)^c = H_X$, then the composition

$$T_\alpha(X) \xrightarrow{\varphi_X} [H_X, T_\alpha] \xrightarrow{v_{H_X}} T_{\alpha+1}(X) \tag{5}$$

(where φ_X is the natural bijection from Theorem 3.19, Ch. 1, and v_{H_X} is the structural component of the colimit (4)), is denoted by $u_X^{\alpha,\alpha+1}$.

We prove that the assignment $X \rightsquigarrow T_{\alpha+1}(X)$ is a presheaf over $\mathscr{C}$. For that let $f\colon Y \to X$ be a morphism in $\mathscr{C}$. Denote by $f^*\colon \tau(X) \to \tau(Y)$ the functor "inverse image", i. e. the functor that associates with $\mathscr{A} \in \tau(X)$, the element $f^*(\mathscr{A})$ of $\tau(Y)$. Furthermore, the canonical morphism $f^*(\mathscr{A}) \to \mathscr{A}$ defines a map

$$[\mathscr{A}^c, T_\alpha] \to [f^*(\mathscr{A})^c, T_\alpha]$$

and finally, these maps define a map $T_{\alpha+1}(f)\colon T_{\alpha+1}(X) \to T_{\alpha+1}(Y)$. We leave it for the reader to check that the diagram

$$\begin{array}{ccc}
T_\alpha(X) & \xrightarrow{u_X^{\alpha,\alpha+1}} & T_{\alpha+1}(X) \\
{\scriptstyle T_\alpha(f)}\downarrow & & \downarrow{\scriptstyle T_{\alpha+1}(f)} \\
T_\alpha(Y) & \xrightarrow{u_Y^{\alpha,\alpha+1}} & T_{\alpha+1}(Y)
\end{array}$$

is commutative. It is easy to show that if $g\colon Z \to Y$ is another morphism of $\mathscr{C}$, then $T_{\alpha+1}(fg) = T_{\alpha+1}(g)T_{\alpha+1}(f)$ and $T_{\alpha+1}(1_X) = 1_{T_{\alpha+1}(X)}$. In this way we get that $T_{\alpha+1}$ is in fact a presheaf over $\mathscr{C}$ and the maps $\{u_X^{\alpha,\alpha+1}\}_{X\in\mathscr{C}}$ define a functorial morphism $u^{\alpha,\alpha+1}\colon T_\alpha \to T_{\alpha+1}$. Finally let us define $u^{\alpha+1}$ to be $u^{\alpha,\alpha+1}u^\alpha$.

If α is a limit ordinal number, and (T_β, u^β) is defined for every $\beta < \alpha$, let us define

$$T_\alpha \text{ to be } \varinjlim_{\beta<\alpha} T_\beta.$$

The canonical morphism $u^\alpha\colon T \to T_\alpha$ is defined to be the structural morphism of the colimit associated with $\beta = 0$.

Note 9.2. Let $v\colon T \to T'$ be a morphism of $[\mathscr{C}^0, \mathscr{S}et]$. We suggest that the reader check that for every ordinal number α a morphism $v^\alpha\colon T_\alpha \to T'_\alpha$ is defined in a canonical way, such that the diagram

$$\begin{array}{ccccc}
T & \xrightarrow{u^\alpha} & T_\alpha & \xrightarrow{u^{\alpha,\alpha+1}} & T_{\alpha+1} \\
{\scriptstyle v}\downarrow & & {\scriptstyle v^\alpha}\downarrow & & \downarrow{\scriptstyle v^{\alpha+1}} \\
T' & \xrightarrow{u'^\alpha} & T'_\alpha & \xrightarrow{u'^{\alpha,\alpha+1}} & T'_{\alpha+1}
\end{array}$$

is commutative.

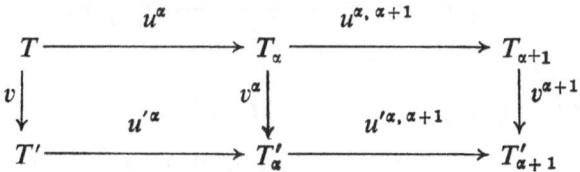

We also suggest that the reader show that for every pair α, α' of ordinal numbers, such that $\alpha < \alpha' \leqslant \theta$, there is a canonical functorial morphism $u^{\alpha,\alpha'}: T_\alpha \to T_{\alpha'}$ such that $u^{\alpha,\alpha'}u^\alpha = u^{\alpha'}$.

LEMMA 9.3. *Let $0 \leqslant \alpha < \theta$ be an ordinal number, $X \in \mathscr{C}$, $\mathscr{A} \in \tau(X)$ and $g: \mathscr{A}^c \to T_\alpha$ a functorial morphism. Then there is a functorial morphism $\bar{g}: H_X \to T_{\alpha+1}$ such that the diagram*

$$\begin{array}{ccc} \mathscr{A}^c & \xrightarrow{\ s\ } & H_X \\ {\scriptstyle g}\downarrow & & \downarrow{\scriptstyle \bar{g}} \\ T_\alpha & \xrightarrow{u^{\alpha,\alpha+1}} & T_{\alpha+1} \end{array} \tag{6}$$

is commutative, where s is the canonical inclusion.

Proof. Let $v_{\mathscr{A}^c}: [\mathscr{A}^c, T_\alpha] \to T_{\alpha+1}(X)$ be the component of the structural morphisms of the colimit (4), associated with $\mathscr{A}^c$, and also let x denote $v_{\mathscr{A}^c}(g)$. Then we can define $\bar{g} = u^x$, where $u^x: H_X \to T_{\alpha+1}$ is the functorial morphism associated with x (see Section 1.3). In order to show that diagram (6) is commutative, let $Y \in \mathscr{C}$ and $f \in \mathscr{A}^c(Y)$. Then $f: Y \to X$ is a monomorphism in $\mathscr{C}$ and $(Y, f) \in \mathscr{A}$. Since s is the canonical inclusion,

$$(\bar{g}s)_Y(f) = \bar{g}_Y(f) = u^x_Y(f) = T_{\alpha+1}(f)(x) \quad \text{(see Section 1.3).}$$

On the other hand, by (5),

$$(u^{\alpha,\alpha+1}g)_Y(f) = u_Y^{\alpha,\alpha+1}(g_Y(f)) = v_{H_Y}\varphi_Y(g_Y(f)) = v_{H_Y}(u^{g_Y(f)}).$$

Furthermore, we have the following cartesian square of $[\mathscr{C}^0, \mathscr{S}et]$ (since $(Y, f) \in \mathscr{A}$)

$$\begin{array}{ccc} H_Y & \xrightarrow{\ 1_{H_Y}\ } & H_Y \\ {\scriptstyle v}\downarrow & & \downarrow{\scriptstyle H(f)} \\ \mathscr{A}^c & \xrightarrow{\ s\ } & H_X \end{array}$$

where v is canonically induced by $H(f)$. Hence, by the construction of $T_{\alpha+1}$, the diagram

$$\begin{array}{ccc} [\mathscr{A}^c, T_\alpha] & \xrightarrow{\ v_{\mathscr{A}^c}\ } & T_{\alpha+1}(X) \\ {\scriptstyle [v, T_\alpha]}\downarrow & & \downarrow{\scriptstyle T_{\alpha+1}(f)} \\ [H_Y, T_\alpha] & \xrightarrow{\ v_{H_Y}\ } & T_{\alpha+1}(Y) \end{array}$$

is commutative.

But then

$$\begin{aligned} (T_{\alpha+1})(f)\, v_{\mathscr{A}^c}(g) &= T_{\alpha+1}(f)\,(v_{\mathscr{A}^c}(g)) = T_{\alpha+1}(f)\,(x) \\ &= v_{H_Y}[v, T_\alpha])\,(g) = v_{H_Y}(gv) = v_{H_Y}(u^{g_Y(f)}) \end{aligned}$$

since $v_Y(1_Y) = f$ and $(gv)_Y(1_X) = g_Y(f)$.

Hence the commutativity of diagram (6) is proven. ✱

LEMMA 9.4. *T_θ is a τ-separated presheaf.*

Proof. We must show that for each $X \in \mathscr{C}$ and each $\mathscr{A} \in \tau(X)$, the canonical morphism

$$[H_X, T_\theta] \to [\mathscr{A}^c, T_\theta]$$

that is induced by the inclusion $\mathscr{A}^c \xrightarrow{s} H_X$ is an injection or, equivalently, if $x, x' \in T_\theta(X)$ are such that $u^x s = u^{x'} s$, then $x = x'$. Indeed, since $T_\theta = \varinjlim_{\alpha<\theta} T_\alpha$, we can find an ordinal $\alpha < \theta$ and elements $y, y' \in T_\alpha(X)$ such that $u_X^{\alpha,\theta}(y) = x$, $u_X^{\alpha,\theta}(y') = x'$ (the morphism $u^{\alpha,\theta}$ is canonically defined). But then

$$u^{\alpha,\theta} u^y s = u^{\alpha,\theta} u^{y'} s.$$

Now let $Y \in \mathscr{C}$ and $f \in \mathscr{A}^c(Y)$. Then

i.e.

$$u_Y^{\alpha,\theta}(u^y s)_Y(f) = u_Y^{\alpha,\theta}(u^{y'} s)_Y(f).$$

Again, by the definition of T_θ, there is an ordinal number $\alpha_{(y,f)}$ such that $\alpha \leqslant \alpha_{(Y,f)} < \theta$, and such that

$$u_Y^{\alpha,\alpha(Y,f)}(u^y s)_Y(f) = u_Y^{\alpha,\alpha(Y,f)}(u^{y'} s)_Y(f).$$

By the choice of the inaccessible ordinal number θ, there is an ordinal number θ' such that $\theta' \geqslant \alpha_{(Y,f)}$ for all objects (Y,f) of $\mathscr{A}$, and $\theta' < \theta$. Finally, for each $(Y,f) \in \mathscr{A}$,

$$u_Y^{\alpha,\theta'}(u^y s)_Y(f) = u_Y^{\alpha,\theta'}(u^{y'} s)_Y(f),$$

i.e.

$$u^{\alpha,\theta'} u^y s = u^{\alpha,\theta'} u^{y'} s.$$

Now let us take $\bar{x} = u_X^{\alpha,\theta'}(y)$ and $\bar{x}' = u_X^{\alpha,\theta'}(y')$. It is clear that $u_X^{\theta',\theta}(\bar{x}) = x$ and $u_X^{\theta',\theta}(\bar{x}') = x'$. By the construction of $T_{\theta'+1}$, it follows that $u_X^{\theta',\theta'+1}(\bar{x}) = u_X^{\theta',\theta'+1}(\bar{x}')$, so that necessarily $x = x'$, i.e. T_θ is a τ-separated presheaf. ✱

LEMMA 9.5. *T_θ is a τ-sheaf.*

Proof. Let $X \in \mathscr{C}$, $\mathscr{A} \in \tau(X)$ and $g: \mathscr{A}^c \to T_\theta$ be a functorial morphism. By the choice of θ, for each object (Y,f) of $\mathscr{A}$, there are an ordinal number $\alpha_{(Y,f)} < \theta$ and an element $x'_{(Y,f)}$ of $T_{\alpha(Y,f)}(Y)$ such that $u^{\alpha(Y,f),\theta}(x'_{(Y,f)}) = g_Y(f)$. Let α be an ordinal number such that $\alpha_{(Y,f)} \leqslant \alpha$ for all $(Y,f) \in \mathscr{A}$, and $\alpha < \theta$. Let $x_{(Y,f)} = u^{\alpha(Y,f),\alpha}(x'_{(Y,f)})$. It is clear that $u^{\alpha,\theta}(x_{(Y,f)}) = g_Y(f)$.

Consider the equivalence relation R on the right ideal $\mathscr{A}$ defined as follows: $(Y,f) \sim (Y',f')\,(R) \Leftrightarrow Y = Y'$ and $g_Y(f) = g_Y(f')$. If $(Y,f) \in \mathscr{A}$, let $\widehat{(Y,f)}$ denote the equivalence class of (Y,f) modulo R. Since θ is an inaccessible ordinal number and $\mathrm{Card}(\theta) > \mathrm{Card}(\widehat{Y,f})$ we obtain an ordinal number $\alpha_{\widehat{(Y,f)}}$, such that $\alpha \leqslant \alpha_{\widehat{(Y,f)}} < \theta$ and

$$u_Y^{\alpha,\,\alpha_{\widehat{(Y,f)}}}(x_{(Y,\,f)}) = u_Y^{\alpha,\,\alpha_{\widehat{(Y,f)}}}(x_{(Y',\,f')})$$

for every $(Y',f') \in \widehat{(Y,f)}$.

Furthermore, let θ' be an ordinal number, such that $\alpha_{\widehat{(Y,f)}} \leqslant \theta'$ for every $(Y,f) \in \mathscr{A}$, and $\theta' < \theta$. Also, let $g'\colon \mathscr{A}^c \to T_{\theta'}$, be the functorial morphism defined as follows. If $(Y,f) \in \mathscr{A}$, then $g'_Y(f) = u_Y^{\alpha,\,\theta'}(x_{(Y,f)})$.

By the definition of θ', it is easy to see that g is well defined.

Now, by Lemma 9.3, there is a functorial morphism $\bar{g}'\colon H_X \to T_{\theta'+1}$ such that $\bar{g}'s = u^{\theta',\,\theta'+1}g'$. But then $u^{\theta'+1,\,\theta}\bar{g}'s = u^{\theta'+1,\,\theta}u^{\theta',\,\theta'+1}g' = u^{\theta',\,\theta}g' = g$. Hence, by Lemma 9.4 we see that T_θ is a τ-sheaf, as required. ✳

LEMMA 9.6. *If T is a τ-sheaf, then for every ordinal number α, with $\alpha \leqslant \theta$, the canonical morphism $u^\alpha\colon T \to T_\alpha$ is an isomorphism.*

The proof is left for the reader. ✳

THEOREM 9.7. *Let τ be a Grothendieck pretopology on the category $\mathscr{C}$. Then for every object T of $[\mathscr{C}^0, \mathscr{S}et]$, the morphism $u^\theta\colon T \to T_\alpha$ defined above is an arrow of reflection, relative to the full subcategory $\tilde{\mathscr{C}}_\tau$.*

Proof. By Lemma 9.5, T_θ is a τ-sheaf. Furthermore, let $v\colon T \to T'$ be a functorial morphism where T' is a τ-sheaf. Then we have the commutative square (see Note 9.2)

$$\begin{array}{ccc} T & \xrightarrow{\;u^\theta\;} & T_\theta \\ {\scriptstyle v}\downarrow & & \downarrow{\scriptstyle v^\theta} \\ T' & \xrightarrow[\sim]{\;u'^\theta\;} & T'_\theta \end{array} .$$

Now the proof follows since u'^θ is an isomorphism, by Lemma 9.6. ✳

THEOREM 9.8. *Let τ be a cofiltered Grothendieck pretopology on $\mathscr{C}$. Then the canonical functor $A\colon [\mathscr{C}^0, \mathscr{S}et] \to \tilde{\mathscr{C}}_\tau$ is left exact.*

Proof. First, we observe that the presheaf F that associates with each object X of $\mathscr{C}$ a one point set, is a final object of $[\mathscr{C}^0, \mathscr{S}et]$, and also that it is a τ-sheaf for every Grothendieck pretopology τ over $\mathscr{C}$. Hence the object $A(F)$ becomes canonically a final object of $\tilde{\mathscr{C}}_\tau$. Therefore, in proving that the functor A is left exact it will suffice to check that it commutes with fibered products (see Exercise 12.2, Ch. 1).

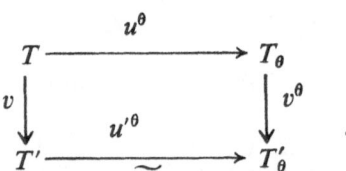

Consequently, consider in $[\mathscr{C}^0, \mathscr{S}et]$ a cartesian square

$$\begin{array}{ccc} F & \xrightarrow{\bar{p}} & G \\ {\scriptstyle \bar{g}}\downarrow & & \downarrow{\scriptstyle g} \\ P & \xrightarrow{p} & T \end{array} \quad .$$

Then for each $X \in \mathscr{C}$ and for each $\mathscr{A} \in \tau(X)$, we have in $\mathscr{S}et$ the following cartesian square:

$$\begin{array}{ccc} [\mathscr{A}^c, F] & \xrightarrow{[\mathscr{A}^c, \bar{p}]} & [\mathscr{A}^c, G] \\ {\scriptstyle [\mathscr{A}^c, \bar{g}]}\downarrow & & \downarrow{\scriptstyle [\mathscr{A}^c, g]} \\ [\mathscr{A}^c, P] & \xrightarrow{[\mathscr{A}^c, p]} & [\mathscr{A}^c, T] \end{array} \quad .$$

Now the proof follows from the construction of Theorem 12, Ch. 1, and the definition of the associated sheaf. ✳

Exercises

9.1. Show that every Grothendieck topology on a category $\mathscr{C}$ is a cofiltered Grothendieck pretopology. (Hint: Show that if $\mathscr{A}, \mathscr{A}' \in \tau(X)$, then $\mathscr{A} \cap \mathscr{A}' \in \tau(X)$.)

9.2. Show that the ordered sets $\tau_0(\mathscr{C})$ and $\tau(\mathscr{C})$ are complete and cocomplete lattices.

9.3. Let $\tau \in \tau(\mathscr{C})$, $X \in \mathscr{C}$ and let $\mathscr{A}$ and $\mathscr{A}'$ be right ideals over X. Prove that if $\mathscr{A} \subseteq \mathscr{A}'$ and $\mathscr{A} \in \tau(X)$ then $\mathscr{A}' \in \tau(X)$.

9.4. Using the notation of Proposition 9.1, show that $\tau \in \tau_0(\mathscr{C})$ belongs to $\tau(\mathscr{C})$ if and only if $\tau = \bar{\tau}$.

9.5. Let Σ be a multiplicative system of morphisms in $\mathscr{C}$. For each $X \in \mathscr{C}$, let $\tau_\Sigma(X)$ denote the set of all right ideals $\mathscr{A}$ over X which have the following property: for each morphism $f\colon Y \to X$ there is a morphism $s\colon Z \to Y$, $s \in \Sigma$, such that $(Z, fs) \in \mathscr{A}$.

a) Show that the assignment $X \rightsquigarrow \tau_\Sigma(X)$ defines a Grothendieck topology on $\mathscr{C}$. This topology is denoted by τ_Σ, and called the *Grothendieck topology associated with the multiplicative system* Σ.

b) Prove that the system Σ is right permutable if and only if for each $s\colon X \to Y$, $s \in \Sigma$, the right ideal $\langle s \rangle$ generated by s belongs to $\tau_\Sigma(Y)$.

9.6. Let $\{T_i\}_{i\in\mathscr{I}}$ be a class of presheaves over $\mathscr{C}$. For each $X\in\mathscr{C}$, let $\tau(X)$ denote the set of all right ideals over X defined as follows. The right ideal $\mathscr{A}$ over X belongs to $\tau(X)$ if and only if for every morphism $f\colon Y\to X$ of $\mathscr{C}$ and for every $i\in\mathscr{I}$ the canonical map

$$[H_Y,\ T_i]\to[f^*(\mathscr{A})^c,\ T_i],$$

induced by the inclusion $s\colon f^*(\mathscr{A})\to H_Y$ is a bijection. Show that the assignment $X\rightsquigarrow\tau(X)$ defines a Grothendieck topology on $\mathscr{C}$. It is the largest Grothendieck topology for which each T_i, $i\in\mathscr{I}$ is a sheaf. In particular, the largest Grothendieck topology on $\mathscr{C}$, for which all H_X, $X\in\mathscr{C}$ are sheaves, is called the *canonical Grothendieck topology* on $\mathscr{C}$ and denoted by k.

Show that a right ideal $\mathscr{A}$ over X belongs to $k(X)$ if and only if for every morphism $f\colon Y\to X$ the pair (s, Y) is a colimit of the underlying functor $\sigma\colon f^*(\mathscr{A})\to\mathscr{C}$, and $s\colon\sigma\to Y$ is canonically defined, i.e. if $(Z, g)\in f^*(\mathscr{A})$, then $s_{(Y,g)}=g$.

9.7. Let τ be a Grothendieck topology on $\mathscr{C}$ and let Σ_τ, or simply Σ, be the set of all morphisms $\{s_X^{\mathscr{A}}\}_{X\in\mathscr{C},\ \mathscr{A}\in\tau(X)}$. It is clear that an object T of $[\mathscr{C}^0,\mathscr{S}et]$ is a τ-sheaf (a τ-separated presheaf) if and only if it is a Σ-sheaf (Σ-separated presheaf) (see Theorem 5.7). Using the notation of Section 2.4, Lemma 4.8, show that:

a) A morphism $u\colon T\to K$ of $[\mathscr{C}^0,\mathscr{S}et]$ belongs to $\overline{\Sigma}$ if and only if for each $X\in\mathscr{C}$ and for each $v\colon H_X\to K$, the image of the canonical morphism $v'\colon T\prod_K H_X\to H_X$ belongs to $\tau(X)$, i.e. is of the form $\mathscr{A}^c$ with $\mathscr{A}\in\tau(X)$.

b) A morphism $u\colon T\to K$ of $[\mathscr{C}^0,\mathscr{S}et]$ belongs to $\tilde{\Sigma}$ if and only if $u\in\overline{\Sigma}$. Furthermore, if $ut=ut'$ with $t, t'\colon T'\to T$, then there is a morphism $s\colon T''\to T'$ such that $ts=t's$ and $s\in\overline{\Sigma}$ (or equivalently, the canonical morphism $s\colon \mathrm{Ker}(t, t')\to T'$ belongs to $\overline{\Sigma}$).

The elements of $\overline{\Sigma}$ (resp. $\tilde{\Sigma}$) are called *covering (bicovering) morphisms*. By Corollary 4.9, $\overline{\Sigma}$ and $\tilde{\Sigma}$ are left calculable and saturated systems of morphisms.

c) If $T\xrightarrow{v}T'\xrightarrow{u}T''$ are morphisms of $[\mathscr{C}^0,\mathscr{S}et]$ and uv is a covering, then u is a covering. Moreover, if u and uv are bicoverings, then v is a bicovering.

d) If in the cartesian square of $[\mathscr{C}^0,\mathscr{S}et]$

$$\begin{array}{ccc} T & \xrightarrow{u'} & T' \\ \downarrow & & \downarrow \\ T'' & \xrightarrow{u} & T''' \end{array}$$

u is a covering (a bicovering) then so is u'. Show that $\overline{\Sigma}$ (resp. $\tilde{\Sigma}$) is also a right calculable system of morphisms of $[\mathscr{C}^0,\mathscr{S}et]$.

e) Any covering monomorphism is a bicovering. Since every morphism of $[\mathscr{C}^0,\mathscr{S}et]$ has a unique triangular decomposition, and $[\mathscr{C}^0,\mathscr{S}et]$ is a locally small category, then using Theorem 14.1, Ch. 1, give a new description of the categories $\overline{\mathscr{C}}_\tau$ and $\tilde{\mathscr{C}}_\tau$.

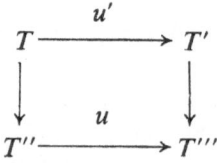

9.8. Let τ be a cofiltered Grothendieck pretopology on $\mathscr{C}$ and let T be an object object of $[\mathscr{C}^0, \mathscr{S}et]$. Show that T_2 is the τ-sheaf associated with T, and $u^2: T \to T_2$ is the arrow of adjunction. Moreover T_1 is the separated τ-presheaf associated with T.

9.9. Let τ and τ' be two Grothendieck pretopologies such that $\tau \leqslant \tau'$. Then $\tilde{\mathscr{C}}_\tau \supseteq \tilde{\mathscr{C}}_{\tau'}$. Show that:

a) If τ and τ' are Grothendieck topologies, then $\tilde{\mathscr{C}}_\tau \subseteq \tilde{\mathscr{C}}_{\tau'}$ if and only if $\tau \geqslant \tau'$. Hence the equality $\tilde{\mathscr{C}}_\tau = \tilde{\mathscr{C}}_{\tau'}$ implies $\tau = \tau'$.

b) Let $\tau \in \tau_0(\mathscr{C})$, and let $\bar{\tau}$ be its associated Grothendieck topology. The following are equivalent:

i) $\tilde{\mathscr{C}}_\tau \equiv \tilde{\mathscr{C}}_{\bar{\tau}}$.

ii) The canonical functor $A: [\mathscr{C}^0, \mathscr{S}et] \to \tilde{\mathscr{C}}_\tau$ is left exact.

iii) τ is a cofiltered Grothendieck pretopology.

(Hint: Use Exercise 9.11.f, below.)

9.10. (Giraud [1]). Let Gir($\mathscr{C}$) be the class of all full reflective and replete subcategories of $[\mathscr{C}^0, \mathscr{S}et]$ such that the reflection functor is left exact. Show that the assignment $\tau \rightsquigarrow \tilde{\mathscr{C}}_\tau$ gives a one-to-one correspondence between $\tau(\mathscr{C})$ and Gir($\mathscr{C}$).

9.11. Show that the following are equivalent for a category $\mathscr{D}$:

1) There is a small category $\mathscr{C}$ and a Grothendieck topology τ on $\mathscr{C}$ such that $\mathscr{D}$ is equivalent to $\tilde{\mathscr{C}}_\tau$.

2) There is a small category $\mathscr{C}$, and a full and faithful functor $I: \mathscr{D} \to [\mathscr{C}^0, \mathscr{S}et]$ such that I has an adjoint which is a left exact functor.

3) $\mathscr{D}$ has the following properties:

i) $\mathscr{D}$ is finitely complete and has an initial object $\emptyset$.

ii) $\mathscr{D}$ has coproducts. The coproducts in $\mathscr{D}$ are universal.

In addition, if $\{D_i\}_i$ is a set of objects of $\mathscr{D}$, then the following squares are cartesian, for all $i, j, k; j \neq k$:

$$\begin{array}{ccc} D_i & = & D_i \\ \| & & \downarrow u_i \\ D_i & \xrightarrow{u_i} & \coprod_i D_i \end{array} \qquad\qquad \begin{array}{ccc} \emptyset & \longrightarrow & D_k \\ \downarrow & & \downarrow u_k \\ D_j & \xrightarrow{u_j} & \coprod_i D_i \end{array}$$

(where the u_i are structural morphisms).

iii) If $(f, g): A \to B$ is a pair of morphisms of $\mathscr{D}$ such that for each $X \in \mathscr{D}$, $(H^X(f), H^X(g))$ is a kernel pair of $\mathscr{S}et$, then (f, g) is a kernel pair and it has a coequalizer that is universal.

iv) $\mathscr{D}$ has a set of generators.

A category $\mathscr{D}$ that satisfies the equivalent conditions 1)—3) is called a *topos*.

9.12. Let $\mathscr{D}$ be a topos. Show that:

a) $\mathscr{D}$ is a complete and cocomplete category.

b) Every monomorphism of $\mathscr{D}$ is an equalizer and every epimorphism is a coequalizer.

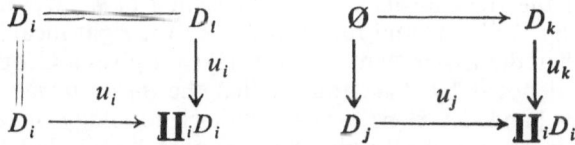

c) The monomorphisms of $\mathscr{D}$ are couniversal and the epimorphisms are universal.

d) $\mathscr{D}$ is a balanced category and every morphism of $\mathscr{D}$ has a unique triangular decomposition.

e) $\mathscr{D}$ contains a small full and dense subcategory.

f) The filtered colimits in $\mathscr{D}$ commute with finite limits. The colimits in $\mathscr{D}$ are universal.

g) Every morphism whose codomain is an initial object is an isomorphism.

9.13. Let Σ be a multiplicative system of morphisms of $\mathscr{C}$, and let τ be the Grothendieck topology associated with Σ. Show that the following are equivalent:

a) The canonical functor $A:[\mathscr{C}^0,\mathscr{S}et] \to \tilde{\mathscr{C}}_\tau$ makes the elements of Σ invertible.

b) The system Σ is right calculable.

c) The category $\mathscr{C}^r(\Sigma^{-1})$ (the right fractional category of $\mathscr{C}$ with respect to Σ) is defined.

Under these conditions, let $P:\mathscr{C} \to \mathscr{C}^r(\Sigma^{-1})$ be the canonical functor (see Section 1.13, Ch. 1) and let $L:\mathscr{C}^r(\Sigma^{-1}) \to \tilde{\mathscr{C}}_\tau$ be the unique functor such that $LP=AH$. Then L is an equivalence of categories.

9.14. Let $T:\mathscr{C} \to \mathscr{C}'$ be a functor between small categories. If $X \in \mathscr{C}$ and $\mathscr{A}$ is a right ideal over X, let $\langle T(A)\rangle$ denote the right ideal over $T(X)$ generated by all pairs $(T(Y), T(f))$ where (Y,f) runs over $\mathscr{A}$. $\langle T(\mathscr{A})\rangle$ is called the *direct image of* $\mathscr{A}$ *relative to* T. On the other hand, if $\mathscr{B}$ is a right ideal over an object X' of $\mathscr{C}'$ and $X \in \mathscr{C}$, let $T_X(\mathscr{B})$ denote the right ideal over X consisting of all pairs (Y,f) such that $(T(Y), T(f)) \in \mathscr{B}$. $T_X(\mathscr{B})$ is called the *inverse image of B relative to T*.

a) Let $X \in \mathscr{C}$, $X' \in \mathscr{C}'$, $\mathscr{A} \in \mathrm{Idr}(X)$ and $\mathscr{B} \in \mathrm{Idr}(X')$. Show that:

i) $\mathscr{A} \subseteq T_X(\langle T(\mathscr{A})\rangle)$.

ii) $T_X(\mathscr{B}) = T_X(\langle T_X(\mathscr{B})\rangle)$.

iii) $\langle T_X(\mathscr{B})\rangle \subseteq \mathscr{B}$.

iv) $\langle T(\mathscr{A})\rangle = \langle T_X(\langle T(\mathscr{A})\rangle)\rangle$.

b) Let τ be a Grothendieck topology on $\mathscr{C}$. For each $X' \in \mathscr{C}'$ let $T_*(\tau)(X')$ denote the set of the right ideals over X' consisting of $\mathscr{C}'/X'$, or all right ideals $\mathscr{B}$ such that for each $f': Y' \to X'$ and for each $X \in \mathscr{C}$, the right ideal $T_X(f'^*(B))$ belongs to $\tau(X)$. Show that the assignment $X' \rightsquigarrow T^*(\tau)(X')$ gives a Grothendieck topology on $\mathscr{C}'$, which is denoted by $T_*(\tau)$ and called the *direct image* of τ relative to T.

On the other hand, let τ' be a Grothendieck topology on $\mathscr{C}'$. For each $X \in \mathscr{C}$ let $T^*(\tau')(X)$ denote the set of all right ideals $\mathscr{A}$ over X such that $\langle T(\mathscr{A})\rangle \in \tau'(T(X))$. Show that the assignment $X \rightsquigarrow T^*(\tau')(X)$ gives a Grothendieck topology on $\mathscr{C}$, denoted $T^*(\tau')$ and called the *inverse image of* τ' relative to T.

c) Let $\tau \in \tau(\mathscr{C})$. Show that there is a functor $T_1:\tilde{\mathscr{C}}'_{T_*(\tau)} \to \tilde{\mathscr{C}}_\tau$ such that the following diagram of categories and functors is commutative

$$\begin{array}{ccc} [(\mathscr{C}')^0,\mathscr{S}et] & \xrightarrow{\quad T_* \quad} & [\mathscr{C}^0,\mathscr{S}et] \\ {\scriptstyle A'}\downarrow & & \downarrow{\scriptstyle A} \\ \tilde{\mathscr{C}}'_{T_*(\tau)} & \xrightarrow{\quad T_1 \quad} & \tilde{\mathscr{C}}_\tau \end{array} .$$

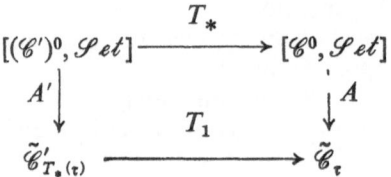

On the other hand if $\tau' \in \tau(\mathscr{C}')$, then there is a functor $T^1\colon \tilde{\mathscr{C}}_{T^\bullet(\tau')} \to \tilde{\mathscr{C}}'_{\tau'}$ such that the diagram

$$\begin{array}{ccc} [\mathscr{C}^0, \mathscr{S}et] & \xrightarrow{\quad T^* \quad} & [(\mathscr{C}')^0, \mathscr{S}et] \\ {\scriptstyle A}\downarrow & & \downarrow{\scriptstyle A'} \\ \tilde{\mathscr{C}}_{T^\bullet(\tau')} & \xrightarrow{\quad T^1 \quad} & \tilde{\mathscr{C}}'_{\tau'} \end{array}$$

is commutative.

9.15. Let $T\colon \mathscr{C} \to \mathscr{C}'$ be a functor between small categories. Show that the following are equivalent:

i) The canonical functor $T^*\colon [\mathscr{C}^0, \mathscr{S}et] \to [(\mathscr{C}')^0, \mathscr{S}et]$ is left exact and the restriction functor T_* is full and faithful.

ii) If for each $X \in \mathscr{C}$, $\tau(X)$ is the set of all right ideals $\mathscr{A}$ over X, such that $\langle T(\mathscr{A})\rangle = \mathscr{C}'/T(X)$, then the assignment $X \rightsquigarrow \tau(X)$ gives a Grothendieck topology τ on $\mathscr{C}$, and there that is an equivalence of categories $S\colon \tilde{\mathscr{C}}_\tau \to [(\mathscr{C}')^0, \mathscr{S}et]$ such that $SA = T^*$.

9.16. (Roos [1]). Let τ be a Grothendieck topology on $\mathscr{C}$. Show that the following are equivalent:

a) The canonical functor $A\colon [\mathscr{C}^0, \mathscr{S}et] \to \tilde{\mathscr{C}}_\tau$ commutes with limits.

b) The canonical functor $A\colon [\mathscr{C}^0, \mathscr{S}et] \to \tilde{\mathscr{C}}_\tau$ has an adjoint.

c) For each $X \in \mathscr{C}$ the set $\tau(X)$ (ordered relative to inclusion) has a minimal element.

d) If $\{f_i\}_i$ is a set of epimorphisms in $\tilde{\mathscr{C}}_\tau$, then the morphism $\prod_i f_i$ is also an epimorphism and furthermore products in $\tilde{\mathscr{C}}_\tau$ commute with coproducts.

e) The products in $\tilde{\mathscr{C}}_\tau$ commute with colimits.

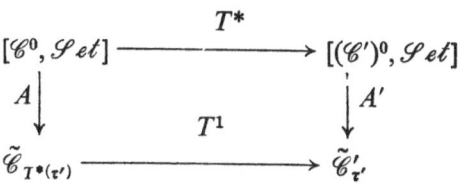

Chapter 3

Algebraic categories

3.1. Algebraic theories

In their work [2], Eilenberg and MacLane suggested the possibility of "functorizing" the study of general algebraic systems. For this purpose, Lawvere [1] has defined the notion of an algebraic theory. We remark that these ideas appear in a non-explicit form in Birkhoff [1], Isbell [5], Mal'cev [2], etc.

In this section, we define algebraic theory and make some observations about this notion.

References: Birkoff [1], Cohn [1], Eilenberg—MacLane [2], Freyd [2], Grätzer [1], Isbell [3], [5], Lawvere [1], Mal'cev [2], Pareigis [1], Schubert [1], Wraith [1].

An *algebraic theory* is a small category $\mathscr{A}$ whose objects are $\mathbf{0}, \mathbf{1}, \mathbf{2}, \ldots, \mathbf{n}, \ldots$, $n \in \mathbf{N}$, in which each object $\mathbf{n}$, $n \in \mathbf{N}$, is the product of the object $\mathbf{1}$ with itself n times. In particular, $\mathbf{0}$ is a final object of $\mathscr{A}$. For each $n \in \mathbf{N}$, denote the only element of $\mathscr{A}(\mathbf{n}, \mathbf{0})$ by $p_0^n \colon \mathbf{n} \to \mathbf{0}$. For each $n \geqslant 1$ let $p_1^n, \ldots, p_n^n \colon \mathbf{n} \to \mathbf{1}$ denote the structural projections. The elements of $\mathscr{A}(\mathbf{n}, \mathbf{1})$ are called the *n-ary operations.* It is clear that the projections p_i^n, $i = 1, \ldots, n$, are n-ary operations, usually called *trivial operations*. The nontrivial n-ary operations are often called *consistent operations.*

Recall that if $\{f_i \colon X \to X_i\}_{i \in \mathscr{I}}$ is a set of morphisms in a category $\mathscr{C}$, $[f_i]_{i \in \mathscr{I}} \colon X \to \prod_i X_i$ denotes the only morphism such that $p_j[f_i]_{i \in \mathscr{I}} = f_j$ for all $j \in \mathscr{I}$, where the $p_j \colon \prod_i X_i \to X_j$ are structural projections.

LEMMA 1.1. *Let $\mathscr{A}$ be an algebraic theory and $n \geqslant 1$ a natural number. Then for every $m \in \mathbf{N}$ and for every $f \in \mathscr{A}(\mathbf{m}, \mathbf{n})$ we have:*

(a) $$f = [p_1^n f, \ldots, p_n^n f]$$

and

(b) $$p_0^n f = p_0^m. \;*$$

The previous lemma says that the set $\mathscr{A}(\mathbf{m}, \mathbf{n})$ is canonically identified with the set $(\mathscr{A}(\mathbf{m}, \mathbf{1})^n)$. Therefore, the morphisms of the algebraic theory $\mathscr{A}$ are completely determined by the sets $\{\mathscr{A}(\mathbf{m}, \mathbf{1})\}_{m \in \mathbf{N}}$, i.e. the morphisms of $\mathscr{A}$ are determined by the m-ary operations, where m runs through $\mathbf{N}$.

By a *theory morphism* from the algebraic theory $\mathscr{A}$ into the algebraic theory $\mathscr{B}$, we mean a functor $F: \mathscr{A} \to \mathscr{B}$ such that

a) $F(\mathbf{n}) = \mathbf{n}$ for all $n \in \mathbf{N}$,

and

b) $F(p_i^n) = p_i^n$, i.e. F preserves the structural projections.

It is clear that the composition of two theory morphisms is again a theory morphism.

Let $\mathscr{T}h$ denote the category whose objects are algebraic theories and whose morphisms are theory morphisms.

Example 1.2. Let $\mathscr{N}$ denote the algebraic theory defined as follows: for each $\mathbf{m}$ the set $\mathscr{N}(\mathbf{m}, \mathbf{1})$ contains only the structural projections $p_1^m, \ldots, p_m^m$, i.e. all m-ary operations are trivial. $\mathscr{N}$ is called the *trivial algebraic theory*. It is easy to verify that $\mathscr{N}$ is an initial object of $\mathscr{T}h$. For each algebraic theory $\mathscr{A}$, let $I^{\mathscr{A}}: \mathscr{N} \to \mathscr{A}$ denote the only theory morphism.

1.3. Let $\mathscr{P}$ denote the *final algebraic theory*, which is defined by $p_0^n: \mathbf{n} \to \mathbf{0}$ is an isomorphism for all $n \in \mathbf{N}$. Since $p_0^n = p_0^i p_i^n$ for all $1 \leqslant i \leqslant n$, we can see that all projections p_i^n coincide and that for every $n \geqslant 0$ there is only one n-ary operation. $\mathscr{P}$ is a final object of $\mathscr{T}h$.

Let $\mathbf{N}$ be the discrete category with a countable set of objects denoted by $0, 1, 2, \ldots, n, \ldots$. Let $K: \mathscr{T}h \to [\mathbf{N}, \mathscr{S}et]$ be the functor defined by:

$$K(\mathscr{A}) = \{\mathscr{A}(\mathbf{n}, \mathbf{1})\}_n, \quad \mathscr{A} \in \mathscr{T}h,$$

$$K(F) = \{F(\mathbf{n}, \mathbf{1}): \mathscr{A}(\mathbf{n}, \mathbf{1}) \to \mathscr{B}(\mathbf{n}, \mathbf{1})\}_n,$$

where $F: \mathscr{A} \to \mathscr{B}$ is a theory morphism.

THEOREM 1.4. *The functor* $K: \mathscr{T}h \to [\mathbf{N}, \mathscr{S}et]$ *defined above has an adjoint which we denote by* L.

Proof. Let $X = \{X_n\}_n$ be an object of $[\mathbf{N}, \mathscr{S}et]$ and $\mathscr{N}$ the trivial algebraic theory. For every pair (r, s) of natural numbers, we define a set $M(r, s)$. First let:

$M_1(r, 0) = \{p_0^r\}$,

$M_1(r, 1) = X_r \coprod \mathscr{N}(\mathbf{r}, \mathbf{1})$, (the coproduct of sets), and

$M_1(r, s) = M_1(r, 1)^s$ for $s > 1$.

Hence an element of $M_1(r, s)$ will be an s-tuple $(f_1, \ldots, f_s)$ where $f_i \in M_1(r, 1)$. Furthermore, we define

$$M_1'(r, s) = [\textstyle\coprod_{t \in \mathscr{N}} (M_1(t, s) \coprod M_1(r, t)] \coprod M_1(r, s).$$

In contrast to the s-tuples in $M_1(r, s)$ we shall write an element of $M_1'(r, s)$ as a pair $[f, h]$ with brackets.

If the sets $M_{i-1}(r, s)$ and $M'_{i-1}(r, s)$ are defined, then we continue as follows:

$$M_i(r, 0) = M'_{-1}(r, 0),$$
$$M_i(r, 1) = M'_{i-1}(r, 1),$$
$$M_i(r, s) = M_i(r, 1)^s \coprod M'_{i-1}(r, s) \quad \text{for } s > i, \text{ and}$$
$$M'_i(r, s) = [\times_{t \in \mathbf{N}}(M_i(t, s) \coprod M_i(r, t))] \coprod M_i(r, s)$$
$$= \{[f, g], f \in M_i(t, s), g \in (M_i(r, t), t \in \mathbf{N}\} \coprod M_i(r, s).$$

Now for each i we identify $M_i(r, s)$ with its image in $M'_i(r, s)$ and $M'_i(r, s)$ with its image in $M_{i+1}(r, s)$. Thus we have the following inclusions:

$$\{p_0^r\} = M_i(r, 0) \subseteq M'_1(r, 0) \subseteq M_2(r, 0) \subseteq M'_2(r, 0) \subseteq \ldots,$$
$$X_r \coprod \mathcal{N}(\mathbf{r}, \mathbf{1}) = M_i(r, 1) \subseteq M'_1(r, s) \subseteq M_2(r, s) \subseteq M'_2(r, s) \subseteq \ldots,$$
$$M_1(r, s) \subseteq M'_1(r, s) \subseteq M_2(r, s) \subseteq M'_2(r, s) \subseteq \ldots.$$

Finally, let us define $M(r, 1) = \cup_{i \in \mathbf{N}} M_i(r, s)$; also for each i we identify $M_i(r, s)$ with its image in $M(r, s)$.

If confusion is possible, we write $M_i^X(r, s)$ instead of $M_i(r, s)$ and $M^X(r, s)$ instead of $M(r, s)$. The following hold:

a) $\{p_0^r\} \subseteq M(r, 0)$ for all $r \in \mathbf{N}$.

b) $X_r \coprod \mathbf{N}(\mathbf{r}, \mathbf{1}) \subseteq M(r, 1)$.

c) If $f_i \in M(r, 1)$, $i = 1, \ldots, s$ with $s > 1$, then the s-tuple $(f_1, \ldots, f_s) \in M(r, s)$ for all $r \geqslant 0$.

d) If $f \in M(t, s)$ and $g \in M(r, t)$, then $[f, g] \in M(r, s)$ for all $r, s, t \in \mathbf{N}$.

Denote by $R(r, s)$ the equivalence relation on the set $M(r, s)$ that is generated by the following pairs of elements:

1) If $f, g \in M(r, 0)$, then $(f, g) \in R(r, 0)$.

2) If $f_j \in M(r, 1)$, for $j = 1, \ldots, s$, then $([p_i^s, (f_i, \ldots, f_s)], f_i) \in R(r, 1)$, $i = 1, \ldots, s$.

3) If $f \in M(r, s)$, then $(([p_1^s, f], \ldots, [p_s^s, f]), f) \in R(r, s)$.

4) If $f \in M(r, s)$, then

$$([(p_1^s, \ldots, p_s^s), f]) \text{ and } ([f, (p_1^r, \ldots, p_r^r)], f) \in R(r, s).$$

5) If $f \in M(r, s)$, $h \in M(s, t)$, $k \in M(t, u)$, then

$$([[k, h, f], [k, [h, f]]) \in R(r, u).$$

6) If $f_i, g_i \in M(r, 1)$, and $(f_i, g_i) \in R(r, 1)$ for $i = 1, \ldots, s$, then $((f_1, \ldots, f_s), (g_1, \ldots, g_s)) \in R(r, s)$.

7) If $f, f' \in M(r, t)$, $h, h' \in M(t, s)$ and $(f, f') \in R(r, t)$, $(h, h') \in R(t, s)$, then $([h, f], [h', f']) \in R(r, s)$.

Now define the algebraic theory $L(X)$ as follows: for every $r, s \in \mathbf{N}$, we have $L(X)(\mathbf{r}, \mathbf{s}) = M(r, s)/R(r, s)$. If $f \in M(r, s)$ denote by $\bar{f}$ the class of f in $L(X)(r, s)$.

Now, if $f: \mathbf{r} \to \mathbf{t}$ and $g: \mathbf{t} \to \mathbf{s}$ are morphisms of $L(X)$, then we define $\bar{g}\bar{f}$ to be $\overline{[g, f]}$. By 7), this class is independent of the choice of the representative for $\bar{f}$ and $\bar{g}$. By 5) the composition is associative, and by 4) the equivalence class of $(p_1^r, \ldots p_r^r)$ gives an identity element of $L(X)(\mathbf{r}, \mathbf{r})$. Thus $L(X)$ is a category. Furthermore, we claim that $L(X)$ is an algebraic theory. Indeed, it follows from 1) that $\mathbf{0}$ is a final object of $L(X)$. Now we assert that for each $r > 1$, $\mathbf{r}$ is equal to $(\mathbf{1})^r$, and the p_i^r are structural projections. To see this, let $s \in \mathbf{N}$ and $f_i \in L(X)(\mathbf{s}, \mathbf{1})$, $i = 1, \ldots, r$; then, by 2), $\bar{p}_i^r(\overline{f_i, \ldots, f_r}) = \bar{f}_i$, $i = 1, \ldots, n$ and, by 3), we see that $\overline{(f_1, \ldots, f_r)}$ is the unique morphism h such that $\bar{p}_i^r h = \bar{f}_i$ for all $i = 1, \ldots, r$. That $\overline{(f_1, \ldots, f_r)}$ is independent of the choice of representatives follows from 6). Hence $L(X)$ is an algebraic theory, as claimed. The element $\bar{p}_i^r$ will also be denoted by p_i^r since $\bar{p}_i^r \neq \bar{p}_{i'}^{r'}$, whenever $r \neq r'$ or $i \neq i'$.

Now let $f: X \to Y$ be a morphism of $[\mathbf{N}, \mathcal{S}et]$. Then $f = (f_n)_{n \in \mathbf{N}}$ where $f_n: X_n \to Y_n$ is a map for each n.

For every $r, s \in \mathbf{N}$, and $i \geqslant 1$ define a map $f_i^{(r,s)}: M_i^X(r, s) \to M_i^Y(r, s)$ in the following manner:

$$f_1^{(r,0)}(p_0^r) = p_0^r,$$

$$f_1^{(r,1)} = f_r \textstyle\coprod 1_{\mathbf{N}(\mathbf{r},\mathbf{1})}, \quad \text{and}$$

$$f_1^{(r,s)} = (f_1^{(r,1)})^s.$$

If $i \geqslant 1$ and the $f_i^{(r,s)}$ are already defined, then we let:

$$f_i'^{(r,s)} = \{[\textstyle\coprod_{t \in \mathbf{N}}(f_i^{(t,s)} \coprod f_i^{(r,t)})] \coprod f_i^{(r,s)}\}: M_i'^X(r, s) \to M_i'^Y(r, s)$$

and furthermore:

$$f_{i+1}^{(r,0)} = f_i'^{(r,0)},$$

$$f_{i+1}^{(r,1)} = f_i'^{(r,1)},$$

$$f_{i+1}^{(r,s)} = (f_{i+1}^{(r,1)})^s f_i'^{(r,s)}.$$

In this way the maps $\{f_i^{(r,s)}\}_{i \geqslant 1}$ define a map $f(r, s): M^X(r, s) \to M^Y(r, s)$. Also it is straightforward to verify that if $f, g \in M^X(r, s)$ and $(f, g) \in R^X(r, s)$ then $(f^{(r,s)}(f), f^{(r,s)}(g)) \in R^Y(r, s)$ (the "exponents" X and Y have an obvious meaning. Now we denote by $L(f): L(X) \to L(Y)$ the functor defined by:

$$L(f)(\mathbf{n}) = \mathbf{n} \quad \text{for all } n \in \mathbf{N}, \text{ and}$$

$$L(f)(\mathbf{r}, \mathbf{s}): L(X)(\mathbf{r}, \mathbf{s}) \to L(Y)(\mathbf{r}, \mathbf{s})$$

is the map induced by $f^{(r,s)}$. It is clear that:

$$L(f)(p_i^r) = p_i^r \quad \text{for all } r \in \mathbf{N} \text{ and all } i = 1, \ldots, r,$$

$$L(f)([\bar{f}_1, \ldots, \bar{f}_s]) = [L(f)(\bar{f}_1), \ldots, L(f)(\bar{f}_s)]$$

so that $L(f)$ is a theory morphism. We can easily verify that $L(fg) = L(f)L(g)$ and $L(1_X) = 1_{L(X)}$. Thus $L: [\mathbf{N}, \mathcal{S}et] \to \mathcal{T}h$ is a functor. It remains to show that L

is an adjoint of K. To see this, let $X = (X_n)_n$ be an object of $[\mathbf{N}, \mathcal{S}et]$. For each $n \in \mathbf{N}$, denote by $(u_X)_n\colon X_n \to KL(X)_n = L(X)(\mathbf{n}, \mathbf{1})$ the composition of canonical maps

$$X_n \to M(n, 1) \to L(X)(\mathbf{n}, \mathbf{1})$$

and let $u_X\colon X \to KL(X)$ denote the morphism $\{u_X\}_{n\in\mathbf{N}}$ of $[\mathbf{N}, \mathcal{S}et]$. It is easy to see that the morphisms $\{u_X\}_{X\in[\mathbf{N}, \mathcal{S}et]}$ define a functorial morphism $u\colon \mathrm{Id}[\mathbf{N}, \mathcal{S}et] \to KL$. Also, it is easy to see that u_X is an argumentwise monomorphism, i.e. u is a functorial monomorphism.

Furthermore, let $\mathcal{A}$ be an algebraic theory; we define a functorial morphism $q_{\mathcal{A}}\colon LK(\mathcal{A}) \to \mathcal{A}$. Firstly, for every $r, s, i \in \mathbf{N}$, $i \geqslant 1$, define a map $q_i^{(r,s)}\colon M_i(r, s) \to \mathcal{A}(\mathbf{r}, \mathbf{s})$ as follows:

$$q_1^{(r,0)}\colon M_1(r, 0) = \{p_0^r\} \to \{p_0^r\} = \mathcal{A}(\mathbf{r}, \mathbf{0}),$$

$$q_1^{(r,1)}\colon M_1(r, 1) = \mathcal{A}(\mathbf{r}, \mathbf{1}) \coprod \mathcal{N}(\mathbf{r}, \mathbf{1}) \to \mathcal{A}(\mathbf{r}, \mathbf{1})$$

is induced by the identity map of $\mathcal{A}(\mathbf{r}, \mathbf{1})$ and $I^{\mathcal{A}}(\mathbf{r}, \mathbf{1})$, and

$$q_1^{(r,s)}\colon M_1(r, 1)) = (M_1(r, 1))^s \to \mathcal{A}(\mathbf{r}, \mathbf{s}), \quad s > 1$$

is defined by $q_1^{(r,s)}(f_1, \ldots, f_s) = [q_1^{(r,1)}(f_1), \ldots, q_1^{(r,1)}(f_s)]$.

Consider

$$q_1'^{(r,s)}\colon M_1'(r, s) = [\textstyle\coprod_{t\in\mathbf{N}}(M_1(t, s) \prod M_1(r, t))] \coprod M_1(r, s) \to \mathcal{A}(\mathbf{r}, \mathbf{s})$$

to be the map induced by $q_1^{(r,s)}$, and the map

$$g\colon [\textstyle\coprod_{t\in\mathbf{N}}(M_1(r, s) \prod M_1(r, t))] \to \mathcal{A}(\mathbf{r}, \mathbf{s})$$

which assigns to the pair $(f, h) \in M_1(t, s) \prod M_1(r, t)$ the composition

$$\mathbf{r} \xrightarrow{q_1^{(r,t)}(h)} \mathbf{t} \xrightarrow{q_1^{(t,s)}(f)} \mathbf{s} \text{ in } \mathcal{A}.$$

If the map $q_{i-1}^{(r,s)}$ and $q_{i-1}'^{(r,s)}$ have already been defined, then we define:

$$q_i^{(r,0)} = q_{i-1}'^{(r,0)},$$

$$q_i^{(r,} \;= q'^{(r,1)}_{-1}, \text{ and}$$

$$q_i^{(r,s)} \text{ is induced by } [q_i^{(r,1)}(?), \ldots, q_i^{(r,1)}(?)] \text{ and } q_{i-1}'^{(r,s)}.$$

The map $q_i'^{(r,s)}$ is defined in the obvious fashion. It is clear that the maps $\{q_i^{(r,s)}\}_{i\geqslant 1}$ define a map $q^{(r,s)}\colon M(r, s) \to \mathcal{A}(\mathbf{r}, \mathbf{s})$. Now, since $\mathcal{A}$ is an algebraic theory, for every $f, h \in M(r, s)$ we have $q^{(r,s)}(f) = q^{(r,s)}(h)$, whereas $(f, h) \in R(r, s)$. Also, it is clear that the assignments $\bar{f} \rightsquigarrow q^{(r,s)}(f)$, where $\bar{f} \in M(r, s)/R(r, s) = L(K(\mathcal{A}))(\mathbf{r}, \mathbf{s})$, define a functor $v_{\mathcal{A}}\colon LK(\mathcal{A}) \to \mathcal{A}$. We leave it for the reader to check that the morphisms $\{v_{\mathcal{A}}\}_{\mathcal{A}\in\mathcal{T}h}$ define a functorial coarrow $v\colon LK \to \mathrm{Id}\,\mathcal{T}h$ that is a quasi-inverse of the arrow $u\colon \mathrm{Id}\,[\mathbf{N}, \mathcal{S}et] \to KL$. The proof of Theorem 1.4 is due to Pareigis [1]. Another proof of this theorem, based on the same idea, is given by Schubert [1]. ✳

If X is an object of $[\mathbf{N}, \mathcal{S}et]$, then the algebraic theory $L(X)$ is called the *free algebraic theory* generated by X.

Let $\mathscr{A}$ be an algebraic theory and R an equivalence relation on categories in $\mathscr{A}$ (see Section 1.5). We say that R is an *equivalence relation on algebraic theories* if given m pairs $(f_i, g_i)_{1 \leqslant i \leqslant m}$ of elements of $\mathscr{A}(\mathbf{n}, \mathbf{1})$ such that $f_i \sim g_i(R(\mathbf{n}, \mathbf{1}))$ for all $1 \leqslant i \leqslant m$, then $[f_1, \ldots, f_m] \sim [g_1, \ldots, g_m](R(\mathbf{n}, \mathbf{m}))$.

We thus see that the equivalence relation R is determined by the equivalence relations $\{R(\mathbf{n}, \mathbf{1}))\}_{n \in \mathbf{N}}$. The proof of the folowing result is easy and is left for the reader.

LEMMA 1.5. *Let $\mathscr{A}$ be an algebraic theory and let R be an equivalence relation on the algebraic theories in $\mathscr{A}$. Then the factor category $\mathscr{A}/R$ is again an algebraic theory and the canonical functor $P: \mathscr{A} \to \mathscr{A}/R$ is a theory morphism. Moreover, $\{P(p_i^n)\}_{1 \leqslant i \leqslant n}$ are structural projections of the product $\mathbf{1}^n = \mathbf{n}$.* ✳

$\mathscr{A}/R$ is called the *factor algebraic theory* of $\mathscr{A}$ by R.

COROLLARY 1.6. *Let $F: \mathscr{A} \to \mathscr{B}$ be a theory morphism and assume that, for each $n \in \mathbf{N}$, the map $F(\mathbf{n}, \mathbf{1})$: in $\mathscr{A}(\mathbf{n}, \mathbf{1}) \to \mathscr{B}(\mathbf{n}, \mathbf{1})$ is surjective. Then the map $F(\mathbf{n}, \mathbf{m}): \mathscr{A}(\mathbf{n}, \mathbf{m}) \to \mathscr{B}(\mathbf{n}, \mathbf{m})$ is also surjective for all $n, m \in \mathbf{N}$, i.e. the functor F is full. Moreover, there is an equivalence relation R on the algebraic theories in $\mathscr{A}$ such that $\mathscr{B}$ is canonically isomorphic to $\mathscr{A}/R$, i.e. F is a coequalizer.*

Proof. The relation R is defined as follows. If $f, g \in \mathscr{A}(\mathbf{n}, \mathbf{1})$, then $f \sim g(R(\mathbf{n}, \mathbf{1})) \Leftrightarrow F(f) = F(g)$. Actually R is an equivalence relation on algebraic theories since F is a morphism of algebraic theories. ✳

Let $\mathscr{A}$ be an algebraic theory; it is easy to see that for each $n \in \mathbf{N}$ the map $v_{\mathscr{A}}(\mathbf{n}, \mathbf{1}): LK(\mathscr{A})(\mathbf{n}, \mathbf{1}) \to \mathscr{A}(\mathbf{n}, \mathbf{1})$ is surjective, so that the functor $v_{\mathscr{A}}$ is full. By Corollary 1.6 there exists an equivalence relation R on algebraic theories in $LK(\mathscr{A})$ such that $LK(\mathscr{A})/R = \mathscr{A}$. Hence we have the following result.

COROLLARY 1.7. *Every algebraic theory is isomorphic to a factor algebraic theory of a free algebraic theory.* ✳

Note 1.8. Let X be an object of $[\mathbf{N}, \mathcal{S}et]$, and for each $n \in \mathbf{N}$, choose a subset $E(n)$ of the product $L(X)(\mathbf{n}, \mathbf{1}) \prod L(X)(\mathbf{n}, \mathbf{1})$. By Exercise 1.5, there is a smallest equivalence relation R on algebraic theories in $L(X)$, such that $f \sim g(R(\mathbf{n}, \mathbf{1}))$ for each pair $(f, g) \in E(n)$.

Now let $\mathscr{A}$ be an algebraic theory. For each $n \in \mathbf{N}$, let $C(n)$ (or $C_{\mathscr{A}}(n)$, if there is danger of confusion) denote the set of all consistent n-ary operations on $\mathscr{A}$. Also, let $C = \{C(n)\}_{n \in \mathbf{N}}$ and denote by P the composition

$$P: L(C) \xrightarrow{\;U\;} LK(\mathscr{A}) \xrightarrow{\;v_{\mathscr{A}}\;} \mathscr{A}$$

where U is defined by the canonical inclusion $C \to K(\mathscr{A})$ and $v_{\mathscr{A}}$ is the coarrow of adjunction defined in Theorem 1.4. Since U is a morphism of algebraic theories (i.e. U preserves projections), P is a full functor. Let R denote the equivalence relation on $L(C)$ defined by P, i.e. for $f, g \in L(C)(\mathbf{n}, \mathbf{m})$ we have $f \sim g(R(\mathbf{n}, \mathbf{m})) \Leftrightarrow$ $\Leftrightarrow P(f) = P(g)$ and let $E(n) = \{(f, g), f, g \in L(C)(\mathbf{n}, \mathbf{1})$, such that $P(f) = P(g)\}$

(i.e. $E(n)$ is the subset of $L(C)(\mathbf{n}, \mathbf{1}) \coprod L(C)(\mathbf{n}, \mathbf{1})$ associated with the equivalence relation $R(\mathbf{n}, \mathbf{1})$). It is clear that $L(C)/R = \mathscr{A}$, and that R is the equivalence relation on the algebraic theories in $L(C)$ generated by the sets $\{E(n)\}_{n\in\mathbf{N}}$.

Consequently, every algebraic theory $\mathscr{A}$ is defined by:

a) A family $C = \{C(n)\}_{n\in\mathbf{N}}$ of sets. The elements of $C(n)$ (or $C_{\mathscr{A}}(n)$ if there is danger of confusion) are called *generators* of the n-th order for $\mathscr{A}$.

b) A family $\{E(n)\}_{n\in\mathbf{N}}$ of sets, where $E(n)$ is a subset of $L(C)(\mathbf{n}, \mathbf{1}) \coprod L(C)(\mathbf{n}, \mathbf{1})$ for every n. The elements of $E(n)$ (or $E_{\mathscr{A}}(n)$ if there is danger of confusion) are called the *identities* (or *relations*) of the n-th order for $\mathscr{A}$.

1.9. Obviously, the family $C = \{C(n)\}_n$ is not uniquely defined by $\mathscr{A}$. However, we can often choose C in a canonical way. The same observations are valid for the sets $\{E(n)\}_{n\in\mathbf{N}}$.

Example 1.10. *(Algebraic theory of premonoids)*. Denote by $\mathscr{A}$ the algebraic theory defined by:

$$C(n) = \emptyset \quad \text{if } n \neq 2, \quad \text{and}$$
$$C(2) = \{m\}.$$

The identity relations are:

$$E(n) = \emptyset \quad \text{if } n \neq 3, \quad \text{and}$$
$$E(3) = \{m((m[p^3_1, p^3_2]) \coprod p^3_3), \quad m(p^3_1 \coprod (m[p^3_2, p^3_3]))\},$$

i.e. in $\mathscr{A}$ we have the following commutative diagram:

$$\begin{array}{ccc} \mathbf{3} & \xrightarrow{(m[p^3_1,\, p^3_2]) \coprod p^3_3} & \mathbf{2} \\ {\scriptstyle p^3_1 \coprod (m[p^3_2,\, p^3_3])}\big\downarrow & & \big\downarrow{\scriptstyle m} \\ \mathbf{2} & \xrightarrow{\quad m \quad} & \mathbf{1} \end{array} \qquad (1)$$

The diagram (1), or equivalently the set $E(3)$, expresses the *associativity* of the operation m.

Example 1.11. *(Algebraic theory of monoids)*. Consider the algebraic theory $\mathscr{A}$ whose primitive operations are:

$$C(n) = \emptyset \quad \text{if } n \neq 0,\ n \neq 2,$$
$$C(2) = \{m\}, \quad \text{and}$$
$$C(0) = \{e\}.$$

The identities of $\mathscr{A}$ are:

$$E(3) = \text{the same as in the previous example,}$$
$$E(1) = \{(m[ep^1_0, p^1_1], p^1_1);\ (m[p^1_1, ep^1_0], p^1_1)\}, \quad \text{and}$$
$$E(n) = \emptyset \quad \text{for } n \neq 1,\ n \neq 3.$$

The identities $E(1)$ express the fact that e is the *identity operation* or the *identity of m*.

Example 1.12. *(Algebraic theory of groups)*. The primitive operations are:

$$C(n) = \emptyset \quad \text{if } n \geqslant 3,$$
$$C(2) = \{m\},$$
$$C(1) = \{s\}, \quad \text{and}$$
$$C(0) = \{e\}.$$

The identities are:

$$F(3) = \text{the same as in Example 1.10},$$
$$E(1) = \{(m[ep_0^1, p_1^1], p_1^1); (m[s, p_1^1], ep_0^1)\}, \quad \text{and}$$
$$E(n) = \emptyset \quad \text{for } n \neq 1,\ n \neq 3.$$

Here s is called the *inverse*, or *symmetric, operation* for m.

The following result will be useful later.

PROPOSITION 1.13. *$\mathcal{T}h$ is a cocomplete category.*

Proof. By Theorem 9.2, Ch. 1, it will suffice to check that $\mathcal{T}h$ is a category with coproducts and cokernels.

Indeed, let $\{\mathscr{A}_i\}_{i \in \mathscr{I}}$ be a set of algebraic theories. Consider the coproduct $X = \coprod_i K(A_i)$ in the category $[\mathbf{N}, \mathscr{S}et]$, and let $u_i\colon K(A_i) \to X$ be the structural injections. In the algebraic theory $L(X)$, consider the equivalence relation R on algebraic theories of $L(X)$ where R is generated by the pairs:

$$(L(u_i)(a),\ L(u_i)(b))$$

when i runs over $\mathscr{I}$ and $a, b \in LK(A_i)(\mathbf{n}, \mathbf{m})$ such that $v_{\mathscr{A}_i}(a) = v_{\mathscr{A}_i}(b)$. Let $\mathscr{B}$ be the algebraic theory $L(X)/R$ and $P\colon L(X) \to \mathscr{B}$ the canonical theory morphism. By the definition of R, there is a unique theory morphism $P_i\colon \mathscr{A}_i \to \mathscr{B}$ such that $P_i v_{\mathscr{A}_i} = PL(u_i)$. We leave it for the reader to show that $\mathscr{B}$ and the morphisms P_i, $i \in \mathscr{I}$ give a coproduct of this set of algebraic theories $\{\mathscr{A}_i\}_{i \in \mathscr{I}}$. Hence $\mathcal{T}h$ is a category with coproducts.

Finally, if $(U, V)\colon \mathscr{A} \to \mathscr{B}$ is a pair of morphisms of $\mathcal{T}h$ denote by R the equivalence relation on the algebraic theories in $\mathscr{B}$ generated by the pairs $(U(\mathbf{n}, \mathbf{m})(f), V(\mathbf{n}, \mathbf{m})(f))$ when f runs over $\mathscr{A}(\mathbf{n}, \mathbf{m})$ and $n, m \in \mathbf{N}$. If $P\colon \mathscr{B} \to \mathscr{B}/R$ is the canonical morphism, then $(P, \mathscr{B}/R)$ is a cokernel of (U, V). The details are left for the reader. ✳

Note 1.14. It is easy to see that a morphism $F\colon \mathscr{A} \to \mathscr{B}$ of algebraic theories is a monomorphism of $\mathcal{T}h$ if and only if F is a faithful functor. A subcategory $\mathscr{A}'$ of an algebraic theory $\mathscr{A}$ is called an *algebraic subtheory* if, for every n and every $1 \leqslant i \leqslant n$, $p_i^n \in \mathscr{A}'(\mathbf{n}, \mathbf{1})$ and, furthermore, if $g_1, \ldots, g_m \in \mathscr{A}'(\mathbf{n}, \mathbf{1})$, then $[g_1, \ldots, g_m]\colon \mathbf{n} \to \mathbf{m}$ belongs to $\mathscr{A}'(\mathbf{n}, \mathbf{m})$. It is easy to see that $\mathscr{A}'$ is itself an algebraic theory, and every subobject of $\mathscr{A}$ is completely determined by a subtheory of $\mathscr{A}$.

Note 1.15. Let R be an equivalence relation on the algebraic theories of an algebraic theory $\mathscr{A}$. Denote by $\mathscr{R}$ the subcategory of $\mathscr{A} \prod \mathscr{A}$ (see Exercise 1.7) whose morphisms are $\mathscr{R}(\mathbf{n}, \mathbf{m}) = \{(f, g) \in (\mathscr{A} \prod \mathscr{A})(\mathbf{n}, \mathbf{m})$ such that $f \sim g(R)\}$. Then $\mathscr{R}$ is a subcategory of $\mathscr{A} \prod \mathscr{A}$ and the assignment $R \rightsquigarrow \mathscr{R}$ gives a one-to-one correspondence between the equivalence relations on the algebraic theories of $\mathscr{A}$ and all subtheories of $\mathscr{A} \prod \mathscr{A}$.

Exercises

1.1. Let $F: \mathscr{A} \to \mathscr{B}$ a morphism of $\mathscr{T}h$. Prove:

a) F commutes with finite products.

b) If $f_i \in \mathscr{A}(\mathbf{n}, \mathbf{1})$, $i = 1, \ldots, n$, then $F([f_1, \ldots, f_n]) = [F(f_1), \ldots, F(f_n)]$.

1.2. Prove that the functor $K: \mathscr{T}h \to [\mathbf{N}, \mathscr{S}et]$ is faithful, and, moreover, that for each $\mathscr{A} \in \mathscr{T}h$ and each $n \in \mathbf{N}$, the map $v_{\mathscr{A}}: LK(\mathscr{A}) \to \mathscr{A}$ is a full functor, i.e. v is an argumentwise epimorphism.

1.3. Let $\mathscr{A}$ be an algebraic theory; denote by $\mathscr{A}_n$ the full subcategory of $\mathscr{A}$ whose objects are $\mathbf{0}, \mathbf{n}, \mathbf{2n}, \ldots, \mathbf{kn}, \ldots$. Prove that $\mathscr{A}_n$ is an algebraic subtheory of $\mathscr{A}$.

1.4. Let $U: \mathscr{A} \to \mathscr{B}$ be a theory morphism in $\mathscr{T}h$ such that $U(\mathbf{n}, \mathbf{1}): \mathscr{A}(\mathbf{n}, \mathbf{1}) \to$ $\to \mathscr{B}(\mathbf{n}, \mathbf{1})$ is a surjective map for every $n \in \mathbf{N}$. Show that $U(\mathbf{n}, \mathbf{m})$ is surjective for every $n, m \in \mathbf{N}$.

1.5. Show that a theory morphism $F: \mathscr{A} \to \mathscr{B}$ is an extremal epimorphism if and only if it is a coequalizer, i.e. if and only if for each natural number n, the map $F(\mathbf{n}, \mathbf{1}): \mathscr{A}(\mathbf{n}, \mathbf{1}) \to \mathscr{B}(\mathbf{n}, \mathbf{1})$ is a surjection. Also show that the class $Q_e(\mathscr{A})$ of all extremal quotient objects of $\mathscr{A}$ is canonically in one-to-one correspondence with the set $R(\mathscr{A})$ of all equivalence relations on algebraic theories in $\mathscr{A}$. Further, show that $Q_e(\mathscr{A})$ is a complete and cocomplete lattice.

Assume that for every pair (m, n) of natural numbers there is given a set $\{f_i, g_i\}_{i \in I(n,m)}$ of pairs of elements in $\mathscr{A}(\mathbf{n}, \mathbf{m})$; prove that there exists a smallest equivalence relation R on the algebraic theories in $\mathscr{A}$ such that $f_i \sim g_i R(n, m))$ for all $n, m \in \mathscr{N}$ and all $i \in I(n, m)$. (This relation is called the *equivalence relation of algebraic theories generated by the given set of pairs.*) Show that every morphism of $\mathscr{T}h$ has a coequalizer decomposition.

1.6. Show that $\mathscr{T}h$ is a complete category. Products and kernels in $\mathscr{T}h$ are formed as in $\mathscr{C}at$.

1.7. For each natural number n, let X^n denote the object of $[\mathbf{N}, \mathscr{S}et]$ such that $\{X^n\}_m = \emptyset$ if $n \neq m$ and $\{X^n\}_n$ contains only one element. Show that the objects $\{L(X^n)\}_{n \in \mathbf{N}}$ give a set of strong generators for $\mathscr{T}h$.

1.8. Let $\mathscr{A}$ be an algebraic theory and $f: \mathbf{n} \to \mathbf{1}$ an n-ary operation. Let σ be a permutation of the set $1, 2, \ldots, n$. We say that f is *σ-commutative* if $f[p^n_{\sigma(1)}, \ldots, p^n_{\sigma(n)}] = f$. f is called *commutative* if f is σ-commutative for every permutation σ of $1, 2, \ldots, n$. Prove: If $\{f_i: \mathbf{n}_i \to \mathbf{1}\}_i$ is a set of operations of $\mathscr{A}$ and σ_i is a permutation of $1, 2, \ldots, n_i$, for all i, then there exists a smallest equivalence relation R on the algebraic theories in $\mathscr{A}$ such that the image of f_i in $\mathscr{A}/R$ is σ_i-commutative for all i.

1.9. Show that the trivial algebraic theory is equivalent to the dual of $\mathscr{S}et\,f$. Show that every algebraic theory is a category with finite products.

1.10. Let α be a regular cardinal number. By an *α-algebraic theory* we mean a category $\mathscr{A}$ such that every object of $\mathscr{A}$ is a product of $\beta < \alpha$ copies of a fundamental object **1**. If $\mathscr{A}$ and $\mathscr{B}$ are α-algebraic theories, then an *α-theory morphism* is defined to be a functor $F: \mathscr{A} \to \mathscr{B}$ which preserves the fundamental objects and commutes with α-products. Let $\mathscr{T}h_\alpha$ denote the category whose objects are α-algebraic theories and whose morphisms are α-theory morphisms.

Show that $\mathscr{T}h\,\alpha$ is a complete and cocomplete category, and, moreover, that all definitions and results about $\mathscr{T}h$ can be carried over to $\mathscr{T}h_\alpha$ — in particular, $\mathscr{T}h = \mathscr{T}h_{\alpha_0}$, where α_0 is the first regular cardinal number.

We can also define the category $\mathscr{T}h_\infty$, whose objects are *infinitary algebraic theories*. i.e. categories $\mathscr{A}$ whose objects are products of a fundamental object **1**. The morphisms are defined in an obvious way. $\mathscr{T}h_\infty$ has almost all the properties of $\mathscr{T}h$.

1.11. Denote by $\mathscr{T}h_0$ the full subcategory of $\mathscr{T}h$ whose objects are all algebraic theories $\mathscr{A}$ such that $\mathscr{A}(\mathbf{0}, \mathbf{1}) = \emptyset$ (i.e. $\mathscr{A}$ has no 0-ary, or *nullary*, operations). Show that $\mathscr{T}h_0$ is a coreflective subcategory of $\mathscr{T}h$.

1.12. Let $\mathscr{A}$ be an algebraic theory, such that $\mathscr{A}$ is not a final object of $\mathscr{T}h$ and $\mathscr{A}(\mathbf{0}, \mathbf{1}) \neq \emptyset$. Show that the monoid $\mathscr{A}(\mathbf{1}, \mathbf{1})$ is not trivial.

1.13. Let $M: \mathscr{T}h \to \mathscr{M}on$ be the functor $M(\mathscr{A}) = \mathscr{A}(\mathbf{1}, \mathbf{1})$. Show that M has an adjoint. (Hint: If R is a monoid, denote by $\tilde{R}$ the algebraic theory defined as follows: $\tilde{R}(\mathbf{1}, \mathbf{1}) = R$ and $\tilde{R}(\mathbf{n}, \mathbf{1})$ is the set of all compositions $p^n_i g$ where $g \in R$ and $1 \leqslant i \leqslant n$. The assignment $R \rightsquigarrow \tilde{R}$ gives an adjoint of M.)

1.14. Let $f \in \mathscr{A}(\mathbf{n}, \mathbf{1})$, $g \in \mathscr{A}(\mathbf{m}, \mathbf{1})$. We shall say that f and g *commute* if the diagram

$$\begin{array}{ccc} \mathbf{nm} & \xrightarrow{f^n} & \mathbf{m} \\ {\scriptstyle g^n}\downarrow & & \downarrow{\scriptstyle g} \\ \mathbf{n} & \xrightarrow{f} & \mathbf{1} \end{array}$$

is commutative.

a) Show that two nullary operations commute if and only if they are equal.
b) Show that a unary operation always commutes with itself.

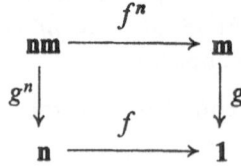

c) Let H be a set of operations of $\mathscr{A}$. Let $H^c(\mathbf{n}, \mathbf{1})$ denote $\{f \in \mathscr{A}(\mathbf{n}, \mathbf{1}),$ f commutes with all $g \in H\}$. Show that H^c is a subtheory of $\mathscr{A}$. In particular, $\mathscr{A}^c$ is denoted by $Z(\mathscr{A})$ and is called the center of $\mathscr{A}$. If $\mathscr{A} = Z(\mathscr{A})$ we say that $\mathscr{A}$ is a *commutative theory*.

d) Show that $Z(\mathscr{A})$, where $\mathscr{A}$ is the algebraic theory of groups, is the free algebraic theory on one nullary operation.

1.15. Let $\mathscr{A}$ be an algebraic theory and let $\{(n_i, m_i)\}$ be a set of pairs of natural numbers. Assume that for each i there are given the operations $f_i \in \mathscr{A}(\mathbf{n}_i, \mathbf{1})$ and $g_i \in \mathscr{A}(\mathbf{m}_i, \mathbf{1})$. Let R denote the equivalence relation on algebraic theories generated by all pairs $(f_i g_i^{n_i}, g_i f_i^{m_i})$. Prove that:

a) For each i, the image of f_i in $\mathscr{A}/R$ commutes with the image of g_i.

b) If $F: \mathscr{A} \to \mathscr{B}$ is a morphism of algebraic theories such that $F(f_i)$ commutes with $F(g_i)$ for all i, then there exists a unique morphism $F: \mathscr{A}/R \to \mathscr{B}$ such that $FP = F$, where $P: \mathscr{A} \to \mathscr{A}/\mathscr{R}$ is canonically defined.

c) Denote by $\mathscr{T}\!hc$ the full subcategory of $\mathscr{T}\!h$ whose objects are commutative algebraic theories. $\mathscr{T}\!hc$ is a reflective subcategory of $\mathscr{T}\!h$.

3.2. Algebraic categories

We continue the program begun in the previous section. Here we define algebraic categories and give some examples.

References: Cohn [1], Eckman—Hilton [1], [2], Freyd [2], Isbell [5], Lawvere [1], Linton [2], MacLane [3], Pareigis [1], Schubert [1] Słomiński [1], Wraith [1].

Let $\mathscr{A}$ be an algebraic theory. An *$\mathscr{A}$-algebra in a category* $\mathscr{C}$ is a functor $F: \mathscr{A} \to \mathscr{C}$ that commutes with finite products. An $\mathscr{A}$-algebra in the category $\mathscr{S}et$ is simply called an *$\mathscr{A}$-algebra*. Generally, we let $[\mathscr{A}, \mathscr{C}]^a$ denote the full subcategory of $[\mathscr{A}, \mathscr{C}]$, whose objects are $\mathscr{A}$-algebras in $\mathscr{C}$ (and whose morphisms are functorial morphisms). Usually a morphism in $[\mathscr{A}, \mathscr{C}]^a$ is called a *morphism of $\mathscr{A}$-algebras*. The category $[\mathscr{A}, \mathscr{S}et]^a$ is simply denoted by $\mathscr{A}^a$. If $f: F \to G$ is a morphism of $\mathscr{A}$-algebras then to simplify the notation we denote $F(\mathbf{n})$ by $F_\mathbf{n}$ and $f_\mathbf{n}$ by f_n for $n \in \mathbf{N}$.

We have the following result concerning morphisms of $\mathscr{A}$-algebras.

PROPOSITION 2.1. *Let $F, G \in [\mathscr{A}, \mathscr{C}]^a$ and let $f: F \to G$ be a morphism of $\mathscr{A}$-algebras. Then, for each $n \in \mathbf{N}$, the morphism $f_n: F_n \to G_n$ is completely determined by f_1, in particular, $f_n = f_1 \coprod f_1 \coprod \cdots \coprod f_1 = f_1^n$,* where the product is n-fold.

Conversely, assume that $f: F_1 \to G_1$ is a morphism in $\mathscr{C}$ such that for every n-ary operation $h: \mathbf{n} \to \mathbf{1}$ of $\mathscr{A}$ we have $G(h)f^n = fF(h)$. Then there is a unique morphism of $\mathscr{A}$-algebras $\bar{f}: F \to G$ such that $\bar{f}_1 = f$.

The proof is left for the reader. ✳

For each category $\mathscr{C}$, let $U_{\mathscr{C}}^{\mathscr{A}}:[\mathscr{A},\mathscr{C}]^a \to \mathscr{C}$ denote the *underlying*, or *forgetful, functor* defined by $U_{\mathscr{C}}^{\mathscr{A}}(F)=F_1$, and $U_{\mathscr{C}}^{\mathscr{A}}(f)=f_1$. The underlying functor $U_{\mathscr{S}et}^{\mathscr{A}}$ is denoted by $U^{\mathscr{A}}$ or simply by U if there is no danger of confusion.

Example 2.2. Let $\mathscr{N}$ be the trivial algebraic theory. An $\mathscr{N}$-algebra in a category $\mathscr{C}$ is given as follows: let X be an object of $\mathscr{C}$ such that for all $n\in\mathscr{N}$, the product $X^n = X\coprod X\coprod\dots\coprod X$ is defined in $\mathscr{C}$. In particular, X^0 is a final object of $\mathscr{C}$. Hence if $\mathscr{C}$ is a category with finite products, then the underlying functor $U_{\mathscr{C}}^{\mathscr{N}}:[\mathscr{N},\mathscr{C}]^a\to\mathscr{C}$ is an equivalence of categories In particular, the assignments $F \rightsquigarrow F_1$ and $f\rightsquigarrow f_1$ define an equivalence between the categories $\mathscr{N}^a$ and $\mathscr{S}et$.

Example 2.3. Let $\mathscr{A}$ be the algebraic theory of premonoids (see Example 1.10). An $\mathscr{A}$-algebra in $\mathscr{C}$ is usually called a *premonoid in* $\mathscr{C}$. A premonoid in $\mathscr{C}$ is defined by choosing an object X of $\mathscr{C}$ such that:

a) X^n is defined for every $n\in\mathscr{N}$.

b) There is a morphism $m: X\coprod X\to X$ such that the following diagram is commutative:

$$\begin{array}{ccc} X\coprod X\coprod X & \xrightarrow{m\coprod 1} & X\coprod X \\ {\scriptstyle 1\coprod m}\downarrow & & \downarrow{\scriptstyle m} \\ X\coprod X & \xrightarrow{m} & X \end{array}$$

By a *premonoid*, we mean a premonoid in $\mathscr{S}et$. It is clear that a premonoid is in fact a pair (M, m), where M is a set and $m: M\coprod M\to M$ is a map (called *multiplication*), that is associative. A morphism of premonoids from the premonoid (M, m) into the premonoid (M', m') is in fact a map $f: M\to M'$ such that the diagram

$$\begin{array}{ccc} M\coprod M & \xrightarrow{f\coprod f=f^2} & M'\coprod M' \\ {\scriptstyle m}\downarrow & & \downarrow{\scriptstyle m'} \\ M & \xrightarrow{f} & M' \end{array}$$

is commutative.

The category of premonoids is denoted by $\mathscr{P}rem$.

2.4. Let $\mathscr{A}$ be the algebraic theory of monoids (see Example 1.11). An $\mathscr{A}$-algebra in $\mathscr{C}$ is called a *monoid in* $\mathscr{C}$, or simply a *monoid*, when $\mathscr{C}=\mathscr{S}et$. Let $F:\mathscr{A}\to\mathscr{S}et$ be a monoid; by the definition of $\mathscr{A}$, $F(m)$: $F_2=F_1\coprod F_1\to F_1$ is an associative multiplication. Also, since F_0 is a final object of $\mathscr{S}et$, i.e. $F_0=\{*\}$, a set consisting of one point, then $F(e)$: $F_0=\{*\}\to F_1$; $F(e)(*)$ is usually denoted by e and called the *identity element of the multiplication* $F(m)$. The conditions imposed in the definition of $\mathscr{A}$ (see Example 1.11) yield that the compositions:

$$F_1 \xrightarrow{[F(e)F(p_0^1),\ F(p_1^1)]} F_2 \xrightarrow{F(m)} F_1,$$

$$F_1 \xrightarrow{[F(p_1^1),\ F(e)F(p_0^1)]} F_2 \xrightarrow{F(m)} F_1$$

are both the identity on F_1. For example,

$$\{F(m)\,[F(e)F(p_0^1),\ F(p_1^1)]\}\,(x) = F(m)F(e)F(p_0^1)\,(x);\ F(p_1^1)\,(x)) = F(m)\,(e, x) = x$$

(since $p_1^1 = 1_1$ and $F(p_0^1)\,(x) = *$).

Similarly,

$$\{F(m)\,[F(p_1^1),\ F(e)F(p_0^1)]\}\,(x) = F(m)\,(x, e) = x.$$

Hence the set F_1 is endowed with an associative law of composition, namely, $F(m)$, whose neutral element is e. Therefore F_1 is in fact a monoid in the sense of Section 1.1.

Conversely, if M is a monoid, as in Section 1.1, whose law of composition is denoted by $f\colon M \coprod M \to M$ and which has a neutral element α, then this monoid defines a functor $F\colon \mathscr{A} \to \mathscr{S}et$ by:

$$F_n = M^n \quad \text{for every } n \in \mathbf{N}$$

(in particular, $F_0 = \{*\}$), and, furthermore,

$$F(m) = f,$$

$$F(e)\colon F_0 = \{*\} \to M = F_1,\ F(e)\,(*) = \alpha,$$

$$F(p_1^n)\colon F_n = M^n \to M \quad \text{are structural projections.}$$

It is easy to see that the category $\mathscr{A}^a$ of monoids defined here is canonically equivalent to the category $\mathscr{M}on$ defined in Section 1.1.

2.5. Let $\mathscr{A}$ be the algebraic theory of groups (see Example 1.12). An $\mathscr{A}$-algebra in $\mathscr{C}$ is called a *group* in $\mathscr{C}$. A group in $\mathscr{S}et$ is simply called a ***group***. It is easy to verify that *tn* this case the category $\mathscr{A}^a$ is actually equivalent to the category $\mathscr{G}r$ defined in Section 1.2. To see this, simply observe that if $s\colon \mathbf{1} \to \mathbf{1}$ is the inverse operation and if $F\colon \mathscr{A} \to \mathscr{S}et$ is a group, then $F(s)\colon F_1 \to F_1$ is a map such that the composition of maps:

$$F_1 \xrightarrow{[F(s),\ F(p_1^1)]} F_2 \xrightarrow{F(m)} F_1$$

carries the element x of F_1 into the neutral element e, since

$$\{F(m)\,[F(s),\ F(p_1^1)]\}\,(x) = F(m)\,(F(s)\,(x), x) = e.$$

The element $F(s)\,(x)$ is in fact the inverse of x, i.e. $F(s)\,(x) = x^{-1}$.

2.6. Let $\mathscr{A}$ be the algebraic theory which is obtained from the algebraic theory of groups by making the only 2-ary operation, m, commutative. (According to Exercise 1.8, $\mathscr{A}$ is obtained as a factor category of the algebraic theory of groups by a suitable equivalence relation on algebraic theories.) We specify that $\mathscr{A}$ has the same operations and identities as in Example 1.12, and furthermore $E(2) = \{(m[p_2^2, p_1^2], m)\}$. We leave it for the reader to check that the category $\mathscr{A}^a$ is equivalent to the category $\mathscr{Ab}$ of abelian groups.

2.7. The category $\mathscr{Rg}$ is equivalent to a category of $\mathscr{A}$-algebras. To see this, let $\mathscr{A}$ be the algebraic theory whose primitive n-ary operations are:

$$C(0) = \{e, 0\},$$

$$C(1) = \{s\},$$

$$C(2) = \{m, a\},$$

$$C(n) = \emptyset \quad \text{for } n > 2.$$

Assume identities of n-th order that make m and a associative. Further, assume that e is an identity for m, 0 is an identity for a, and s is a symmetric operation for a. In addition, assume that $E(3)$ contains the pairs $(m[p_1^3, a[p_2^3, p_3^3]], a[m[p_1^3, p_2^3], m[p_1^3, p_3^3]])$ and $(m[a[p_1^3, p_2^3], p_3^3], a[m[p_1^3, p_3^3], m[p_2^3, p_3^3]])$. The last two identities express the distributivity of m (also called *multiplication*) relative to a (which we now call *addition*). 0 is called the *zero operation* or simply *zero*. We leave it for the reader to check that $\mathscr{A}^a$ is canonically equivalent to the category $\mathscr{Rg}$. Moreover, the algebraic theory $\mathscr{A}$ is called the *algebraic theory of rings*.

Note 2.8. Usually we understand a ring to be a unitary ring and usually a ring that is not necessarily unitary is called a *prering*. It is easy to see that the *category of prerings* is equivalent to $\mathscr{A}^a$, where $\mathscr{A}$ is obtained from the algebraic theory of rings by leaving out the axiom for the existence of the unit element e, for multiplication.

Let $\mathscr{A}$ and $\mathscr{B}$ be two objects of $\mathscr{Th}$. We define a new algebraic theory $\mathscr{A} \otimes \mathscr{B}$, called the *tensor*, or *Kronecker*, *product* of $\mathscr{A}$ with $\mathscr{B}$, as follows. Consider the coproduct $\mathscr{A} \coprod \mathscr{B}$ and let $U: \mathscr{A} \to \mathscr{A} \coprod \mathscr{B}$ and $V: \mathscr{B} \to \mathscr{A} \coprod \mathscr{B}$ be the structural injections. Let R denote the equivalence relation of algebraic theories defined on $\mathscr{A} \coprod \mathscr{B}$, such that $R(\mathbf{n}, \mathbf{1})$ is generated by all pairs

$$(V(k)U(h)^r, \; U(h)V(k)^m),$$

where $h \in \mathscr{A}(\mathbf{m}, \mathbf{1})$, $k \in \mathscr{B}(\mathbf{r}, \mathbf{1})$ and $rm = n$. Define $\mathscr{A} \otimes \mathscr{B}$ to be $\mathscr{A} \coprod \mathscr{B}/R$. If $P: \mathscr{A} \coprod \mathscr{B} \to \mathscr{A} \otimes \mathscr{B}$ is the canonical functor, then the functors $U^* = PU$ and $V^* = PV$ are called the *structural morphisms of the tensor product*.

In general, if $h \in \mathscr{A}(\mathbf{n}, \mathbf{m})$ ($k \in \mathscr{B}(\mathbf{n}, \mathbf{m})$), denote $U^*(h)$ by h^*($V^*(k)$ by k^*).

It follows from the definition of the tensor product that if $h \in \mathscr{A}(\mathbf{m}, \mathbf{1})$ and $k \in \mathscr{B}(\mathbf{r}, \mathbf{1})$ we have the following commutative diagram in $\mathscr{A} \otimes \mathscr{B}$:

$$\begin{array}{ccc} \mathbf{mr} & \xrightarrow{(k^*)^m} & \mathbf{m} \\ {\scriptstyle (h^*)^r}\downarrow & & \downarrow{\scriptstyle h^*} \\ \mathbf{r} & \xrightarrow{k^*} & \mathbf{1} \end{array}$$

i.e. h^* commutes with k^*.

In particular, if $m = r = 0$, then $(k^*)^0 = (h^*)^0 = p_0^0 = 1_0$. Hence $h^* = k^*$.

COROLLARY 2.9. *Let $\mathscr{A}$ and $\mathscr{B}$ be two algebraic theories. Then for each $h \in \mathscr{A}(\mathbf{0}, \mathbf{1})$ and $k \in \mathscr{B}(\mathbf{0}, \mathbf{1})$ we have that $h^* = k^*$ in $\mathscr{A} \otimes \mathscr{B}$. Hence, if $\mathscr{A}(\mathbf{0}, \mathbf{1})$ and $\mathscr{B}(\mathbf{0}, \mathbf{1})$ are nonempty, then $(\mathscr{A} \otimes \mathscr{B})(\mathbf{0}, \mathbf{1})$ contains exactly one element.* ✳

PROPOSITION 2.10. *Let $\mathscr{A}_i$, $i = 1, 2$, be algebraic theories with $m_i: \mathbf{2} \to \mathbf{1}$, and $e_i: \mathbf{0} \to \mathbf{1}$, such that e_i is an identity of m_i (i.e. $m_i[1_1, e_i p_0^1] = 1_1$). Then if we let m_i^* denote the induced operation in $\mathscr{A}_1 \otimes \mathscr{A}_2$ we have:*

a) $m_1^* = m_2^*$,

b) *m_1^* is commutative, and*

c) *m_1^* is associative.*

Proof. Consider the commutative diagram in $\mathscr{A}_1 \otimes \mathscr{A}_2$:

$$\begin{array}{ccc} \mathbf{4} & \xrightarrow{m_2^* \coprod m_2^*} & \mathbf{2} \\ {\scriptstyle m_1^* \coprod m_1^*}\downarrow & & \downarrow{\scriptstyle m_1^*} \\ \mathbf{2} & \xrightarrow{m_2^*} & \mathbf{1} \end{array} .$$

Then for every $\mathbf{n}$, we have the following commutative diagram of sets

$$\begin{array}{ccc} (\mathscr{A}_1 \otimes \mathscr{A}_2)(\mathbf{n}, \mathbf{4}) = (\mathscr{A}_1 \otimes \mathscr{A}_2)(\mathbf{n}, \mathbf{1})^4 & \xrightarrow{m_2' \coprod m_2'} & (\mathscr{A}_1 \otimes \mathscr{A}_2)(\mathbf{n}, \mathbf{1})^2 \\ {\scriptstyle m_1' \coprod m_1'}\downarrow & & \downarrow{\scriptstyle m_1'} \\ (\mathscr{A}_1 \otimes \mathscr{A}_2)(\mathbf{n}, \mathbf{1})^2 & \xrightarrow{m_2'} & (\mathscr{A}_1 \otimes \mathscr{A}_2)(\mathbf{n}, \mathbf{1}) \end{array}$$

where $m_i' = (\mathscr{A}_1 \otimes \mathscr{A}_2)(\mathbf{n}, m_i^*)$. Now let $\alpha = (x, y, z, w)$ be an element of $(\mathscr{A}_1 \otimes \otimes \mathscr{A}_2)(\mathbf{n}, \mathbf{4})$ and put $m_1'(a, b) = a \cdot b$ and $m_2'(a, b) = a + b$. Then:

$$(m_1'(m_2' \coprod m_2'))(\alpha) = m_1'(x * y, z * w) = (x * y) \cdot (z * w)$$
$$= (m_2'(m_1' \coprod m_1'))(\alpha) = m_2'(x \cdot y, z \cdot w) = (x \cdot y) + (z \cdot w).$$

If $0 = e_1^* p_0^n = e_2^* p_0^n$, then $x \cdot 0 = 0 \cdot x = x * 0 = 0 * x = x$. Hence $x \cdot w = (x * 0) \cdot \cdot (0 * w) = (x \cdot 0) + (0 \cdot w) = x * w$, i.e. $m_1^* = m_2^*$. Furthermore,

$$x \cdot y = (0 \cdot x) * (y \cdot 0) = (0 \cdot y) \cdot (x \cdot 0) = y \cdot x,$$

i.e. m_1^* is commutative. Finally:

$$x\cdot(y\cdot z)=(x\cdot 0)\cdot(y\cdot z)=(x\cdot y)\cdot(0\cdot z)=(x\cdot y)\cdot z,$$

so that m_1^* is associative. ✳

COROLLARY 2.11. *Let $\mathscr{A}$ be the algebraic theory of monoids and let $\mathscr{B}$ be the algebraic theory of commutative monoids. Then $\mathscr{B}\simeq\mathscr{A}\otimes\mathscr{A}$. Moreover, if $\mathscr{A}$ is the algebraic theory of groups, then $\mathscr{A}\otimes\mathscr{A}$ is the algebraic theory of commutative groups.*

Proof. By Corollary 2.9 and Proposition 2.10, $\mathscr{A}\otimes\mathscr{A}$ has exactly one commutative 2-ary operation and exactly one 0-ary operation (which is an identity of the 2-ary operation) under consideration. The proof follows from the definition of the tensor product. ✳

THEOREM 2.12. *Let $\mathscr{A}$ and $\mathscr{B}$ be two algebraic theories. Then for each category $\mathscr{C}$ with finite products, we have the canonical isomorphisms:*

$$[\mathscr{A}\otimes\mathscr{B},\mathscr{C}]^a\simeq[\mathscr{A},[\mathscr{B},\mathscr{C}]^a]^a\simeq[\mathscr{B},[\mathscr{A},\mathscr{C}]^a]^a.$$

Proof. Let $F\in[\mathscr{A}\otimes\mathscr{B},\mathscr{C}]^a$; let $T(F):\mathscr{A}\to[\mathscr{B},\mathscr{C}]^a$ denote the functor defined as follows. If $\mathbf{n}\in\mathscr{A}$, then $T(F)(\mathbf{n}):\mathscr{B}\to\mathscr{C}$ is the functor defined by

$$T(F)(\mathbf{n})(\mathbf{m})=F_{nm}.$$

Furthermore, if $k\in\mathscr{B}(\mathbf{m},\mathbf{s})$, then $T(F)(\mathbf{n})(k)=F(k^{*n})$.

Now, if $h:\mathbf{n}\to\mathbf{n}'$ is a morphism of $\mathscr{A}$, then $T(F)(h):T(F)(\mathbf{n})\to T(F)(\mathbf{n}')$ will be the functorial morphism whose components are

$$T(F)(h)_{\mathbf{m}}=F(h^*)^m.$$

It is easy to see that $T(F)$ is an $\mathscr{A}$-algebra in $[\mathscr{B},\mathscr{C}]^a$. If $f:F\to G$ is a morphism of $[\mathscr{A}\otimes\mathscr{B},\mathscr{C}]^a$, let $T(f):T(F)\to T(G)$ denote the morphism of $\mathscr{A}$-algebras (i.e. the functorial morphism) whose components are

$$T(f)_n:T(F)(\mathbf{n})\to T(G)(\mathbf{n})$$

such that $(T(f)_n)_m=f_{nm}$. We have thus defined the functor

$$T:[\mathscr{A}\otimes\mathscr{B},\mathscr{C}]^a\to[\mathscr{A},[\mathscr{B},\mathscr{C}]^a]^a.$$

Conversely, let $G\in[\mathscr{A},[\mathscr{B},\mathscr{C}]^a]$. By Exercise 4.6, Ch. 1, we can define a functor $S'(G):\mathscr{A}\coprod\mathscr{B}\to\mathscr{C}$ (here $\mathscr{A}\coprod\mathscr{B}$ is the product in $\mathscr{C}at$) by:

$$S'(G)(\mathbf{n},\mathbf{m})=G(\mathbf{n})(\mathbf{m}),$$

$$S'(G)(h,k)=G(h)_{m'}\,G(\mathbf{n})(k)=G(\mathbf{n}')(k)G(h)_m,$$

where $h:\mathbf{n}\to\mathbf{n}'$, $k:\mathbf{m}\to\mathbf{m}'$.

According to the definition of $[\mathscr{A},[\mathscr{B},\mathscr{C}]^a]^a$, the functor $S'(G)$ commutes with products in each argument separately. Then we may define

$$S(G):\mathscr{A}\otimes\mathscr{B}\to\mathscr{C}$$

by:

$$S(G)(\mathbf{n}) = S'(G)(\mathbf{1},\mathbf{1})^n,$$

$$S(G)(h^*) = S'(G)(h, 1_1),\ h \in \mathscr{A}(\mathbf{n},\mathbf{1}),$$

$$S(G)(k^*) = S'(k)(1_1, k),\ k \in \mathscr{B}(\mathbf{n},\mathbf{1}).$$

By use of the definition of $\mathscr{A} \otimes \mathscr{B}$, it is easy to check that $S(G)$ is an $(\mathscr{A} \otimes \mathscr{B})$-algebra in $\mathscr{C}$ and that the assignment $G \rightsquigarrow S(G)$ gives a functor $S\colon [\mathscr{A}, [\mathscr{B}, \mathscr{C}]^a]^a \to [\mathscr{A} \otimes \mathscr{B}, \mathscr{C}]^a$ such that $TS = \mathrm{Id}$ and $ST = \mathrm{Id}$. The details are left for the reader. ✳

COROLLARY 2.13. *The only group objects (see Exercise 2.2) in the category of groups are the commutative groups.*

The proof follows from Theorem 2.12, and Corollary 2.11. ✳

COROLLARY 2.14. *The only group object in $\mathscr{R}g$ is the pointed ring (or zero ring).*

Proof. By Corollary 2.9 and Theorem 2.12, for a group object F in $\mathscr{R}g$, the zero element and the identity are necessarily equal. Thus for every $a \in F$, $1 \cdot a = a = 0 \cdot a = 0$. ✳

Note 2.15. Let α be a regular cardinal number and let $\mathscr{A}$ be an α-algebraic theory. By an *α-$\mathscr{A}$-algebra in a category $\mathscr{C}$*, we mean a functor $F\colon \mathscr{A} \to \mathscr{C}$ that commutes with α-products. As before the α-$\mathscr{A}$-algebras in $\mathscr{S}et$ are simply called *α-$\mathscr{A}$-algebras*. Almost all results valid for algebras can be transferred to α-$\mathscr{A}$-algebras.

2.16. Let $\mathscr{A}$ be an algebraic theory and f a zero-ary operation of $\mathscr{A}$. Then for every $\mathscr{A}$-algebra X, $X(f)\colon X_0 = \{*\} \to X_1$ is a constant function and the element $X(f)(*)$ is called the *constant associated with the zero-ary operation f*. We remark that generally a constant is not the identity (or neutral, or zero) element for a suitable operation.

Exercises

2.1. Let $\mathscr{A}$ be an algebraic theory. Show that a functor $F\colon \mathscr{A} \to \mathscr{C}$ is an $\mathscr{A}$-algebra in $\mathscr{C}$ if and only if $H^X F\colon \mathscr{A} \to \mathscr{S}et$ is an $\mathscr{A}$-algebra for every $X \in \mathscr{C}$. Prove: If $f\colon X' \to X$ is a morphism in $\mathscr{C}$ and F an $\mathscr{A}$-algebra in $\mathscr{C}$, then $H^0(f)F\colon H^X F \to H^{X'} F$ is a morphism of $\mathscr{A}$-algebras.

2.2. Let $\mathscr{A}$ be an algebraic theory. An *$\mathscr{A}$-object* of a category $\mathscr{C}$ is an object X of $\mathscr{C}$ such that there exists a cofunctor $F\colon \mathscr{C} \to \mathscr{A}^a$, with $U^{\mathscr{A}} F \to H_X$. Prove that an object X of $\mathscr{C}$ is an $\mathscr{A}$-object if and only if for each $Y \in \mathscr{C}$, the set $\mathscr{C}(Y, X)$ is an $\mathscr{A}$-algebra and for each morphism $f\colon Y' \to Y$ of $\mathscr{C}$, the map $H_X(f)\colon \mathscr{C}(Y, X) \to \mathscr{C}(Y', X)$ is a morphism of $\mathscr{A}$-algebras.

Let $\mathscr{C}^{\mathscr{A}}$ denote the full subcategory of $\mathscr{C}$ whose objects are $\mathscr{A}$-objects. If $\mathscr{C}$ is a category with finite products, show that $\mathscr{C}^{\mathscr{A}}$ is canonically equivalent to $[\mathscr{A}, \mathscr{C}]^a$.

2.3. Let $T: \mathscr{C} \to \mathscr{C}'$ be a functor that commutes with finite products. Show that for each algebraic theory $\mathscr{A}$ there is a commutative diagram:

$$\begin{array}{ccc} [\mathscr{A}, \mathscr{C}]^a & \xrightarrow{[\mathscr{A}, T]} & [\mathscr{A}, \mathscr{C}']^a \\ {\scriptstyle U_{\mathscr{C}}^{\mathscr{A}}}\downarrow & & \downarrow{\scriptstyle U_{\mathscr{C}'}^{\mathscr{A}}} \\ \mathscr{C} & \xrightarrow{T} & \mathscr{C}' \end{array} .$$

2.4. Prove that for every category, $\mathscr{C}$, the underlying functor $U_{\mathscr{C}}^{\mathscr{A}}: [\mathscr{A}, \mathscr{C}]^a \to \mathscr{C}$ is faithful and commutes with limits. If $\mathscr{C}$ is complete, show that $[\mathscr{A}, \mathscr{C}]^a$ is also complete. Show that, moreover, a morphism $f: F \to G$ in $[\mathscr{A}, \mathscr{C}]^a$ is a monomorphism if and only if $f_1: F_1 \to G_1$ is a monomorphism in $\mathscr{C}$.

2.5. Let $\mathscr{A}$ be the algebraic theory of rings. Prove that the 2-ary operation a (i.e. addition) is commutative.

2.6. Let $\mathscr{A}$ be an algebraic theory. Prove that there is an $\mathscr{A}$-algebra F such that F_1 consists of exactly one element. All $\mathscr{A}$-algebras with one element are isomorphic. Such an algebra is called a point algebra.

2.7. Let $\mathscr{A}$ be an algebraic theory. Prove that there exists an empty $\mathscr{A}$-algebra if and only if the set $\mathscr{A}(\mathbf{0}, \mathbf{1})$ is empty. Show that the category $\mathscr{A}^a$ has a zero object if and only if there is precisely one nullary operation.

2.8. Prove: If $\mathscr{A}, \mathscr{B}, \mathscr{C} \in \mathscr{T}\!h$, then we have the canonical isomorphisms

$$\mathscr{A} \otimes \mathscr{B} \simeq \mathscr{B} \otimes \mathscr{A}, \quad \mathscr{A} \otimes (\mathscr{B} \otimes \mathscr{C}) \simeq (\mathscr{A} \otimes \mathscr{B}) \otimes \mathscr{C}$$

and $\mathscr{A} \otimes \mathscr{N} \simeq \mathscr{A}$.

2.9. Show that the tensor product of any two commutative algebraic theories is canonically isomorphic to their coproduct in $\mathscr{T}hc$.

2.10. Let $\mathscr{A}$ be an algebraic theory, $f \in \mathscr{A}(\mathbf{n}, \mathbf{1})$, $F: \mathscr{A} \to \mathscr{C}$ an $\mathscr{A}$-algebra in $\mathscr{C}$ and σ a permutation of the set $1, 2, \ldots, n$. We shall say that F *is σ-commutative for f*, if $F(f)[F(p_{\sigma(1)}^n), \ldots, F(p_{\sigma(n)}^n)] = F(f)$. We shall say that F *is commutative for f* if F is σ-commutative for all permutations of $1, 2, \ldots, n$. Prove that the following are equivalent:

a) If $\mathscr{C}$ is a category, then every $\mathscr{A}$-algebra $F: \mathscr{A} \to \mathscr{C}$ is σ-commutative (commutative) for f.

b) Every $\mathscr{A}$-algebra is σ-commutative (commutative) for f.

c) The $\mathscr{A}$-algebra $H^{\mathbf{n}}: \mathscr{A} \to \mathscr{S}et$ is σ-commutative (commutative) for f.

d) The operation f is σ-commutative (commutative).

2.11. Give a direct proof of Corollary 2.13 (without using either Theorem 2.12 or Corollary 2.11).

2.12. Show that the only monoid object in the category of monoids is a commutative one. Give a direct proof of this result without using Corollary 2.11 or Theorem 2.12.

2.13. Let $\mathscr{A}$ be an algebraic theory. An object $(F, \mathscr{B})$ of $\mathscr{A}/\mathscr{T}h$ is simply called an *$\mathscr{A}$-algebraic theory*. It is clear that an $\mathscr{A}$-algebraic theory is in fact a pair $(F, \mathscr{B})$, where $F: \mathscr{A} \to \mathscr{B}$ is a theory morphism. If there is no chance of doubt, we shall write $\mathscr{B}$ instead of $(F, \mathscr{B})$.

Let $(F, \mathscr{B})$ and $(G, \mathscr{C})$ be two $\mathscr{A}$-algebraic theories. Denote by R the equivalence relation on $\mathscr{B} \otimes \mathscr{C}$ generated by all pairs $(F(f)^*, G(f)^*)$ where f runs over all operations of $\mathscr{A}$. The algebraic theory $\mathscr{B} \otimes_{\mathscr{A}} \mathscr{C}$ is defined to be $\mathscr{B} \otimes \mathscr{C}/R$ (which is in a canonical way an $\mathscr{A}$-algebraic theory), and is called the *tensor*, or *Kronecker, product of $\mathscr{B}$ and $\mathscr{C}$ relative to $\mathscr{A}$.*

a) Let $\mathscr{B}, \mathscr{C}$ and $\mathscr{D}$ be $\mathscr{A}$-algebraic theories. Show that the following isomorphisms exist:

i) $\mathscr{B} \otimes_{\mathscr{A}} \mathscr{C} \simeq \mathscr{C} \otimes_{\mathscr{A}} \mathscr{B}$,

ii) $\mathscr{A} \otimes_{\mathscr{A}} \mathscr{B} \simeq \mathscr{B}$,

iii) $(\mathscr{B} \otimes_{\mathscr{A}} \mathscr{C}) \otimes_{\mathscr{A}} \mathscr{D} \simeq \mathscr{B} \otimes_{\mathscr{A}} (\mathscr{C} \otimes_{\mathscr{A}} \mathscr{D})$.

b) Let $F: \mathscr{B} \to \mathscr{B}'$ be a morphism of $\mathscr{A}$-algebraic theories (i.e. a morphism of $\mathscr{A}/\mathscr{T}h$). Show that, for every $\mathscr{A}$-algebraic theory $\mathscr{D}$, F defines in a canonical way a morphism of $\mathscr{A}$-algebraic theories:

$$F \otimes \mathscr{D}: \mathscr{B} \otimes_{\mathscr{A}} \mathscr{D} \to \mathscr{B}' \otimes_{\mathscr{A}} \mathscr{D}.$$

Moreover, if $G: \mathscr{D} \to \mathscr{D}'$ is another morphism of $\mathscr{A}$-algebraic theories, show that the diagram

$$\begin{array}{ccc} \mathscr{B} \otimes_{\mathscr{A}} \mathscr{D} & \xrightarrow{F \otimes \mathscr{D}} & \mathscr{B}' \otimes_{\mathscr{A}} \mathscr{D} \\ {\scriptstyle \mathscr{B} \otimes G}\downarrow & & \downarrow{\scriptstyle \mathscr{B}' \otimes G} \\ \mathscr{B} \otimes_{\mathscr{A}} \mathscr{D}' & \xrightarrow{F \otimes \mathscr{D}'} & \mathscr{B}' \otimes_{\mathscr{A}} \mathscr{D}' \end{array}$$

is commutative.

Usually the diagonal morphism given by thisdiagram is denoted by $F \otimes G$, and is called the *tensor*, or *Kronecker, product of F* and G. It is obvious that $F \otimes \mathscr{D} = F \otimes 1_{\mathscr{D}}$.

c) Show that a theory morphism $F: \mathscr{A} \to \mathscr{B}$ is an epimorphism in the category $\mathscr{T}h$ if and only if the canonical morphism

$$\mathscr{B} \otimes_{\mathscr{A}} \mathscr{A} \xrightarrow{\mathscr{B} \otimes F} \mathscr{B} \otimes_{\mathscr{A}} \mathscr{B}$$

is an isomorphism.

3.3. Algebraic functors

Let $F: \mathscr{A} \to \mathscr{B}$ be a morphism in $\mathscr{T}h$. Since F commutes with finite products, the restriction functor $F_*: [\mathscr{B}, \mathscr{S}et] \to [\mathscr{A}, \mathscr{S}et]$ induces a functor $F_a: \mathscr{B}^a \to \mathscr{A}^a$. Functors of this kind are called *algebraic* functors. In this section we prove that every algebraic functor has an adjoint. Some problems concerning limits and colimits in algebraic categories are studied.

References: Cohn [1], Freyd [2], Lawvere [1], Pareigis [1], Schubert [1], Wraith [1].

For each algebraic theory $\mathscr{A}$, let $I_{\mathscr{A}}$ (or simply I, if no confusion is possible) denote the inclusion functor $\mathscr{A}^a \to [\mathscr{A}, \mathscr{S}et]$. As a consequence of the result of Section 2.4 we obtain the following theorem.

THEOREM 3.1. *If $\mathscr{A}$ is an algebraic theory, then the inclusion functor $I: \mathscr{A}^a \to$ $\to [\mathscr{A}, \mathscr{S}et]$ has an adjoint.*

Proof. Let $H^0: \mathscr{A} \to [\mathscr{A}, \mathscr{S}et]$ be the canonical embedding, i.e. $H^0(\mathbf{n}) = H^{\mathbf{n}}$. Let Σ be the set of all morphisms $\{\langle H^0(p_1^n), \ldots, H^0(p_n^n)\rangle: (H^1)^{(n)}, \to H^{\mathbf{n}}\}_{n \in \mathbf{N}}$. It is clear that a functor $F: \mathscr{A} \to \mathscr{S}et$ is an $\mathscr{A}$-algebra if and only if it is a Σ-sheaf over the category $\mathscr{A}^0$, the dual of $\mathscr{A}$. Then, by Theorem 4.17, Ch. 2, the inclusion functor I has an adjoint. ✳

Let $B_{\mathscr{A}}$ (or simply B) be an adjoint of I.

For each $n \in \mathbf{N}$, the functor $H^{\mathbf{n}}: \mathscr{A} \to \mathscr{S}et$ commutes with products so that if defines an $\mathscr{A}$-algebra, usually denoted by $M_n^{\mathscr{A}}$. In particular, $M_1^{\mathscr{A}}$ is called the *fundamental $\mathscr{A}$-algebra*. It is clear that the assignment $\mathbf{n} \rightsquigarrow M_n^{\mathscr{A}}$ defines a cofunctor $M^{\mathscr{A}}: \mathscr{A} \to \mathscr{A}^a$, called the *fundamental cofunctor* (we simply write M and M_n if there is no risk of confusion).

PROPOSITION 3.2. *Let $\mathscr{A}$ be an algebraic theory. Then:*

a) *There is a canonical isomorphism*

$$BH^0 \simeq M.$$

Moreover, we can choose B so that $BH^0 = M$.

b) The cofunctor $M: \mathscr{A} \to \mathscr{A}^a$ is full and faithful, commutes with finite products (i.e. transforms finite products into finite coproducts), and is dense.

The proof follows from Theorem 7.1, Ch. 2, and is left for the reader. ✳

COROLLARY 3.3. *For each algebraic theory $\mathscr{A}$ the category $\mathscr{A}^a$ is complete and cocomplete. Moreover, limits in $\mathscr{A}^a$ are formed argumentwise.*

The proof follows from Theorem 3.1 and Proposition 13.12, Ch. 1. ✳

THEOREM 3.4. *Let $F: \mathscr{A} \to \mathscr{B}$ be a theory morphism. Then the algebraic unctor $F_a: \mathscr{B}^a \to \mathscr{A}^a$ has an adjoint, denoted F^a. Moreover, F^a can be chosen so* *that the diagram*

$$\begin{array}{ccc} \mathscr{A} & \xrightarrow{\;F\;} & \mathscr{B} \\ {\scriptstyle M^{\mathscr{A}}}\downarrow & F^a & \downarrow{\scriptstyle M^{\mathscr{B}}} \\ \mathscr{A}^a & \longrightarrow & \mathscr{B}^a \end{array}$$

is commutative.

Proof. Consider the following diagram of categories and functors

$$\begin{array}{ccc} [\mathscr{B}, \mathscr{S}et] & \underset{F^*}{\overset{F_*}{\rightleftarrows}} & [\mathscr{A}, \mathscr{S}et] \\ {\scriptstyle I_{\mathscr{B}}}\uparrow\downarrow{\scriptstyle B_{\mathscr{B}}} & & {\scriptstyle I_{\mathscr{A}}}\uparrow\downarrow{\scriptstyle B_{\mathscr{A}}} \\ \mathscr{B}^a & \xrightarrow{\;F_a\;} & \mathscr{A}^a \end{array}$$

where F^* is adjoint to F_* (see Corollary 2.10, Ch. 2). Denote $B_B F^* I_{\mathscr{A}}$ by F^a. Then F^a is adjoint to F_a. Indeed, if $X \in \mathscr{A}^a$ and $Y \in \mathscr{B}^a$ then we have the following canonical functorial isomorphisms:

$$\mathscr{A}^a(X, F_a(Y)) \simeq [I_{\mathscr{A}}(X), I_{\mathscr{A}} F_a(Y)] \simeq [I_{\mathscr{A}}(X), F_* I_{\mathscr{B}}(Y)]$$
$$\simeq [F^* I_{\mathscr{A}}(X), I_{\mathscr{B}}(Y)] \simeq \mathscr{B}^a(B_{\mathscr{B}} F^* I_{\mathscr{A}}(X), Y).$$

The last part of the theorem follows from Corollary 2.10, Ch. 2. ⚹

COROLLARY 3.5. *Let $\mathscr{A}$ be an algebraic theory. Then:*

a) *The underlying functor $U^{\mathscr{A}}: \mathscr{A}^a \to \mathscr{S}et$ has an adjoint, denoted $L^{\mathscr{A}}$ (we simply write U and L if there is no danger of confusion).*

b) *The underlying functor is canonically isomorphic to H^{M_1}, where M_1 is the fundamental $\mathscr{A}$-algebra, and it commutes with filtered colimits.*

c) *The fundamental $\mathscr{A}$-algebra is a generator and a separator of $\mathscr{A}^a$. Moreover, if M is a set, then $L(M) \simeq M_1^{(M)}$.*

Proof. a) By Example 2.2, the category $\mathscr{S}et$ is canonically equivalent to $\mathscr{N}^a$, where $\mathscr{N}$ is the trivial algebraic theory. It is easy to see that the underlying functor $U^{\mathscr{A}}$ is equal to the functor $I_a^{\mathscr{A}}$, where $I_a^{\mathscr{A}}: \mathscr{N} \to \mathscr{A}$ is defined in Example 1.2.

b) This follows from Theorem 3.1, Proposition 3.2 of this chapter and Theorem 12.4 of Ch. 1.

c) The proof is left for the reader. ⚹

The functor $L: \mathscr{S}et \to \mathscr{A}^a$ is called the *free $\mathscr{A}$-algebra functor*. If M is a set then $L(M)$ is called the *free $\mathscr{A}$-algebra generated by M.*

If X is an $\mathscr{A}$-algebra, then $X_1 = X(\mathbf{1})$ is called the *carrier* of X. Likewise, if $u: X \to Y$ is a morphism of $\mathscr{A}$-algebras, then $u_1: X_1 \to Y_1$ is called the *carrier* of u. It is obvious that X is completely determined by its carrier and the map $X(f): X_n = X_1^n \to X_1$, where f runs all through the operations of $\mathscr{A}$. Generally, when we speak of an $\mathscr{A}$-algebra X we shall understand its carrier to be X_1. For example, the expression "x is an element of X" means that x is an element of X_1.

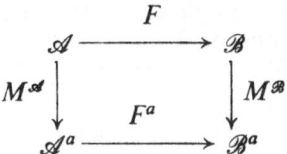

A subobject of an $\mathscr{A}$-algebra X is called an *$\mathscr{A}$-subalgebra*. It is clear that every $\mathscr{A}$-subalgebra of X is completely determined by a subset X' of X_1 that is closed under all the operations of X and that contains all constants. In particular, we obtain the notions of: submonoid, subgroup, subring, etc.

By an *equivalence relation of $\mathscr{A}$-algebras* on an $\mathscr{A}$-algebra X we mean an equivalence relation R on X_1, the carrier of X, such that the following condition is satisfied. If f is an n-ary operation of $\mathscr{A}$, and $x_1, \ldots, x_n, y_1, \ldots, y_n$ are elements of X_1 such that $x_i \sim y_i(R)$, $1 \leqslant i \leqslant n$, then

$$X(f)(x_1, \ldots, x_n) \sim X(f)(y_1, \ldots, y_n)(R).$$

(We note that by an equivalence relation on a set M we mean a subset R of the product $M \coprod M$ that is reflexive, symmetric and transitive.)

If $u: X \to Y$ is a morphism of $\mathscr{A}$-algebras, then the equivalence relation on X_1, R_u, defined by the carrier of u, is an equivalence relation of $\mathscr{A}$-algebras on X. We can obtain a partial converse.

PROPOSITION 3.6. *Let X be an $\mathscr{A}$-algebra and let R be an equivalence relation of $\mathscr{A}$-algebras on X. Let X/R denote the $\mathscr{A}$-algebra defined by $(X/R)_n = (X_1/R)^n$. If $f \in A(\mathbf{n}, \mathbf{1})$, then $X(f): X_n \to X$ defines a unique map $(X/R)(f): (X_1/R)^n \to X_1/R$ such that the diagram*

$$\begin{array}{ccc} X_n & \xrightarrow{X(f)} & X_1 \\ {\scriptstyle p_1^n}\downarrow & & \downarrow{\scriptstyle p_1} \\ (X_1/R)^n & \xrightarrow{(X/R)(f)} & X_1/R \end{array}$$

is commutative, where p_1 is the canonical map. In this way X/R becomes an $\mathscr{A}$-algebra and p_1 defines a morphism of $\mathscr{A}$-algebras $p: X \to X/R$. Moreover $R_p = R$. ✳

Usually X/R is called the *factor algebra* of X relative to R and p is called the *canonical morphism*.

The equivalence relations of $\mathscr{A}$-algebras on an $\mathscr{A}$-algebra X are closely related to kernel pairs $(u, v): Y \to X$ as well as to some subalgebras of $X \coprod X$. To see this, let $(u, v): Y \to X$ be a kernel pair. We observed in Proposition 8.7, Ch. 1, that the morphism

$$[u, v]: Y \to X \coprod X$$

is a monomorphism and so it defines an $\mathscr{A}$-subalgebra, denoted $\mathrm{Im}[u, v]_1$. It is easy to see that $\mathrm{Im}[u, v]_1$ is an equivalence relation of $\mathscr{A}$-algebra on X.

PROPOSITION 3.7. *Let X be an $\mathscr{A}$-algebra and R a subset of $X_1 \coprod X_1$. The following are equivalent:*

a) *R is an equivalence relation of $\mathscr{A}$-algebras on X.*

b) *R is an equivalence relation on the set X_1 and it defines an $\mathscr{A}$-subalgebra of $X \coprod X$ (i.e. it is closed under the operations and contains all the constants).*

c) *There exists a kernel pair $(u, v): Y \to X$ such that $R = \mathrm{Im}[u, v]_1$.*

The proof is left for the reader. ✳

COROLLARY 3.8. *The set of equivalence relations of $\mathscr{A}$-algebras on an $\mathscr{A}$-algebra X is canonically in one-to-one correspondence with the set of all those $\mathscr{A}$-subalgebras X' of $X \prod X$ such that X'_1 is an equivalence relation on X_1.* ✳

The following result gives a characterization of coequalizers in algebraic categories.

PROPOSITION 3.9. *Let $u: X \to Y$ be a morphism in the category $\mathscr{A}^a$. The following are equivalent:*

1) *u is an extremal epimorphism.*

2) *u is a coequalizer.*

3) *The unique morphism $u': X/R_u \to Y$ such that $u'p = u$ (where $p: X \to X/R_u$ is canonical) is an isomorphism.*

4) *u_1 is a surjective map.* ✳

COROLLARY 3.10 a) *A double arrow (u, v) in $\mathscr{A}^a$ is a kernel pair if and only if $(U(u), U(v))$ is a kernel pair of $\mathscr{S}et$.*

b) *A morphism p of $\mathscr{A}^a$ is an equalizer if and only if $U(p)$ is an equalizer of $\mathscr{S}et$, i.e. a surjective map.*

Proof. a) follows from Proposition 3.7, and b) from Proposition 3.9,4. ✳

COROLLARY 3.11. *Every $\mathscr{A}$-algebra is a factor algebra of a free $\mathscr{A}$-algebra.*

Proof. Since the underlying functor $U: \mathscr{A}^a \to \mathscr{S}et$ is faithful, we have (see Proposition 13.7, Ch. 1) that the coarrow of adjunction $p: LU \to \mathrm{Id}\mathscr{A}^a$ is a functorial epimorphism. We remark that for each $X \in \mathscr{A}^a$, p_X is in fact a coequalizer (see Exercise 3.13). ✳

Corollary 3.11 can be strengthened in the following way. Let M be a set and $\{x_m\}_{m \in M}$ a set of elements of an $\mathscr{A}$-algebra X. Let $f: M \to U(X)$ denote the map defined by $f(m) = x_m$, and let $\bar{f} = p_X L(f): L(M) \to X$. We say that the set $\{x_m\}_{m \in M}$ of elements of X is a *set of generators* (a *set of free generators*) of X if the morphism $\bar{f}$ is a coequalizer (an isomorphism). If α is a cardinal number, we say that X is *α-generated* (*α-free generated*) if it has a set $\{x_m\}_{m \in M}$ of generators (of free generators) such that $\mathrm{Card}(M) = \alpha$. In particular, X is called *finitely generated (finitely free generated)* if it is n-generated (n-free generated) where n is a finite cardinal or equivalently a natural number. We sometimes call a finitely generated (finitely free generated) $\mathscr{A}$-algebra an *$\mathscr{A}$-algebra of finite type* (a *free $\mathscr{A}$-algebra of finite type*).

If M is a subset of an $\mathscr{A}$-algebra X, let $\langle M \rangle$ denote the image of the canonical morphism

$$L(M) \to X$$

that is defined by the inclusion $M \subseteq X_1$. $\langle M \rangle$ is called the subalgebra of X *generated* by M.

Now we reconsider algebraic functors. If $F: \mathscr{A} \to \mathscr{B}$ is a morphism of algebraic theories, then the diagram of categories and functors

$$\begin{array}{ccc} \mathscr{B}^a & \xrightarrow{\quad F_a \quad} & \mathscr{A}^a \\ & {\scriptstyle U^{\mathscr{B}}} \searrow \quad \swarrow {\scriptstyle U^{\mathscr{A}}} & \\ & \mathscr{S}et & \end{array}$$

is commutative.

We show the converse.

THEOREM 3.12. *Let $\mathscr{A}$ and $\mathscr{B}$ be algebraic theories. Assume that there is a functor $R: \mathscr{B}^a \to \mathscr{A}^a$ such that the diagram of categories and functors*

$$\begin{array}{ccc} \mathscr{B}^a & \xrightarrow{\quad R \quad} & \mathscr{A}^a \\ & {\scriptstyle U^{\mathscr{B}}} \searrow \quad \swarrow {\scriptstyle U^{\mathscr{A}}} & \\ & \mathscr{S}et & \end{array} \tag{2}$$

is commutative.

Then there is a unique theory morphism $F: \mathscr{A} \to \mathscr{B}$ such that $F_a \simeq R$.

Proof. We will define a theory morphism $F: \mathscr{A} \to \mathscr{B}$ such that $F_a \simeq R$.

We first observe that the functor R commutes with limits. Indeed, that follows from the fact that underlying functors commute and reflect limits (see Exercise 3.1).

Furthermore, by hypothesis and Corollary 3.5.b, for each pair of natural numbers m and n we have the following functorial isomorphisms:

$$\begin{aligned} \mathscr{B}(\mathbf{m}, \mathbf{n}) = \mathscr{B}(\mathbf{m}, \mathbf{1})^n = U^{\mathscr{B}}(M_m^{\mathscr{B}})^n \simeq U^{\mathscr{A}}R(M_m^{\mathscr{B}})^n = \mathscr{A}^a\,(M_1^{\mathscr{A}}, R(M_m^{\mathscr{B}})^n \\ \simeq \mathscr{A}^a\,(M_n^{\mathscr{A}}, R(M_m^{\mathscr{B}})). \end{aligned} \tag{3}$$

Now let $f \in \mathscr{A}(\mathbf{n}, \mathbf{1})$ and let $M(f): M_1^{\mathscr{A}} \to M_n^{\mathscr{A}}$ denote the induced functorial morphism (see the definition of $M_n^{\mathscr{A}}$). Then there is a unique map $F(f)_m: \mathscr{B}(\mathbf{m}, \mathbf{n}) \to$ $\to \mathbf{B}(\mathbf{m}, \mathbf{1})$ such that the diagram

$$\begin{array}{ccc} \mathscr{B}(\mathbf{m}, \mathbf{n}) & \xrightarrow{\quad \sim \quad} & \mathscr{A}^a\,(M_n^{\mathscr{A}}, R(M_m^{\mathscr{B}})) \\ {\scriptstyle F(f)_m} \downarrow & & \downarrow {\scriptstyle \mathscr{A}^a(M(f), R(M_m^{\mathscr{B}}))} \\ \mathscr{B}(\mathbf{m}, \mathbf{1}) & \xrightarrow{\quad \sim \quad} & \mathscr{A}^a\,(M_1^{\mathscr{A}}, R(M_m^{\mathscr{B}})) \end{array} \tag{4}$$

is commutative (the horizontal isomorphisms are obtained from (3)).

A trivial computation shows that for each morphism $g \in \mathscr{B}(\mathbf{m}', \mathbf{m})$ the diagram

$$\begin{array}{ccc} \mathscr{B}(\mathbf{m}, \mathbf{n}) & \xrightarrow{\quad \mathscr{B}(g, n) \quad} & \mathscr{B}(\mathbf{m}', \mathbf{n}) \\ {\scriptstyle F(f)_m} \downarrow & & \downarrow {\scriptstyle F(f)_{m'}} \\ \mathscr{B}(\mathbf{m}, \mathbf{1}) & \xrightarrow{\quad \mathscr{B}(g, \mathbf{1}) \quad} & \mathscr{B}(\mathbf{m}', \mathbf{1}) \end{array}$$

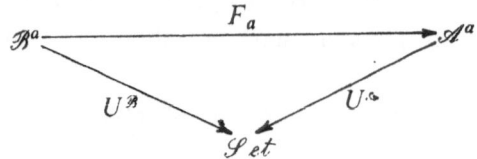

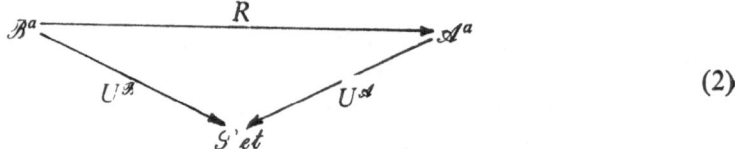

is commutative, i.e. the maps $\{F(f)_m\}_{m\in\mathbf{N}}$ define a functorial morphism from $H_{\mathbf{n}}$ into $H_{\mathbf{1}}$. But then, by Theorem 3.19, Ch. 1, there is a unique morphism $F(f)\in\mathscr{B}(\mathbf{n},\mathbf{1})$ such that $H(F(f))_m = F(f)_m$ for all $m\in\mathbf{N}$.

Again an easy computation shows that the assignments

$$\mathbf{n}\rightsquigarrow\mathbf{n},\ f\rightsquigarrow F(f)$$

define a functor $F:\mathscr{A}\to\mathscr{B}$, and this functor is in fact a theory morphism.

Our aim is now to show that $F_a\simeq R$. For this, let X be a $\mathscr{B}$-algebra. By Exercise 3.3, for each n-ary operation f of $\mathscr{A}$, we have the following commutative diagrams:

$$\begin{array}{ccc} R(X)_n & \xrightarrow{\sim} & \mathscr{A}^a(M_n^{\mathscr{A}}, R(X)) \\ {\scriptstyle R(X)(f)}\downarrow & & \downarrow{\scriptstyle \mathscr{A}^a(M(f),R(X))} \\ R(X)_1 & \xrightarrow{\sim} & \mathscr{A}^a(M_1^{\mathscr{A}}, R(X)), \end{array}$$

and (5)

$$\begin{array}{ccc} F_a(X)_n & \xrightarrow{\sim} & \mathscr{A}^a(M_n^{\mathscr{A}}, F_a(X)) \\ {\scriptstyle F_a(X)(f)}\downarrow & & \downarrow{\scriptstyle \mathscr{A}^a(M(f),\, F_a(X))} \\ F_a(X)_1 & \xrightarrow{\sim} & \mathscr{A}^a(M_1^{\mathscr{A}}, F_a(X)). \end{array}$$

But $F_a(X)(f) = XF(f)$ so that using the definition of $F(f)$ and the diagrams (5), we obtain the commutative diagram

$$\begin{array}{ccc} R(X)_n & \xrightarrow{\sim} & F_a(X)_n \\ {\scriptstyle R(X)(f)}\downarrow & & \downarrow{\scriptstyle F_a(X)(f)} \\ R(X)_1 & \xrightarrow{\sim} & F_a(X)_1 \end{array}$$

where the horizontal isomorphisms are functorial in X.

Finally, we leave it for the reader to show that F is uniquely defined (up to isomorphism) by R. ✳

Recall that we have denoted the illegitimate category of all categories and functors by $\mathscr{CAT}$ (see Section 1.9). Now we shall consider another illegitimate category, namely $\mathscr{CAT}/\mathscr{S}et$: the objects are pairs $(\mathscr{C}, U)$ where $U:\mathscr{C}\to\mathscr{S}et$ is a functor, and the morphisms from $(\mathscr{C}, U)$ into $(\mathscr{C}', U')$ are functors $F:\mathscr{C}\to\mathscr{C}'$ such that $U'F\simeq U$. As we already noted in Section 1.9, $\mathscr{CAT}$ is an illegitimate category in which every functor has a limit and a colimit where these notions are defined in the obvious manner.

Let Sem: $\mathscr{T}h\to\mathscr{CAT}/\mathscr{S}et$ denote the (illegitimate) cofunctor defined by:

$$\mathrm{Sem}(\mathscr{A}) = (\mathscr{A}^a, U^{\mathscr{A}}) \text{ and } \mathrm{Sem}(F) = F_a.$$

Sem is usually called the *semantic cofunctor.*

THEOREM 3.13. *The semantic cofunctor takes colimits to limits*

Proof. We wish to show that for every functor $F:\mathscr{I}\to\mathscr{T}h$,

$$(\varinjlim F,\ U^{\varinjlim F}) = \varprojlim \mathrm{Sem}(F).$$

Clearly, to prove this equality, it will suffice to show that Sem takes coproducts (of $\mathscr{T}h$) to products (of $\mathscr{CAT}|\mathscr{S}et$) and coequalizer diagrams to equalizer diagrams.

Let $\{\mathscr{A}_i\}_i$ be a set of algebraic theories and let $F_i\colon \mathscr{A}_i \to \coprod_i \mathscr{A}_i$ be the structural morphisms. By Proposition 1.13, if f_i is an operation of $\mathscr{A}_i$ and f_j is an operation of $\mathscr{A}_j$, then $F_i(f_i)$ and $F_j(f_j)$ are independent operations, whenever $i \neq j$. Hence a $\coprod_i \mathscr{A}_i$-algebra is in fact a set X that is simultaneously an $\mathscr{A}_i$-algebra for every i, and such that the $\mathscr{A}_i$-operations and $\mathscr{A}_j$-operations act independently, whenever $i \neq j$. A morphism of $\coprod_i \mathscr{A}_i$-algebras is a function f which is also an $\mathscr{A}_i$-morphism for all i. In this way we observe that the following diagram of $\mathscr{CAT}$:

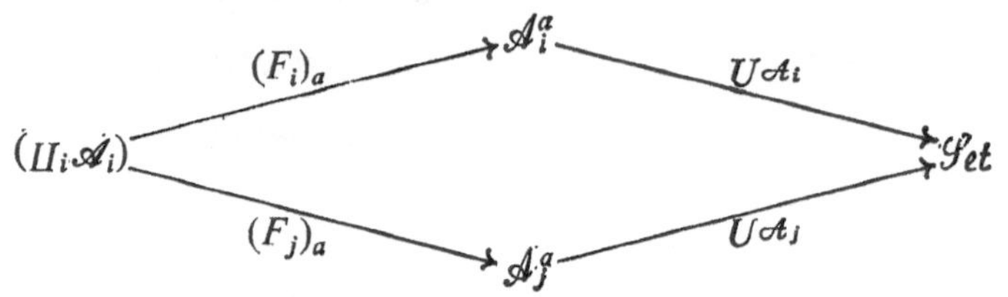

is a generalized fibered product or equivalently that Sem takes coproducts to products.

Next we consider the following coequalizer diagram of $\mathscr{T}h$:

$$\mathscr{A} \underset{G}{\overset{F}{\rightrightarrows}} \mathscr{B} \xrightarrow{P} \mathscr{C}.$$

Recall that P is full and the equivalence relation associated with P is the smallest equivalence relation on algebraic theories on $\mathscr{B}$, generated by all pairs $(F(f), G(f))$ where f runs through all the operations of $\mathscr{A}$. Now, if $X\colon \mathscr{B} \to \mathscr{S}et$ is a $\mathscr{B}$-algebra, such that $XF \simeq XG$, then the equivalence relation associated with X is finer than the equivalence relation on algebraic theories generated by all pairs $(F(f), G(f))$, so that there is a unique functor $X\colon \mathscr{C} \to \mathscr{S}et$, such that $XP = X$, and X commutes with finite products. Therefore, the following diagram of $\mathscr{CAT}|\mathscr{S}et$

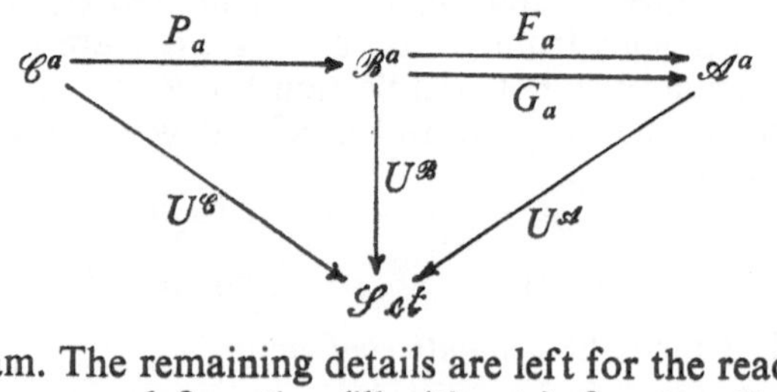

is an equalizer diagram. The remaining details are left for the reader. ⚹

In a similar way, we define the (illegitimate) functor Cosem: $\mathscr{T}h \to \mathscr{CAT}$ as follows:

$$\text{Cosem}(\mathscr{A}) = \mathscr{A}^a \text{ and } \text{Cosem}(F) = F^a.$$

We suggest that the reader prove the following result.

THEOREM 3.14. *The functor Cosem takes limits to limits.*

We note only that a $\coprod_i \mathscr{A}_i$-algebra is given by a family $\{X_i\}_i$ where X_i is an $\mathscr{A}_i$-algebra and a morphism $\{X_i \xrightarrow{f_i} X_i'\}_i$ of $\coprod_i \mathscr{A}_i$-algebras is given by a family $\{X_i \xrightarrow{f_i} X_i'\}_i$ of functions such that f_i is a morphism of $\mathscr{A}_i$-algebras for all i. ⚹

Exercises

3.1. Let $F: \mathscr{A} \to \mathscr{B}$ be a theory morphism. Then we have the following commutative diagram of categories and functors:

$$\begin{array}{ccc} [\mathscr{B}, \mathscr{S}et] & \xrightarrow{[F, \mathscr{S}et] = F_*} & [\mathscr{A}, \mathscr{S}et] \\ {\scriptstyle I_{\mathscr{B}}}\uparrow & & \uparrow{\scriptstyle I_{\mathscr{A}}} \\ \mathscr{B}^a & \xrightarrow{F_a} & \mathscr{A}^a. \end{array}$$

Show that F_a is faithful and that it commutes with and reflects isomorphisms, limits, monomorphisms and filtered colimits. (Hint: Observe that filtered colimits in $\mathscr{A}^a$ are formed as in $[\mathscr{A}, \mathscr{S}et]$ or equivalently $I_{\mathscr{A}}$ commutes with filtered colimits (see Theorem 12.1, Ch. 1).)

3.2. Let $\mathscr{A}$ be an algebraic theory such that there exists an $\mathscr{A}$-algebra F, with F_1 containing more than one element. Prove that the free $\mathscr{A}$-algebra functor $L: \mathscr{S}et \to \mathscr{A}^a$ is faithful. This is the case precisely when $\mathscr{A}$ is not a final object of $\mathscr{T}h$ or $\mathscr{A}$ is not the algebraic theory which contains only a nullary operation.

3.3. Let $\mathscr{A}$ be an algebraic theory, f an n-ary operation of $\mathscr{A}$, and X an $\mathscr{A}$-algebra. Show that the diagram

$$\begin{array}{ccc} X_n & \xrightarrow{\sim} & \mathscr{A}^a(M_n, X) \\ {\scriptstyle X(f)}\downarrow & & \downarrow{\scriptstyle \mathscr{A}^a(M(f), X)} \\ X_1 & \xrightarrow{\sim} & \mathscr{A}^a(M_1, X) \end{array}$$

is commutative, where the horizontal isomorphisms are canonical (see Theorem 3.19 and Proposition 3.2 of Ch. 1).

3.4. Show that the inclusion functor $\mathscr{M}on \to \mathscr{P}re$ is algebraic and describe its adjoint. Show that this functor is not full. (Hint: the desired adjoint associates each premonoid M with the monoid $M \coprod \{e\}$ (the coproduct of sets), where e is an element that does not belong to M and it becomes a neutral element for the multiplication in M.)

3.5. Prove: The inclusion functor $\mathscr{G}r \to \mathscr{M}on$ is algebraic and its adjoint associates with each monoid M (a category with one object) the category of fractions of M relative to all its morphisms (see Section 1.13). This inclusion functor also has a coadjoint and is full.

3.6. Every set M can be viewed as a diagram scheme with one vertex and whose arrows are the elements of M. Show that the free monoid generated by M is just the free category generated by this diagram scheme. Using Exercise 3.5 describe the free group generated by M.

3.7. Show that the inclusion functor $\mathscr{Ab} \to \mathscr{Gr}$ is algebraic and that the abelianization functor (see Section 1.11) is its adjoint.

3.8. The inclusion functor $\mathscr{Rg} \to \mathscr{Mon}(\mathscr{Rg} \to \mathscr{Ab})$ which associates with each ring its multiplicative monoid (its additive group) is algebraic. Describe its adjoint.

3.9. Let $\mathscr{A}$ be an algebraic theory. Using only the definition of $\mathscr{A}^a$ and Theorem 12.12, Ch. 1, observe that $\mathscr{A}^a$ is a category with filtered colimits and the underlying functor commutes with these colimits. Furthermore, let $[n] = \{1, 2, \ldots \ldots, n\}$ and $[0] = \emptyset$, and let $\mathscr{F}$ denote the full subcategory of $\mathscr{Set}$ generated by all sets $[n]$, $n \in \mathbf{N}$. Denote by $F\colon \mathscr{F} \to \mathscr{A}$ the cofunctor defined by: $F([n]) = \mathbf{n}$ and if $f\colon [n] \to [m]$ is a map then $F(f) = [p^m_{f(1)}, \ldots, p^m_{f(n)}]\colon \mathbf{m} \to \mathbf{n}$. Let $G' = MF$; we may extend G' to a functor $G\colon \mathscr{Set} \to \mathscr{A}^a$, since every set is the filtered colimit of its finite subsets. Then G is an adjoint of the underlying functor $U\colon \mathscr{A}^a \to \mathscr{Set}$. Now, using this result, show that $\mathscr{A}^a$ is a cocomplete category. These considerations can be used to show that $\mathscr{A}^a$ is cocomplete when $\mathscr{A}$ is an infinitary algebraic theory. In this case, the cofunctor F is directly defined from $\mathscr{Set}$ into $\mathscr{A}$.

3.10. a) Show that the set $R(X)$ of all equivalence relations of $\mathscr{A}$-algebras on an $\mathscr{A}$-algebra X is in a canonical way a complete and cocomplete lattice. Moreover, $R(X)$ is an ordered subset of the (complete and cocomplete) lattice of all equivalence relations on X_1, and it is closed under limits and colimits.

b) Show that $\mathscr{A}^a$ is a category with cokernels without using Theorem 3.1.

c) Show that every morphism u of $\mathscr{A}^a$ has an image, a coimage and a coequalizer decomposition.

d) Prove that a morphism u of $\mathscr{A}^a$ is an isomorphism if and only if u_1 is a bijective map.

e) Show that generally $\mathscr{A}^a$ is not a balanced category (see Ch. 1, Exercise 2.9)

3.11. Let $X \in \mathscr{A}^a$ and $R \in R(X)$ (see Exercise 3.10.a)) and let $p\colon X \to X/R$ be the canonical morphism. Prove the following:

a) For $R' \in R(X/R)$, let $p^{-1}(R')$ denote the equivalence relation on X defined as follows: $x \sim x'(p^{-1}(R')) \Leftrightarrow p(x) \sim p(x')(R')$. Then $p^{-1}(R') \in R(X)$ and $R \subseteq p^{-1}(R')$. Moreover, the assignment $R' \rightsquigarrow p^{-1}(R')$ gives a one-to-one correspondence between $R(X/R)$ and the subset of $R(X)$ consisting by all elements R'' such that $R \subseteq R''$.

b) If $R' \in R(X/R)$, then we have the following commutative diagram of $\mathscr{A}^a$:

$$\begin{array}{ccc} X & \xrightarrow{\;p\;} & X/R \\ {\scriptstyle p_1}\downarrow & & \downarrow{\scriptstyle p_2} \\ X/p^{-1}(R') & \xrightarrow{\;\overline{p}\;} & X/(R/R') \end{array}$$

where $p, \overline{p}, p_1, p_2$ are canonically defined and $\overline{p}$ is an isomorphism.

c) Let $R \in R(X)$ and let X' be a subalgebra of X. Let R' denote the restriction of R to X'. Then we have the following commutative diagram in $\mathscr{A}^a$:

$$\begin{array}{ccc} X' & \xrightarrow{\quad u \quad} & X \\ {\scriptstyle p'}\big\downarrow & & \big\downarrow{\scriptstyle p} \\ X'/R' & \xrightarrow{\quad u' \quad} & X/R \end{array}$$

where all morphisms are canonical and u' is a monomorphism. Also, every subalgebra of X/R is of the form X'/R' where X' is a suitable subalgebra of X.

3.12. Show that the fundamental $\mathscr{A}$-algebra M_1 is generated by one element. If M is a nonempty set, then $L(M) = M_1^{\{M\}}$, i.e. it is a coproduct of M-copies of M_1.

3.13. Show that for every object X of $\mathscr{A}^a$ the canonically defined diagram

$$LULU(X) \underset{(LUp)_X}{\overset{p_{LU(X)}}{\rightrightarrows}} LU(X) \xrightarrow{\;p_X\;} X$$

(where $p: LU \to \mathrm{Id}\, \mathscr{A}^a$ is a coarrow of adjunction) is a coequalizer diagram.

3.14. Let $\mathscr{A}$ be an algebraic theory such that there is a natural number n for which every nontrivial operation of $\mathscr{A}$ is at most n-ary. Assume that $\mathscr{C}$ is a full subcategory of $\mathscr{A}^a$ that contains an n-freely generated $\mathscr{A}$-algebra X. Show that then the full subcategory of $\mathscr{C}$ generated by X is dense in $\mathscr{C}$.

3.15. An object P of a category $\mathscr{C}$ is called *projective* if for every extremal epimorphism $g: X \to Y$ and every morphism $f: P \to Y$, there exists a morphism $\bar{f}: P \to X$ such that $g\bar{f} = f$.

a) Show that a coproduct of projective objects is also a projective object.

b) Show that an $\mathscr{A}$-algebra X is projective in $\mathscr{A}^a$ if and only if there is a free $\mathscr{A}$-algebra A' and a right invertible morphism $X' \to X$ and, in particular, that every free $\mathscr{A}$-algebra is projective.

c) Show that the projective objects in $\mathscr{G}r$ and $\mathscr{A}b$ are precisely the free ones. (Hint: See Kuroš's book [1].)

3.16. Prove Theorem 3.1 using Theorem 6.2, Ch. 2.

3.17. a) Prove: If, in the cartesian square

$$\begin{array}{ccc} X & \xrightarrow{\qquad} & Y \\ {\scriptstyle p'}\big\downarrow & & \big\downarrow{\scriptstyle p} \\ Z & \xrightarrow{\qquad} & T \end{array}$$

of $\mathscr{A}^a$, p is an extremal epimorphism, then so is p'.

b) Show that the extremal epimorphisms of $\mathscr{A}^a$ form a left and right permutable multiplicative system of morphisms.

c) Prove that a product of extremal epimorphisms is again an extremal epimorphism.

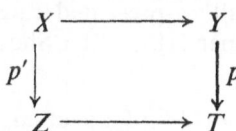

3.18. Let $\{X_i\}_i$ be a set of $\mathscr{A}$-algebras, $M = \coprod_i U(X_i)$ (the coproduct of sets), $u_i: U(X_i) \to M$ the structural injections and $v_i = s_M u_i$, where $s: \mathrm{Id}\,\mathscr{S}et \to UL$ is an arrow of adjunction. Denote by R the equivalence relation of $\mathscr{A}$-algebras on $L(M)$, such that for each $n \in \mathbf{N}$, for each $f \in \mathscr{A}(\mathbf{n}, 1)$, for each i and for each set $(x_1^i, \ldots, x_n^i)$ of elements of X_i we have that

$$v_i[X_i(f)(x_1^i, \ldots, x_n^i)] \sim L(M)(f)[v^i(x_1^i, \ldots, v_i(x_n^i)](R).$$

Let $q: L(M) \to L(M)/R$ denote the canonical morphism. Show that $qv_i: X_i \to$ $\to L(M)/R$ is a morphism of $\mathscr{A}$-algebras and these morphisms define $L(M)/R$ as a coproduct of X_i's in $\mathscr{A}^a$.

3.19. Let α be a regular cardinal number. An object X of a category $\mathscr{C}$ is called *α-presentable (α-generated)* if the functor $H^X: \mathscr{C} \to \mathscr{S}et$ commutes with the colimits of all α-filtered systems (with the colimits of all α-monofiltered systems) which exist in $\mathscr{C}$. The object X is called *finitely presented (finitely generated)* if it is α_0-presented (α_0-generated), where α_0 is the first regular cardinal number.

a) Show that an $\mathscr{A}$-algebra X is α-presented if and only if there exists an equalizer diagram of $\mathscr{A}^a$:

$$L_1 \rightrightarrows L_0 \longrightarrow X$$

where L_1 and L_0 are free algebras generated by the sets M_1 and M_0 respectively and such that $\mathrm{Card}(M_1) < \alpha$, $\mathrm{Card}(M_0) < \alpha$.

b) Show that an $\mathscr{A}$-algebra X is α-generated if and only if it is generated by a set M of elements, such that $\mathrm{Card}(M) < \alpha$, and, in particular, that X is α_0-generated if and only if it is generated by a finite set of elements.

c) Show that the set of all finitely generated subalgebras of an $\mathscr{A}$-algebra X forms a monofiltered system relative to inclusion and that X is the colimit of this system.

3.20. Let α be a regular cardinal number. A category $\mathscr{C}$ is called *locally α-presented* if it is cocomplete and has a set of strong generators which are simultaneously α-presented. A category $\mathscr{C}$ is called *locally finitely presented* if it is locally α_0-presented.

a) Show that the category $\mathscr{A}^a$ is locally finitely presented.

b) Show that if $\mathscr{A}$ is an α-algebraic theory, then the category $\mathscr{A}^a$ is locally α-presented.

c) Prove that the category $\mathscr{T}h$ is locally finitely presented.

3.21. Show that every locally α-presented category is a well copowered category. (Hint: See Gabriel — Ulmer [1], p. 81.) Show that, in particular, $\mathscr{A}^a$ is a wellcopowered category.

3.22. As usual, a subalgebra X' of an $\mathscr{A}$-algebra X is called *proper* if it is not equal to X. Show that if X is a finitely generated algebra, then every proper subalgebra can be embedded in a maximal proper subalgebra.

3.23. Let X be a set and let $P(X)$ be the power set of X. By a *closure operator* on X, we mean a map $J: P(X) \to P(X)$ such that:

1) $X' \subseteq J(X')$ for all $X' \in P(X)$.

2) If $X' \subseteq X''$, then $J(X') \subseteq J(X'')$.

3) $JJ(X) = J(X)$.

A closure operator J on X is called *algebraic* if for every $X' \subseteq X$ and every $x \in J(X')$ there is a finite subset X'' of X such that $x \in J(X'')$.

a) Let X be an $\mathscr{A}$-algebra. Show that the assignment $M \rightsquigarrow \overline{M}$ (where $\overline{M}$ is the subalgebra of X generated by M) defines an algebraic closure operator on X, called the *closure operator associated with the $\mathscr{A}$-algebra X*.

b) Prove that, conversely, if J is an algebraic closure operator on a set X, then there exists a suitable algebraic theory $\mathscr{A}$ such that X becomes in a canonical way an $\mathscr{A}$-algebra and J is precisely the closure operator associated with the $\mathscr{A}$-algebra X. (Hint: See Cohn [1], Theorem 5.2.)

3.4. Coalgebras

The concept of a coalgebra is the dualization of the notion of algebra. In this section we define coalgebra and show that coalgebras arise in the study of functors between algebraic categories, as well as elsewhere.

References: Eckmann—Hilton [1], [2], [3], Freyd [2], Hilton [4], Isbell [5], Lawvere [1], Linton [3], Wraith [1].

Let $\mathscr{A}$ be an algebraic theory. An *$\mathscr{A}$-coalgebra*, or *$\mathscr{A}^0$-algebra*, in a category $\mathscr{C}$, is a cofunctor $F: \mathscr{A} \to \mathscr{C}$ that commutes with finite products, i.e. takes finite products to finite coproducts. Obviously, an $\mathscr{A}$-coalgebra in $\mathscr{C}$ is defined by a suitable $\mathscr{A}$-algebra in $\mathscr{C}^0$, the dual of $\mathscr{C}$, and conversely.

If $F: \mathscr{A} \to \mathscr{C}$ is an $\mathscr{A}$-coalgebra in $\mathscr{C}$, then $F(\mathbf{n}) = F(\mathbf{1})^{(n)}$ so that F is determined by $F(\mathbf{1})$. Usually we denote $\mathbf{F(n)}$ by F_n. We assume that this notation will not lead to error, since it will be clear from the context whether we are dealing with algebras or coalgebras.

A *morphism of $\mathscr{A}$-coalgebras* is defined in the obvious manner: it is a functorial morphism. If $f: F \to G$ is a morphism of $\mathscr{A}$-coalgebras, denote $f_{\mathbf{n}}$ by f_n. It is clear that

$$f_n = \langle f_1, \ldots, f_1 \rangle = f_1^{(n)}.$$

The category of all $\mathscr{A}$-coalgebras in $\mathscr{C}$ is denoted by $[\mathscr{A}^0, \mathscr{C}]^a$. Also, let $U_{\mathscr{A}}^{\mathscr{C}}: [\mathscr{A}^0, \mathscr{C}]^a \to \mathscr{C}$ denote the underlying functor, defined by $U_{\mathscr{A}}^{\mathscr{C}}(F) = F_1$.

Usually, to design an $\mathscr{A}$-coalgebra for some known algebraic theory, we use the notions of *cogroup*, *coring*, *comonoid*, etc.

Let $\mathscr{A}$ and $\mathscr{B}$ be two algebraic theories. An $\mathscr{A}$-algebra in $\mathscr{B}^a$ is called an *$(\mathscr{A}, \mathscr{B})$-bialgebra*. Similarly, an $\mathscr{A}$-coalgebra in $\mathscr{B}^a$ is called an $(\mathscr{A}^0, \mathscr{B})$-bialgebra. We hope that the reader does not confuse the $(\mathscr{A}^0, \mathscr{B})$-bialgebras with $(\mathscr{A}, \mathscr{B})$-bialgebras. We note only that those notions are partially dual.

PROPOSITION 4.1. *The cofunctor $F: \mathscr{A} \to \mathscr{C}$ is an $\mathscr{A}$-coalgebra in $\mathscr{C}$ if and only if for each $X \in \mathscr{C}$ the functor $H_X F$ is an $\mathscr{A}$-algebra.* ✳

Example 4.2. Every free $\mathscr{A}$-algebra has a canonical structure as an $(\mathscr{A}^0, \mathscr{A})$-bialgebra. To see this, let M be a set and f an n-ary operation of $\mathscr{A}$. We want to define a morphism $\bar{f}: L(M) \to L(M^{(n)}) = L(M)^{(n)}$ of $\mathscr{A}$-algebras. By the definition of a free algebra, it will suffice to find a map $f': M \to L(M)_1^{(n)}$. To this end, let $x \in M$ and define f' by:

$$f'(x) = L(M^{(n)})(f)(u_1(x), \ldots, u_n(x)) \in L(M)_1^{(n)}$$

where the $u_i: M \to M^{(n)}$ are the structural injections of the coproduct. Set $\bar{f} = L(f'): L(M) \to L(M^{(n)}) = L(M)^{(n)}$.

If $p_i^n: \mathbf{n} \to \mathbf{1}$, $i = 1, \ldots, n$, are the structural projections of a product, then $(p_i^n)'(x) = L(M^{(n)})(p_i^n)(u_1(x), \ldots, u_n(x)) = u_i(x)$ and hence $\bar{p}_i^n = L(u_i)$, $i = 1, \ldots, n$, i.e. they are the structural injections of the coproduct $L(M^{(n)}) = L(M)^{(n)}$. We thus see that the assignments

$$\mathbf{n} \rightsquigarrow L(M)^{(n)}, \quad f \rightsquigarrow \bar{f}$$

define a cofunctor from $\mathscr{A}$ into $\mathscr{A}^a$ and this cofunctor commutes with products, i.e. it is an $\mathscr{A}$-coalgebra in $\mathscr{A}^a$.

As a direct consequence of Corollary 7.7, Ch. 2, we have the following result.

THEOREM 4.3. *Let $\mathscr{A}$ be an algebraic theory, $\mathscr{C}$ a cocomplete category and $T: \mathscr{A} \to \mathscr{C}$ an $\mathscr{A}$-coalgebra. Then there is a unique functor $\tilde{T}: \mathscr{A}^a \to \mathscr{C}$ such that:*

a) $\tilde{T}M \simeq T$, where $M: \mathscr{A} \to \mathscr{A}^0$ is the canonical cofunctor (see Section 3.3).

b) $\tilde{T}$ commutes with colimits and has a coadjoint. Actually, every functor $R: A^a \to \mathscr{C}$ that commutes with colimits has a coadjoint. ✳

Note 4.4. The functor $\tilde{T}$ defined in Theorem 4.3 is called "the *tensor product associated with* T". Moreover, if $X \in \mathscr{A}^a$, then we shall write:

$$\tilde{T} = T \otimes_A ? \quad \text{and} \quad \tilde{T}(X) = T \otimes_{\mathscr{A}} X.$$

By Theorem 7.6, Ch. 2, a coadjoint of $T \otimes_{\mathscr{A}} ?$ is the functor $\mathscr{C}(T, ?): \mathscr{C} \to \mathscr{A}^a$ defined by

$$\mathscr{C}(T, ?)(Y) = \mathscr{C}(T, Y) = H_Y T.$$

We hope that this notation does not lead to confusion. Although T is not an object of $\mathscr{C}$, it is determined by the object $T_1 = T(\mathbf{1})$ (sometimes called the *carrier* of T).

With the same hypothesis as in Theorem 4.3, let $f: T \to T'$ be a morphism of $\mathscr{A}$-coalgebras in $\mathscr{C}$. Then f defines in a canonical way the functorial morphisms:

$$F \otimes_{\mathscr{A}} ?: T \otimes_{\mathscr{A}} ? \to T' \otimes_{\mathscr{A}} ?, \quad \text{and}$$

$$\mathscr{C}(f, ?): \mathscr{C}(T, ?) \to \mathscr{C}(T, ?).$$

Moreover, if (u, v) and (u', v') are the corresponding arrow and coarrow of adjunction, then the diagrams

$$\begin{array}{ccc} \mathrm{Id}\mathscr{A}^a & \xrightarrow{\quad u \quad} & \mathscr{C}(T, ?)T \otimes_{\mathscr{A}} ? \\ {\scriptstyle u'}\downarrow & & \downarrow{\scriptstyle \mathscr{C}(T, ?)f \otimes_{\mathscr{A}} ?} \\ \mathscr{C}(T', ?)T' \otimes_{\mathscr{A}} ? & \xrightarrow{\mathscr{C}(f, ?)T' \otimes_{\mathscr{A}} ?} & \mathscr{C}(T, ?)T' \otimes_{\mathscr{A}} ? \end{array}$$

and

$$\begin{array}{ccc} (T\otimes_{\mathscr{A}}?)\mathscr{C}(T',?) & \xrightarrow{(T\otimes_{\mathscr{A}}?)\mathscr{C}(f,?)} & (T\otimes_{\mathscr{A}}?)\mathscr{C}(T,?) \\ {\scriptstyle (f\otimes_{\mathscr{A}}?)\mathscr{C}(T',?)}\downarrow & & \downarrow{\scriptstyle v} \\ (T'\otimes_{\mathscr{A}}?)\mathscr{C}(T',?) & \xrightarrow{\quad v' \quad} & \mathrm{Id}\mathscr{C} \end{array}$$

are commutative.

Thus, as a direct consequences of Exercise 7.7, Ch. 2, we obtain the following result.

COROLLARY 4.5. *Let $\mathscr{A}$ be an algebraic theory and $\mathscr{C}$ a cocomplete category. Then the assignments*

$$T \rightsquigarrow (T\otimes_{\mathscr{A}}?, \mathscr{C}(T, ?)), \quad \textit{and}$$

$$f \rightsquigarrow (f\otimes_{\mathscr{A}}?, \mathscr{C}(f, ?))$$

define an equivalence between the category of all $\mathscr{A}$-coalgebras and $\mathrm{Ad}(\mathscr{A}^a, \mathscr{C})$, *the category of all adjoint pairs.* ✳

For every algebraic theory $\mathscr{A}$, the canonical cofunctor $M: \mathscr{A} \to \mathscr{A}^a$ defines an $(\mathscr{A}^0, \mathscr{A})$-bialgebra, called the *fundamental $(\mathscr{A}^0, \mathscr{A})$-algebra* and also denoted by $M^{\mathscr{A}}$ (or by M if there is no danger of confusion).

COROLLARY 4.6. *If $\mathscr{A}$ is an algebraic theory, then the tensor product associated with the fundamental $(\mathscr{A}^0, \mathscr{A})$-bialgebra is isomorphic to the identity functor of $\mathscr{A}^a$.* ✳

Exercises

4.1. Let $\mathscr{A}$ be an algebraic theory. An *$\mathscr{A}$-coobject* of a category $\mathscr{C}$ is an object X of $\mathscr{C}$ for which there is a functor $F: \mathscr{C} \to \mathscr{A}^a$ such that $U^{\mathscr{A}}F \simeq H^X$.

a) Show that an object X of $\mathscr{C}$ is an $\mathscr{A}$-coobject if and only if for each $Y \in \mathscr{C}$ the set $\mathscr{C}(X, Y)$ is an $\mathscr{A}$-algebra and, for each morphism $f: Y \to Y'$ of $\mathscr{C}$, the map $H^X(f): \mathscr{C}(X, Y) \to \mathscr{C}(X, Y')$ is a morphism of $\mathscr{A}$-algebras.

b) Let $\mathscr{C}_{\mathscr{A}}$ denote the full subcategory of $\mathscr{C}$ whose objects are $\mathscr{A}$-coobjects. Show that if $\mathscr{C}$ is a category with finite coproducts, then $\mathscr{C}_{\mathscr{A}}$ is canonically equivalent to $[\mathscr{A}^0, \mathscr{C}]^a$.

c) A functor $F: \mathscr{C} \to \mathscr{A}^a$ is called *$\mathscr{A}$-representable* if there is an $\mathscr{A}$-coobject X of $\mathscr{C}$ such that $U^{\mathscr{A}}F \simeq H^X$. Show that if $\mathscr{C}$ is a complete and cocomplete category, then F is $\mathscr{A}$-representable if and only if it has an adjoint. (Hint: Note that a functor $F: \mathscr{C} \to \mathscr{A}^a$ is $\mathscr{A}$-representable if and only if $U^{\mathscr{A}}F$ is representable.)

4.2. Let $\mathscr{A}$ be an algebraic theory and $f \in \mathscr{A}(\mathbf{n}, \mathbf{1})$. Show that f belongs to the center of $\mathscr{A}$ if and only if for every $\mathscr{A}$-algebra X, the map

$$X(f): X_n \to X_1$$

defines a morphism $\bar{f}\colon X^n \to X$ of $\mathscr{A}$-algebras. In particular, if $\mathscr{A}$ is a commutative theory, then every object of $\mathscr{A}^a$ is an $\mathscr{A}$-coobject.

4.3. An operation $f \in \mathscr{A}(\mathbf{n}, \mathbf{1})$ is called *associative* if for each integer $k \geqslant 1$ and for each sequence $k+1, \ldots, k+n$, the diagram

$$\begin{array}{ccc} \mathbf{2n-1} & \xrightarrow{p_1^{2n+1} \amalg \cdots \amalg p_k^{2n+1} \amalg (f[p_{k+1}^{2n+1}, \ldots, p_{k+n}^{2n+1}]) \amalg p_{k+n+1}^{2n+1} \amalg \cdots \amalg p_{2n-1}^{2n+1}} & \mathbf{n} \\ \downarrow & & \downarrow \\ \mathbf{n} & \xrightarrow{p_1^{2n+1} \amalg \cdots \amalg p_{k+1}^{2n+1} \amalg (f[p_{k+2}^{2n+1}, \ldots, p_{k+n+1}^{2n+1}]) \amalg p_{k+n+2}^{2n+1} \amalg \cdots \amalg p_{2n-1}^{2n+1}} & \mathbf{p} \end{array}$$

is commutative.

a) Show that every n-ary operation which is simultaneously associative and commutative commutes with itself.

b) Show that every object of $\mathscr{A}b$ is a cogroup.

4.4. Let X and Y be two objects of a category $\mathscr{C}$ such that X s an $\mathscr{A}$-object and Y is a $\mathscr{B}$-coobject. Show that the set $\mathscr{C}(Y, X)$ is canonically a $(\mathscr{B} \otimes \mathscr{A})$ object of $\mathscr{S}et$ and, in particular, that if m is a binary operation of $\mathscr{A}$ and m' a binary operation of $\mathscr{B}$, then they induce on $\mathscr{C}(Y, X)$ the same binary law of composition which is associative and commutative (see Proposition 2.10).

4.5. Prove that every cogroup in $\mathscr{G}r$ is free (see Hilton [4]).

4.6. Let $G\colon \mathscr{R}g \to \mathscr{G}r$ be the functor that associates with each ring A, its group of units. Show that this functor has a coadjoint P. Hence P is of the form $P = T \otimes_{\mathscr{A}} ?$ where $\mathscr{A}$ is the algebraic theory of groups and T a cogroup-object in $\mathscr{R}g$. What is this cogroup-object? (Hint: A coadjoint of G is the functor P that associates with each group X its *group-algebra*, i.e. the ring whose elements are formal sums $\Sigma_i n_i x_i$, where $n_i \in \mathbf{Z}$ and $x_i \in X$, and the multiplication is given by $(nx)(n'x') = (nn')(xx')$ and extended by linearity. To determine T use Theorem 4.3. a.)

4.7. Let $F\colon \mathscr{A} \to \mathscr{B}$ be a theory morphism. If $M\colon \mathscr{B} \to \mathscr{B}^a$ is the canonical cofunctor, then MF is an $(\mathscr{A}^0, \mathscr{B})$-bialgebra.

a) Show that we have the following canonical functorial isomorphism:

$$MF \otimes_{\mathscr{A}} ? \simeq F^a, \quad \mathscr{B}^a(MF, ?) \simeq F_a.$$

b) Let T be an $(\mathscr{A}^0, \mathscr{B})$-bialgebra. Show that the following are equivalent:

i) There exists a theory morphism $F\colon \mathscr{A} \to \mathscr{B}$ such that $T \simeq MF$.

ii) If $U\colon [\mathscr{A}^0, \mathscr{B}^a]^a \to B^a$ is the underlying functor then $U(T) \simeq M_1^{\mathscr{B}}$.

(Hint: Use Theorem 3.11.) We remark that part b) of this exercise is equivalent to Theorem 3.11.

4.8. Let $F\colon \mathscr{A} \to \mathscr{B}$ be a theory morphism. Show that the following are equivalent:

a) F is an epimorphism in the category $\mathscr{T}h$.

b) The functor $F_a\colon \mathscr{B}^a \to \mathscr{A}^a$ is full and faithful.

c) If $\overline{F}: M_1^{\mathscr{A}} \to F_a(M_1^{\mathscr{B}})$ is the canonical morphism of $\mathscr{A}$-algebras defined by F, then the morphism of $\mathscr{B}$-algebras

$$MF \otimes_{\mathscr{A}} M_1^{\mathscr{A}} \xrightarrow{MF \otimes \overline{F}} MF \otimes_{\mathscr{A}} F_a(M_1^{\mathscr{B}})$$

is an isomorphism.

4.9. Let T be an $(\mathscr{A}^0, \mathscr{B})$-bialgebra. Show that the functor $\mathscr{B}^a(T, ?): \mathscr{B}^a \to \mathscr{A}^a$ commutes with monofiltered colimits if and only if T is a finitely generated $\mathscr{B}$-algebra.

4.10. Let $\mathscr{A}$ and $\mathscr{B}$ be algebraic theories. Show that $[\mathscr{A}^0, \mathscr{B}^a]^a$ is a complete and cocomplete category.

4.11. A theory morphism $F: \mathscr{A} \to \mathscr{B}$ is called *good* if the restriction functor F_a has a coadjoint. Show that F is a good morphism if and only if F_a commutes with coproducts.

4.12. Two algebraic theories $\mathscr{A}$ and $\mathscr{B}$ are called *Morita equivalent* if the categories $\mathscr{A}^a$ and $\mathscr{B}^a$ are equivalent.

a) Show that there are nonisomorphic algebraic theories that are Morita equivalent.

b) Show that Morita equivalent theories have isomorphic centers.

3.5. Characterization of algebraic categories

In this section we give some necessary and sufficient conditions for a category $\mathscr{C}$ to be equivalent to an algebraic category. These results are essentially due to Isbell [5], Lawvere [1], and Mal'cev [3].

References: Bourbaki [1], Gabriel—Ulmer [1], Isbell [5], Lawvere [1], Linton [1], [2], [3], MacLane [3], Mal'cev [2], [3], Schubert [1], Calenko—Šulgeifer [1], Ulmer [3], Wraith [1].

Let $S: \mathscr{C} \to \mathscr{S}et$ be a functor. For each natural number n, let $S^n: \mathscr{C} \to \mathscr{S}et$ denote the functor defined by:

$$S^n(X) = S(X)^n, \quad X \in \mathscr{C}, \quad \text{and}$$
$$S^n(f) = S(f)^n, \quad f \in \mathscr{C}.$$

In particular, S^0 is the constant functor associated with a set consisting of one point.

The functor S is called *tractable* if for each pair of natural numbers n, m, the family $[S^n, S^m]$ of all functorial morphisms from S^n into S^m is a set. Since S^n is in fact the product functor of n-copies of S (see Section 1.10, Ch. 1), S is tractable if and only if for each natural number n, the family $[S^n, S]$ is a set.

The proof of the following result is easy and is left for the reader.

PROPOSITION 5.1. *Let X be an object of a category $\mathscr{C}$ such that $X^{(n)}$ is defined for each $n \in \mathbf{N}$. Then the functor $H^X: \mathscr{C} \to \mathscr{S}et$ is tractable. In particular, every functor $S: \mathscr{C} \to \mathscr{S}et$ that has an adjoint is tractable.* ⁂

Let $S: \mathscr{C} \to \mathscr{S}et$ be a tractable functor. Let $\mathscr{A}_S$ denote the algebraic theory defined by

$$\mathscr{A}_S(\mathbf{n}, \mathbf{m}) = [S^n, S^m],$$

where the composition of morphisms is the composition of functorial morphisms. $\mathscr{A}_S$ is called the *algebraic theory associated with S*. An $\mathscr{A}_S$-algebra is also called an *S-algebra*. It is clear that an S-algebra X is in fact defined by a set, also denoted by X, and a family of functions

$$X(f): X^n \to X^m,$$

one for each functorial morphism $f: S^n \to S^m$, all subjected to the following two requirements:

a) if $p_i^n: S^n \to S$, $1 \leqslant i \leqslant n$, are the structural projections, then $X(p_i^n): X^n \to X$ are also the structural projections, and

b) if $f_i: S^n \to S$, $1 \leqslant i \leqslant m$, and $g: S^m \to S$ are functorial morphisms, then, for all $x \in X^n$,

$$X(g[f_1, \ldots, f_m])(x) = X(g)(X(f_1)(x), \ldots, X(f_m)(x)).$$

If X is an object of $\mathscr{C}$, then the set $S(X)$ determines an S-algebra, denoted $\tilde{S}(X)$, as follows:

$$\tilde{S}(X)_1 = S(X)$$

and

$$\tilde{S}(X)(f) = f_X \quad \text{for all } f \in [S^n, S^m].$$

If $h: X \to Y$ is a morphism of $\mathscr{C}$, then the map $S(h)$ defines a morphism of S-algebras, denoted $\tilde{S}(h): \tilde{S}(X) \to \tilde{S}(Y)$.

In this way the assignments

$$X \rightsquigarrow \tilde{S}(X) \quad \text{and} \quad f \rightsquigarrow \tilde{S}(f)$$

define a functor

$$\tilde{S}: \mathscr{C} \to \mathscr{A}_S^a.$$

This is called the *Mal'cev functor associated with S*.

The proof of the following result is easy and is left for the reader.

PROPOSITION 5.2. *Let $S: \mathscr{C} \to \mathscr{S}et$ be a tractable functor. Then:*

a) *If $U: \mathscr{A}_S^a \to \mathscr{S}et$ is the underlying functor, then $U\tilde{S} \simeq S$.*

b) *$\tilde{S}$ is faithful if and only if S is faithful.*

c) *$\tilde{S}$ commutes with filtered (monofiltered) colimits if and only if S commutes with filtered (monofiltered) colimits.* ⁂

If P is an object of $\mathscr{C}$, then the algebraic theory $\mathscr{A}_H^P$ associated with the functor H^P will be denoted by $\mathscr{A}_P$ and called the *algebraic theory associated with the object P*.

LEMMA 5.3. *Let P be an object of $\mathscr{C}$. Assume that the coproduct $P^{(n)}$ is defined for each natural number n. If $\tilde{H}^P: \mathscr{C} \to \mathscr{A}^a_P$ is the Mal'cev functor associated with H^P, then:*

a) *For each natural number n, we have the following isomorphism of $\mathscr{A}_P$-algebras:*

$$\tilde{H}^P(P^{(n)}) \simeq M_1^{(n)}, \tag{6}$$

where M_1 is the fundamental $\mathscr{A}_P$-algebra.

b) *For each pair of natural numbers (n, m) the canonical map*

$$\tilde{H}^P(P^{(n)}, P^{(m)}): \mathscr{C}(P^{(n)}, P^{(m)}) \to \mathscr{A}^a_P(\tilde{H}^P(P^{(n)}), \tilde{H}^P(P^{(m)})) \tag{7}$$

is a bijection.

Proof. a) For each natural number n, we have a canonical functorial isomorphism $(H^P)^n \simeq H^{P(n)}$ such that $[(H^P)^n, H^P] \simeq [H^{P(n)}, H^P] \simeq \mathscr{C}(P, P^{(n)})$. To obtain the isomorphism (6) we observe that we also have the following functorial isomorphisms. For each pair of natural numbers n and m,

$$\tilde{H}^P(P^{(n)}) = (H^P(P^{(n)}))^m \simeq H^{P^{(m)}}(P^{(n)}) \simeq [H^{P^{(n)}}, HP^{(m)}] = (M_1^{(n)})_m.$$

We leave it for the reader to show that these isomorphisms (of sets) determine the required isomorphism of $\mathscr{A}_P$-algebras.

b) First we observe that it is clear that the map (7) is an injection. Now, for the remainder, let $f: \tilde{H}^P(P^{(n)}) \to \tilde{H}^P(P^{(m)})$ be a morphism of $\mathscr{A}_P$-algebras. Then in particular we have that

$$f_1: \tilde{H}^P(P^{(n)})_1 = \mathscr{C}(P, P^{(n)}) \to \mathscr{C}(P, P^{(m)}) = \tilde{H}^P(P^{(m)})_1.$$

Let $g: P^{(n)} \to P^{(m)}$ denote the unique morphism such that $gu_i = f_1(u_i)$, where the $u_i: P \to P^{(n)}$, $i = 1, \ldots, n$ are the structural morphisms. It is left for the reader to show that $H^P(g) = f$. ∸

Note 5.4. a) Let P be an object of $\mathscr{C}$ such that $P^{(n)}$ is not necessarily defined for every natural number n. However, it is easy to see that $\tilde{H}^P(P)$ is canonically isomorphic to the fundamental $\mathscr{A}_P$-algebra.

b) The object $\tilde{H}^P(P^{(n)})$ will be denoted by $\tilde{P}^{(n)}$.

COROLLARY 5.5. *Let $S: \mathscr{C} \to \mathscr{S}et$ be a functor and let F be adjoint to S. If S commutes with monofiltered colimits, then $\tilde{S}F$ is adjoint to the underlying functor $U: \mathscr{A}^a_S \to \mathscr{S}et$.*

The proof follows from Proposition 5.2. c and Lemma 5.3.a, since every set is the colimit of the monofiltered system of its finite subsets. ⋇

Now we are interested in seeing under what conditions a category $\mathscr{C}$ is equivalent to an algebraic category. The question reduces to the study of the Mal'cev functor associated with a given functor. We first have the following result.

THEOREM 5.6. *Let $\mathscr{C}$ be a category with cokernels and P an object of $\mathscr{C}$. Assume that for each set $\mathscr{I}$ the object $P^{(\mathscr{I})}$ is defined in $\mathscr{C}$. Then the Mal'cev functor*

$$\tilde{H}^P: \mathscr{C} \to \mathscr{A}^a_P$$

has an adjoint (usually denoted $P \otimes ?$).

Proof. The proof can be obtained by use of Theorem 4.3; however, for the sake of greater clarity we give a direct proof.

Let X be a free $\mathscr{A}_P$-algebra; since $\tilde{H}^P(P) = \tilde{P}$ is the fundamental $\mathscr{A}_P$-algebra (see Lemma 5.3), there is a set $\mathscr{I}$ such that $X = \tilde{P}^{(\mathscr{I})}$. Let us denote $P \otimes X$ by $P^{(\mathscr{I})}$.

Now, let $\mathscr{I}$ and $\mathscr{J}$ be two sets and let $f\colon \tilde{P}^{(\mathscr{I})} \to \tilde{P}^{(\mathscr{J})}$ be a morphism of $\mathscr{A}_P$-algebras. We define a morphism

$$P \otimes f\colon P^{(\mathscr{I})} \to P^{(\mathscr{J})}$$

in the following way:

For each $i \in \mathscr{I}$ ($j \in \mathscr{J}$), let $u_i : \tilde{P} \to \tilde{P}^{(\mathscr{I})}$ and $u_i' : P \to P^{(\mathscr{I})}$ ($v_j\colon \tilde{P} \to \tilde{P}^{(\mathscr{J})}$ and $v_j'\colon P \to P^{(\mathscr{J})}$) denote the corresponding structural morphisms. Furthermore, if $\mathscr{F}$ is a finite subset of $\mathscr{J}$, let $v_F\colon \tilde{P}^{(\mathscr{F})} \to \tilde{P}^{(\mathscr{J})}$ ($v_F'\colon P^{(\mathscr{F})} \to P^{(\mathscr{J})}$) denote the unique morphism such that $v_F \bar{v}_j = v_j$ ($v_F' \bar{v}_j' = v_j'$) where $\bar{v}_j\colon P \to P^{(\mathscr{F})}$ ($\bar{v}_j'\colon P \to P^{(\mathscr{F})}$), $j \in \mathscr{F}$, are also the structural morphisms.

Furthermore, since P is a finitely generated $\mathscr{A}_P$-algebra (see Exercise 3.12), for each $i \in \mathscr{I}$ there is a finite subset $\mathscr{F}_i$ of $\mathscr{J}$ such that fu_i can be factored as follows:

$$\tilde{P} \xrightarrow{g_i} \tilde{P}^{(\mathscr{F}_i)} \xrightarrow{v_{F_i}} \tilde{P}^{(\mathscr{J})}.$$

By Lemma 5.3. b, there is a unique morphism $g_i'\colon P \to P^{(\mathscr{F}_i)}$ such that $\tilde{H}^P(g_i') = g_i$. Define $P \otimes f$ to be the unique morphism such that, for each $i \in \mathscr{I}$, $(P \otimes f)u_i$ is equal to the composition:

$$P \xrightarrow{g_i'} P^{(\mathscr{F}_i)} \xrightarrow{v'_{\mathscr{F}_i}} P^{(\mathscr{J})}.$$

We must show that $P \otimes f$ does not depend on the subset $\mathscr{F}_i$ of $\mathscr{J}$. To this end, we shall prove that the following diagram of $\mathscr{A}_P^a$ is commutative:

$$\begin{array}{ccc}
\tilde{P}^{(\mathscr{I})} & \xrightarrow{\quad f \quad} & \tilde{P}^{(\mathscr{J})} \\
{\scriptstyle u}\downarrow & & \downarrow{\scriptstyle v} \\
\tilde{H}^P(P^{(\mathscr{I})}) & \xrightarrow{\tilde{H}^P(P \otimes f) = \tilde{f}} & \tilde{H}^P(P^{(\mathscr{J})})
\end{array}$$

(where $uu_i = \tilde{H}^P(u_i')$ for every $i \in \mathscr{I}$, and $vv_j = \tilde{H}^P(v_j')$ for every $j \in \mathscr{J}$). Indeed, for every $i \in \mathscr{I}$,

$$\tilde{f}uu_i = \tilde{f}\tilde{H}^P(u_i') = \tilde{H}^P((P \otimes f)u_i') = \tilde{H}^P(v'_{\mathscr{F}_i})\tilde{H}^P(g_i') = \tilde{H}^P(v'_{\mathscr{F}_i})g_i.$$

Since it is easy to see that $vv_{F_i} = \tilde{H}^P(v'_{\mathscr{F}_i})$, we have:

$$\tilde{f}uu_i = vv_{\mathscr{F}_i}g_i = vfu_i \quad \text{for each } i \in \mathscr{I},$$

so that $\tilde{f}u = vf$. Hence, if for each $i \in \mathscr{I}$ we choose another subset $\mathscr{F}_i'$ of $\mathscr{J}$ and a morphism $\bar{g}_i\colon \tilde{P} \to \tilde{P}^{(\mathscr{F}_i')}$ such that $fu_i = v_{\mathscr{F}_i'}\bar{g}_i$, and as above define a morphism $f'\colon P^{(\mathscr{I})} \to P^{(\mathscr{J})}$ such that $f'u_i' = v'_{\mathscr{F}_i'}\bar{g}_i'$ (where $\tilde{H}^P(\bar{g}_i') = \bar{g}_i$), then we necessarily have that $\tilde{H}^P(f')u = vf$. But then for each i,

$$\tilde{f}uu_i = \tilde{f}\tilde{H}^P(u_i') = \tilde{H}^P((P \otimes f)u_i') = \tilde{H}^P(f'u') = \tilde{H}^P(f')uu_i$$

or, equivalently,

$$(P \otimes f)u_i' = f'u_i', \quad i \in \mathscr{I}.$$

Finally $P \otimes f = f'$. Hence $P \otimes ?$ has been defined for all free algebras and all morphisms between them.

Finally, if X is any object of $\mathscr{A}_P^q$, then there is a coequalizer diagram (see Corollary 3.10):

$$\tilde{P}^{(\mathscr{I})} \underset{g}{\overset{f}{\rightrightarrows}} \tilde{P}^{(\mathscr{I})} \xrightarrow{p} X.$$

Then in $\mathscr{C}$ we may define the diagram:

$$P^{(\mathscr{I})} \underset{P \otimes g}{\overset{P \otimes f}{\rightrightarrows}} P^{(\mathscr{I})} \xrightarrow{q} P \otimes X$$

where q is a coequalizer of the morphisms $(P \otimes f, P \otimes g)$ defined as above. $P \otimes X$ is thus defined. It is left for the reader to show that the functor $P \otimes ?$ is well defined and that it is adjoint to $\tilde{H}^P$. ✳

Let $\mathscr{C}$ be a category, X an object of $\mathscr{C}$ and $\{X_i\}_{i \in \mathscr{I}}$ a family of subobjects of X. This family is called a *filtered family of subobjects* if, for each $i, j \in \mathscr{I}$, there is a $k \in \mathscr{I}$ such that $X_i \subseteq X_k$ and $X_j \subseteq X_k$. For example, the finitely generated subalgebras of an $\mathscr{A}$-algebra give a filtered family of subobjects (see Exercise 3.19.c)).

It is clear that a filtered family of subobjects of an object canonically gives a monofiltered system of objects. With respect to a filtered family of subobjects of an object, we introduce the following so-called *Grothendieck condition:*

(G.C.) *A filtered family $\{X_i\}_i$ of subobjects of an object X satisfies the Grothendieck condition* if a) the monofiltered system $\{X_i\}_i$ has a colimit, and b) the canonical morphism u: $\varinjlim X_i \to X$ defined so that $uv_i = u_i$ for all i, is a monomorphism (where the u_i: $X_i \to X$ are the canonical inclusions and the v_i: $X_i \to \varinjlim X_i$ are the structural morphisms).

Note 5.7. If a filtered family $\{X_i\}_i$ of subobjects of an object X satisfies the (G.C.)-condition, then $\cup_i X_i$ is defined and is equal to $\varinjlim X_i$. However, a simple example shows that the converse is not true.

We shall say that a category $\mathscr{C}$ is a *(G.C.)-category* if for each object X of $\mathscr{C}$, and for each filtered family $\{X_i\}$ of subobjects of X, $\{X_i\}_{i \in \mathscr{I}}$ satisfies the (G.C.)-condition.

It is easy to see that every algebraic category and every topos is a (G.C.)-category.

THEOREM 5.8. *Let $\mathscr{C}$ be a category such that:*

a) $\mathscr{C}$ *is a* (G.C.)-*category.*

b) *There is an object P of $\mathscr{C}$ such that P is a generator and a separator, and such that for each $\mathscr{I}$, the coproduct $P^{(\mathscr{I})}$ is defined.*

c) *Every morphism of $\mathscr{C}$ has a coequalizer decomposition. Then the Mal'cev functor*

$$\tilde{H}^P: \mathscr{C} \to \mathscr{A}^a_P$$

is full and faithful.

If, moreover, $\mathscr{C}$ is a category with cokernels, then it is a complete and cocomplete category.

Proof. For an object X of $\mathscr{C}$, for the sake of simplification let $\mathscr{C}(P, X)$ be denoted by $I(X)$ and let

$$p_X: P^{(I(X))} \to X$$

denote the unique morphism such that for each $f \in I(X)$, $p_X u_f = f$ (where $u_f: P \to P^{(I(X))}$ is the structural morphism associated with f). Since P is a separator of $\mathscr{C}$, p_X is an epimorphism and, since P is a generator, it is in fact a coequalizer. Indeed, if $P^{(I(X))} \xrightarrow{q} X' \xrightarrow{v} X$ is a coequalizer decomposition of p_X, then v must be an isomorphism. Otherwise there is a morphism $f: P \to X$ that does not factor through X', which is a contradiction.

Now, for every finite subset $F = \{f_1, \ldots, f_n\}$ of $I(X)$, let

$$u_F \text{ denote } \coprod_i u_{f_i}: P^{(F)} \to P^{(I(X))}$$

and let

$$f_F \text{ denote } \langle f_1, \ldots, f_n \rangle: P^{(F)} \to X.$$

It is clear that $p_X u_F = f_F$. Now let

$$P^{(F)} \xrightarrow{p_F} X_F \xrightarrow{v_F} X$$

be a coequalizer decomposition of f_F. It is left for the reader to show that the family of subobjects $\{X_F\}_F$ of X (where F runs through all finite subsets of $I(X)$), defines a filtered family of subobjects. In fact, $X_F \subseteq X_{F'}$, whenever $F \subseteq F'$.

Furthermore, since $\mathscr{C}$ is (G.C.)-category, the canonical morphism

$$u: \varinjlim X_F \to X$$

is a monomorphism. We claim that this morphism is in fact an isomorphism. For otherwise there would exist a morphism $f: P \to X$ which would not factor through u. This would be a contradiction, since $\{f\}$ is a finite subset of $I(X)$. We keep the notation used above and now show that the functor

$$\tilde{H}^P: \mathscr{C} \to \mathscr{A}^a_P$$

is full and faithful. $H^P: \mathscr{C} \to \mathscr{S}et$ is faithful since P is a separator, so it will suffice to show that $\tilde{H}^P$ is full. For this purpose, let X and Y be any two objects of $\mathscr{C}$ and let $s: \tilde{H}^P(X) \to \tilde{H}^P(Y)$ be a morphism of $\mathscr{A}^a$-algebras. By the definition of $\tilde{H}^P$, we have that s is defined by a map

$$s: \tilde{H}^P(X)_1 = \mathscr{C}(P, X) \to \mathscr{C}(P, Y) = \tilde{H}^P(Y)_1$$

such that for each functorial morphism

$$t' : (H^P)^n \to H^P$$

we have the following commutative diagram:

$$\begin{array}{ccc} (H^P)^n(X) & \xrightarrow{s^n} & (H^P)^n(Y) \\ {\scriptstyle t'_X}\downarrow & & \downarrow{\scriptstyle t'_Y} \\ H^P(X) & \xrightarrow{s} & H^P(Y). \end{array} \tag{8}$$

It is clear that t' is determined by a morphism $t: P \to P^{(n)}$ and $t'_X = H_X(t)$.

It follows from the commutativity of diagram (8) that, for every $\{f_1, \ldots, f_n\} \in \in (H^P)^n(X) = \mathscr{C}(P, X)^n$,

$$s(\langle f_1, \ldots, f_n \rangle t) = \langle s(f_1), \ldots, s(f_n) \rangle t. \tag{9}$$

Suppose that $F_1 = \{f_1, \ldots, f_n\}$ is a finite subset of $H^P(X) = I(X)$. Using the above notation, we have the diagram:

$$\begin{array}{ccccc} & P^{(F)} & \xrightarrow{p_F} & X_F & \xrightarrow{v_F} & X \\ g_F = \langle s(f_1), \ldots, s(f_n) \rangle & \downarrow & & & & \\ & Y & & & & \end{array}$$

where $v_F p_F = \langle f_1, \ldots, f_n \rangle = f_F$. We claim that there is a morphism $s_F: X_F \to Y$ such that $s_F p_F = g_F$. To get that, let $h, k: Z \to P^{(F)}$ be a pair of morphisms, whose equalizer is p_F. The existence of s_F will be shown once we have proven that $g_F h = = g_F k$. To this end, let $r: P \to Z$ be a morphism. Then $f_F hr = f_F kr$, so that by (9),

$$\begin{aligned} g_F hr &= \langle s(f_1), \ldots, s(f_n) \rangle hr = s(\langle f_1, \ldots, f_n \rangle hr) \\ &= s(\langle f_1, \ldots, f_n \rangle kr) = \langle s(f_1), \ldots, s(f_n) \rangle kr = g_F kr. \end{aligned}$$

Hence, since P is a separator of $\mathscr{C}$, we get that $g_F h = g_F k$, as required.

We leave it for the reader to show that the morphisms $\{s_F\}_F$ define a morphism from the monofiltered system $\{X_F\}_F$ to Y. Then there is a unique morphism $s': X \to Y$ such that $s' v_F = s_F$. It is now clear that $\tilde{H}^P(s') = s$. Hence $\tilde{H}^P$ is full. Finally, suppose that $\mathscr{C}$ is a category with coequalizers; then by Theorem 5.6, $\tilde{H}^P$ has an adjoint. Hence $\mathscr{C}$ is a full reflective subcategory of $\mathscr{A}^a_P$ and thus it is complete and cocomplete by Proposition 13.12, Ch. 1. ✳

THEOREM 5.9. *A category $\mathscr{C}$ is equivalent to an algebraic category if and only if the following three conditions are satisfied:*

a) *There is a functor $S: \mathscr{C} \to \mathscr{S}et$ that commutes with monofiltered colimits and has an adjoint F.*

b) *$\mathscr{C}$ is a category with coequalizers and a morphism f in $\mathscr{C}$ is a coequalizer if and only if $S(f)$ is a surjection.*

c) *A double arrow (u, v) of $\mathscr{C}$ is a kernel pair if and only if $(S(u), S(v))$ is a kernel pair of $\mathscr{S}et$.*

Proof. If $\mathscr{C}$ is equivalent to an algebraic category, then the conditions a) — c) are satisfied, by Corollaries 3.5 and 3.10.

Conversely, assume that conditions a) — c) are satisfied. In this case, we shall prove that the Mal'cev functor

$$\tilde{S}: \mathscr{C} \to \mathscr{A}_S^a$$

associated with S is an equivalence of categories.

We divide the proof into several steps:

i) The functor S is representable — precisely, it is isomorphic to H^P, where $P = F(*)$ ($*$ being a one-element set). We shall assume that $S = H^P$.

Since every set M is the coproduct of its one-element subsets, then $F(M) = P^{(M)}$. Hence the coproduct of every set of copies of P is defined in $\mathscr{C}$.

ii) The object P is a strong separator of $\mathscr{C}$ and the functor S reflects monomorphisms.

By b), S reflects epimorphisms, so that by Proposition 13,7, Ch 1, it is a faithful functor, or equivalently P is a separator of $\mathscr{C}$.

Furthermore, let $f: X \to Y$ be a morphism of $\mathscr{C}$ such that $S(f)$ is a monomorphism. If $u, v: Z \to X$ satisfy $fu = fv$, then $S(f)S(u) = S(f)S(v)$, i.e. $S(u) = S(v)$ since $S(f)$ is a monomorphism. Hence $u = v$ since S is faithful.

Now, if $S(f)$ is an isomorphism, then f is a monomorphism and a coequalizer, i.e. an isomorphism. Hence P is a strong separator of $\mathscr{C}$. We also observe that P is a generator of $\mathscr{C}$.

iii) $\mathscr{C}$ is a cocomplete category.

By b), $\mathscr{C}$ is a category with cokernels, so that by Theorem 9.2, Ch. 1 it will suffice to show that $\mathscr{C}$ is a category with coproducts.

An object of $\mathscr{C}$ of the form $P^{(M)}$ is called *P-free*.

By i) every set of free objects of $\mathscr{C}$ has a coproduct. We assert that for every object X of $\mathscr{C}$, there is a coequalizer diagram:

$$L_1 \rightrightarrows L_0 \longrightarrow X \tag{10}$$

where L_0 and L_1 are free objects.

Let

$$P^{(S(X))} \xrightarrow{\ p\ } X$$

denote the unique morphism such that for each $f \in S(X) = \mathscr{C}(P, X)$, $pu_f = f$ (where $u_f: P \to P^{(S(X))}$ is the structural injection associated with f). It is easy to see that $S(p)$ is a surjection, so that by b), p is a coequalizer.

Furthermore, if p is the coequalizer of a pair $(u, v)\colon Y \rightrightarrows P^{(S(X))}$ and $q\colon L \to Y$ is an epimorphism, then p is also a coequalizer of (uq, vq). Hence the existence of a diagram of the form (10) is now proven.

Now let $\{X_i\}_i$ be a set of objects of $\mathscr{C}$. For each i we choose a coequalizer diagram:

$$L_1^i \underset{g_i}{\overset{f_i}{\rightrightarrows}} L_0^i \xrightarrow{p_i} X_i$$

where L_1^i and L_0^i are free. Let L_1 denote $\coprod_i L_1^i$, L_a denote $\coprod_i L_0^i$, f denote $\coprod_i f_i$ and g denote $\coprod_i g_i$. Consider the following coequalizer diagram:

$$L_1 \underset{g}{\overset{f}{\rightrightarrows}} L_0 \xrightarrow{p} X.$$

For each i let $u_i\colon X_i \to X$ denote the unique morphism such that $u_i p_i = p v_i$ (where the $v_i\colon L_0^i \to L_0$ are the structural injections). We leave it for the reader to show that X together with the morphisms u_i give a coproduct of the set of objects $\{X_i\}_i$.

iv) Every morphism of $\mathscr{C}$ has a coequalizer decomposition. Let $f\colon X \to Y$ be a morphism in $\mathscr{C}$. Let R be the subset of $S(X) \coprod S(X)$ that consists of all pairs (x, y) such that $S(f)(x) = S(f)(y)$. Consider the coproduct $P^{(R)}$ and, for each $(x, y) \in R$, let $u_{(x,y)}\colon P \to P^{(R)}$ denote the structural morphisms. Let $s\colon P^{(R)} \to X$ ($t\colon P^{(R)} \to X$) denote the unique morphism such that for each $(x, y) \in R$, $su_{(x,y)} = x$ ($tu_{(x,y)} = y$). It is easy to see that $fs = ft$. Now let $p\colon X \to Z$ be a coequalizer of (s, t) and let $h\colon Z \to Y$ be the unique morphism such that $hp = f$. It is easy to see that $S(h)$ is a monomorphism, so that by ii) h is also a monomorphism. Hence $X \overset{p}{\to} Z \overset{h}{\to} Y$ gives a coequalizer decomposition of f, as required.

v) $\mathscr{C}$ is a (G.C.)-category.

Let $\{X_i\}_i$ be a filtered family of subobjects of an object X, and let

$$u\colon \varinjlim X_i \to X$$

be the canonical morphism such that $uv_i = u_i$ for every i (where $u_i\colon X_i \to X$ is the natural inclusion and $v_i\colon X_i \to \varinjlim X_i$ is the structural morphism). Since S commutes with monofiltered colimits, we see that $S(u)$ is a monomorphism. Hence, by step ii), u is a monomorphism, i.e. $\mathscr{C}$ is a (G.C.)-category.

We conclude by observing that steps i), ii), iv) and v) yield the hypothesis of Theorem 5.8. Hence the Mal'cev functor

$$\tilde{S}\colon \mathscr{C} \to \mathscr{A}_S^a$$

is full and faithful. Then, by Theorem 4.1, Ch. 1, to complete the proof of Theorem 5.9 it will suffice to show that the functor $\tilde{S}$ is representative, i.e. for each object X of $\mathcal{A}_S^q$, there exists an object X' of $\mathscr{C}$ such that $\tilde{S}(X') \tilde{\rightarrow} X$. To that end, we prove the following claim.

vi) If $X \in \mathscr{C}$ and if Y is a subobject of $\tilde{S}(X)$ in $\mathcal{A}_S^q$, then there is a subobject X' of X in $\mathscr{C}$ such that $\tilde{S}(X') = Y$.

Let $X \in \mathscr{C}$ and let $u: Y \rightarrow \tilde{S}(X)$ be a monomorphism of $\mathcal{A}_S^q$. There is an object Z of $\mathscr{C}$ and a coequalizer $p: \tilde{S}(Z) \rightarrow Y$ (for example, $\tilde{S}(Z)$ may be a free $\mathcal{A}_S$-algebra). Since $\tilde{S}$ is full, there is a morphism $g: Z \rightarrow Y$ such that $S(g) = up$. Let

$$Z \xrightarrow{q} Y' \xrightarrow{v} X$$

be a coequalizer decomposition of g. Then $\tilde{S}(v)\tilde{S}(q)$ is a coequalizer decomposition of $\tilde{S}(g)$ so that there is an isomorphism $t: \tilde{S}(Y') \rightarrow Y$ such that $ut = \tilde{S}(v)$ (see Exercise 19.18 a, Ch. 1).

Finally, let X be an object of $\mathcal{A}_S^q$ and let K be an object of $\mathscr{C}$ such that there is a coequalizer $p: \tilde{S}(K) \rightarrow X$. Also let $(u, v): R \rightarrow \tilde{S}(K)$ be a kernel pair of p. Since R is in fact a subobject of $\tilde{S}(K) \coprod \tilde{S}(K)$ we may assume that $R = \tilde{S}(R')$ where $R' \in \mathscr{C}$. Let u', $v': R' \rightarrow K'$ be a pair of morphisms such that $\tilde{S}(u') = u$ and $\tilde{S}(v') = v$. By Corollary 3.10 and c) of the present theorem, (u', v') is a kernel pair in $\mathscr{C}$. Hence, if $q: K \rightarrow X$ is a coequalizer of (u', v') then $\tilde{S}(q)$ is a coequalizer of (u, v), i.e. $\tilde{S}(X')$ is canonically isomorphic to X. The proof of Theorem 5.9 is now complete. ✳

Note 5.10. Condition a) of Theorem 5.9 is equivalent to:

a') There exists a finitely generated object P of $\mathscr{C}$ such that $P^{(M)}$ is defined for every set M.

Exercises

5.1. Let $S: \mathscr{C} \rightarrow \mathcal{S}et$ be a functor. Assume that S commutes with monofiltered colimits and has an adjoint F. Show that the functor FS is dominated by all finite sets and, moreover, that if α is a regular cardinal number and S commutes with α-monofiltered colimits, then SF is dominated by all sets M such that Card $(M) < \alpha$.

5.2. Let $S: \mathscr{C} \rightarrow \mathcal{S}et$ be a tractable functor. For each regular cardinal number α, let $\mathcal{A}_S(\alpha)$ denote the α-algebraic theory defined by

$$\mathcal{A}_S(\alpha)(\beta, \gamma) = [S^\beta,\ S^\gamma], \qquad \beta, \gamma < \alpha.$$

a) Assume that for every $\beta < \alpha$, the object $P^{(\beta)}$ is defined in $\mathscr{C}$. Reformulate Lemma 5.3 in this context.

b) Also reformulate Theorem 5.6.

5.3. Formulate the (G.C.)-condition in terms of a regular cardinal number α, thus obtaining the so-called α-(G.C.)-condition. Reformulate Theorem 5.8 in this context.

5.4 Prove: A category $\mathscr{C}$ is equivalent to the category $\mathscr{A}^a$ where $\mathscr{A}$ is an α-algebraic theory if and only if all of the following two conditions are satisfied:

a) There is a functor $S: \mathscr{C} \to \mathscr{S}et$ that commutes with α-monofiltered colimits and has an adjoint F.

a') Conditions b) and c) of Theorem 5.9 are satisfied.

5.5. Prove: A category $\mathscr{C}$ is equivalent to the category $\mathscr{A}^a$, where $\mathscr{A}$ is an ∞-algebraic theory, if and only if all of the following two conditions are satisfied:

a) There is a functor $S: \mathscr{C} \to \mathscr{S}et$ having an adjoint F.

a') Conditions b) and c) of Theorem 5.9 are satisfied.

(Hint: The only difficult point is in showing that the corresponding Mal'cev functor $S: \tilde{\mathscr{C}} \to \mathscr{A}_S^a(\infty)$ is full. To do this, observe that for every pair of objects X, Y of $\mathscr{C}$, we can define the coequalizers $P^{(\mathscr{I})} \xrightarrow{u} X$, $P^{(\mathscr{I})} \xrightarrow{v} Y$, where $P = F(*)$. Now, if $f: \tilde{S}(X) \to \tilde{S}(Y)$ is a morphism of $A_S(\infty)$-algebras, then there will be a morphism $g: P^{(\mathscr{I})} \to P^{(\mathscr{I})}$ of $\mathscr{C}$ such that $f\tilde{S}(u) = \tilde{S}(vg)$. If $(h, k): Z \to P^{(\mathscr{I})}$ is a pair of morphisms such that u is a coequalizer of (h, k), then it will follow that $vgh = vgk$. Finally, a morphism $f': X \to Y$ can be found such that $\tilde{S}(f') = f$.)

5.6. Let $\mathscr{C}$ be the full subcategory of $\mathscr{T}op$ whose objects are compact Hausdorff spaces. Show that there is an ∞-algebraic theory $\mathscr{A}$ such that $\mathscr{C}$ is equivalent to $\mathscr{A}^a$. (Hint: Consider the underlying functor $S: \mathscr{C} \to \mathscr{S}et$ whose adjoint is provided by the Stone-Čech compactification of discrete spaces (see Schubert [2]). The other conditions are easily verified.)

5.7. Let $\mathscr{A}$ be an algebraic theory and M_n the free $\mathscr{A}$-algebra with n generators. Denote by $\mathscr{A}_n$, the algebraic theory associated with the object M_n, and let $\tilde{U}_n$: $\mathscr{A}^a \to \mathscr{A}_n^a$ denote the Mal'cev functor associated with the functor $U_n = H^{M_n}$: $\mathscr{A}^a \to \mathscr{S}et$ (in particular, $U_1 = U_{\mathscr{A}}$, the underlying functor).

a) Show that U_n is an equivalence of categories (i.e. $\mathscr{A}$ and $\mathscr{A}_n$ are Morita equivalent). ($\mathscr{A}_n$ is called the *matrix theory associated with* $\mathscr{A}$.)

b) Show that $Z(\mathscr{A}_n)$, the center of $\mathscr{A}_n$, is canonically isomorphic to $Z(\mathscr{A})$, the center of $\mathscr{A}$.

c) Let $\mathscr{A}$ and $\mathscr{B}$ be two algebraic theories. Define an $(\mathscr{A} \coprod \mathscr{B})$-algebra that is projective but not free.

5.8. Let X be an object of $\mathscr{A}^a$, and $\mathscr{A}_X$ the algebraic theory associated with X. Show that the Mal'cev functor $\tilde{H}^X$ $\mathscr{A}^a \to \mathscr{A}_X^a$ is an equivalence of categories if and only if X is a finitely generated separator and a projective generator of $\mathscr{A}^a$.

5.9. Prove: An algebraic theory $\mathscr{A}$ is Morita equivalent to the trivial algebraic theory $\mathscr{N}$ if and only if there is a finite set X such that $\mathscr{A}$ is isomorphic to $\mathscr{A}_X$ (the algebraic theory associated with the object X of $\mathscr{S}et$) or, equivalently, if and only if it is a matrix theory associated with $\mathscr{N}$.

5.10. Let $\mathscr{A}$ be an algebraic theory and let $\mathscr{I}$ be a finite category. Show that the category $[\mathscr{I}, \mathscr{A}^a]$ is an algebraic category.

5.11. We say that a category $\mathscr{C}$ is an *algebraic category in the sense of Gabriel* if there is a small category $\mathscr{A}$ and a set, Σ, of morphisms of $[\mathscr{A}, \mathscr{S}et]$ such that $\mathscr{C}$ is equivalent to $\tilde{\mathscr{A}}_\Sigma$ (the category of all Σ-sheaves over $\mathscr{A}$). Prove:

a) If α is a regular cardinal number and $\mathscr{A}$ is an α-algebraic theory, then $\mathscr{A}^a$ is an algebraic theory in the sense of Gabriel.

b) A category $\mathscr{C}$ is algebraic in the sense of Gabriel if and only if a) it is cocomplete and contains a full, small and dense subcategory $\mathscr{A}$, and furthermore b) there exists an infinite cardinal number α such that the objects of $\mathscr{A}$ are α-generated.

5.12. Let $T = (T, u, v)$ be a triple in a category $\mathscr{C}$. We define a *T-algebra* to be a pair (X, h), where $X \in \mathscr{C}$ (an underlying object of the algebra) and $h: T(X) \to X$ (called the *structure morphism* of the algebra), such that the diagrams

$$\begin{array}{ccc} T(T(X)) & \xrightarrow{T(h)} & T(X) \\ {\scriptstyle v_X}\downarrow & & \downarrow{\scriptstyle h} \\ T(X) & \xrightarrow{h} & X \end{array} \qquad \begin{array}{ccc} X & \xrightarrow{u_X} & T(X) \\ & {\scriptstyle 1_X}\searrow & \downarrow{\scriptstyle h} \\ & & X \end{array}$$

are commutative.

The first diagram depicts the associative law, and the second the unit law.

A morphism $f: (X, h) \to (X', h')$ of T-algebras is a morphism $f: X \to X'$ in $\mathscr{C}$ such that the diagram

$$\begin{array}{ccc} T(X) & \xrightarrow{h} & X \\ {\scriptstyle T(f)}\downarrow & & \downarrow{\scriptstyle f} \\ T(X') & \xrightarrow{h'} & X' \end{array}$$

is commutative.

Denote the category of all T-algebras by $\mathscr{C}^T$. Also, let $\tilde{T}: \mathscr{C} \to \mathscr{C}^T$ denote the functor defined by $\tilde{T}(X) = (T(X), v_X)$.

a) Show that the underlying functor $U: \mathscr{C}^T \to \mathscr{C}$ defined by $U(X, h) = X$ is adjoint to $\tilde{T}$ and that $U\tilde{T} = T$.

b) For triples (T, u, v) and (T', u', v') in $\mathscr{C}$, define a morphism t of triples as a suitable functorial morphism $t: T \to T'$. Show that t defines a functor $t_*: \mathscr{C}^{T'} \to \mathscr{C}^T$, such that $Ut_* = U'$, and t defines a functorial morphism $\tilde{T} \to t_*\tilde{T}'$.

c) Let $F: \mathscr{C} \to \mathscr{C}'$ be a functor, G be coadjoint to F, $u: \mathrm{Id}\,\mathscr{C} \to GF$ be an arrow of adjunction and v be a coarrow quasi-inverse of u. Let $T = (GF, u, GvF)$ be the associated triple. Show that there is a unique functor $K: \mathscr{C}' \to \mathscr{C}^T$ such that $UK = G$ and $KF = T$.

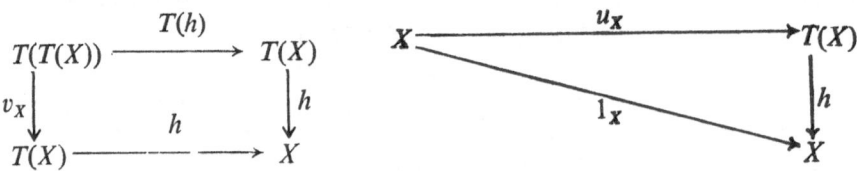

5.13. A morphism $p: X \to Y$ is called an *absolute coequalizer* of the pair $(f, g): Z \to X$ if for every functor $F: \mathscr{C} \to \mathscr{C}'$ the resulting diagram:

$$T(Z) \underset{T(g)}{\overset{T(f)}{\rightrightarrows}} T(X) \xrightarrow{T(p)} T(Y)$$

is a coequalizer diagram.

a) Prove: if the diagram

$$Z \underset{g}{\overset{f}{\rightrightarrows}} X \xrightarrow{p} Y$$

is a contractible coequalizer diagram, then p is an absolute coequalizer of the pair (f, g).

b) We shall say that a functor $F: \mathscr{C} \to \mathscr{C}'$ *creates coequalizers* for a double arrow $(f, g): Z \to X$ in $\mathscr{C}$, if, for each coequalizer $p': F(X) \to Y'$ of $(F(f), F(g))$ in $\mathscr{C}'$, there is a unique object Y and a unique morphism $p: X \to Y$ such that $F(Y) = Y'$ and $F(p) = p'$ and furthermore this unique morphism is a coequalizer of f and g.

Using the notation of part c), Exercise 5.12, show that the following are equivalent:

i) The (unique) functor $K: \mathscr{C}' \to \mathscr{C}^T$ is an equivalence of categories.

ii) The functor $G: \mathscr{C}' \to \mathscr{C}$ creates coequalizers for those double arrows (f, g) in $\mathscr{C}'$ for which $(G(f), G(g))$ has an absolute coequalizer in $\mathscr{C}$.

iii) The functor $G: \mathscr{C}' \to \mathscr{C}$ creates coequalizers for those double arrows (f, g) in $\mathscr{C}'$ for which $(G(f), G(g))$ can be embedded in a contractible coequalizer diagram in $\mathscr{C}$.

This result is due to Beck (see MacLane [3]).

Chapter 4

Abelian categories

4.1. Preadditive and additive categories

The prototype for preadditive category is the category $\mathscr{Ab}$ of all abelian groups. This category plays an important role in mathematics. Preadditive and additive categories are natural generalizations of the category $\mathscr{Ab}$. They play an important role in the theory of categories and elsewhere.

In this section, we define preadditive and additive categories, and prove some of their elementary properties.

References: Buchsbaum [1], Bucur—Deleanu [1], Cartan—Eilenberg [1], Freyd [1], Gabriel [1], Grothendieck [1], MacLane [3], Mitchell [2], N. Popescu [3], Pareigis [1], Schubert [1].

A category $\mathscr{C}$ is *preadditive* if the following condition is satisfied:

(Ad 1) For each pair of objects (X, Y) in $\mathscr{C}$, the set $\mathscr{C}(X, Y)$ has the structure of an abelian group, and for all objects X, Y, and Z of $\mathscr{C}$, the composition map

$$\mathscr{C}(X, Y)\prod\mathscr{C}(Y, Z) \to \mathscr{C}(X, Z)$$

is *bilinear*, i.e. for every $f, f' \in \mathscr{C}(X, Y)$ and for every $g, g' \in \mathscr{C}(X, Z)$,

$$g(f+f') = gf + gf', \text{ and}$$

$$(g + g')f = gf + g'f.$$

Note 1.1. For each pair (X, Y) of objects we denote the zero element of the group $\mathscr{C}(X, Y)$ by O_{XY} (or simply by O if there is no confusion). We suggest that the reader show: for every triple of objects (X, Y, Z) of $\mathscr{C}$,

$$O_{YZ}O_{XY} = O_{XZ}.$$

Note 1.2. It is sometimes convenient to assume that there exists a zero object O in $\mathscr{C}$. Then, for each object X of $\mathscr{C}$, the group $\mathscr{C}(O, X)$ (the group $\mathscr{C}(X, O)$) reduces to the element O_{OX} (O_{XO}). Obviously, for any pair (X, Y) of objects of $\mathscr{C}$,

$$O_{XY} = O_{OY}O_{XO}.$$

Example 1.3. The category $\mathcal{Ab}$ of abelian groups is a typical example of a preadditive category. If X and Y are any two abelian groups, then the set $\mathcal{Ab}(X, Y)$ has canonically an abelian group structure: for $f, g \in \mathcal{Ab}(X, Y)$ we let $f+g: X \to Y$ denote the morphism of groups defined by $(f+g)(x) = f(x) + g(x)$. We see immediately that the composition of morphisms is bilinear.

Example 1.4. Let A be a ring. Then A is a preadditive category which will be denoted by the same symbol A. The category A has a single object $*$ and $A(*, *) = A$. The composition of morphisms is the multiplication of elements of A. The group structure of $A(*, *)$ is that of the underlying additive group A. Conversely, a preadditive category with a single object is in fact a ring. Indeed, if X is an object in a preadditive category $\mathscr{C}$, then the monoid $\mathrm{End}_{\mathscr{C}}(X)$ is canonically a ring.

Example 1.5. Let $\mathscr{C}$ be a preadditive category. A subcategory $\mathscr{C}'$ of $\mathscr{C}$ is preadditive if, for every two objects X and Y of $\mathscr{C}'$, the set $\mathscr{C}'(X, Y)$ is a subgroup of $\mathscr{C}(X, Y)$. In fact, $\mathscr{C}'$ is a preadditive category if and only if for every two objects X and Y of $\mathscr{C}'$ and for every $f, g \in \mathscr{C}'(X, Y)$ the difference $f - g$ which is in $\mathscr{C}(X, Y)$, also belongs to $\mathscr{C}'(X, Y)$. In particular a full subcategory of a preadditive category is also preadditive. If A is a ring, then a preadditive subcategory of A is in fact a subring.

Example 1.6. Let $\mathscr{I}$ be a small category. Then for every preadditive category $\mathscr{C}$, the category $[\mathscr{I}, \mathscr{C}]$ is preadditive: the addition of functorial morphisms is defined argumentwise.

1.7. The dual $\mathscr{C}^0$ of a preadditive category $\mathscr{C}$ is canonically a preadditive category.

Note 1.8. Let $\mathscr{C}$ be a preadditive category and $X \in \mathscr{C}$; then the functor $H^X: \mathscr{C} \to \mathscr{S}et$ can be factored as

$$H^X: \mathscr{C} \xrightarrow{h^X} \mathcal{Ab} \xrightarrow{U} \mathscr{S}et \tag{1}$$

where U is the underlying functor. This factorization expresses the fact that X is an abelian group-object of $\mathscr{C}$. A preadditive category $\mathscr{C}$ may thus be defined as a category in which every object is an abelian group-object, or, equivalently, every object is an abelian cogroup-object.

The category $\mathcal{Ab}$ plays the same role for preadditive categories as the category $\mathscr{S}et$ does for arbitrary categories.

The object h^X is called the *additive-representable functor associated with* X. Dually, we may define the functor $h_X: \mathscr{C} \to \mathcal{Ab}$ which is called the *additive-representable cofunctor associated with* X.

Let $\mathscr{C}$ and $\mathscr{C}'$ be any two preadditive categories and let $F: \mathscr{C} \to \mathscr{C}'$ be a functor. F is said to be an *additive functor* if for each pair of morphisms $f, g \in \mathscr{C}(X, Y)$, $F(f+g) = F(f) + F(g)$.

Additive cofunctors are defined by duality.

Note 1.9. Unless stated explicitly otherwise, whenever we consider a functor F from a preadditive category into a preadditive category, we will assume F to be additive.

Example 1.10. For each object X of a preadditive category $\mathscr{C}$, the functor $h^X : \mathscr{C} \to \mathscr{A}b$ is additive. Also the cofunctor $h_X : \mathscr{C} \to \mathscr{A}b$ is additive.

Example 1.11. By a *left (right) ideal of a preadditive category* $\mathscr{C}$ we mean a preadditive subcategory $\mathscr{I}$ of $\mathscr{C}$ which is also a left (right) ideal of $\mathscr{C}$ (see Section 1.16). A *two-sided ideal of a preadditive category* is a preadditive subcategory that is both a left and right ideal.

If $\mathscr{I}$ is a two-sided ideal of the preadditive category $\mathscr{C}$, we let R denote the equivalence relation on $\mathscr{C}$ defined as follows. If $f, g \in \mathscr{C}(X, Y)$ then $f \sim g(R(X, Y)) \Leftrightarrow$ $\Leftrightarrow f - g \in \mathscr{I}(X, Y)$. (Observe that $\mathscr{I}(X, Y)$ is always nonempty.) The factor category $\mathscr{C}/R$ is usually denoted by $\mathscr{C}/\mathscr{I}$. $\mathscr{C}/\mathscr{I}$ is canonically a preadditive category and the canonical functor $T: \mathscr{C} \to \mathscr{C}/\mathscr{I}$ is additive.

If A is a ring, denote by $\mathbf{a}$, $\mathbf{b}$, $\mathbf{c}$, ... the left, right or two-sided ideals of the preadditive category A; these preadditive ideals are simply called *ideals.* If $\mathbf{a}$ is a two-sided ideal of A, then the ring $A/\mathbf{a}$ is called *the factor ring of* A *by* $\mathbf{a}$.

Example 1.12. If A and B are rings, then an additive functor $f: A \to B$ is, in fact, a ring morphism.

Example 1.13. Let $\mathscr{C}$ and $\mathscr{C}'$ be preadditive categories. Assume that $\mathscr{C}$ is small. Let $\mathrm{Mod}(\mathscr{C}, \mathscr{C}')$ denote the preadditive category whose objects are all additive functors from $\mathscr{C}$ into $\mathscr{C}'$, and whose morphisms are functorial morphisms (addition of morphisms is defined argumentwise). Let $\mathrm{Mod}(\mathscr{C}^0, \mathscr{C}')$ denote the preadditive category whose objects are all additive cofunctors from $\mathscr{C}$ into $\mathscr{C}'$. Furthermore, $\mathrm{Mod}(\mathscr{C}, \mathscr{A}b)$ ($\mathrm{Mod}(\mathscr{C}^0, \mathscr{A}b)$) is denoted by $\mathrm{Mod}\mathscr{C}$ ($\mathrm{Mod}\mathscr{C}^0$). An object of $\mathrm{Mod}\mathscr{C}$ ($\mathrm{Mod}\mathscr{C}^0$) is called a *left (right) module* over $\mathscr{C}$.

Using the ideas occurring in the proof of Theorem 3.19, Ch. 1, we suggest that the reader prove the following.

THEOREM 1.14. *Let* $\mathscr{C}$ *be a preadditive category, and let* $F: \mathscr{C} \to \mathscr{A}b$ *be an (additive) functor. If* $X \in \mathscr{C}$, *let*

$$\varphi: [h^X, F] \to F(X)$$

denote the map defined by $\varphi(u) = u_X(1_X)$. *Then* φ *is an isomorphism of abelian groups (the structure of the domain of* φ *being defined argumentwise). Moreover,* φ *is functorial in* X *and* F. ✳

COROLLARY 1.15. *For every small preadditive category* $\mathscr{C}$, *the assignment* $X \rightsquigarrow h^X (X \rightsquigarrow h_X)$ *defines a full and faithful additive cofunctor* $h^0: \mathscr{C} \to \mathrm{Mod}\mathscr{C}$ *(functor* $h: \mathscr{C} \to \mathrm{Mod}\mathscr{C}^0$). ✳

The cofunctor h^0 (the functor h) is called *canonical.* Sometimes we shall write $h^0_{\mathscr{C}}(h_{\mathscr{C}})$ if there is any danger of confusion.

We obtain the following result concerning finite coproducts and finite products in a preadditive category.

THEOREM 1.16. *Let $\{X_i\}_{1\leqslant i\leqslant n}$ be a finite set of objects of a preadditive category $\mathscr{C}$. The following are equivalent:*

1) *$\{X_i\}_{1\leqslant i\leqslant n}$ has a coproduct.*

2) *$\{X_i\}_{1\leqslant i\leqslant n}$ has a product.*

3) *There is an object X of $\mathscr{C}$ and there are morphisms $u_i\colon X_i \to X$, $p_i\colon X \to X_i$, $1 \leqslant i \leqslant n$, such that:*

$$\text{a)}\ \sum_{i=1}^{n} u_i p_i = 1_X,$$

$$\text{b)}\ p_i u_j = \begin{cases} 0 & \text{if } i \neq j \\ 1_{X_i} & \text{if } i = j. \end{cases} \tag{2}$$

Moreover, the coproduct and the product of $\{X_i\}_{1\leqslant i\leqslant n}$ are canonically isomorphic.

Proof. 1) ⇒ 3). Let $X = \coprod_i X_i$ and let $u_i\colon X_i \to X$, $1 \leqslant i \leqslant n$ be the structural morphisms. Then, for each i, $1 \leqslant i \leqslant n$, there is a unique morphism $p_i\colon X \to X_i$ such that $p_i u_j = 0$ if $i \neq j$ and $p_i u_i = 1_{X_i}$. These morphisms satisfy conditions (2). Let $f = \Sigma_i u_i p_i$. Then for each j, $1 \leqslant j \leqslant n$ we have that $fu_j = (\Sigma_i u_i p_i)u_j = \Sigma_i u_i p_i u_j = u_j$. Hence from the definition of coproduct we see that $f = 1_X$.

3) ⇒ 2). We show that X and the morphisms $p_i\colon X \to X_i$, $1 \leqslant i \leqslant n$ define a product of $\{X_i\}_{1\leqslant i\leqslant n}$. In order to do this let Y be an object of $\mathscr{C}$ and consider, for any i, $1 \leqslant i \leqslant n$, a morphism $g_i\colon Y \to X_i$. Let $g = \Sigma_i u_i g_i$. Then by relations b) we have that

$$p_j g = p_j \sum_i u_i g_i = \sum_i p_j u_i g_i = g_j.$$

We shall now prove the uniqueness of g. Let $g'\colon Y \to X$ be another morphism such that $p_j g' = g_j$, $1 \leqslant j \leqslant n$. Then we have that $u_j p_j g' = u_j g_j$. Now using (2), a), we get that

$$g' = \sum_j u_j p_j g' = \sum_j u_j g_j = g.$$

The implication 2)⇒1) is obtained by considering $\mathscr{C}^0$ and noting that conditions (2) are self-dual. ✳

Note 1.17. In conditions (2), the p_i are called the *projections associated with u_i* and, dually, the u_i are called the *injections associated with p_i*. We observe that these notions are meaningful even for an infinite set of objects. Of course, in that case, condition a) is devoid of meaning.

Note 1.18. Let $\{X_i\}_{i\in\mathscr{I}}$ be a set of objects of a preadditive category $\mathscr{C}$. Assume that their product and coproduct are defined in $\mathscr{C}$. Then there is a unique morphism $t\colon \coprod_i X_i \to \prod_i X_i$ such that $p_i' t = p_i$, where the p_i are the projections associated with the structural morphisms of the coproduct, and the p_i' are the structural projections of the product. We observe that generally t is neither a monomorphism nor an epimorphism. However, from Theorem 1.16 we see that if $\mathscr{I}$ is a finite set, then t is an isomorphism. In this way we shall identify the coproduct with the product. This will be done for any finite set of objects of a preadditive category (at least whenever these notions have a meaning).

A preadditive category $\mathscr{C}$ is said to be *additive* if the following condition is satisfied:

(Ad 2) For each pair (X, Y) of objects of $\mathscr{C}$, the coproduct $X \coprod Y$ is defined.

From Theorem 1.16 we deduce that an additive category is in fact a preadditive category with finite products and finite coproducts.

Note 1.19. We shall usually tacitly assume that an additive category has a zero object.

Example 1.20. The typical example of an additive category is also the category $\mathscr{Ab}$.

Example 1.21. For every small preadditive category $\mathscr{C}$, the categories $\mathrm{Mod}\mathscr{C}$ and $\mathrm{Mod}\mathscr{C}^0$ are additive.

Note 1.22. Just as in every category with zero morphisms (see Section 1.18), in a preadditive category we have the notions kernel and cokernel of a morphism. Namely, the kernel (cokernel) of a morphism f is defined to be the kernel (cokernel) of the double arrow $(f, 0)$.

Note 1.23. By Exercise 2.3 we see that if the morphism $f: X \to Y$ has a kernel (cokernel), then $-f$, the opposite element of f, in the group $\mathscr{C}(X, Y)$ has the same kernel (cokernel).

Note 1.24. Let $\mathscr{C}$ be a small preadditive category. Then for every $X, Y \in \mathrm{Mod}\mathscr{C}$, the group $\mathrm{Mod}\mathscr{C}(X, Y)$ will be denoted by $\mathrm{Hom}_{\mathscr{C}}(X, Y)$.

Exercises

1.1. Prove that every additive functor commutes with finite products and finite coproducts.

1.2. Let $\mathscr{C}$ be a preadditive category, $\mathscr{I}$ a two-sided ideal of $\mathscr{C}$ and $T: \mathscr{C} \to \mathscr{C}/\mathscr{I}$ the canonical functor.

a) Prove that if $\mathscr{C}$ is additive then $\mathscr{C}/\mathscr{I}$ is also additive.

b) Let $\mathscr{I}'$ be a left (a right) ideal of $\mathscr{C}/\mathscr{I}$. Let $T^{-1}(\mathscr{I}')$ denote the subcategory of $\mathscr{C}$ whose objects are those objects X of $\mathscr{C}$ such that $T(X) \in \mathscr{I}'$ and whose morphisms are those morphisms f such that $T(f) \in \mathscr{I}'$. Show that $T^{-1}(\mathscr{I}')$ is a left (a right) ideal of $\mathscr{C}$ and the assignment $\mathscr{I}' \rightsquigarrow T^{-1}(\mathscr{I}')$ defines a bijection from the class of all left (right) ideals of $\mathscr{C}/\mathscr{I}$ to the class of all left (right) ideals of $\mathscr{C}$ which contain $\mathscr{I}$. Also show that this correspondence preserves two-sided ideals.

1.3. For any category $\mathscr{C}$ define a category $\mathrm{Add}(\mathscr{C})$ as follows. The objects of $\mathrm{Add}(\mathscr{C})$ are the same as the objects of $\mathscr{C}$. The set $\mathrm{Add}(\mathscr{C})(X, Y)$ is the free abelian group generated by $\mathscr{C}(X, Y)$, that is, the set of all finite formal linear combinations of the form $\Sigma_i n_i f_i$ where the n_i are integers and $f_i \in \mathscr{C}(X, Y)$. Composition of morphisms in $\mathrm{Add}(\mathscr{C})$ is defined as follows:

$$(\Sigma_i\, n_i f_i)\,(\Sigma_j\, m_j g_j) = \Sigma_{i,j}(n_i m_j)\, f_i g_j.$$

Prove:

a) $\mathrm{Add}(\mathscr{C})$ is a preadditive category and contains $\mathscr{C}$ as a subcategory.

b) If $U: \mathscr{C} \to \mathrm{Add}(\mathscr{C})$ is the inclusion functor, then for every preadditive category $\mathscr{C}'$ and every functor $T: \mathscr{C} \to \mathscr{C}'$ there is a unique additive functor $\bar{T}: \mathrm{Add}(\mathscr{C}) \to \mathscr{C}'$ such that $\bar{T}U = T$.

c) If $\mathscr{C}$ is a category with zero object, then there is a two-sided ideal $\mathscr{I}$ in $\mathrm{Add}(\mathscr{C})$ such that the functor TU: $\mathscr{C} \to \mathrm{Add}(\mathscr{C})/\mathscr{I}$ is faithful and the zero object of $\mathscr{C}$ becomes a zero object of $\mathrm{Add}(\mathscr{C})/\mathscr{I}$.

1.4. Let $(f, g): X \to Y$ be a double arrow of a preadditive category $\mathscr{C}$. Prove that the following are equivalent:

a) The double arrow (f, g) has a kernel (a cokernel).

b) The morphism $f - g$ has a kernel (a cokernel).

c) The morphism $g - f$ has a kernel (a cokernel).

1.5. Prove that the following are equivalent for a category $\mathscr{C}$:

a) $\mathscr{C}$ is an additive category.

b) $\mathscr{C}$ is a category with finite products and finite coproducts and every object of $\mathscr{C}$ is simultaneously a group and cogroup object.

c) The following conditions hold:

(A_1) $\mathscr{C}$ has a zero object.

(A_2) $\mathscr{C}$ has finite products and finite coproducts.

(A_3) For each pair (X, Y) of objects of $\mathscr{C}$, the uniquely defined morphism $t(X, Y): X \coprod Y \to X \prod Y$, such that

$$pt(X, Y)\,u = 1_X, \quad pt(X, Y)v = 0,$$

$$qt(X, Y)u = 0, \quad qt(X, Y)v = 1_Y$$

(where $X \xrightarrow{u} X \coprod Y \xleftarrow{v} Y$, $X \xleftarrow{p} X \prod Y \xrightarrow{q} Y$ are respectively the structural morphisms), is an isomorphism.

(A_4) For each object X of $\mathscr{C}$, there is a morphism $c(X): X \to X$ such that the diagram.

$$\begin{array}{ccc} X & \xrightarrow{\quad c(X) \quad} & X \\ {\scriptstyle \Delta}\downarrow & & \uparrow{\scriptstyle \nabla} \\ X \coprod X & \xrightarrow{\; c(X) \coprod 1_X \;} & X \coprod X \end{array}$$

is commutative, where Δ and ∇ are as usual the diagonal and codiagonal morphisms.

1.6. Prove that a morphism f in a preadditive category is a monomorphism (an epimorphism) if and only if $\mathrm{Ker}\, f = 0$ ($\mathrm{Coker}\, f = 0$).

1.7. A category $\mathscr{C}$ is called *semi-preadditive* if for every $X, Y \in \mathscr{C}$, the set $\mathscr{C}(X, Y)$ has the structure of a commutative monoid, and the composition of morphisms is bilinear.

a) Reformulate the results of Note 1.8, Theorem 1.14 and Corollary 1.15 for a semi-preadditive category. In this case the category $\mathscr{Ab}$ is replaced by the category $\mathscr{Mon}c$ (the category of commutative monoids).

b) Show that Theorem 1.16 is valid in a semi-preadditive category.

c) Let $\mathscr{C}$ be a semi-preadditive category. Assume that for every $X \in \mathscr{C}$, the coproduct $X \coprod X$ is defined. Prove that if in addition $\mathscr{C}$ is a balanced category, then it is a preadditive category.

1.8. Let A be a ring. Show that the category Mod$A$ is an algebraic category. (Hint: Let $\mathscr{A}$ be the algebraic theory of abelian groups. Define a new algebraic theory $\mathscr{A}(A)$ as follows. The generators and relations of $\mathscr{A}(A)$ (see Section 3.1) are those of $\mathscr{A}$ and in addition the elements of $\mathscr{A}$ are unary operations. Let $m: \mathbf{2} \to \mathbf{1}$ be the "addition" in $\mathscr{A}$. Assume that the following new relations are satisfied in $\mathscr{A}(A)$:

(1) $am = m[a, a]$, $a \in A$;
(2) $(ab)m = a(bm)$, $a, b \in A$, ab being the multiplication of a with b in A;
(3) the unity of A is equal to 1_1, the identity of $\mathbf{1}$;
(4) $a + b = m[a, b]\Delta$, where "$a + b$" means the sum of a with b in A.)

1.9. Show that the following are equivalent for an object X of an additive category $\mathscr{C}$:

1) For each morphism $f: X \to \coprod_{i \in \mathscr{I}} X_i$ from X into a coproduct, there exists a factorization:

$$X \xrightarrow{f'} \coprod_{j \in F} X_j \xrightarrow{u_F} \coprod_{i \in \mathscr{I}} X_i$$

where F is a finite subset of $\mathscr{I}$ and u_F is the unique morphism such that $u_F u'_j = u_j$ for every $j \in F$ (where $u_j: X_j \to \coprod_{i \in \mathscr{I}} X_i$, $u'_j: X_j \to \coprod_{j \in F} X_j$ are the structural morphisms).

2) For each morphism $f: X \to \coprod_{i \in \mathscr{I}} X_i$ from X into a coproduct there is a finite subset F of $\mathscr{I}$ such that $f = \Sigma_{j \in F} u_j p_j f$, where the u_j are the structural morphisms and the p_j are associated projections.

3) The functor $h^X: \mathscr{C} \to \mathscr{Ab}$ commutes with coproducts.

An object X which satisfies any one of these equivalent conditions is called a *small* object.

1.10. Prove: a) For every small preadditive category $\mathscr{C}$, the objects $\{h^X\}_{X \in \mathscr{C}}$ of Mod$\mathscr{C}$ give a set of small projective and strong generators.

b) Every finite coproduct of small objects is again a small object.

4.2. Abelian categories

Abelian category is the most important notion in the theory of additive categories. In this section we define abelian category and study some elementary properties.

References: Andreotti [1], Buchsbaum [1], Bucur—Deleanu [1], Freyd [1], Gabriel [1], Grothendieck [1], MacLane [1], [3], Mitchell [2], Pereigis [1], N. Popescu [3], Schubert [1].

An additive category that is finitely complete and finitely cocomplete is called *preabelian.*

Let $\mathscr{C}$ be a preabelian category and $f: X \to Y$ a morphism of $\mathscr{C}$. Let $(\mathrm{Ker} f, i)$ be a kernel of f and $(p, \mathrm{Coker}(\mathrm{Ker} f))$ a cokernel of i. Also let $(q, \mathrm{Coker} f)$ be a cokernel of f and $(\mathrm{Ker}(\mathrm{Coker} f), j)$ a kernel of j. Then there is a unique morphism $\bar{f}: \mathrm{Coker}(\mathrm{Ker} f) \to \mathrm{Ker}(\mathrm{Coker} f)$ such that the diagram

$$\begin{array}{ccccccc} \mathrm{Ker} f & \xrightarrow{\ i\ } & X & \xrightarrow{\ f\ } & Y & \xrightarrow{\ q\ } & \mathrm{Coker} f \\ & & {\scriptstyle p}\downarrow & & \uparrow{\scriptstyle j} & & \\ & & \mathrm{Coker}(\mathrm{Ker} f) & \xrightarrow{\ \bar{f}\ } & \mathrm{Ker}(\mathrm{Coker} f) & & \end{array} \qquad (3)$$

is commutative.

We say that $\bar{f}$ is *parallel* to f and the equality $f = j\bar{f}p$ is the *canonical factorization* of f. In general, $\bar{f}$ need not be an isomorphism as the following example shows.

Example 2.1. Let $\mathscr{A}$ be the algebraic theory of abelian groups (see Section 3.1). If $\mathscr{H}d$ is the category of all Hausdorff spaces, then the category $[\mathscr{A}, \mathscr{H}d]^a$ is preabelian. However, if: $\mathbf{Q} \to \mathbf{R}$ is the natural inclusion ($\mathbf{Q}$ is the topological group of rational numbers and $\mathbf{R}$ is the topological additive group of real numbers) then $\bar{f} = f$, and $\bar{f}$ is not an isomorphism. Moreover, $\bar{f}$ is a biomorphism (see Exercise 2.7, Ch. 1) in $[\mathscr{A}, \mathscr{H}d]^a$,

THEOREM 2.2. *Let $\mathscr{C}$ be a preabelian category. The following are equivalent:*

a) The parallel $\bar{f}$ of each morphism $f: X \to Y$ in $\mathscr{C}$ is an isomorphism.

b) $\mathscr{C}$ is a normal and conormal category and every morphism has a triangular decomposition.

Proof. b) $\Rightarrow$ a). Consider diagram (3). Let $X \xrightarrow{h} Z \xrightarrow{u} Y$ be a triangular decomposition of f. It is easy to verify that there is a unique morphism $r: \mathrm{Coker}(\mathrm{Ker} f) \to Z$ such that $rp = h$ and a unique morphism $v: Z \to \mathrm{Ker}(\mathrm{Coker} f)$ such that $iv = u$. Now let (X', i') be a kernel of h. Since $f = uh$ then $fi' = 0$ so that there is a unique morphism $s: X' \to \mathrm{Ker} f$ such that $is = i'$. But then $pi' = pis = 0$. Hence there is a unique morphism $r': Z \to \mathrm{Coker}(\mathrm{Ker} f)$ such that $r'h = p$. But it is easy to see that r' is an inverse of r, or, equivalently, that r is an isomorphism. In a dual fashion, we see that v is an isomorphism.

We leave it for the reader to check that $\bar{f} = vr$ and also to show that a)$\Rightarrow$b). $\divideontimes$

A preabelian category $\mathscr{C}$ that satisfies either a) or b) of Theorem 2.2 is called *abelian*.

Example 2.3. The most important example of an abelian category is the category $\mathscr{Ab}$ of abelian groups. Indeed, $\mathscr{Ab}$ is an algebraic category so that it is preabelian and every morphism of $\mathscr{Ab}$ has a triangular decomposition (see Section 3.3). Since every subgroup of an abelian group is normal, it follows that $\mathscr{Ab}$ is a normal category. Finally, by Example 16.14, Ch. 1, it follows that $\mathscr{Ab}$ is a balanced category and thus a conormal category (see also Section 3.3). Hence $\mathscr{Ab}$ is an abelian category.

2.4. Let $\mathscr{C}$ be an abelian category. Then for every small category $\mathscr{I}$ the category $[\mathscr{I}, \mathscr{C}]$ becomes in a canonical way (working argumentwise) an abelian category. Similarly, for every small preadditive category $\mathscr{I}$, the category Mod$(\mathscr{I}, \mathscr{C})$ is abelian. In particular, for every ring A the category Mod A is abelian.

Let $\mathscr{C}$ be an abelian category. It is easy to see that every morphism of $\mathscr{C}$ has a unique triangular decomposition that is also an equalizer and a coequalizer decomposition. Moreover, every morphism of $\mathscr{C}$ has a coimage and an image and these objects are canonically isomorphic. Consider diagram (3). Coker(Ker f) is in fact a coimage of f and Ker(Coker$f$) is an image of f. In general, when working with a morphism of $\mathscr{C}$, we shall identify in a canonical way its image and its coimage.

We say that a sequence

$$X \xrightarrow{f} Y \xrightarrow{g} Z$$

of objects and morphisms of $\mathscr{C}$ is *exact at* Y if Ker$g$ = Imf. In other words, in the canonical factorization of the morphism f

$$\begin{array}{ccccc} X & \xrightarrow{\quad f \quad} & Y & \xrightarrow{\quad g \quad} & Z \\ {\scriptstyle p}\downarrow & & \uparrow{\scriptstyle j} & & \\ \mathrm{Coim} f & \xrightarrow{\quad \bar{f} \quad} & \mathrm{Im} f & & \end{array}$$

(Imf, j) is a kernel of g. The above sequence is said to be a *zero sequence at* Y if $gf = 0$. It is clear that every exact sequence is a zero sequence. However, simple examples show that there exist zero sequences which are not exact. We shall say that the sequence of objects and morphisms of $\mathscr{C}$:

$$\cdots \to X_{n-2} \to X_{n-1} \to X_n \to X_{n+1} \to \cdots$$

is *exact* (*zero*) if it is exact (zero) at X_n for each n. In particular, the sequence

$$0 \longrightarrow X' \xrightarrow{f} X$$

is exact at X' if and only if f is a monomorphism. Dually, the sequence

$$X \xrightarrow{p} X'' \longrightarrow 0$$

is exact at X'' if and only if p is an epimorphism. Finally, we shall say that the sequence

$$0 \longrightarrow X' \longrightarrow X \longrightarrow X'' \longrightarrow 0$$

is a *short exact sequence* if it is exact at each of its objects.

We have the following result concerning cartesian and cocartesian squares in an abelian category.

PROPOSITION 2.5. *Consider the square:*

$$\begin{array}{ccc} X & \xrightarrow{f} & Y \\ {\scriptstyle k}\downarrow & & \downarrow{\scriptstyle h} \\ T & \xrightarrow{g} & Z \end{array} \qquad (4)$$

Then: a) This square is commutative if and only if the following is a zero sequence:

$$X \xrightarrow{[f,k]} Y \coprod T \xrightarrow{\langle -h, g\rangle} Z.$$

b) This square is cartesian if and only if the following sequence is exact:

$$0 \longrightarrow X \xrightarrow{[f,k]} Y \coprod T \xrightarrow{\langle -h, g\rangle} Z.$$

c) *This square is cocartesian if and only if the following sequence is exact:*

$$X \xrightarrow{[f,k]} Y \coprod T \xrightarrow{\langle -h, g\rangle} Z \longrightarrow 0.$$

Proof. a) Let $u: Y \to Y \coprod T$ and $v: T \to Y \coprod T$ be structural injections and p and q the corresponding associated projections.

Square (4) is commutative if and only if $gk + (-h)f = 0$. But this last relation is equivalent to the following:

$$\langle -h, g\rangle vq[f,k] + \langle -h, g\rangle up[f,k] = 0. \qquad (5)$$

Since $up + vq = 1$, relation (5) is equivalent to the zeroness of the sequence given in a).

b) Let (4) be a cartesian square and let $r: U \to Y \coprod T$ be a morphism such that $\langle -h, g\rangle r = 0$. Then $r = (up + vq)r$ and so $0 = \langle -h, g\rangle r = \langle -h, g\rangle upr +$

$+\langle -h, g\rangle vqr = -hpr + gqr$, i.e. $hpr = gqr$. Then there exists a unique morphism $m\colon U \to X$ such that $fm = pr$ and $km = qr$. Now, from the definition of $[f, k]$,

$$p[f, k]m = pr \quad \text{and} \quad q[f, k]m = qr.$$

Hence by the definition of product we get that $[f, k]m = r$.

Conversely, if $(X, [f, k])$ is a kernel of $\langle -h, g\rangle$, then by a) the square (4) is commutative. Now, let $m\colon U \to Y$ and $n\colon U \to T$ be morphisms such that $hm = gn$. Then $\langle -h, g\rangle [m, n] = 0$. In fact, by the definition of $\langle -h, g\rangle$ we have that $\langle -h, g\rangle um = -\langle -h, g\rangle vn$, or, $0 = \langle -h, g\rangle up[m, n] + \langle -h, g\rangle vq[m, n] = \langle -h, g\rangle\, [m, n]$.

Let $j\colon U \to X$ be the unique morphism such that $[f, k]j = [m, n]$. Then $p[f, k]j = p[m, n] = m$ and $q[f, k]j = q[m, n] = n$. Finally, we get that $kj = n$ and $fj = m$.

Assertion c) is obtained by duality. ✳

COROLLARY 2.6. *If in the cartesian (cocartesian) square* (4) *g is an epimorphism (f is a monomorphism) then f is an epimorphism (g is a monomorphism).*

Proof. Let (4) be a cocartesian square and let f be a monomorphism. By assertion c) of Proposition 2.5, we have that $(\langle -h, g\rangle, Z)$ is a cokernel of $[f, k]$. Also we have $p[f, k] = f$ and hence $[f, k]$ is a monomorphism. By b) of Proposition 2.5 we get that (4) is cartesian. Now, if $m\colon U \to T$ satisfies $gm = 0$, then there is a unique morphism $n\colon U \to X$ such that $fn = 0$ and $kn = m$. But then $n = 0$, so that $m = 0$. Hence g is a monomorphism. ✳

Note 2.7. Corollary 2.6 says that in an abelian category the epimorphisms (and hence the coequalizers) are universal, and the monomorphisms (hence the equalizers) are couniversal.

Note 2.8. If (4) is a cartesian (cocartesian) square and if g is an epimorphism (f is a monomorphism) then (4) is bicartesian.

Note 2.9. Let (X', u) be a subobject of an object X in an abelian category. The quotient object Coker$u$ is denoted by X/X' and is called the *quotient object of* X *relative to* X'. It follows from the definition of an abelian category that if $p\colon X \to X/X'$ is the canonical epimorphism, then (X', u) is a kernel of p. Hence we have the exact sequence

$$0 \to X' \xrightarrow{u} X \xrightarrow{p} X/X' \to 0 \quad .$$

Clearly, this result can be dualized. If (p, X'') is a quotient object of X, then the subobject Ker$p$ (which is denoted by $X \diagdown X''$) is called the *subobject of* X *relative to* X''. If $u\colon X \diagdown X'' \to X$ is the canonical inclusion, then we get the exact sequence

$$0 \to X \diagdown X'' \xrightarrow{u} X \xrightarrow{p} X'' \to 0.$$

Exercises

2.1. Show that a category $\mathscr{C}$ is abelian if and only if $\mathscr{C}$ is a finitely complete, finitely cocomplete, normal and conormal category.

2.2. Prove that a preabelian category $\mathscr{C}$ is abelian if and only if $\mathscr{C}$ is a balanced category and the parallel $\bar{f}$ of every morphism f of $\mathscr{C}$ is a bimorphism.

2.3. Let $\mathscr{C}$ be a normal preabelian category. Show that there is a natural injective function from the class of subobjects of an object X to the class of quotient objects of X. Hence, if $\mathscr{C}$ is wellcopowered then it is also wellpowered. If $\mathscr{C}$ is normal and conormal, then the above function is a bijection.

2.4. Prove that a family $\{U_i\}_i$ of objects of an abelian category $\mathscr{C}$ is a family of generators if and only if it is a family of separators.

2.5. Let X_1 and X_2 be objects in an abelian category. Let $u_1: X_i \to X_1 \coprod X_2$, and let $p_i: X_1 \coprod X_2 \to X_i$, $i = 1, 2$ be the structural injections and structural projections respectively. Show that when $i \neq j$ sequence

$$0 \to X_i \xrightarrow{u_i} X_1 \coprod X_2 \xrightarrow{p_j} X_j \to 0$$

is exact.

2.6. Assume that the following diagram is commutative and has exact rows, in the category $\mathscr{A}b$:

$$\begin{array}{ccccccccc}
X_1 & \longrightarrow & X_2 & \longrightarrow & X_3 & \longrightarrow & X_4 & \longrightarrow & X_5 \\
{\scriptstyle f_1}\downarrow & & {\scriptstyle f_2}\downarrow & & {\scriptstyle f_3}\downarrow & & {\scriptstyle f_4}\downarrow & & {\scriptstyle f_5}\downarrow \\
Y_1 & \longrightarrow & Y_2 & \longrightarrow & Y_3 & \longrightarrow & Y_4 & \longrightarrow & Y_5
\end{array} .$$

Prove the following:

i) If f_1 is an epimorphism and if f_2 and f_4 are monomorphisms, then f_3 is a monomorphism.

ii) If f_5 is a monomorphism and f_2 and f_4 are epimorphisms, then f_3 is an epimorphism.

iii) If f_1 is an epimorphism, f_5 is a monomorphism and f_2 and f_4 are isomorphisms, then f_3 is an isomorphism.

2.7. Consider the following commutative diagram with exact rows, in the category $\mathscr{A}b$:

$$\begin{array}{ccccccccc}
 & & X' & \xrightarrow{u_1} & X & \xrightarrow{u_2} & X'' & \longrightarrow & 0 \\
 & & {\scriptstyle f'}\downarrow & & {\scriptstyle f}\downarrow & & {\scriptstyle f''}\downarrow & & \\
0 & \longrightarrow & Y' & \xrightarrow{v_1} & Y & \xrightarrow{v_2} & Y'' & &
\end{array} .$$

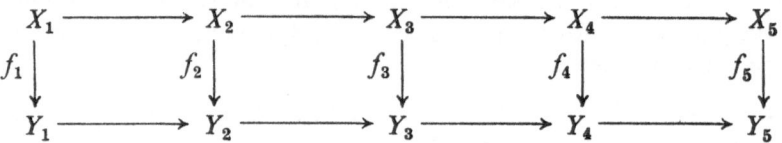

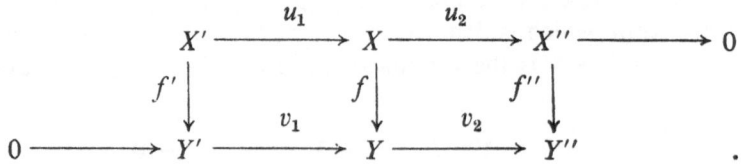

a) Show that the following induced sequences:
$\mathrm{Ker} f' \to \mathrm{Ker} f \to \mathrm{Ker} f''$ and $\mathrm{Coker} f' \to \mathrm{Coker} f \to \mathrm{Coker} f''$
are exact.

b) Define a function $d\colon \mathrm{Ker} f'' \to \mathrm{Coker} f'$ as follows: for every $x \in \mathrm{Ker} f''$, let $u_2(y) = x$. Then $v_2 f(y) = 0$ so $f(y) = v_1(z)$ for some $z \in Y'$. Define $d(x) = \bar{z}$ (the image of z in $\mathrm{Coker} f'$). Show that d is a well-defined morphism of groups and that the sequence

$$\mathrm{Ker} f \to \mathrm{Ker} f'' \to \mathrm{Coker} f' \to \mathrm{Coker} f$$

is exact.

Assertions a), b) hold in the category Mod $\mathscr{C}$, where $\mathscr{C}$ is a small preadditive category.

2.8. Prove the following: a) Let $T\colon \mathscr{C} \to \mathscr{C}'$ be an additive functor, where $\mathscr{C}$ is an abelian category. Then T is a left exact functor if and only if, for every exact sequence

$$0 \to X' \xrightarrow{u} X \xrightarrow{f} X''$$

of $\mathscr{C}$ in the induced sequence

$$T(X') \xrightarrow{T(u)} T(X) \xrightarrow{T(f)} T(X''),$$

$T(u)$ is an equalizer of $T(f)$.

This result can be dualized for right exact functors in the obvious manner.

b) If $\mathscr{C}'$ is also abelian, then the functor T is exact if and only if for every exact sequence:

$$X' \to X \to X''$$

of $\mathscr{C}$ the corresponding sequence of $\mathscr{C}'$ is also exact.

2.9. An object X of a category $\mathscr{C}$ is called *injective* if its dual X^0 is a projective object of $\mathscr{C}^0$ (the dual of $\mathscr{C}$). Show that an object X of an abelian category $\mathscr{C}$ is projective (injective) if and only if the functor h^X (the cofunctor h_X) is exact.

4.3. The isomorphism theorems

In this section, the well-known "isomorphism theorems" that hold in the category $\mathscr{A}b$, are proven in an arbitrary abelian category. The first part of the section contains some technical results which the isomorphism theorems have as corollaries. The results of this section, as well as those of the previous section, will be used extensively in the following sections.

References: Andreotti [1], Buchsbaum [1], Freyd [1], Grothendieck [1], Mitchell [2], MacLane [3], Pareigis [1], N. Popescu [3], Schubert [1].

Here, as well as in the next section (unless the contrary is expressly specified), $\mathscr{C}$ will be an abelian category.

LEMMA 3.1. *Consider the commutative diagram*

$$\begin{array}{ccccccc} & & X & \xrightarrow{f} & Y & & \\ & & \downarrow{\scriptstyle u} & & \downarrow{\scriptstyle v} & & \\ 0 & \longrightarrow & X' & \xrightarrow{g} & Y' & \xrightarrow{h} & Z \end{array}$$

where the bottom row is exact. The square is cartesian if and only if the sequence

$$0 \to X \xrightarrow{f} Y \xrightarrow{hv} Z \tag{6}$$

is exact.

Proof. Assume that the square is cartesian. Then $hvf = hgu = 0$. Let $s: U \to Y$ be a morphism such that $hvs = 0$. Then there is a unique morphism $t: U \to X'$ such that $gt = vs$. Hence there exists a unique morphism $r: U \to X$ such that $fr = s$, i.e., (X,f) is a kernel of hv, or equivalently (6) is exact.

Conversely, assume that (6) is exact and let $s: U \to Y$ and $t: U \to X'$ be morphisms such that $vs = gt$. Then $hvs = hgt = 0$ and there is a unique morphism $r: U \to X$ for which $fr = s$. Also, $gur = vfr = vs = gt$. Finally, $ur = t$, since g is a monomorphism. Consequently the square is cartesian. ✳

LEMMA 3.2. *If the following is a commutative diagram*

$$\begin{array}{ccccccccc} & & X & \xrightarrow{f} & Y & \xrightarrow{p} & \operatorname{Coker} f & \longrightarrow & 0 \\ & & \downarrow{\scriptstyle g} & & \downarrow{\scriptstyle k} & & \downarrow{\scriptstyle s} & & \\ 0 & \longrightarrow & T & \xrightarrow{h} & Z & \xrightarrow{q} & \operatorname{Coker} h & \longrightarrow & 0 \end{array}$$

in which the first square is cartesian, then s is a monomorphism. Moreover, if k is an epimorphism, then s is an isomorphism.

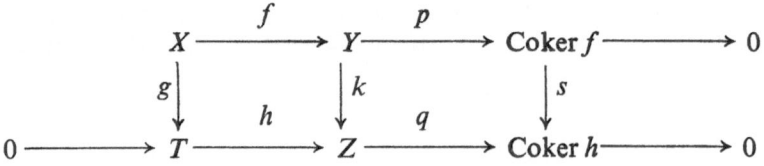

Proof. Let $r: U \to \operatorname{Coker} f$ be a morphism such that $sr = 0$. Let (K, u, v) be a fibered product for the angle $Y \xrightarrow{p} \operatorname{Coker} f \xleftarrow{r} U$, $u: K \to U$ and $v: K \to Y$. By Lemma 3.1, (X, f) is a kernel of $sp = qk$. Now we have that $sru = spv = 0$. Hence there is a unique morphism $t\colon K \to X$ such that $ft = v$. Thus $pv = pft = ru = 0$. Since u is an epimorphism (see Corollary 2.6) we obtain $r = 0$. It follows that s is a monomorphism. If k is an epimorphism, then sp is an epimorphism and finally s is an epimorphism, hence an isomorphism. ✳

LEMMA 3.3. *Consider the following commutative diagram*

$$\begin{array}{ccccccc}
0 & \longrightarrow & \operatorname{Ker} f & \xrightarrow{\;i\;} & X & \xrightarrow{\;f\;} & Y \\
 & & \Big\downarrow{\scriptstyle u} & & \Big\downarrow{\scriptstyle k} & & \Big\downarrow{\scriptstyle h} \\
0 & \longrightarrow & \operatorname{Ker} g & \xrightarrow{\;j\;} & Z & \xrightarrow{\;g\;} & T
\end{array}$$

in which the last square is cartesian. If h is a monomorphism, then u is a monomorphism. If h is a monomorphism and g is an epimorphism, then u is an isomorphism.

Proof. If h is a monomorphism, then k is a monomorphism since the square is cartesian. Then ju is a monomorphism, hence u is also a monomorphism. Assume that g is an epimorphism. Then the square is cocartesian (see Note 2.8). Furthermore we can use the dual of the previous lemma. ✳

LEMMA 3.4. *Consider the commutative diagram*

$$\begin{array}{ccccccccc}
0 & \longrightarrow & X' & \xrightarrow{\;f'\;} & X & \xrightarrow{\;g\;} & X'' & \longrightarrow & 0 \\
 & & \Vert & & \Vert & & \Big\downarrow{\scriptstyle u} & & \\
0 & \longrightarrow & X' & \xrightarrow{\;f'\;} & X & \xrightarrow{\;g'\;} & Y & &
\end{array}$$

in which the upper row is exact. The lower row is exact if and only if u is a monomorphism.

Proof. Assume that the lower row is exact. Let (K, i) be a kernel of u and (U, t, s) a fibered product of the angle $X \xrightarrow{g} X'' \xleftarrow{i} K$, $t\colon U \to K$ and $s\colon U \to X$. By Corollary 2.6, t is an epimorphism and s is a monomorphism. Now we have that $uit = ugs = 0$, i.e. $g's = 0$. Hence there is a unique morphism $r\colon U \to X$ such that $f'r = s$. From this we deduce that $it = gs = gf'r = 0$, so that $i = 0$, since t is an epimorphism. Hence u is a monomorphism.

Conversely, if u is a monomorphism, and $r\colon U \to X$ is a morphism such that $g'r = 0$, then $ugr = 0$ and $gr = 0$. Hence a unique morphism $h\colon U \to X'$ can be found such that $f'h = r$, i.e. f' is a kernel of g'. ✳

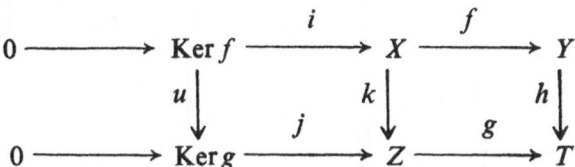

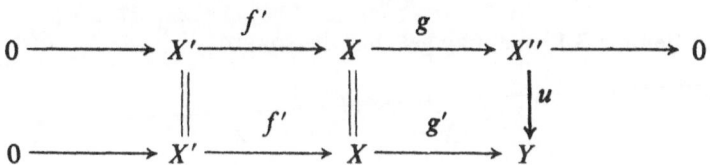

LEMMA 3.5. *Consider the commutative diagram*

$$\begin{array}{ccccccc}
 & & 0 & & 0 & & 0 \\
 & & \downarrow & & \downarrow & & \downarrow \\
0 & \longrightarrow & X' & \xrightarrow{f'} & X & \xrightarrow{f} & X'' \\
 & & \downarrow{\scriptstyle u'} & & \downarrow{\scriptstyle u} & & \downarrow{\scriptstyle u''} \\
0 & \longrightarrow & Y' & \xrightarrow{g'} & Y & \xrightarrow{g} & Y'' \\
 & & \downarrow{\scriptstyle v'} & & \downarrow{\scriptstyle v} & & \\
0 & \longrightarrow & Z' & \xrightarrow{h} & Z & & \\
 & & \downarrow & & & & \\
 & & 0 & & & &
\end{array}$$

in which all columns and the middle row are exact. The upper row is exact if and only if the lower row is exact (i.e., h is a monomorphism).

Proof. Assume that the upper row is exact. Then $\operatorname{Ker}(u''f) = \operatorname{Ker} f$ (see Exercise 3.1). It follows that $\operatorname{Ker}(gu) = \operatorname{Ker} f$, i.e., the sequence:

$$0 \to X' \xrightarrow{f'} X \xrightarrow{gu} Y''$$

is exact. By Lemma 3.1, one may get that the square

$$\begin{array}{ccc}
X' & \xrightarrow{f'} & X \\
{\scriptstyle u'}\downarrow & & \downarrow{\scriptstyle u} \\
Y' & \xrightarrow{g'} & Y
\end{array} \tag{7}$$

is cartesian. Then it follows from Lemma 3.2 that h is a monomorphism.

Conversely, if h is a monomorphism, then we get the exact sequence

$$0 \to X' \xrightarrow{u'} Y' \to Z,$$

and it follows that the square (7) is cartesian.

Let $r: U \to X$ be a morphism such that $fr = 0$. Then $gur = 0$ and there exists a unique morphism $t: U \to Y'$ such that $g't = ur$. Since the square (7) is cartesian, there exists a unique morphism $s: U \to X'$ such that $f's = r$. Finally, (X', f') is a kernel of f. ✳

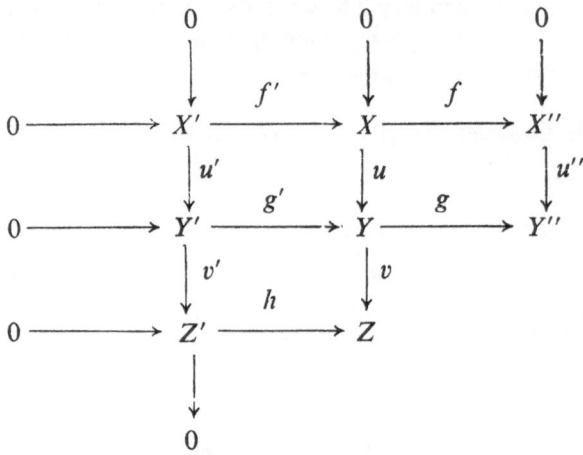

Lemma 3.6 (3×3 Lemma). *Let us consider the commutative diagram:*

$$\begin{array}{ccccccccc}
 & & 0 & & 0 & & 0 & & \\
 & & \downarrow & & \downarrow & & \downarrow & & \\
0 & \longrightarrow & X' & \longrightarrow & X & \longrightarrow & X'' & \longrightarrow & 0 \\
 & & \downarrow & & \downarrow & & \downarrow & & \\
0 & \longrightarrow & Y' & \longrightarrow & Y & \longrightarrow & Y'' & \longrightarrow & 0 \\
 & & \downarrow & & \downarrow & & \downarrow & & \\
0 & \longrightarrow & Z' & \longrightarrow & Z & \longrightarrow & Z'' & \longrightarrow & 0 \\
 & & \downarrow & & \downarrow & & \downarrow & & \\
 & & 0 & & 0 & & 0 & &
\end{array}$$

in which the middle row and the columns are exact. Then the upper row is exact if and only if the lower row is exact.

Proof. This lemma follows from Lemma 3.5 and its dual. ✳

Lemma 3.7. (First isomorphism theorem). *Let X' be a subobject of X and let Y be a subobject of X'. Then we have the following commutative diagram with exact rows:*

$$\begin{array}{ccccccccc}
0 & \longrightarrow & X' & \xrightarrow{\ i\ } & X & \xrightarrow{\ p\ } & X/X' & \longrightarrow & 0 \\
 & & \downarrow{\scriptstyle u'} & & \downarrow{\scriptstyle u} & & \downarrow{\scriptstyle u''} & & \\
0 & \longrightarrow & X'/Y & \xrightarrow{\ j\ } & X/Y & \xrightarrow{\ q\ } & X/X' & \longrightarrow & 0
\end{array}$$

where all the morphisms are canonically defined, and u'' is an isomorphism.

Proof. Consider the commutative diagram:

$$\begin{array}{ccccccccc}
 & & 0 & & 0 & & & & \\
 & & \downarrow & & \downarrow & & & & \\
0 & \longrightarrow & Y & \xrightarrow{\ 1_Y\ } & Y & \longrightarrow & 0 & & \\
 & & \downarrow & & \downarrow & & \downarrow & & \\
0 & \longrightarrow & X' & \xrightarrow{\ i\ } & X & \xrightarrow{\ p\ } & X/X' & \longrightarrow & 0 \\
 & & \downarrow{\scriptstyle u'} & & \downarrow{\scriptstyle u} & & \downarrow{\scriptstyle u''} & & \\
0 & \longrightarrow & X'/Y & \xrightarrow{\ j\ } & X/Y & \xrightarrow{\ q\ } & X/X' & & \\
 & & \downarrow & & \downarrow & & \downarrow & & \\
 & & 0 & & 0 & & 0 & &
\end{array}$$

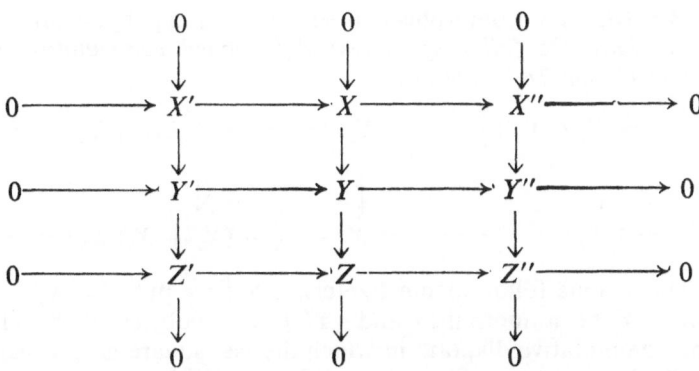

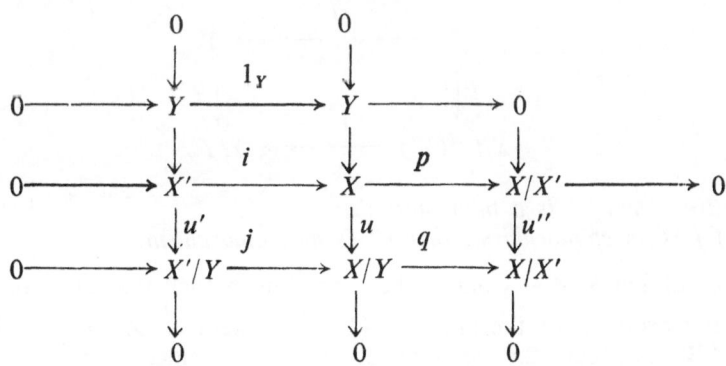

The proof follows from Lemma 3.6. ✱

LEMMA 3.8. (Second isomorphism theorem). *Let X_1, X_2 be any two subobjects of X. Then we have the following canonically defined commutative diagram with exact rows and u'' an isomorphism:*

$$\begin{array}{ccccccccc} 0 & \longrightarrow & X_1 \cap X_2 & \longrightarrow & X_2 & \longrightarrow & X_2/X_1 \cap X_2 & \longrightarrow & 0 \\ & & \downarrow{\scriptstyle u'} & & \downarrow{\scriptstyle u} & & \wr\downarrow{\scriptstyle u''} & & \\ 0 & \longrightarrow & X_1 & \longrightarrow & X_1 + X_2 & \rightarrow & (X_1 + X_2)/X_1 & \longrightarrow & 0. \end{array}$$

Proof. This lemma follows from Exercise 3.5 if we put $X = X_1 + X_2$. ✱

Let $f: X \to Y$ be a morphism and (Y', j) a subobject of Y. Then we have the following commutative diagram in which the last square is cartesian (by Proposition 18.3, Ch. 1, this square is used to define $f^{-1}(Y')$):

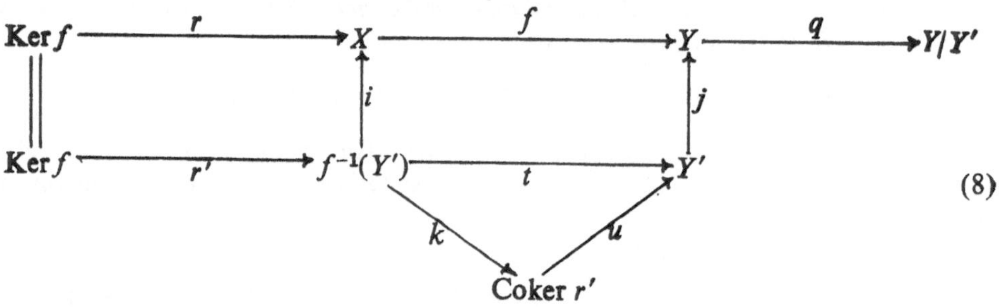

(8)

and all morphisms are canonically defined.

LEMMA 3.9. (Third isomorphism theorem). *Let $f: X \to Y$ be a morphism and let (Y', j) be a subobject of Y. Then the following hold:*

a) *The morphism u in diagram* (8) *is a monomorphism.*

b) *If f is an epimorphism, then u is an isomorphism.*

c) *Let $f'': X/f^{-1}(Y') \to Y/Y'$ denote the unique morphism which makes the diagram*

$$\begin{array}{ccc} X & \xrightarrow{\quad f \quad} & Y \\ {\scriptstyle p}\downarrow & & \downarrow{\scriptstyle q} \\ X/f^{-1}(Y') & \xrightarrow{\quad f'' \quad} & Y/Y' \end{array}$$

commutative. Then f'' is a monomorphism.

d) *If f is an epimorphism, then f'' is an isomorphism.*

Proof. a) Let $h: Z \to \operatorname{Coker} r'$ be a morphism such that $uh = 0$. Let (K, m, n) be a fibered product for the angle $f^{-1}(Y') \overset{k}{\to} \operatorname{Coker} r' \overset{h}{\leftarrow} Z$, $m: K \to Z$ and $n: K \to f'(Y')$. By Corollary 2.6, m is an epimorphism, since k is a coequalizer. We have that $jukn = juhm = 0$. Now, $fi = juk$ and thus $fin = 0$. Hence there is a unique morphism $w: K \to \operatorname{Ker} f$ such that $rw = in$. Also $ir'w = in$, i.e., $r'w = n$, since i is a monomorphism. Then, $hm = kn = kr'w = 0$. In other words, $h = 0$, since m is an epimorphism. Hence u is a monomorphism.

b) Assume that f is an epimorphism. Let $h: Y' \to U$ be a morphism such that $hu = 0$. Let (m, n, K) be a fibered coproduct of the coangle $Y \overset{j}{\leftarrow} Y' \overset{h}{\to} U$, $m: U \to K$ and $n: Y \to K$. By Corollary 2.6 we see that m is a monomorphism, since j is an equalizer. Then $nfi = njuk = mhuk = 0$. Hence, there exists a unique morphism $w: Y/Y' \to K$ such that $wqf = nf$ since $(qf, Y/Y')$ is a cokernel of i (see Exercise 3.1). Then $wq = n$, since f is an epimorphism. Also $mh = nj = wqj = 0$, and thus $h = 0$, since m is a monomorphism, hence u is an epimorphism, and by a) an isomorphism (see Exercise 2.2).

c) Consider the following commutative diagram with exact rows:

$$\begin{array}{ccccccccc}
0 & \longrightarrow & f^{-1}(Y') & \overset{i}{\longrightarrow} & X & \overset{p}{\longrightarrow} & X/f^{-1}(Y') & \longrightarrow & 0 \\
 & & \downarrow{\scriptstyle t} & & \downarrow{\scriptstyle f} & & \downarrow{\scriptstyle f''} & & \\
0 & \longrightarrow & Y' & \overset{j}{\longrightarrow} & Y & \overset{q}{\longrightarrow} & Y/Y' & \longrightarrow & 0
\end{array} .$$

Since the first square is cartesian, it follows from Lemma 3.2 that f'' is a monomorphism.

d) This also follows from Lemma 3.2. ✳

Exercises

3.1. Let $X \overset{f}{\to} Y \overset{g}{\to} Z$ be any two morphisms of $\mathscr{C}$. Prove: If g is a monomorphism (f an epimorphism), then $\operatorname{Ker} f = \operatorname{Ker}(gf)$ ((Coker $g = \operatorname{Coker}(gf)$).

3.2. Let $\{X_i, u_i\}_i$ be a (not necessarily finite) set of subobjects of X. Assume that the object $\coprod_i X_i$ is defined. The morphism $\langle u_i \rangle_i : \coprod_i X_i \to X$ is called the *sum morphism*. The image of this morphism is denoted by $\Sigma_i X_i$ and is called the *sum* of $\{X_i\}_i$. Show that this subobject is the union as well as the upper bound of $\{X_i\}_i$.

3.3. Let $\{p_i, X_i\}_i$ be a (not necessarily finite) set of quotient objects of X. Assume that $\prod_i X_i$ is defined. The morphism $[p_i]_i : X \to \prod_i X_i$ is called the *product morphism*. Show that the quotient object $\{p, X/(\cap_i \operatorname{Ker} p_i)\}$ is the counion as well as the lower bound of $\{X_i\}_i$.

3.4. Let $\{X_i\}_i$ be a (not necessarily finite) set of subobjects of X. Let $p_i: X \to X/X_i$ be the canonical morphism. Assume that $\prod_i X/X_i$ is defined. Prove that $\cap_i X_i = \operatorname{Ker}([p_i]_i)$.

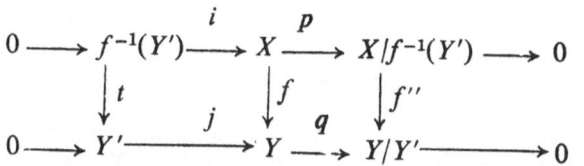

3.5. Let X_1 and X_2 be any two subobjects of X. Prove that the following canonically defined diagram with exact rows and columns:

$$
\begin{array}{ccccccccc}
 & & 0 & & 0 & & 0 & & \\
 & & \downarrow & & \downarrow & & \downarrow & & \\
0 & \longrightarrow & X_1 \cap X_2 & \longrightarrow & X_2 & \longrightarrow & X_2/X_1 \cap X_2 & \longrightarrow & 0 \\
 & & \downarrow & & \downarrow & & \downarrow & & \\
0 & \longrightarrow & X_1 & \longrightarrow & X & \longrightarrow & X/X_1 & \longrightarrow & 0 \\
 & & \downarrow & & \downarrow & & \downarrow & & \\
0 & \longrightarrow & X_1/X_1 \cap X_2 & \longrightarrow & X/X_2 & \longrightarrow & X/(X_1 + X_2) & \longrightarrow & 0 \\
 & & \downarrow & & \downarrow & & \downarrow & & \\
 & & 0 & & 0 & & 0 & &
\end{array}
$$

is commutative. (Hint: Use Lemma 3.6.)

3.6. Let $0 \to X' \to X \xrightarrow{g} X'' \to 0$ be an exact sequence. Prove that if X_1 and X_2 are subobjects of X satisfying the relation $X' \subseteq X_1 \subseteq X_2$, then $X_1 = X_2$ if and only if $g(X_1) = g(X_2)$.

3.7. Let $\{X_i\}_{i \in \mathscr{I}}$ be a family of subobjects of an object X of $\mathscr{Ab}$. Prove that its union (denoted by $\Sigma_i X_i$) is the set of elements in X of the form $\Sigma_i a_i$ where $a_i \in X_i$ for all $i \in \mathscr{I}$ and $a_i = 0$ for all but a finite number of i.

3.8. (Modular law). Let X_1, X_2 and X_3 be subobjects of an object X such that $X_1 \subseteq X_2$. Prove that $X_2 \cap (X_1 + X_3) = X_1 + (X_2 \cap X_3)$.

3.9. Let $f\colon X \to Y$ be a morphism, X' a subobject of X, and Y' a subobject of Y. Show that we have the following isomorphisms:

$$f(X') \overset{\sim}{\longrightarrow} X'/(X' \cap \operatorname{Ker} f) \overset{\sim}{\longrightarrow} (X' + \operatorname{Ker} f)/\operatorname{Ker} f,$$

$$X/f^{-1}(Y') \overset{\sim}{\longrightarrow} \operatorname{Im} f/(\operatorname{Im} f \cap Y'),$$

$$f(f^{-1}(Y') \cap X') \overset{\sim}{\longrightarrow} f(X') \cap Y'.$$

3.10. Show that the results of Exercises 2.6 and 2.7 hold in every abelian category (see Corollary 8.8 below).

3.11. Let X_1 and X_2 be subobjects of X.

a) Prove that the following are equivalent:

1) $X_1 \cap X_2 = 0$.
2) The sum morphism $X_1 \coprod X_2 \to X$ is a monomorphism.
3) The product morphism $X \to X/X_1 \prod X/X_2$ is a monomorphism.

b) Also, prove that, dually, the following are equivalent:

1) $X_1 + X_2 = X$.

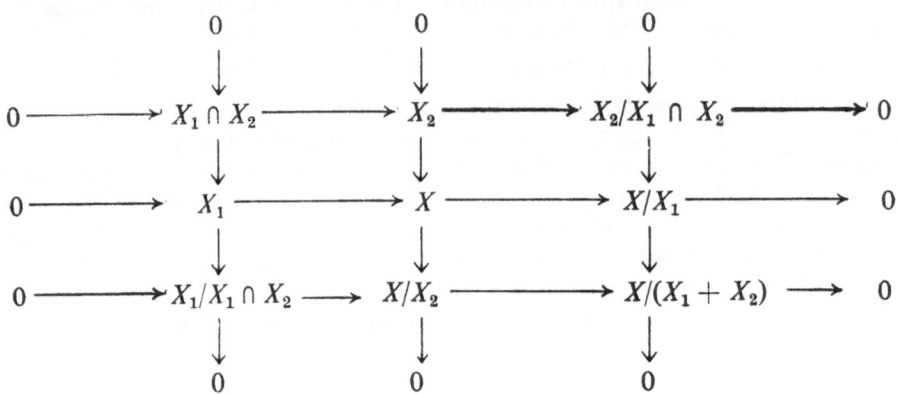

2) The product morphism is an epimorphism.
3) The sum morphism is an epimorphism.

Two subobjects X_1, X_2 of X which satisfy condition b) of Exercise 3.11 are called *comaximal.* Two subobjects X_1 and X_2 are called *supplementary* if they satisfy conditions a) and b) simultaneously. In this case, we say that X_1 is a *supplement* of X_2. A subobject X having a supplement is called a *direct summand.*

Let X be a nonzero object and let $u_i: X \to X \coprod X$, $i = 1, 2$ be the structural injections. Let $X_i = u_i(X)$ and $X_3 = \Delta(X)$, where $\Delta: X \to X \coprod X$ is the diagonal morphism. Show that X_1 and X_2 are nonzero supplements of X_3 and $X_1 \cap X_2 = \emptyset$. Hence a subobject may have many distinct supplements. However, any two supplements of X_1 are isomorphic.

3.12. Prove that a subobject (X_1, u) of X is a direct summand if and only if u is a section or, equivalently, if and only if there is a morphism $p: X \to X_1$ such that up is an idempotent.

3.13. Let $0 \to X' \xrightarrow{u} X \xrightarrow{p} X'' \to 0$ be an exact sequence. Prove then that u is a section if and only if p is a retraction. Thus X is canonically isomorphic to the coproduct $X' \coprod X''$.

An exact sequence such as in Exercise 3.13 is called a *split short exact sequence.*

3.14. Prove that the following are equivalent for an endomorphism u of X:
1) u is an idempotent.
2) $\operatorname{Ker} u$ is a direct summand of X.
3) $\operatorname{Im} u$ is a direct summand.

3.15. Show that assertions 2)—3) of a) and 2)—3) of b) in Exercise 3.11 are valid for a finite set $X_1, \ldots, X_n$ of subobjects of X. Also show that, moreover, these assertions are equivalent to the following one:
1) For every i_0, $1 \leqslant i_0 \leqslant n$, $X_{i_0} \cap (\Sigma_{i \neq i_0} X_i) = 0$.

4.4. Limits and colimits in abelian categories

Limits and colimits in abelian categories have some special properties. In this section we define the so-called *Ab*-conditions due to Grothendieck [1] and investigate some properties of categories in which these conditions hold.

References: Freyd [1], Gabriel [1], Grothendieck [1], Mitchell [2], Pareigis [1], N. Popescu [3], Roos [4], Schubert [1].

By an *Ab* 3-*category* we mean an (abelian) category with coproducts. An (abelian) category with products is said to be an *Ab* 3*-*category.* These two conditions are obviously dual. By Theorem 9.2, Ch. 1, (resp. its dual), we see that *Ab* 3-(resp. *Ab* 3*-) categories are just the cocomplete (complete) categories.

Let $\mathscr{C}$ be an *Ab* 3-category, $\{X_i\}_{i \in \mathscr{I}}$ a set of objects of $\mathscr{C}$ and $u_i: X_i \to \coprod_i X_i$ the structural injections. Since $\mathscr{C}$ has a zero object, for each $i \in \mathscr{I}$ there is a unique morphism $p_i: \coprod_i X_i \to X_i$ such that $p_i u_i = 1_{X_i}$ and $p_i u_j = 0$ if $j \neq i$. As we remarked in Note 1.18, the morphisms $\{p_i\}_{i \in \mathscr{I}}$ are called the projections associated with the structural injections. Dually, we can speak of the injections associated with the structural projections.

Note 4.1. The category $\mathscr{Ab}$ is clearly an *Ab* 3- and an *Ab* 3*-category (see Section 3.3). If $\{X_i\}_{i \in \mathscr{I}}$ is a set of abelian groups, let $\oplus_i X_i$ denote the subgroup of the product $\prod_i X_i$ that consists of all elements $\{x_i\}_{i \in \mathscr{I}}$ such that $x_i = 0$ for all but a finite number of indices.

Let $p_i: \prod_i X_i \to X_i$ be the structural projections and $u_i: X_i \to \prod_i X_i$ their associated injections. It is easy to see that u_i can be factored as $X_i \xrightarrow{u_i'} \oplus_i X_i \xrightarrow{u} \prod_i X_i$ where u is the inclusion. We leave it for the reader to check that $\oplus_i X_i$ and the morphisms $\{u_i'\}_i$ give a coproduct of $\{X_i\}_i$ groups. Since $\oplus_i X_i$ is a subset of $\prod_i X_i$, u_i can be identified with u_i'. Hence we may conclude that the injections associated with the projections of the product are actually the injections for the coproduct. Note however that the coproduct and the product of an infinite set of objects of $\mathscr{Ab}$ are generally not isomorphic.

4.2. A family $\{X_i, u_i\}_i$ of subobjects of X is called *independent* (*complete*) if the sum morphism $\langle u_i \rangle_i: \coprod_i X_i \to X$ is a monomorphism (an epimorphism). A family of subobjects is called *direct* if it is both independent and complete.

Using Proposition 18.6, Ch. 1, we obtain the following result.

LEMMA 4.3. *Let $F: \mathscr{I} \to \mathscr{C}$ be a small functor into an Ab 3-category and let (f, X) be a colimit of F. If $g: F \to Y$ is a morphism and $u: X \to Y$ is the unique morphism such that $uf = g$, then $u(X) = \Sigma_i g_i(F_i)$, $i \in \mathscr{I}$.* ✳

We say that an *Ab* 3-category is an *Ab 4-category* if the coproduct of each set of monomorphisms is also a monomorphism. Dually, we obtain the concept of an *Ab 4*-category*. The conditions *Ab* 3 and *Ab* 4 are independent (see Exercise 4.1).

By Exercise 4.2 we will see that the coproduct (product) is "right (left) exact". The following result shows that the Ab 4- (Ab 4*-) categories are just those categories in which the coproduct (product) is exact.

PROPOSITION 4.4. *Let $\mathscr{C}$ be an Ab 3-(resp. Ab 3*-)category. The following are equivalent:*

1) *$\mathscr{C}$ is an Ab 4-(resp. Ab 4*-)category.*

2) *For each set of exact sequences*

$$X_i' \xrightarrow{u_i} X_i \xrightarrow{p_i} X_i''$$

the sequence

$$\coprod_i X_i' \xrightarrow{\coprod u_i} \coprod_i X_i \xrightarrow{\coprod p_i} \coprod_i X_i''$$

$$(\text{resp. } \prod_i X_i' \xrightarrow{\prod u_i} \prod_i X_i \xrightarrow{\prod p_i} \prod_i X_i'')$$

is exact.

The proof is left for the reader. ✳

PROPOSITION 4.5. *Let $\mathscr{C}$ be an Ab 3- and Ab 3*-category. Assume that for every set $\{X_i\}_i$ of objects of $\mathscr{C}$, the canonical morphism $t\colon \coprod_i X_i \to \prod_i X_i$ (see Note 1.18) is a monomorphism. Then $\mathscr{C}$ is an Ab 4-category.*

Proof. Let $\{f_i\colon X_i' \to X_i\}_i$ be a set of monomorphisms. Then we have the following commutative diagram:

$$\begin{array}{ccc} \coprod_i X_i' & \xrightarrow{\coprod_i f_i} & \coprod_i X_i \\ {\scriptstyle t'}\downarrow & & \downarrow{\scriptstyle t} \\ \prod_i X_i' & \xrightarrow{\prod_i f_i} & \prod_i X_i \end{array} \quad .$$

It follows that $\coprod_i f_i$ is a monomorphism since $\prod_i f_i$ is always a monomorphism (see Exercise 4.2). ✳

A category $\mathscr{C}$ that satisfies the hypothesis of Proposition 4.5 is called a *C_2-category.*

Now we define a class of abelian categories that play an important role, particularly in ring theory. We begin with some definitions.

A *filtered system of morphisms* in a category $\mathscr{C}$ is a filtered system of objects in the category Morf($\mathscr{C}$), the category of morphisms in $\mathscr{C}$. A filtered system $F\colon \mathscr{I} \to$ $\to$ Morf($\mathscr{C}$) is called a *filtered system of monomorphisms* if for each $i \in \mathscr{I}$, the morphism $F_1 = \{X_i' \xrightarrow{f_i} X_i\}$ is a monomorphism. The notions *directed system of morphisms and direct system of monomorphisms* have an obvious meaning (see Section 1.7). Likewise the notions *filtered system of (short) exact sequences* and *directed system of (short) exact sequences* are obvious.

THEOREM 4.6. *Let $\mathscr{C}$ be an Ab 3-category. The following are equivalent:*

1) *For every direct system $\{X_i' \xrightarrow{f_i} X_i\}_i$ of monomorphisms of $\mathscr{C}$, the morphism:*

$$\varinjlim_i f_i\colon \varinjlim_i X_i' \to \varinjlim_i X_i$$

is a monomorphism (i.e., the directed colimits are exact).

2) *$\mathscr{C}$ is a (G.C.)-category.*

3) *Let X be any object, $\{X_i\}_i$ a directed set of subobjects and X' a subobject of X. Then*

$$\Sigma_i(X_i \cap X') = (\Sigma_i X_i) \cap X'.$$

4) *Let $f\colon Y \to X$ be a morphism and $\{X_i\}_i$ a directed set of subobjects of X. Then*

$$f^{-1}(\Sigma_i X_i) = \Sigma_i f^{-1}(X_i).$$

5) *Let $\{X_i\}_{i\in\mathscr{I}}$ be a set of objects. For each finite subset F of $\mathscr{I}$ let X_F denote the image of the canonical morphism $u_F\colon \coprod_{i'\in F} X_{i'} \to \coprod_i X_i$, defined such that*

$u_F u'_{i'} = u_{i'}$, *(where* $u'_{i'}$ *and* $u_{i'}$ *are respectively the structural morphisms). Then for every subobject* X *of* $\coprod_i X_i$,

$$X = \Sigma_{F \in \mathcal{T}}(X \cap X_F),$$

where $\mathcal{T}$ *is the set of all finite subsets of* $\mathcal{I}$.

Proof. 1) ⇒ 2). Let $\{X_i\}_i$ be a direct system of subobjects of X and let $j_i\colon X_i \to X$ be the canonical inclusions. We consider the following direct system of monomorphisms:

$$\{X_i \xrightarrow{j_i} X\}_i.$$

Then by 1) the morphism

$$\varinjlim_i X_i \xrightarrow{\varinjlim j_i = j} \varinjlim_i X = X$$

is a monomorphism.

(We observe that the morphism j is in fact an isomorphism from $\varinjlim_i X_i$ onto $\Sigma_i X_i$. This follows since in this case $\varinjlim_i X_i$ is a union of the X_i's.)

2) ⇒ 3). Let $\{X_i\}_i$ be a direct set of subobjects of X. Then for each subobject X' of X, $X_i \cap X' \subseteq X_i$ and $X_i \cap X' \subseteq X'$. Hence $\Sigma_i(X_i \cap X') \subseteq (\Sigma_i X_i) \cap X'$.

By the definition of the sum or subobjects we have the equality.

$$\Sigma_i(X_i + X') = (\Sigma_i X_i) + X'.$$

We shall now consider the following canonically defined directed system of short exact sequences, t_i being the inclusions:

$$0 \to X' \xrightarrow{t_i} (X_i + X') \to (X_i + X')/X' \to 0,$$

where the t_i's are the inclusions.

By 2),

$$\varinjlim_i (X_i + X') = \Sigma_i(X_i + X') = (\Sigma_i X_i) + X'.$$

Since the colimits are right exact (see Exercise 4.2) we get the exact sequence

$$X' \xrightarrow{t = \varinjlim t_i} \varinjlim_i (X_i + X') \to \varinjlim_i ((X_i + X')/X') \to 0.$$

It is clear by 2) that t is a monomorphism.

Hence we obtain the canonical isomorphisms

$$\varinjlim_i ((X_i + X')/X') \simeq (\varinjlim_i (X_i + X'))/X' \simeq ((\Sigma_i X_i) + X')/X'. \qquad (9)$$

On the other hand, we have the following directed set of canonically defined (see Lemma 3.8) short exact sequences:

$$0 \to X_i \cap X' \xrightarrow{k_i} X_i \to (X_i + X')/X' \to 0.$$

Passing to colimits, we get the exact sequence

$$\Sigma_i(X_i \cap X') \xrightarrow{k = \varinjlim k_i} \Sigma_i X_i \to \varinjlim_i ((X_i + X')/X') \to 0.$$

Again using 2) we see that k is a monomorphism, so that we have the isomorphism:

$$\varinjlim ((X_i + X')/X') \simeq (\Sigma_i X_i)/(\Sigma_i(X_i \cap X')). \tag{10}$$

Hence by Lemma 3.8 we have the isomorphism

$$((\Sigma_i X_i) + X')/X' \simeq (\Sigma_i X_i)/((\Sigma_i X_i) \cap X').$$

Now by (9) and (10), we deduce the inclusion

$$(\Sigma_i X_i) \cap X' \subseteq \Sigma_i(X_i \cap X').$$

3)⇒2). Let $\{X_i, j_i\}_i$ be a directed system of subobjects of X and let $j\colon \varinjlim_i X_i \to X$ be the unique morphism such that $ju_i = j_i$, where the $u_i\colon X_i \to \varinjlim_i X_i$ are the structural morphisms. We must prove that j is a monomorphism.

Let $K = \operatorname{Ker} j$ and $X_i' = u_i(X_i)$. The set $\{X_i'\}_i$ of subobjects of $\varinjlim_i X_i$ is directed and complete (i.e. $\Sigma_i X_i' = \varinjlim_i X_i$). Hence $K = (\Sigma_i X_i') \cap K = \Sigma_i(X_i' \cap K)$. If $K \neq 0$ there is an index i such that $u_i^{-1}(K) = u_i^{-1}(K \cap X_i) \neq 0$, and thus $u_i^{-1}(K) = \operatorname{Ker}(ju_i) = \operatorname{Ker} j_i = 0$, which is a contradiction. Hence $K = 0$, or equivalently, j is a monomorphism.

3) ⇒ 4). Let $\{X_i\}_i$ be a directed system of subobjects of X and let $f\colon Y \to X$ be a morphism. Then the system $\{f^{-1}(X_i)\}_i$ of subobjects of Y is directed. For any i we have the exact sequence:

$$0 \to f^{-1}(K_i) \xrightarrow{r_i} Y \longrightarrow (\operatorname{Im} f/(X_i \cap \operatorname{Im} f)) \to 0.$$

Passing to colimits, we obtain the exact sequence

$$\varinjlim_i f^{-1}(X_i) \xrightarrow{r = \varinjlim r_i} Y \to \varinjlim_i (\operatorname{Im} f/(X_i \cap \operatorname{Im} f)) \to 0 \tag{11}$$

where r is a monomorphism given by the equivalence 2) ⇔ 3).

Now, we apply Exercise 4.4 to the exact sequence (11) and get that

$$\varinjlim_i (\operatorname{Im} f/(X_i \cap \operatorname{Im} f)) \simeq \operatorname{Im} f/(\Sigma_i(X_i \cap \operatorname{Im} f)) \simeq \operatorname{Im} f/(\operatorname{Im} f \cap (\Sigma_i X_i)).$$

From Exercises 3.9 and 4.4 we deduce that

$$\operatorname{Im} f/(\operatorname{Im} f \cap (\Sigma_i X_i)) \simeq Y/(f^{-1}(\Sigma_i X_i)) \simeq Y/(\Sigma_i f^{-1}(X_i)).$$

Now 4) follows immediately.

5) $\Rightarrow$ 3). Let $\{X_i\}_i$ be a directed system of subobjects and X' a subobject of X. Let $s: \coprod_i X_i \to X$ be the sum morphism. It is easy to see that

$$X' \cap (\Sigma_i X_i) = X' \cap \operatorname{Im} s = s(s^{-1}(X') \cap (\Sigma_i X_i)). \tag{12}$$

Let $\mathscr{T}$ be the set of all finite subset of $\mathscr{I}$. It is clear that $\{X_F\}_{F \in \mathscr{T}}$ is a directed and complete system of subobjects of $\coprod_i X_i$.

Hence

$$s(s^{-1}(X') \cap \textstyle\coprod_i X_i) = s(s^{-1}(X') \cap \Sigma_F X_F) = s(\Sigma_F(s^{-1}(X') \cap X_F)) =$$
$$= \Sigma_F s(s^{-1}(X') \cap X_F) = \Sigma_F(X' \cap s(X_F)). \tag{13}$$

Now it is easy to see that, for each $F \in \mathscr{T}$, there is an $i \in \mathscr{I}$ such that $s(X_F) \subseteq X_i$ and thus

$$\Sigma_F(X' \cap s(X_F)) \subseteq \Sigma_i(X' \cap X_i).$$

Since the reverse inclusion is obvious, we deduce that 3) holds (use relations (12) and (13)).

The proof of the implication 4) $\Rightarrow$ 3) is easy and is left for the reader.

We now prove that 2) + 3) + 4) + 5) $\Rightarrow$ 1).

For that, let $\{f_i: X_i' \to X_i\}_{i \in \mathscr{I}}$ be a directed system of monomorphisms of $\mathscr{C}$. For each i, consider the commutative diagram

$$\begin{array}{ccccc} & & X_i' & \xrightarrow{\ f_i\ } & X_i \\ & & {\scriptstyle u_i'}\downarrow & & \downarrow{\scriptstyle u_i} \\ 0 \to K \to & & \varinjlim X_i' & \xrightarrow{\ f = \varinjlim f_i\ } & \varinjlim X_i \end{array}$$

where K is the kernel of f, and the u_i' and u_i are respectively the structural morphisms. Let $Y_i = \operatorname{Im} u_i'$; we have by Lemma 4.3 that $\varinjlim_i X_i' = \Sigma_i Y_i$, and obviously $\{Y_i\}_i$ is a directed system of subobjects. Then by 3), $K = (\Sigma_i Y_i) \cap K = \Sigma_i(Y_i \cap K)$ and so if $K \neq 0$ then $K \cap Y_{i_0} \neq 0$ for some index $i_0 \in \mathscr{I}$. Thus $M = u_{i_0}'^{-1}(K \cap Y_{i_0}) \neq 0$. On the other hand, $u_{i_0}(f_{i_0}(M)) = f(u_{i_0}'(M)) = f(u_{i_0}'(u_{i_0}'^{-1}(K \cap Y_{i_0}))) \subseteq f(K \cap Y_{i_0}) \subseteq f(K) = 0$.

Hence, by Exercise 4.6, $f_{i_0}(M)$ is a subobject of $\Sigma_{i_0 \leqslant j} \operatorname{Ker}(f_{i_0 j})$. Again using condition 3),

$$f_{i_0}(M) = \Sigma_{i_0 \leqslant j}(\operatorname{Ker}(f_{i_0 j}) \cap f_{i_0}(M)).$$

$$M = f_{i_0}^{-1}(f_{i_0}(M)) = f_{i_0}^{-1}(\Sigma_{i_0 \leqslant j}(\operatorname{Ker} f_{i_0 j} \cap f_{i_0}(M)) =$$
$$= \Sigma_{i_0 \leqslant j} f_{i_0}^{-1}(\operatorname{Ker}(f_{i_0 j}) \cap f_{i_0}(M)). \tag{14}$$

(The first equality holds because f_{i_0} is a monomorphism and the third equality holds by condition 4) of Theorem 4.6.)

Since f_j is a monomorphism it follows that

$$f'_{i_0 j}(f_{i_0}^{-1}(\operatorname{Ker}(f_{i_0 j}) \cap f_{i_0}(M))) = 0$$

for each $j \geqslant i_0$ and so

$$u'_{i_0}(f_{i_0}^{-1}(\operatorname{Ker}(f_{i_0 j}) \cap f_{i_0}(M))) = 0 \text{ for all } j \geqslant i_0.$$

Therefore by (14), $u'_{i_0}(M) = 0$. But then, using Lemma 3.9, we see that $K \cap Y_{i_0} = 0$. This contradiction implies that $K = 0$, and so f is a monomorphism. ✳

Note 4.7. Observe that Theorem 4.6 still holds if we change the word "directed" to the word "filtered".

An *Ab* 3-category which satisfies the conditions of the above theorem is called an *Ab 5-category*. Dually, we have the notion of an *Ab 5*-category*.

Example 4.8. It is easy to see that $\mathscr{Ab}$ is an *Ab* 5-category, since $\mathscr{Ab}$ is an algebraic category, and hence also a (G.C.)-category.

4.9. If $\mathscr{C}$ is an *Ab* 5-category, then for every small (preadditive) category $\mathscr{I}$, the category $[\mathscr{I}, \mathscr{C}]$ ($\operatorname{Mod}(\mathscr{I}, \mathscr{C})$) is an *Ab* 5-category. In particular, for each ring A, the category Mod A is an *Ab* 5-category.

PROPOSITION 4.10. *An Ab 5-category which is also a C_2^*-category consists only of zero objects.*

Proof. By Exercise 4.8, $\mathscr{C}$ is a C_2-category. Since it is also C_2^* we find that for any set $\{X_i\}_i$ of objects, the canonical morphism $t: \coprod_i X_i \to \prod_i X_i$ (see Note 1.18) is an isomorphism. For every $X \in \mathscr{C}$, $X^{(\mathbf{N})} = X^{\mathbf{N}}$. For each natural number n, the object X^n can be viewed as a subobject of $X^{\mathbf{N}}$. It is clear that $\{X^n\}_{n \in \mathbf{N}}$ gives a directed and total system of subobjects. Let $\Delta: X \to X^{\mathbf{N}}$ be the diagonal morphism. Then by Theorem 4.6, $X = \Sigma_{n \in \mathbf{N}} \Delta^{-1}(X^n)$. We claim that $\Delta^{-1}(X^n) = 0$ for all n. Indeed, let q be the composition

$$X \xrightarrow{\Delta} X^{\mathbf{N}} \xrightarrow{p} X^{\mathbf{N}}/X^n.$$

It is clear that $\Delta^{-1}(X_n) = \operatorname{Ker} q$. Now if $u_n: X^n \to X^{\mathbf{N}}$ is the canonical inclusion, then $p_{n+1}u_n = 0$ ($p_n: X^{\mathbf{N}} \to X$ is the structural projection). Hence we have a morphism $\bar{p}_{n+1}: X^{\mathbf{N}}/X^n \to X$ such that $\bar{p}_{n+1}p = p_{n+1}$. Then $\bar{p}_{n+1}q = 1_X$, so that q must be a monomorphism. Therefore, $\Delta^{-1}(X^n) = 0$, so that $X = 0$ by Theorem 4.6, part 4. ✳

We say that an *Ab* 3-category is an *Ab 6-category* if, for each object X and for each set $\mathscr{J}$ such that for each $j \in \mathscr{J}$ a directed set $\{X_{j(i_j)}\}_{i \in \mathscr{I}_j}$ of subobjects of X is given, we have that

$$\bigcap_{j \in \mathscr{J}} (\Sigma_{i \in \mathscr{I}_j} X_{j(i_j)}) = \Sigma_{\{j(i_j)\} \in \Pi_j \mathscr{I}_j} (\bigcap_{\{j(i_j)\}} X_{j(i_j)}).$$

Dually, we have the notion of an *Ab 6*-category.*

Note 4.11. An *Ab* 6-category is obviously an *Ab* 5-category.

4.12. We leave it for the reader to verify that the category $\mathscr{A}b$ as well as the category of modules over an *Ab* 6-category are examples of *Ab* 6-categories.

An *Ab* 5-category having a set of generators is called a *Grothendieck category*

Exercises

4.1. An abelian group X is called a *torsion* group if for each $x \in X$ there is a nonzero integer n such that $na = a + \ldots + a$ (n-fold) $= 0$. Let $\mathscr{T}ors$ be the full subcategory of $\mathscr{A}b$ consisting of all torsion groups.

a) Show that $\mathscr{T}ors$ is a coreflective abelian subcategory of $\mathscr{A}b$.

b) Show that the product in $\mathscr{T}ors$ of a family $\{X_i\}_{i \in \mathscr{I}}$ of torsion groups is given by the subgroup of the product in $\mathscr{A}b$, consisting by all elements $\{x_i\}_{i \in \mathscr{I}}$ such that there is a nonzero integer n with $nx_i = 0$ for all i (i.e. all elements x_i have bounded order).

c) For every positive integer m, let $f_m \colon \mathbf{Z}_{2m} \to \mathbf{Z}_2$ be the morphism, which takes the coset r modulo 2^m into the coset r modulo 2. Show that each f_m is an epimorphism in $\mathscr{T}ors$, but $\prod_m f_m$ is not an epimorphism (in $\mathscr{T}ors$).

Thus an *Ab* 3- and *Ab* 3*- need not be an *Ab* 4*-category.

4.2. Let $\mathscr{C}$ be an *Ab* 3-(resp. an *Ab* 3*-) category. Prove that for every small category $\mathscr{I}$, the colimit (limit) functor $\underrightarrow{\lim}_{\mathscr{I}} \colon [\mathscr{I}, \mathscr{C}] \to \mathscr{C}$ ($\underleftarrow{\lim}_{\mathscr{I}} \colon [\mathscr{I}, \mathscr{C}] \to \mathscr{C}$) is right (left) exact. (Hint: Use the Theorem 12.8, Ch. 1.)

4.3. Let $\mathscr{C}$ be an *Ab* 3- and *Ab* 3*-category. Prove that the following are equivalent:

1) $\mathscr{C}$ is a C_2-category.

2) If $\{X_i\}_i$ is a set of objects of $\mathscr{C}$ and $f \colon Y \to \coprod_i X_i$ is a morphism such that $p_i f = 0$ for each i (p_i being the canonical projection associated with the structural injections) then $f = 0$.

4.4. Let $\mathscr{C}$ be an *Ab* 3-category and $\{X_i\}_i$ a directed set of subobjects of X. Prove that $\underrightarrow{\lim}_i (X/X_i) \simeq X/\Sigma_i X_i$.

4.5. Let $\{X_i, f_{ij}\}_{i\in\mathscr{I}}$ be a direct system in an *Ab* 3-category. Let R be the subset of $\mathscr{I} \coprod \mathscr{I}$ consisting of all ordered pairs (i,j) such that $i \leqslant j$. Let $v_i\colon X_i \to \coprod_i X_i$ be the structural morphisms and $X_R = \Sigma_{(i,j)\in R} \operatorname{Im}(v_i - v_j f_{ij})$. Show that $\coprod_i X_i/X_R$ with the canonical morphisms $\{pv_i\}_i$ ($p\colon \coprod_i X_i \to (\coprod_i X_i)/X_R$ being canonical) defines a colimit of this direct system.

4.6. Let $\mathscr{C}$ be an *Ab* 3-category in which any one of the parts 2) to 5) of Theorem 4.6 are satisfied, and let $\{X_i, f_{ij}\}_i$ be a direct system of $\mathscr{C}$. Let K_{ij} denote the kernel of $f_{ij}\colon X_i \to X_j$, $i \leqslant j$, and let K_i be the kernel of $u_i\colon X_i \to \varinjlim X_i$, where the u_i are the structural morphisms. Prove that $K_i = \Sigma_{i<j} K_{ij}$.

4.7. Show that every *Ab* 5-category is also an *Ab* 4-category.

4.8. Show that every *Ab* 5- and *Ab* 3*-category is a C_2-category.

4.9. Let $\{X_i, f_{ij}\}_i$ be a direct system in an *Ab* 5-category, and let $f\colon Y \to X_i$ be a morphism for some i. Show that $\operatorname{Ker}(u_i f) = \Sigma_{j>i} \operatorname{Ker}(f_{ij} f)$, where $u_i\colon X_i \to \varinjlim_i X_i$ are the structural morphisms.

4.10. Prove that a set $\{X_i\}_{i\in\mathscr{I}}$ of subobjects of an object X in an *Ab* 5-category is independent if and only if for every $i_0 \in \mathscr{I}$,

$$(\Sigma_{i\neq i_0} X_i) \cap X_{i_0} = 0.$$

4.11. Prove that an object X in an *Ab* 5-category is finitely generated if and only if every filtered and complete set $\{X_i\}_i$ of subobjects is stationary (i.e. $X = X_{i_0}$ for some index i_0).

4.12. An abelian category $\mathscr{C}$ is called *locally finitely generated* if each object X of $\mathscr{C}$ is the sum of its finitely generated subobjects.

a) Show that every object of a locally small and locally finitely generated *Ab* 5-category is the sum of the directed set of its finitely generated subobjects.

b) Show that an object X of a locally finitely generated and locally small *Ab* 5-category is finitely presented if and only if it is finitely generated and, furthermore, if each $p\colon X \to Y$ is an epimorphism, where Y is finitely generated, then $\operatorname{Ker} p$ is also finitely generated.

4.13. Let $\mathscr{C}$ be an *Ab* 5-category. We say that a subobject X' of X is *finitely generated relative to* X if for each complete and direct system $\{X_i\}_i$ of subobjects of X, the direct system $\{X' \cap X_i\}_i$ is stationary. Assuming that $\mathscr{C}$ is locally small, show that the following are equivalent:

a) $\mathscr{C}$ is an *Ab* 6-category.

b) Each object X of $\mathscr{C}$ is a sum of subobjects which are finitely generated relative to X.

4.5. The extension theorem in the additive case. A characterization of functor categories

The extension theorem stated in Section 2.2 has a different form in the additive case. That and some related questions will be examined in this section.

References: Bourbaki [1], Cartan – Eilenberg [1], Freyd [1], [2], Gabriel [1], Gabriel—Ulmer [1], Mitchell [2], Morita [1], Pareigis [1], N. Popescu [3], Ulmer [1], [4], Watts [1], Wraith [1].

LEMMA 5.1. *Consider the following diagram in an abelian category $\mathscr{C}$:*

$$\begin{array}{ccccccc} P_1 & \xrightarrow{d} & P_0 & \xrightarrow{q} & X & & \\ & & & & \downarrow{\scriptstyle f} & & \\ P_1' & \xrightarrow{d'} & P_0' & \xrightarrow{q'} & X' & \to & 0, \end{array} \tag{15}$$

where P_1 and P_0 are projective, the bottom row is exact, and $qd = 0$. Then there are morphisms $f_0\colon P_0 \to P_0'$ and $f_1\colon P_1 \to P_1'$ such that the diagram resulting when these are inserted in (15) *is commutative. Furthermore, let $T\colon \mathscr{C}_1 \to \mathscr{C}'$ be an additive functor into an additive category with cokernels, where $\mathscr{C}_1$ is a full subcategory of $\mathscr{C}$ that contains P_0, P_1, P_0' and P_1'.*

Then the induced morphism

$$\mathrm{Coker}\, T(d) \to \mathrm{Coker}\, T(d')$$

does not depend on the choice of f_0 and f_1.

Proof. Making use of the projectivity of P_0, we can find $f_0\colon P_0 \to P_0'$ such that $fq = q'f_0$. Then $q'f_0d = fqd = 0$.

Thus the image of f_0d is a subobject of $\mathrm{Ker}\, q' = \mathrm{Im}\, d'$. Hence by the projectivity of P_1, we can find a morphism $f_1\colon P_1 \to P_1'$ such that $f_0d = d'f_1$. Let $p\colon T(P_0) \to Y$ and $p'\colon T(P_0') \to Y'$ be the cokernels of $T(d)$ and $T(d')$, respectively. Assume that $g_0\colon P_0 \to P_0'$ is another pair of morphisms which when inserted in (15) yield a commutative diagram. Let $u, v\colon Y \to Y'$ be the morphisms induced by f_0, f_1 and g_0, g_1, respectively. Now $q'(f_0 - g_0) = 0$ and again using the projectivity of P_0 and the exactness of the bottom row, we obtain a morphism $h\colon P_0 \to P_1'$ such that $d'h = f_0 - g_0$. Then $(u - v)p = up - vp = p'T(f_0) - p'T(g_0) = p'T(f_0 - g_0) = p'T(d'h) = p'T(d')T(h) = 0$. Since p is an epimorphism, then we have that $u = v$. ✳

THEOREM 5.2. *Let $\mathscr{P}$ be a full subcategory of an Ab 3-category $\mathscr{C}$ and suppose that the objects of $\mathscr{P}$ form a set of small projective generators of $\mathscr{C}$. Let $T\colon \mathscr{P} \to \mathscr{C}'$ be an additive (covariant) functor into a cocomplete additive category. Then T can be uniquely extended to a functor $\overline{T}\colon \mathscr{C} \to \mathscr{C}'$ that commutes with colimits.*

Proof. We first extend T to the full subcategory of $\mathscr{C}$ which consists of all *free objects*, that is, the objects of the form $\coprod_i \mathscr{P}_i$, where $P_i \in \mathscr{P}$ for all i. We define

$\bar{T}(\coprod_i P_i) = \coprod_i T(P_i)$. For a morphism $t: \coprod_{i\in\mathscr{I}} P_i \to \coprod_{j\in\mathscr{J}} P_j$ of $\mathscr{C}$, P_i, $P_j \in \mathscr{P}$ for all i and j, let t_i denote tu_i (where $u_i: P_i \to \coprod_i P_i$ are the structural morphisms). By Exercise 1.9, and since P_i is small, we may write $t_i = \Sigma_{j\in\mathscr{J}} u_j p_j t_i$ (where u_j are also structural morphisms and the p_j are the associated projections of the coproduct $\coprod_j P_j$). Then define $\bar{T}(t): \coprod_i T(P_i) \to \coprod_j T(P_j)$ as the morphism which when composed with $\bar{u}_i$ (the structural injection of the coproduct $\coprod_i T(P_i)$ gives $\coprod_j \bar{u}_j T(p_j t_i)$ (where the $\bar{u}_j$ are also the structural injections of the coproduct $\coprod_j T(P_j)$). The additive functorial properties of $\bar{T}$ follow from the additive functorial properties of T.

Now let X be an object of $\mathscr{C}$. By Exercise 5.1, we get the exact sequence

$$P_1 \xrightarrow{d} P_0 \xrightarrow{q} X \longrightarrow 0 \tag{16}$$

where P_0 and P_1 are free objects. Then define $\bar{T}(X)$ to be $\operatorname{Coker} T(d)$.

For a morphism $f: X \to X'$, let

$$P_1' \xrightarrow{d'} P_0' \xrightarrow{q'} X' \longrightarrow 0$$

be an exact sequence used to define $\bar{T}(X')$. Then by Lemma 5.1 there are morphisms $f_0: P_0 \to P_0'$ and $f_1: P_1 \to P_1'$ such that $fq = q'f_0$ and $f_0 d = d'f_1$. Thus define $\bar{T}(f)$ as the morphism that makes the following diagram commutative:

$$\begin{array}{ccccccc}
\bar{T}(P_1) & \xrightarrow{\bar{T}(d)} & \bar{T}(P_0) & \xrightarrow{p} & \bar{T}(X) & \longrightarrow & 0 \\
\downarrow{\scriptstyle \bar{T}(f_1)} & & \downarrow{\scriptstyle \bar{T}(f_0)} & & \downarrow{\scriptstyle \bar{T}(f)} & & \\
\bar{T}(P_1') & \xrightarrow{\bar{T}(d')} & \bar{T}(P_0') & \xrightarrow{p'} & \bar{T}(X') & \longrightarrow & 0,
\end{array}$$

where p and p' are the canonical epimorphisms.

Again by Lemma 5.1, we see that $\bar{T}(f)$ is independent of the choice of f_0 and f_1, and thus $\bar{T}$ is seen to be an additive functor from $\mathscr{C}$ to $\mathscr{C}'$. Setting $X = X'$ and $f = 1_X$ in the above construction, we see at this point that (up to isomorphism) $\bar{T}$ is independent of the choice of the sequence (16) that was used to define it. Furthermore, the isomorphism is natural with respect to the morphisms to and from X.

Now we show that $\bar{T}$ commutes with colimits. We first show that $\bar{T}$ is right exact. To this end, let

$$0 \longrightarrow X' \xrightarrow{i} X \longrightarrow X'' \longrightarrow 0 \tag{17}$$

be an exact sequence of $\mathscr{C}$, and take the free resolutions

$$P_1' \longrightarrow P_0' \xrightarrow{h} X' \longrightarrow 0,$$

and

$$P_1'' \longrightarrow P_0'' \longrightarrow X'' \longrightarrow 0$$

for X' and X''. We construct the following commutative diagram:

$$\begin{array}{ccccccc} P_1' & \longrightarrow & P_0' & \xrightarrow{h} & X' & \longrightarrow & 0 \\ \downarrow & & \downarrow & & \downarrow{\scriptstyle i} & & \\ P_1' \amalg P_1'' & \xrightarrow{s} & P_0' \amalg P_0'' & \xrightarrow{q} & X & \longrightarrow & 0 \\ \downarrow & & \downarrow & & \downarrow{\scriptstyle p} & & \\ P_1'' & \longrightarrow & P_0'' & \longrightarrow & X'' & \longrightarrow & 0 \end{array} \qquad (18)$$

The middle column is a split exact sequence and the morphism q is defined by the projectivity of P_0'' and the composition ih. Taking the kernels of the morphisms from the middle column to the right-hand column, we obtain by Lemma 3.5 a short exact sequence. If we repeat the above construction with the sequence of kernels, replacing P_0' and P_0'' by P_1' and P_1'' respectively, we obtain the commutative diagram (18), the rows and columns being short exact sequences. Hence we can use the middle row to define $\overline{T}(X)$. Applying $\overline{T}$ to (18), the two left columns will still be split exact sequences by the additivity of $\overline{T}$. It follows from Lemma 3.6 that the sequence

$$\overline{T}(X') \longrightarrow \overline{T}(X) \longrightarrow \overline{T}(X'') \longrightarrow 0$$

is exact. In other words $\overline{T}$ is right exact.

Now, in order to show that $\overline{T}$ commutes with colimits, it will suffice to show that it commutes with coproducts.

Given a set $\{X_i\}_i$ of objects of $\mathscr{C}$, for each i choose a free resolution:

$$P_1^i \xrightarrow{d_i} P_0^i \xrightarrow{q_i} X_i \longrightarrow 0.$$

Then, using the fact that coproducts are right exact (see Exercise 4.2), we obtain the sequence

$$\amalg_i P_1^i \xrightarrow{\amalg_i d_i} \amalg_i P_0^i \xrightarrow{\amalg_i q_i} \amalg_i X_i \longrightarrow 0.$$

We obtain directly from this that $\overline{T}$ commutes with coproducts. Since at each stage the construction of $\overline{T}$ was subject to the requirement that it should commute with colimits, we have the uniqueness of $\overline{T}$. ⁎

THEOREM 5.3. *Let $\mathscr{C}$ be a preadditive small category and $T:\mathscr{C}\to\mathscr{C}'$ an (additive) cofunctor into a cocomplete additive category. Then there is a unique functor $T^*: \mathrm{Mod}\,\mathscr{C} \to \mathscr{C}'$ such that:*

1) *$T^*h^0 \simeq T$, where $h^0:\mathscr{C}\to \mathrm{Mod}\,\mathscr{C}$ is the canonical cofunctor.*

2) *T^* has a coadjoint.*

Moreover, every functor F: Mod $\mathscr{C} \to \mathscr{C}'$ *which commutes with colimits has a coadjoint.*

Proof. Let $\mathscr{P}$ be the full subcategory of Mod $\mathscr{C}$, generated by the objects h^X, when X runs through $\mathscr{C}$. Let $T': \mathscr{P} \to \mathscr{C}$ denote the functor defined by $T'(h^X) = T(X)$, and for every $f: h^X \to h^Y$ define $T'(f) = T(f')$ (where $f': Y \to X$ is the unique morphism of $\mathscr{C}$ such that $h^0(f') = f$).

By Theorem 5.2, there is a functor T^*: Mod $\mathscr{C} \to \mathscr{C}'$ such that T^* extends T' and T^* commutes with colimits. Define a functor $T_*: \mathscr{C}' \to$ Mod $\mathscr{C}$ as follows. For each object X' of $\mathscr{C}'$ define $T_*(X')$ to be $h_X\, T$. Similarly, if f' is a morphism of $\mathscr{C}'$, then define $T_*(f')$ to be $h(f')T$.

It is easy to see that T^* is an additive functor from $\mathscr{C}'$ into Mod $\mathscr{C}$ and we assert that it is coadjoint to T^*. The idea of this proof is the same as that of the proof of Corollary 2.8, Ch. 1, and is left for the reader. ✳

The functor T^* defined above is usually called the *additive tensor product* associated with the additive functor T.

Note 5.4. Theorem 5.3 has a dual which is obtained by replacing Mod $\mathscr{C}$ by Mod $\mathscr{C}^0$ and T by a functor.

The following result, due to Freyd, shows that Theorem 5.3 in fact characterizes the category Mod $\mathscr{C}$, where $\mathscr{C}$ is a small preadditive category.

THEOREM 5.5 (Freyd) [1]). *Let $\mathscr{C}$ be an abelian category. The following are equivalent:*

1) *$\mathscr{C}$ is an Ab 3-category and has a set $\{U_i\}_i$ of small projective generators.*

2) *There is a small preadditive category $\mathscr{P}$ such that $\mathscr{C}$ is equivalent to Mod $\mathscr{P}^0$.*

Proof. 1) ⇒ 2). Let $\mathscr{P}$ be the full subcategory of $\mathscr{C}$ generated by the objects $\{U_i\}_i$ and let $T: \mathscr{P} \to \mathscr{C}$ be the inclusion functor. Then there exists a unique functor T^*: Mod $\mathscr{P}^0 \to \mathscr{C}$ such that

$$T^*(h_{U_i}) = T(U_i) = U_i$$

for each i, and T^* has a coadjoint T_*. We show that T_* is an equivalence of categories. First note that for each i we have $T_*(U_i) = h_{U_i}$. Furthermore, since the objects $\{U_i\}_i$ are small and projective, they give a set of strong generators of $\mathscr{C}$. It also follows that the functor T_* commutes with colimits.

Let u: Id Mod $\mathscr{P}^0 \to T_*T^*$ be an arrow of adjunction and v a quasi-inverse coarrow. It is easy to see that $u_{h_{U_i}}$ is an isomorphism for every i, and since the objects $\{h_{U_i}\}_i$ give a set of small and projective generators, we see that u is a functorial isomorphism.

It is also easy to see that for every i the morphism v_{U_i} is a functorial isomorphism. As above it follows that v is a functorial isomorphism. The details are left for the reader.

The implication 2) ⇒ 1) is obvious (see Exercise 10). ✳

Let A be a ring and $\mathscr{C}$ a preadditive category. A (covariant) additive functor from A into $\mathscr{C}$ consists of an object X of $\mathscr{C}$ and a ring morphism from A into $\text{End}_{\mathscr{C}}(X)$.

In this case, X is called a *left A-object of $\mathscr{C}$*. In particular, any left A-module is a left A-object of $\mathscr{Ab}$. The definition of a right A-object is obtained by duality.

Theorems 5.3 and 5.5 can be reformulated for modules.

THEOREM 5.6. *Let A be a ring, $\mathscr{C}$ a cocomplete additive category and X a right A-object of $\mathscr{C}$. Then there exists a unique functor $X \otimes_A ?$: Mod $A \to \mathscr{C}$ such that:*

1) $X \otimes_A A = X$.
2) $X \otimes_A ?$ *has a coadjoint.*

Moreover, any functor from Mod A into $\mathscr{C}$ that commutes with colimits has a coadjoint. ⚹

THEOREM 5.7. *An abelian category $\mathscr{C}$ is equivalent to the category of modules over a ring if and only if $\mathscr{C}$ is an Ab 3-category and has a small projective generator.* ⚹

LEMMA 5.8. *Let $\mathscr{C}$ be a preadditive category and let U be an object of $\mathscr{C}$. Then the category $\mathscr{A}_U^a$ of algebras over the algebraic theory associated with U (see Section 3.5) is equivalent to the category of right modules over the ring $A_U = \mathrm{End}_{\mathscr{C}}(U)$.*

Proof. Let M be an $\mathscr{A}_U^a$-algebra. Then M_1, the carrier of M, is a right A_U-module. Indeed, by Section 3.5, an n-ary operation of $\mathscr{A}_U$, i.e. an element of $\mathscr{A}_U(\mathbf{n}, \mathbf{1})$, is defined by a morphism $t: U \to U^{(n)} = U^n$ in $\mathscr{C}$. Let $s: U \to U \coprod U$ denote the morphism $u_1 + u_2$, where the $u_i: U \to U \coprod U$ are the structural morphisms of the coproduct. It is easy to see that the binary operation induced by s is commutative and associative and has a neutral element induced by the zero morphism in U. Also the morphism $-1_U: U \to U$ defines the opposite of each element relative to the above binary operation. Hence M_1 becomes in a canonical way an abelian group. Furthermore, if $a \in \mathscr{A}_U(\mathbf{1}, \mathbf{1}) = A_U$, then the map $M(a)_1: M_1 \to M_1$ is a morphism of abelian groups, and the rules of associativity and addition of elements of the ring A_U show that M_1 is an A_U-module.

It is easy to see that a right A_U-module can be interpreted in a canonical way as an $\mathscr{A}_U$-algebra. The details are left for the reader. ⚹

By Theorem 5.9, Ch. 3, and Lemma 5.8, Ch. 3, we obtain the following result.

COROLLARY 5.9. *An abelian category $\mathscr{C}$ is an algebraic category if and only if it is equivalent to the category of (right) modules over a ring.* ⚹

Let $\mathscr{C}$ be an abelian category and $U \in \mathscr{C}$. Usually, the category $\mathscr{A}_U^a$, the category of $\mathscr{A}_U$-algebras, where, as above, $\mathscr{A}_U$ is the algebraic theory associated with U, will be systematically substituted for its equivalent category Mod A_U^0, the category of right modules over the ring $A_U^0 = \mathrm{End}_{\mathscr{C}}(U)$. In this case the Mal'cev functor $\mathscr{C} \to$ Mod $\mathscr{A}_U^0$ will be denoted by S_U. If there is no danger of confusion we shall write A and S instead of A_U and S_U.

By Lemmas 5.3 and 5.8 of Ch. 3, we have the following result.

LEMMA 5.10. *Let U be an object of an abelian category $\mathscr{C}$. Then:*

a) *The Mal'cev functor $S: \mathscr{C} \to$ Mod A^0 is left exact.*
b) *For every pair of natural numbers m, n, the canonical map:*

$$S(U^n, U^m): \mathscr{C}(U^n, U^m) \to \mathrm{Hom}_A(S(U^n), S(U^m))$$

is an isomorphism of abelian groups. ⚹

If $\mathscr{C}$ is an Ab 3-category, then by Theorem 5.6, Ch. 3, as well as by Theorem 5.6 of this chapter we see that for each object U of $\mathscr{C}$, the functor $S_U: \mathscr{C} \to \operatorname{Mod} A_U^0$ has an adjoint, namely the functor? $\otimes_{A_U} U$: $\operatorname{Mod} A_U^0 \to \mathscr{C}$. This functor will be denoted by T_U or simply by T, if there is no risk of confusion.

THEOREM 5.11. (Ulmer [4]). *Let $\mathscr{C}$ be a Grothendieck category and $U \in \mathscr{C}$. The following are equivalent:*

1) *The canonical functor T: $\operatorname{Mod} A^0 \to \mathscr{C}$ is left exact.*

2) *For every finite set $\{f_i\}_{1 \leqslant i \leqslant n}$ of elements of A, there is a set K and for every $k \in K$ a set $\{g_{ki}\}_{1 \leqslant i \leqslant n}$ of elements of A such that $\Sigma_i f_i g_{ki} = 0$ for all $k \in \mathscr{K}$, and the sequence*

$$U^{(K)} \xrightarrow{\coprod_k [g_{k1}, \ldots, g_{kn}] = g} U^n \xrightarrow{\langle f_1, \ldots, f_n \rangle = f} U$$

is exact.

Proof. 1) $\Rightarrow$ 2). Let $\{f_i : U \to U\}_{1 \leqslant i \leqslant n}$ be a finite set of elements of A. Let f denote $\langle f_1, \ldots, f_n \rangle$: $U^n \to U$, and let (X, t) be the kernel of f. Since the Mal'cev functor $S: \mathscr{C} \to \operatorname{Mod} A$ is left exact, we get the following exact sequence:

$$0 \to S(X) \xrightarrow{S(t)} S(U^n) \xrightarrow{S(f)} S(U) \quad . \qquad (*)$$

Now, since $S(U) = A$ is a generator of $\operatorname{Mod} A^0$, there is a set K and an epimorphism p: $S(U)^{(K)} \to S(X)$. Furthermore, since T is left exact, by applying T to $(*)$ we obtain the sequence

$$T(S(U)^{(K)}) = U^{(K)} \xrightarrow{T(p)} TS(X) \xrightarrow{TS(t)} TS(U^n) \xrightarrow{TS(f)} TS(U).$$

Now let v: $TS \to \operatorname{Id} \mathscr{C}$ be a coarrow of adjunction. Then we have the commutative diagram:

$$\begin{array}{ccccccc}
U^{(K)} & \xrightarrow{T(p)} & TS(X) & \xrightarrow{TS(t)} & TS(U^n) & \xrightarrow{TS(f)} & TS(U) \\
 & & \downarrow{\scriptstyle v_X} & & \wr\downarrow{\scriptstyle v_{U^n}} & & \wr\downarrow{\scriptstyle v_U} \\
0 & \longrightarrow & X & \xrightarrow{t} & U^n & \xrightarrow{f} & U
\end{array}$$

where v_{U^n} and v_U are isomorphisms since S and T commute with finite coproducts and since by hypothesis $TS(t)$ is a monomorphism. It follows that v_X is an isomorphism.

Let h_k: $U \to U^{(K)}$ and p_i: $U^n \to U$ be the obvious structural morphisms. Let g_{ki} denote $p_i t v_X T(p) h_k$. These morphisms are those required for 2).

2) $\Rightarrow$ 1). We divide the proof into several steps:

a) We first show that for each natural number n and each finitely generated submodule (X, t) of A, the morphism $T(t)$: $T(X) \to T(A)$ is a monomorphism (here t is the natural inclusion). To this end, let $(f_1, \ldots, f_n)$ be a set of generators of X and $f = \langle f_1, \ldots, f_n \rangle$: $U^n \to U$. It is easy to see that $\operatorname{Im} S(f) = X$.

Now let K be a set and for each $k \in K$ let $\{g_{ki}\}_{1 \leq i \leq n}$ be a set of endomorphisms of U such that condition 2) holds. Consider the morphisms:

$$A^{(K)} \xrightarrow{q = \coprod_k [S(g_{k1}), \ldots, S(g_{kn})]} A^n \quad .$$

We may assume that $\operatorname{Im} q = \operatorname{Ker} S(f)$, otherwise this can be obtained by adding finite families $\{g_{k'1}, \ldots, g_{k'n}\}_{k' \in K'}$, which satisfy $\Sigma_i f_i g_{k'i} = 0$ for every k', and that relative to $\overline{K} = K \cup K'$ satisfy condition 2). Hence, we may consider the diagram:

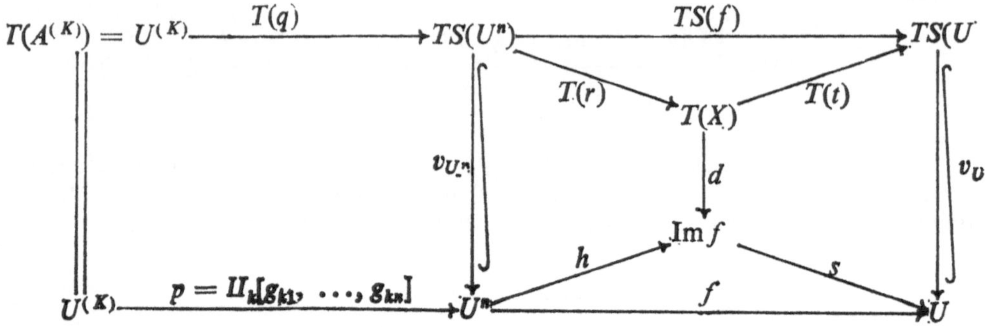

where $T(q) = p$, $tr = S(f)$ is the canonical decomposition of $S(f)$, and $T(r)$ is a coequalizer of $T(q)$. The above holds since T is right exact by adjunction. In a canonical manner the existence of the morphism $d = T(X) \to \operatorname{Im} f$ follows, which makes all diagrams commutative. Now it is easy to see that d is an isomorphism, and hence $T(f)$ is a monomorphism.

b) If n is a natural number and (X, t) is a finitely generated submodule of A^n, then $T(t)$ is a monomorphism.

This assertion is obtained by induction on n and is left for the reader.

c) If M is any set and (X, t) is a submodule of $A^{(M)}$, then $T(t)$ is a monomorphism. The following considerations yield this part. Every submodule is the union of the direct system of its finitely generated submodules. Every finitely generated submodule of $A^{(M)}$ is contained in A^n for a suitable natural number n, and T commutes with colimits.

d) Finally, let $f: X \to Y$ be a monomorphism in $\operatorname{Mod} A^0$. Then we obtain the commutative diagram:

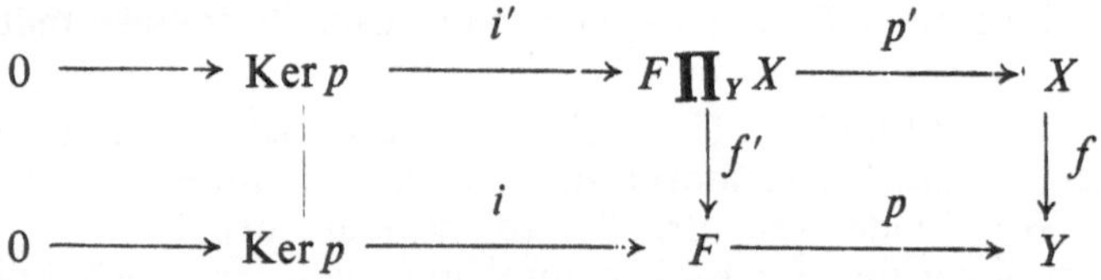

which is canonically constructed, and where the bottom row is exact and F is a free A^0-module. Since f is a monomorphism and the last square is cartesian, f' is a monomorphism. Then, by c), $T(f')$ and $T(i)$ are monomorphisms. Therefore

$T(i')$ is an equalizer of $T(p')$ and hence also of $T(f)T(p')$. Thus $T(f)$ is a monomorphism. ✳

THEOREM 5.12 (Gabriel—Popescu [1]). *Let $\mathscr{C}$ be a Grothendieck category, $U \in \mathscr{C}$, $S: \mathscr{C} \to$ Mod A_U^0 the Mal'cev functor, and T an adjoint to S. The following are equivalent:*

1) *U is a generator of $\mathscr{C}$.*

2) *S is full and faithful, and T is exact.*

Proof. 1) ⇒ 2). It is easy to see that the conditions of Theorem 5.8, Ch. 3, are verified, so that S is full and faithful. Also since U is a generator of $\mathscr{C}$, condition 2) of Theorem 5.11 also holds. The details are left for the reader.

The implication 2) ⇒ 1) follows from Exercise 5.6. ✳

COROLLARY 5.13. *Every Grothendieck category is an Ab 3*-category.*

The proof follows by Exercise 5.6 since every category of modules is an *Ab* 3*-category (see Proposition 13.12, Ch. 1). ✳

Exercises

5.1. Assume the hypotheses of Theorem 5.2. Show that for each object X of $\mathscr{C}$ we get an exact sequence:

$$\ldots \to P_n \to P_{n-1} \to \ldots \to P_1 \to P_0 \to X \to 0$$

where the P_i are free objects. This exact sequence is called a *free resolution* of X.

5.2. Let $T: \mathscr{C} \to \mathscr{C}'$ be an additive functor, where $\mathscr{C}$ and $\mathscr{C}'$ are small preadditive categories. Show that the restriction functor T: Mod $\mathscr{C}' \to$ Mod $\mathscr{C}$ has a coadjoint ${}^*T$ and an adjoint T_*.

5.3. Let A, B be two rings. An A-object of Mod B is simply called an *(A, B)-bimodule*. Prove: If X is an (A, B)-bimodule, then there exists a functor $? \otimes_A X$: Mod $A^0 \to$ Mod B such that:

1) $A \otimes_A X = X$.

2) $? \otimes_A X$ has a coadjoint.

Also prove that, moreover, for any functor T: Mod $A^0 \to$ Mod B which commutes with colimits, the object $T(A)$ is canonically an (A, B)-bimodule and $T \simeq\, ? \otimes_A T(A)$.

5.4. Let A and B be two rings and S: Mod $B \to$ Mod A^0 a functor. Prove that the following are equivalent:

1) S is an equivalence of categories.

2) There is a finitely generated projective generator U of Mod B and a ring isomorphism: $A \xrightarrow{\sim} \mathrm{End}_B(U)$ such that S is the Mal'cev functor associated with U.

5.5. (Morita [1]). Let A and B be any two rings and U an (A, B^0)-bimodule. Prove that the following are equivalent:

1) The functor $?\otimes_A U$: Mod $A^0 \to$ Mod B^0 is an equivalence of categories.

2) a) U is a projective and finitely generated A-module, b) U is a projective and finitely generated B^0-module, and c) the canonical morphisms of rings: $A \to \mathrm{End}_{B^0}(U)$ and $B^0 \to \mathrm{End}_A(U)$ are isomorphisms.

5.6. Let $T: \mathscr{C} \to \mathscr{C}'$ be a functor and S a coadjoint of T. Assume that T is exact and S is full and faithful. Prove:

a) If $\mathscr{C}$ is an *Ab* 3-, *Ab* 3*-, *Ab* 4- or *Ab* 5-category, then so is $\mathscr{C}'$.

b) If U is a generator of $\mathscr{C}$, then $T(U)$ is a generator of $\mathscr{C}'$.

5.7. Let $\mathscr{C}$ be a preadditive category. Show that the center of $\mathscr{C}$ is a commutative ring. Prove that if $\mathscr{C}$ is a Grothendieck category and U a generator of $\mathscr{C}$, then the center of $\mathscr{C}$ is canonically isomorphic to the center of the ring $\mathrm{End}_{\mathscr{C}}(U)$. (Hint: If we make use of Theorem 5.12, then this exercise may be reduced to the case of ModA^0, for a suitable ring A.)

5.8. An object U of an additive category $\mathscr{C}$ is called a *dense generator* if the full subcategory of $\mathscr{C}$ generated by $U \coprod U$ is dense.

a) Show that a dense generator is always a separator.

b) Show that in an *Ab* 5-category a generator is also a dense generator.

c) Prove: An additive category $\mathscr{C}$ is equivalent to a full reflective subcategory of a category of modules (over a small preadditive category) if and only if it is cocomplete and has a dense generator.

5.9. Let A be a ring and let $\mathscr{A}$ be the algebraic theory associated with the object A which is viewed as an A^0-module (as usual we identify A with $h(A)$, where $h: A \to$ Mod A^0 is the canonical functor). Such an algebraic theory is called an *annular theory*. Show that:

a) If U is an object in an additive category $\mathscr{C}$, then the algebraic theory associated with U is an annular theory.

b) An algebraic theory $\mathscr{A}$ is annular if and only if the category $\mathscr{A}^a$ is an additive category.

c) If $f: A \to B$ is a ring morphism, then it defines in a canonical way a theory morphism: $\tilde{f}: \mathscr{A} \to \mathscr{B}$. In fact, the assignments $A \rightsquigarrow \mathscr{A}$, $f \rightsquigarrow \tilde{f}$ define a full and faithful functor $R: \mathscr{R}g \to \mathscr{T}h$. This functor has an adjoint.

d) If A is a ring and $\mathscr{B}$ an algebraic theory, then the tensor product $\mathscr{A} \otimes \mathscr{B}$ is an annular theory. Moreover, an adjoint of the functor R defined in c) is the functor which assigns to each algebraic theory $\mathscr{B}$ the ring $\mathscr{B} \otimes \mathscr{Z}$, where Z is the ring of integers.

e) If A, B, are rings, then the ring associated with the tensor product $\mathscr{A} \otimes \mathscr{B}$ is canonically isomorphic to the usual tensor product of rings (see Bourbaki [2], Chevalley [1]).

5.10. Let $f: A \to B$ be a ring morphism. Let f_*: Mod $B \to$ Mod A denote the restriction functor, associated with f. Show that f is an epimorphism of rings if and only if f_* is a full functor.

4.6. Injective objects in abelian categories

The notion of an injective object plays a central role in the theory of abelian categories and in homological algebra. This section is devoted to showing that the most important abelian categories are categories with enough injective objects. The notion of an injective envelope is also discussed.

References. Baer [1], Buchsbaum [1], Cartan—Eilenberg [1], Eckmann—Schöpf [1], Freyd [1], Gabriel [1], Grothendieck [1], Kuroš [1], Mitchell [2], N. Popescu [3].

LEMMA 6.1. *Let $\mathscr{C}$ be an $\mathscr{Ab}$ 5-category, U a generator and Q an object of $\mathscr{C}$. The following are equivalent:*

1) *Q is an injective object.*

2) *For any monomorphism $j\colon U' \to U$ and for any morphism $f\colon U' \to Q$, there is a morphism $\bar{f}\colon U \to Q$ such that $\bar{f}j = f$.*

Proof. The implication 1) ⇒ 2) follows from the definition of an injective object.

2) ⇒ 1). Let X' be a subobject of some object X and let $f'\colon X' \to X$ be a morphism. Let $\mathscr{M}$ denote the set of all pairs (X'', f'') where X'' is a subobject of X that contains X', and $f''\colon X'' \to Q$ is a morphism which extends f'. It is easy to see that $\mathscr{M}$ is canonically an inductive set and so by Zorn's lemma we deduce that it has a maximal element. We may assume that (X', f') is a maximal element of $\mathscr{M}$. If $X' \neq X$, then, by hypothesis, there is a morphism $t\colon U \to X$ such that $t(U) \nsubseteq X'$, and $X'' = X' + t(U) \neq X'$. Let $U' = t^{-1}(X')$, and let $g'\colon U' \to X'$ be the unique morphism such that $g' = tr$ (where $r\colon U' \to U$ is the inclusion). We consider the exact sequence

$$U' \xrightarrow{h=[-r,\, g']} U \textstyle\coprod X' \xrightarrow{p} X'' \to 0$$

where p is defined by the canonical inclusion $X' \to X''$ and t. Let $g\colon U \to Q$ be a morphism which extends $f'g'$, and $v = \langle g, f\rangle\colon U \coprod X' \to Q$. Then $vh = 0$. Hence there is a unique morphism $f''\colon X'' \to Q$ such that $f''p = v$, i.e., $f''pk = vk = f'$, where $k\colon X' \to U \coprod X'$ is the canonical inclusion, but this is a contradiction. ✳

We now characterize the injective objects in the category $\mathscr{Ab}$. We observed previously that $\mathscr{Ab}$ is equivalent to the category Mod **Z**, so that the group **Z** of integers is a generator of $\mathscr{Ab}$ (see also Examples 4.4 and 4.5, Ch. 1). We assume that the reader is aware that every subgroup of **Z** is of the form $n\mathbf{Z}$, i.e. is generated by an element n of **Z** (see Kuroš [1]).

We say that an abelian group Q is *divisible* if for each $x \in Q$ and each $n \in \mathbf{Z}$ there is an element $x' \in Q$ such that $nx' = x$.

THEOREM 6.2. *An object Q in $\mathscr{Ab}$ is injective if and only if it is divisible.*

Proof. Let Q be an injective object of $\mathscr{Ab}$, $x \in Q$ and $n \in \mathbf{Z}$. Let $f\colon n\mathbf{Z} \to Q$ denote the morphism defined by $f(nm) = mx$. There is a morphism $g\colon \mathbf{Z} \to Q$ that extends f. Hence $g(n) = ng(1) = f(n) = x$, i.e., Q is divisible.

Conversely, we assume that Q is divisible and let $f\colon n\mathbf{Z} \to Q$ be a morphism; there is an element $x' \in Q$ such that $nx' = f(n)$. Then the morphism $\bar{f}\colon \mathbf{Z} \to Q$ defined by $\bar{f}(m) = mx'$ extends f. Hence Q is injective by Lemma 6.1. ✳

We say that an abelian category $\mathscr{C}$ *has enough injective objects* (or *has enough injectives*) if every object of $\mathscr{C}$ is a subobject of an injective one, or equivalently, for every object X of $\mathscr{C}$, there is an injective object Q and there exists a monomorphism $u\colon X \to Q$.

The following result is fundamental.

THEOREM 6.3. *The category* $\mathscr{A}b$ *has enough injectives.*

Proof. For each object X of $\mathscr{A}b$ consider the diagram:

$$\begin{array}{ccccccccc} 0 & \longrightarrow & \operatorname{Ker} p & \xrightarrow{\;i\;} & \mathbf{Z}^{(\mathscr{J})} & \xrightarrow{\;p\;} & X & \longrightarrow & 0 \\ & & & & \downarrow{\scriptstyle u} & & \downarrow{\scriptstyle v} & & \\ & & & & \mathbf{Q}^{(\mathscr{J})} & \xrightarrow{\;q\;} & \mathbf{Q}^{(\mathscr{J})}/R & \longrightarrow & 0 \end{array}$$

in which the top row is exact, u is induced by the natural inclusion $\mathbf{Z} \to \mathbf{Q}$ and $R = u(\operatorname{Ker} p)$. It is clear that there is a unique monomorphism v which makes the diagram commutative. By Exercise 6.1, $\mathbf{Q}^{(\mathscr{J})}/R$ is divisible, and hence injective by Theorem 6.2. ✳

THEOREM 6.4. *For every ring A, the category* Mod A *has enough injectives.*

Proof. Let $U\colon \operatorname{Mod} A \to \mathscr{A}b$ be the underlying functor. It is easy to see that this functor is canonically induced by the $(A^0, \mathbf{Z})$-bimodule A, and so it has a coadjoint S. For each $X \in \mathscr{A}b$, $S(X) = \operatorname{Hom}_{\mathbf{Z}}(A, X)$ and its structure as an A-module is canonically defined as follows. If $a \in A$ and $f \in S(X)$, then $af\colon A \to X$ is the map defined by $(af)(x) = f(xa)$. Since U is an exact functor then by Exercise 6.2, a) the functor S preserves injective objects. Hence, let X be an A-module and $h\colon U(X) \to Q$ be a monomorphism where Q is an injective object of $\mathscr{A}b$. Now let $U\colon \operatorname{Id} \operatorname{Mod} A \to SU$ be an arrow of adjunction. Since U is faithful, u is a functorial monomorphism (see Proposition 13.7, Ch. 1). Therefore the composite morphism

$$X \xrightarrow{\;u_X\;} SU(X) \xrightarrow{\;S(h)\;} S(Q)$$

is a monomorphism, and $S(Q)$ is injective. ✳

Let X' be a subobject of X. A subobject X'' of X is called a *complement* of X' if $X' \cap X'' = 0$ and $X' + X''$ is an essential subobject of X. In this situation we say that X' is *complemented*, or that X' is *complemented by* X''.

LEMMA 6.5. *Let $\mathscr{C}$ be a locally small Ab 5-category. Then every subobject X' of an object X is complemented. Moreover, any complement of X' is contained in a maximal one.*

Proof. Let $\mathscr{M}$ be the set of all subobjects of X that trivially intersect X'. Under the natural ordering of subobjects, $\mathscr{M}$ is an ordered set, and, by the *Ab* 5-condition,

$\mathscr{M}$ is an inductive set. Hence by Zorn's lemma it has a maximal element X''. Then X'' is a complement of X'. Indeed, if Y is a nonzero subobject of X such that $(X' + X'') \cap Y = 0$, then $X' \cap (X'' + Y) = 0$ and $X'' + Y$ strictly contains X'', which is a contradiction. The last part of the lemma follows in a canonical way. ✳

LEMMA 6.6. *Let $\mathscr{C}$ be a category that satisfies the following condition: If X' is a subobject of an object X, then X' is essential, or X' is complemented and any complement of X' is contained in a maximal one. An object Q of $\mathscr{C}$ is injective if and only if Q does not admit proper essential extensions.*

Proof. Let $u\colon Q \to X$ be a monomorphism. If u is not essential, then a maximal complement X' of $u(Q)$ exists and the composite morphism

$$Q \xrightarrow{\;u\;} X \xrightarrow{\;p\;} X/X'$$

is an essential monomorphism (p is the canonical morphism). In fact, if Y is a nonzero subobject of X/X' such that $pu(Q) \cap Y = 0$, then $u(Q) \cap p^{-1}(Y) = 0$ and obviously $p^{-1}(Y) \supset X'$, which is a contradiction. But then pu is an isomorphism and so $u(pu)^{-1}p$ is an idempotent of $\mathrm{End}_{\mathscr{C}}(X)$. Equivalently, u is a section (see Exercise 3.14). The lemma now follows from Exercise 6.8. ✳

An *injective envelope* of an object X is an essential extension $X \to Q$ where Q is injective.

LEMMA 6.7. *Let $\mathscr{C}$ be a category satisfying the condition given in Lemma 6.6. If X is a subobject of an injective object Q, then X has an injective envelope.*

Proof. Let Y be a maximal complement of X in Q, and $\bar{Q}$ a maximal complement of Y that contains X. Then X is an essential subobject of $\bar{Q}$, and $\bar{Q}$ is injective. Indeed, let X' be a subobject of $\bar{Q}$ such that $X' \neq 0$ and $X' \cap X = 0$. Then $X' \cap Y = 0$, and $X' + Y$ is a complement of X that strictly contains Y, which is a contradiction. Hence X is essential in $\bar{Q}$.

In order to get that $\bar{Q}$ is an injective object we shall use Lemma 6.6. To this end, let $f\colon \bar{Q} \to T$ be an essential monomorphism. By the injectivity of Q, there is a morphism $g\colon T \to Q$ such that gf is a monomorphism (in fact gf is the inclusion $\bar{Q} \to Q$). Then, by Exercise 6.6, we get that g is a monomorphism since f is essential. Hence $g(T)$ is a subobject of Q that contains $\bar{Q}$ and $\bar{Q}$ is essential in $g(T)$. Thus we have that $g(T) \cap Y = 0$. Now, since $\bar{Q}$ is a maximal complement of Y, we get that $\bar{Q} = g(T)$, i.e., f is an isomorphism. Hence $\bar{Q}$ is injective by Lemma 6.6. ✳

We say that *a category $\mathscr{C}$ has injective envelopes* if each object of $\mathscr{C}$ has an injective envelope.

COROLLARY 6.8. *For every ring A, the category* Mod A *is a category with injective envelopes.*

The proof follows from Theorem 6.4, Lemma 6.5, and Lemma 6.7. ✳

Corollary 6.8 holds in every Grothendieck category.

We shall first prove the following result.

LEMMA 6.9. *Let $\mathscr{C}$ be an abelian category, U a generator of $\mathscr{C}$ and $S: \mathscr{C} \to \to \operatorname{Mod} A^0$ the Mal'cev functor associated with U. If $t: X' \to X$ is an essential monomorphism, then $S(t)$ is again an essential monomorphism.*

Proof. Assume that $g \in \operatorname{Hom}_{\mathscr{C}}(U, X) = S(X)$ is a nonzero element. We must find an element $a \in A = \operatorname{End}_{\mathscr{C}}(U)$ such that $0 \neq ga \in S(X') = \mathscr{C}(U, X')$, or more precisely such that $\operatorname{Im}(ga) \subseteq \operatorname{Im} t$. Consider the following diagram:

$$\begin{array}{ccccc} V & \xrightarrow{\ k\ } & X' \cap X'' & \xrightarrow{\ w\ } & X' \\ {\scriptstyle h}\downarrow & & \downarrow & & \downarrow{\scriptstyle t} \\ U & \xrightarrow{\ q\ } & X'' & \xrightarrow{\ v\ } & X \end{array}$$

in which the bottom row is the factorization of g through its image and each square is cartesian. Since $g \neq 0$, we have that $X'' \neq 0$ and thus $X' \cap X'' \neq 0$ (see Exercise 6.7, c)). Since q is an epimorphism, k will also be an epimorphism by Corollary 2.6, and consequently $k \neq 0$. Hence there is a morphism $s: U \to V$ such that $ks \neq 0$. Now, since t and w are monomorphisms, $twks \neq 0$. Therefore $vqhs = = ghs \neq 0$, and we can take $a = hs$. ⁎

THEOREM 6.10. *Any Grothendieck category is a category with injective envelopes.*

Proof. The proof of this theorem will be similar to that of Theorem 5.12. Let U be a generator of $\mathscr{C}$, $S: \mathscr{C} \to \operatorname{Mod} A^0$ the al'cev functor associated with U, T an adjoint to S and $u: \operatorname{Id} \operatorname{Mod} A^0 \to ST$ an arrow of adjunction. Let X be an object of $\mathscr{C}$, and let $f: S(X) \to Q$ be an essential monomorphism, where Q is injective (see Theorem 6.4). We get the commutative diagram:

$$\begin{array}{ccc} S(X) & \xrightarrow{\ f\ } & Q \\ {\scriptstyle u_{S(X)}}\downarrow & & \downarrow{\scriptstyle u_Q} \\ STS(X) & \xrightarrow{\ ST(f)\ } & ST(Q) \end{array}$$

where $u_{S(X)}$ is an isomorphism, and $ST(f)$ a monomorphism (see the conditions for adjointness in Section 1.11 and Theorem 5.12).

We claim that $T(Q)$ is an injective object of $\mathscr{C}$, and that $T(f)$ is an essential monomorphism. For this, let us assume that $T(f)$ is essential. Then, by Lemma 6.9, $ST(f)$ is an essential monomorphism. Hence u_Q is also an essential monomorphism; therefore it is an isomorphism. Furthermore, let $r: Y' \to Y$ be a monomorphism of $\mathscr{C}$, and $g: Y' \to T(Q)$. Then there is a morphism $h: S(Y) \to ST(Q)$ such that $hS(r) = S(g)$. But since S is full and faithful, there exists a unique morphism $t: Y \to T(Q)$ such that $S(t) = h$. Hence $tr = g$, and therefore $T(Q)$ is injective.

Now we must show that $T(f)$ is essential. For that, let Y be a nonzero subobject of $T(Q)$ such that $Y \cap X = 0$ (we identify X with $TS(X)$ and X with the image of $T(f)$). Then $S(X) \cap S(Y) = 0$, and $S(Y) \neq 0$. Hence $S(Y) \cap u_Q(Q) = 0$. Therefore, we have the split exact sequence

$$0 \to Q \xrightarrow{u_Q} ST(Q) \to Q' \to 0,$$

and $S(Y) \subseteq Q'$. But $T(u_Q)$ is an isomorphism and thus $T(Q') = 0$, i.e. $TS(Y) \simeq Y = 0$, a contradiction. Therefore $T(f)$ is an essential monomorphism. ✳

Exercises

6.1. a) Show that the additive group of rational numbers **Q** is divisible, hence injective, in $\mathcal{A}b$.

b) Let p be a prime number. Let $\mathbf{Q}_p$ denote the multiplicative group of all complex numbers that satisfy an equation of the form $z^{p^n} = 1$, $n > 0$. Show that $\mathbf{Q}_p$ is an injective object in $\mathcal{A}b$.

c) Show that any product and any coproduct of divisible abelian groups is divisible and that a quotient group of a divisible abelian group is divisible.

6.2. Prove:

a) Let $F: \mathscr{C} \to \mathscr{C}'$ be a functor and G a coadjoint of F. If the functor F is exact, then G transforms any injective object of $\mathscr{C}'$ into an injective object of $\mathscr{C}$.

b) Conversely, if $\mathscr{C}'$ is a category with enough injective objects and G transforms any injective object of $\mathscr{C}'$ into an injective object, then F is exact.

6.3. Show that any *Ab* 3-category with enough injective objects is an *Ab* 4-category.

6.4. Let $\mathscr{C}$ be an *Ab* 3-category with a set of generators and enough injective objects. Show that $\mathscr{C}$ has an injective cogenerator.

6.5. Let A be a ring and T: Mod $A \to \mathcal{A}b$ a functor which commutes with limits. Show that T is representable (Hint: By Exercise 6.4 Mod A has an injective cogenerator. Then we use the same idea as in Theorem 6.11, Ch. 2.)

6.6. Let $u: X' \to X$ be a monomorphism in an abelian category. Prove that the following are equivalent:

1) For each subobject Y of X, $u^{-1}(Y) = 0$ implies $Y = 0$.

2) If $f: X \to Z$ is a morphism such that $\operatorname{Ker}(fu) = \operatorname{Ker} u$, then f is a monomorphism.

Such a morphism is called *essential*.

6.7. Prove that:

a) If $X \xrightarrow{u} Y \xrightarrow{v} Z$ are monomorphisms then vu is an essential monomorphism if and only if u and v are essential monomorphisms.

b) If $u_i: X_i' \to X_i$, $i = 1, \ldots, n$ are essential monomorphisms, then $\coprod_i u_i$ is also an essential monomorphism.

Let X' be a nonzero subobject of X. We say that X' is an *essential subobject* of X, or that X is an *essential extension* of X', if the natural inclusion $X' \subseteq X$ is an essential morphism.

c) A subobject X' of X is essential if and only if for each subobject X'' of X, the relation $X' \cap X'' = 0$ implies $X'' = 0$. In particular, if X is an A-module, then X' is essential if and only if for each $x \in X$, $x \neq 0$, there is an element $a \in A$, such that $ax \neq 0$ and $ax \in X'$.

d) The additive group $\mathbf{Q}$ of rational numbers is an essential extension of every nonzero subgroup of $\mathbf{Q}$. The same is true for the group $\mathbf{Q}_p$ (defined in Exercise 6.1. b).

6.8. Show that an object X is injective (projective) if and only if every monomorphism $f: X \to Y$ (every epimorphism $g: Y \to X$) is a section (a retraction).

6.9. Let $u: X \to Q$ and $u': X \to Q'$ be any two injective envelopes of X. Show that there is an isomorphism $f: Q \to Q'$ (not necessarily unique) such that $fu = u'$.

6.10. Show that the converse of Lemma 6.9 is true. Namely, show that if $S(t)$ is an essential extension, then t is also essential.

6.11. Show that any Grothendieck category has enough injective objects, and consequently it is a category with injective envelopes. In doing this, make specific use of the procedure due to Cartan and Eilenberg [1]. Then obtain a new proof of Theorem 5.12 by proving that the functor S preserves injective objects and using Exercise 6.2.

4.7. Categories of additive fractions

In this section we define and study categories of additive fractions. An important example of a category of additive fractions is the quotient category of an abelian category relative to a dense subcategory. Localizing subcategories and their associated quotient categories are defined similarly.

References: Almkvist [1], Gabriel [1], Gabriel—Oberst [1], Gabriel—Zisman [1], Grothendieck [1], N. Popescu [3], Schubert [1].

Let $\mathscr{C}$ be a preadditive category and Σ a class of morphisms in $\mathscr{C}$. A pair $(P_\Sigma, \mathscr{C}^a(\Sigma^{-1}))$, where $P_\Sigma: \mathscr{C} \to \mathscr{C}^a(\Sigma^{-1})$ is an additive functor, is called a *category of additive fractions of $\mathscr{C}$ relative to* Σ if the following conditions are satisfied:

i) For each $f \in \Sigma$, $P_\Sigma(f)$ is invertible.

ii) If $F: \mathscr{C} \to \mathscr{D}$ is an additive functor, and if for each $f \in \Sigma$, $F(f)$ is invertible, then there is a unique additive functor $\bar{F}: \mathscr{C}^a(\Sigma^{-1}) \to \mathscr{D}$ such that $\bar{F}P_\Sigma = F$.

As in the general case, the functor P_Σ (simply denoted by P if there are no doubts) is called the *canonical functor.*

Note 7.1. As in the general case (see Note 13.1, Ch. 1), we can show that the category $\mathscr{C}^a(\Sigma^{-1})$ need not exist. Moreover, it is possible for the category $\mathscr{C}(\Sigma^{-1})$ to be defined but $\mathscr{C}^a(\Sigma^{-1})$ not to be.

Note 7.2. The conventions used in Note 13.2, Ch. 1, will be used here also.

Example 7.3. The classical example of a category of additive fractions is given next. Let $\mathbf{Z}$ be the ring of integers and $u: \mathbf{Z} \to \mathbf{Q}$ the canonical inclusion. Then $(u, \mathbf{Q})$ is a category of additive fractions of $\mathbf{Z}$ relative to the set Σ of all its nonzero elements.

THEOREM 7.4. *Let $\mathscr{C}$ be a small preadditive category and Σ a set of morphisms of $\mathscr{C}$. Then the category $\mathscr{C}^a(\Sigma^{-1})$ exists.*

Proof. Let $(T, \mathscr{C}(\Sigma^{-1}))$ be a category of fractions of $\mathscr{C}$ relative to Σ (see Theorem 13.3, Ch. 1). Consider the preadditive category Add $(\mathscr{C}(\Sigma^{-1}))$ defined in Exercise 1.3 and let $I: \mathscr{C}(\Sigma^{-1}) \to \mathrm{Add}(\mathscr{C}(\Sigma^{-1}))$ be the canonical functor. Denote by $\mathscr{A}$ the two-sided ideal of $\mathrm{Add}(\mathscr{C}(\Sigma^{-1}))$ generated by elements of the form

$$IT(f+g) - IT(f) - IT(g), \quad f, g \in \mathscr{C}(X, Y)$$

when (X, Y) runs through the set $\mathcal{O}b(\mathscr{C}) \coprod \mathcal{O}b(\mathscr{C})$.

Let $\mathscr{C}'$ be the functor category $\mathrm{Add}(\mathscr{C}(\Sigma^{-1}))/\mathscr{A}$ and $H: \mathrm{Add}(\mathscr{C}(\Sigma^{-1})) \to \mathscr{C}'$ the canonical functor. Let P denote HIT. Then $(P, \mathscr{C}')$ is a category of additive fractions of $\mathscr{C}$ relative to Σ. The details are left for the reader. ✳

Other examples of categories of additive fractions may be constructed by use of the following result.

THEOREM 7.5. *Let Σ be a multiplicative system of a preadditive category $\mathscr{C}$. If the left fractional category of $\mathscr{C}$ relative to Σ is defined, then it is a category of additive fractions.*

Proof. By Theorem 14.1, Ch. 1, Σ is a left calculable system, and for any pair (X, Y) of objects of $\mathscr{C}$ the functor $H^Y\sigma_X: X/\Sigma \to \mathscr{S}et$ has a colimit. By Note 1.8 we see that the functor $H^Y: \mathscr{C} \to \mathscr{S}et$ can be factored as

$$\mathscr{C} \xrightarrow{h^Y} \mathscr{A}b \xrightarrow{U} \mathscr{S}et$$

where U is the underlying functor. Now, since X/Σ is filtered (see Exercise 14.1, Ch. 1) and the underlying functor U commutes with filtered colimits (see Corollary 3.5, Ch. 3), it follows from Theorem 14.1, Ch. 1, that for any $X, Y \in \mathscr{C}^l(\Sigma^{-1})$ the set $\mathscr{C}^l(\Sigma^{-1})(X, Y)$ has canonically an abelian group structure. We leave it for the reader to show that $\mathscr{C}^l(\Sigma^{-1})$ is a preadditive category, the canonical functor $P: \mathscr{C} \to \mathscr{C}^l(\Sigma^{-1})$ is additive and finally, $(P, \mathscr{C}^l(\Sigma^{-1}))$ is a category of additive fractions of $\mathscr{C}$ relative to Σ. ✳

By Propositions 14.4 and 14.5, Ch. 1, we obtain the following result.

PROPOSITION 7.6. *Let $\mathscr{C}$ be a preadditive category and Σ a multiplicative system of morphisms of $\mathscr{C}$. Assume that either $\mathscr{C}^l(\Sigma^{-1})$ or $\mathscr{C}^r(\Sigma^{-1})$ is defined and that Σ is bicalculable. Then, if $\mathscr{C}$ is preabelian (abelian) then so is $\mathscr{C}^a(\Sigma^{-1}) = \mathscr{C}^l(\Sigma^{-1}) = \mathscr{C}^r(\Sigma^{-1})$, and the canonical functor is exact.* ⚹

A full subcategory $\mathscr{A}$ of an abelian category $\mathscr{C}$ is called *thick* if for any exact sequence from $\mathscr{C}$

$$0 \to X' \to X \to X'' \to 0$$

X belongs to $\mathscr{A}$ if and only if X' and X'' belong to $\mathscr{A}$.

If $\mathscr{A}$ is a thick subcategory of $\mathscr{C}$, let $\Sigma_{\mathscr{A}}$ or simply Σ denote the class of all morphisms s of $\mathscr{C}$ such that $\operatorname{Ker} s$ and $\operatorname{Coker} s$ belong to $\mathscr{A}$.

PROPOSITION 7.7. *The system Σ is a bicalculable system of morphisms.*

Proof. Using duality, it will suffice to show that Σ is right calculable.

We first show that Σ is multiplicative. Indeed, if $s: X \to Y$ and $s': Y \to Z$ are morphisms of Σ, then by the canonically defined exact sequence

$$0 \to \operatorname{Ker} s \to \operatorname{Ker}(s's) \to (\operatorname{Im} s) \cap (\operatorname{Ker} s') \to 0$$

we see that $\operatorname{Ker}(s's) \in \mathscr{A}$. In a dual fashion, we get that $\operatorname{Coker}(s's) \in \mathscr{A}$. Hence $s's \in \Sigma$.

Now consider the following angle $X \xrightarrow{s} Y \xleftarrow{f} Z$ with $s \in \Sigma$. Then by the commutative square

$$\begin{array}{ccc} Z \coprod_Y X & \xrightarrow{f'} & X \\ {\scriptstyle s'}\downarrow & & \downarrow{\scriptstyle s} \\ Z & \xrightarrow{f} & Y \end{array}$$

we have that $\operatorname{Ker} s \simeq \operatorname{Ker} s'$ and that $\operatorname{Coker} s'$ is canonically included in $\operatorname{Coker} s$. Hence we get that $s' \in \Sigma$.

Finally, let $X \xrightarrow{f} Y \xrightarrow{s} Z$ be morphisms of $\mathscr{C}$ with $s \in \Sigma$ and such that $sf = 0$. Then $\operatorname{Im} f \subseteq \operatorname{Ker} s$ so we obtain that the equalizer of f belongs to Σ. ⚹

THEOREM 7.8. *Let $\mathscr{C}$ be a wellpowered abelian category, $\mathscr{A}$ a thick subcategory of $\mathscr{C}$, and $\Sigma = \Sigma_{\mathscr{A}}$. Then the category $\mathscr{C}^a(\Sigma^{-1})$ is defined.*

Proof. Let $\Sigma'(\Sigma'')$ be the set of all monomorphisms (epimorphisms) of Σ. It is easy to see that Σ' is a right calculable system. Also, for any object X, the category Σ'/X has a coinitial small subcategory (which consists of all subobjects X' of X such that $X/X' \in \mathscr{A}$). Hence by Theorem 7.5 and Proposition 7.6 we see that the category $\mathscr{C}^a(\Sigma'^{-1})$ is defined, is a finitely complete category, and the canonical functor $P': \mathscr{C}^a(\Sigma'^{-1})$ is left exact.

Let $\bar{\Sigma}''$ be the system of all morphisms of $\mathscr{C}^a(\Sigma'^{-1})$ that can be inserted in the commutative diagram

$$\begin{array}{ccc} P'(X) & \xrightarrow{P'(s)} & P'(X'') \\ \wr & & \wr \\ P'(X_1) & \xrightarrow{\alpha} & P'(X_1'') \end{array}$$

where $s: X \to X''$ belongs to Σ''. Now we show that $\bar{\Sigma}''$ is a left calculable system of $\mathscr{C}^a(\Sigma'^{-1})$. Obviously, we may assume that any morphism in $\bar{\Sigma}''$ has the form $P'(s'')$, with $s'' \in \Sigma''$. Let

$$P'(Z) \xleftarrow{(f/s')} P'(X) \xrightarrow{P'(s'')} P'(Y)$$

be a coangle in $\mathscr{C}^a(\Sigma'^{-1})$ (see the duals of conditions l_1)—l_3) stated in Section 1.14), where $f: X' \to Z$, $s': X' \to X$ and $s' \in \Sigma$. Then $s''s' \in \Sigma$, and the canonical morphisms $u: X' \to \mathrm{Im}(s''s')$ and $v: \mathrm{Im}(s''s') \to Y$ are in Σ'' and Σ' respectively. Thus we obtain the commutative diagram of $\mathscr{C}$

$$\begin{array}{ccc} X & \xrightarrow{s''} & Y \\ {\scriptstyle s'}\uparrow & & \uparrow{\scriptstyle v} \\ X' & \xrightarrow{u} & \mathrm{Im}(s''s') \\ {\scriptstyle f}\downarrow & & \downarrow{\scriptstyle f'} \\ Z & \xrightarrow{u''} & Z \coprod_{X'} (\mathrm{Im}(s''s')) \end{array}$$

and obviously $u'' \in \Sigma''$. It is clear that $P'(s')$ and $P'(v)$ are isomorphisms. Thus

$$P'(u'')P'(f)P'(s')^{-1} = P'(f')P'(u)P'(s')^{-1} = P'(f')P'(v)^{-1}P'(s'') = (f'/v)P'(s'').$$

These relations show that $\bar{\Sigma}''$ is left permutable

Furthermore, we show that $\bar{\Sigma}''$ is left simplifiable. We shall first show that $P'(s'')$ is an epimorphism for all $s'' \in \Sigma''$. Indeed, if $s'': X \to Y$ and $(f/s'): P'(Y) \to$ $\to P'(Z)$ are morphisms such that $(f/s')P'(s'') = 0$, then we may assume that s' is an identity and $(f/s') = P'(f)$. This implies that $P'(fs'') = 0$. Hence, by the dual of condition l_3) in Section 1.14, there is an element u of Σ', $u: X' \to X$, such that $fs''u = 0$. Now since the canonical morphism $v: \mathrm{Im}(s''u) \to Y$ belongs to Σ', we have $fv = 0$. But then we must have that $P'(f) = 0$. Hence, any element of $\bar{\Sigma}''$ is an epimorphism, so that $\bar{\Sigma}''$ is a left simplifiable system and thus a left calculable system.

Now from the definition of the system $\bar{\Sigma}''$ we get that for each object $P'(X)$ of $\mathscr{C}^a(\Sigma'^{-1})$ the category

$$P'(X)/\bar{\Sigma}''$$

has a cofinal small subcategory. Hence by Corollary 14.2, Ch. 1, and Theorem 7.5 we get that the category $\bar{\mathscr{C}} = (\mathscr{C}^a(\Sigma'^{-1}))^a(\bar{\Sigma}''^{-1})$ is defined. Let P'' denote the canonical functor and let P denote $P''P'$. Then the pair $(P, \bar{\mathscr{C}})$ is a category of additive fractions of $\mathscr{C}$ relative to Σ. Indeed, since every element s of Σ can be written as $s = s's''$, with $s' \in \Sigma'$ and $s'' \in \Sigma''$, we see that $P(f)$ is invertible for all elements $f \in \Sigma$. Furthermore, let $H: \mathscr{C} \to \mathscr{C}'$ be an additive functor such that $H(f)$ is invertible for all elements f of Σ. Since $\Sigma' \subseteq \Sigma$, there is a unique functor $H': \mathscr{C}^a(\Sigma'^{-1}) \to \mathscr{C}'$ such that $H'P' = H$. Furthermore, since $\Sigma'' \subseteq \Sigma$, we get that H' makes invertible all elements of $\bar{\Sigma}''$, so that there is a unique functor $H'': \bar{\mathscr{C}} \to \mathscr{C}'$ such that $H''P'' = H'$. But it is clear that H'' is the unique functor such that $H''P = H$. ⁎

Let us consider $\mathscr{C}$ and $\mathscr{A}$ as in Theorem 7.8. Then the category $\mathscr{C}^a(\Sigma_{\mathscr{A}}^{-1})$ is denoted by $\mathscr{C}/\mathscr{A}$ and is called the *quotient category* of $\mathscr{C}$ relative to $\mathscr{A}$. The hypothesis of Theorem 7.8 will be used again.

By Lemma 15.1, Ch. 1, Proposition 15.2, Ch. 1, and by Exercise 7.3, we get the following result. The details of the proof are left for the reader.

LEMMA 7.9. *For an object M of $\mathscr{C}$, the following are equivalent:*

1) *M is left closed for $\Sigma_{\mathscr{A}}$ (we shall say simply $\mathscr{A}$-closed).*

2) *M does not contain subobjects in $\mathscr{A}$, and every morphism $s: M \to X$, $s \in \Sigma_{\mathscr{A}}$ is a section.*

3) *For each object X of $\mathscr{C}$, the morphism of abelian groups*

$$P(X, M): \mathscr{C}(X, M) \to (\mathscr{C}/\mathscr{A})P(X), P(M))$$

is an isomorphism. ⁎

An object X of $\mathscr{C}$ which has no nonzero subobjects in $\mathscr{A}$ is called an *$\mathscr{A}$-torsionfree object*. It is clear from Lemma 7.9,2 that every $\mathscr{A}$-closed object is also $\mathscr{A}$-torsionfree.

We shall now deal with the following question. Under what conditions does the canonical functor $P: \mathscr{C} \to \mathscr{C}/\mathscr{A}$ have a coadjoint? First we get the following result which is the additive variant of Theorem 15.3, Ch. 1.

THEOREM 7.10. *Let $\mathscr{A}$ be a thick subcategory of $\mathscr{C}$. The following are equivalent:*

a) *The canonical functor $P: \mathscr{C} \to \mathscr{C}/\mathscr{A}$ has a coadjoint.*

b) *For each object X of $\mathscr{C}$, the set of all subobjects of X in $\mathscr{A}$ has a greatest element. If in addition, X is an A-torsionfree object, then there is a monomorphism X into an A-closed subobject.* ⁎

A thick subcategory $\mathscr{A}$ or $\mathscr{C}$ is called *localizing* if the canonical functor $P: \mathscr{C} \to \mathscr{C}/\mathscr{A}$ has a coadjoint. By Theorem 15.3, Ch. 1, as well as by Theorem 7.10 it follows that a coadjoint of P is full and faithful.

The following result gives a way for constructing localizing subcategories.

THEOREM 7.11. *Let $T: \mathscr{C} \to \mathscr{C}'$ be an exact functor between abelian categories, and S a full and faithful coadjoint of T. Then $\mathrm{Ker}T$ is a localizing subcategory, and S defines an equivalence between $\mathscr{C}'$ and $\mathscr{C}/\mathrm{Ker}T$.*

Proof. It is easy to see that $\mathrm{Ker}\, T$ is a thick subcategory, and that the system Σ of all morphisms of $\mathscr{C}$ which are made invertible by T is precisely the system of all morphisms whose kernel and cokernel belong to $\mathrm{Ker}T$. Now, by Theorems 13.10 and 14.1, Ch. 1, and Theorem 7.5 of this chapter we see that the category $\mathscr{C}^a(\Sigma^{-1})$ is canonically equivalent to $\mathscr{C}'$. The details are left for the reader. ✳

Now we shall give some examples of localizing subcategories.

Example 7.12. Assume that $\mathscr{C}$ is a category with injective envelopes and that $\mathscr{A}$ is a thick subcategory of $\mathscr{C}$. Then $\mathscr{A}$ is localizing if and only if, for each object X, the set of all subobjects of X in $\mathscr{A}$ has a maximal element. This follows from Exercise 7.4. a and Theorem 7.10.

Example 7.13. Let $\mathscr{C}$ be an *Ab* 3-category with injective envelopes. Then a thick subcategory $\mathscr{A}$ of $\mathscr{C}$ is localizing if and only if $\mathscr{A}$ is closed under coproducts in $\mathscr{C}$. Indeed, if $\mathscr{A}$ is closed under coproducts and $\{X_i\}_i$ is the set of all subobjects (in $\mathscr{A}$) of an object X then the image of the sum morphism $\coprod_i X_i \to X$ belongs to $\mathscr{A}$ and it is the greatest subobject of X in $\mathscr{A}$. Hence, by Example 7.12, $\mathscr{A}$ is localizing. The converse also holds since the canonical functor $P: \mathscr{C} \to \mathscr{C}/\mathscr{A}$ commutes with coproducts, provided $\mathscr{A}$ is localizing.

7.14. By Example 7.13 it follows that a thick subcategory of a Grothendieck category is localizing if and only if it is closed under coproducts (see Theorem 6.19).

7.15. The localizing subcategories in the category $\mathrm{Mod}\, \mathscr{C}^0$ ($\mathscr{C}$ being a small preadditive category) are closely related to so-called localizing systems. We remark that localizing systems are the additive variation of the Grothendieck topologies, defined in Section 2.9.

A *right localizing system* L on the (small) preadditive category $\mathscr{C}$ is given if for each object X, there is given a nonempty set $L(X)$ of additive right ideals over X such that:

L_1). If $\mathscr{A} \subseteq L(X)$ and $\mathscr{B}$ is a right ideal over X such that $\mathscr{A} \subseteq \mathscr{B}$, then $\mathscr{B} \in \mathscr{L}(X)$.

L_2). If $\mathscr{A} \in \mathscr{L}(X)$, then for each morphism $f: Y \to X$, the right ideal $f^*(\mathscr{A})$ belongs to $L(Y)$.

L_3). If $\mathscr{A}$ and $\mathscr{B}$ are right ideals over X such that $\mathscr{A} \in L(X)$, and if for every $(Y,f) \in \mathscr{A}$, $f^*(\mathscr{B}) \in L(Y)$, then $\mathscr{B} \in L(X)$.

Let L be a right localizing system on $\mathscr{C}$.Let $\mathscr{L}$ denote the full subcategory of $\mathrm{Mod}\,\mathscr{C}^0$ whose objects are all $\mathscr{C}$-modules F such that for each $X \in \mathscr{C}$ and for each $x \in F(X)$, the kernel of the canonical morphism associated with x:

$$u^x: h_X \to F,$$

belongs to $L(X)$. Then $\mathscr{L}$ is a localizing subcategory and the assignment

$$L \rightsquigarrow \mathscr{L}$$

gives a bijection between all right localizing systems of $\mathscr{C}$ and all localizing subcategories of Mod $\mathscr{C}^0$. This result is due to Gabriel [1].

Using the same idea as in Section 2.9, the reader can construct the quotient category of Mod $\mathscr{C}^0$, relative to a localizing subcategory. These ideas as well as many related problems are discussed in N. Popescu [3].

Exercises

7.1. Let $\mathscr{C}$ be a preadditive category and let Σ be a system of morphisms of $\mathscr{C}$ such that $\mathscr{C}^a(\Sigma^{-1})$ is defined. Show that the canonical functor commutes with finite products. Hence, if $\mathscr{C}$ is additive, then so is $\mathscr{C}^a(\Sigma^{-1})$.

7.2. Let $\mathscr{C}$ be an abelian category. Let E denote the class of all essential monomorphisms of $\mathscr{C}$. Prove that:

a) E is a right calculable system. (Hint: Show that the essential monomorphisms are universal.)

b) If $\mathscr{C}$ is a wellpowered Ab 5-category, then the category $\mathscr{C}^a(E^{-1})$ exists. Moreover, every monomorphism of this category is a retraction.

7.3. Let $\mathscr{C}$ be a wellpowered category, $\mathscr{A}$ a thick subcategory of $\mathscr{C}$ and $P: \mathscr{C} \to \mathscr{C}/\mathscr{A}$ the canonical functor.

a) Let f be a morphism of $\mathscr{C}$. Show that $P(f) = 0$ if and only if the image of f belongs to $\mathscr{A}$.

b) Let f be a morphism of $\mathscr{C}$. Show that $P(f)$ is a monomorphism (an epimorphism) if and only if $\operatorname{Ker} f \in \mathscr{A}$ ($\operatorname{Coker} f \in \mathscr{A}$).

c) Show that the category $\mathscr{C}/\mathscr{A}$ is preabelian and the canonical functor P is exact.

d) Show that if f is a morphism of $\mathscr{C}$, then $P(f)$ is an isomorphism if and only if $f \in \Sigma_{\mathscr{A}}$.

Use parts a) — d) to show that $\mathscr{C}/\mathscr{A}$ is an abelian category and that the canonical functor P is exact.

e) Show that any morphism of $\mathscr{C}/\mathscr{A}$ can be written (and this form need not be unique) as

$$P(s'')^{-1}\, P(f)\; P(s')^{-1},$$

with $s' \in \Sigma'$ and $s'' \in \Sigma''$.

f) Let $H: \mathscr{C} \to \mathscr{C}'$ be an exact functor such that $H(X) = 0$ for all objects X of $\mathscr{A}$. Show that there exists a unique functor $\bar{H}: \mathscr{C}/\mathscr{A} \to \mathscr{C}'$ such that $\bar{H}P = H$.

g) Let $G: \mathscr{C}/\mathscr{A} \to \mathscr{C}'$ be an additive functor, where $\mathscr{C}'$ is abelian. Show that GP is exact if and only if G is exact.

7.4. Let $\mathscr{A}$ be a thick subcategory of $\mathscr{C}$. Prove that:

a) Every $\mathscr{A}$-torsionfree injective object of $\mathscr{C}$ is $\mathscr{A}$-closed (Hint: Use Lemma 7.8, 2.)

b) If $\mathscr{A}$ is localizing and if $u: X' \to X$ is an essential monomorphism such that X is $\mathscr{A}$-torsionfree, then $P(u)$ is an essential monomorphism.

c) If $\mathscr{A}$ is localizing and if Q is an injective envelope of an $\mathscr{A}$-torsionfree object X, then $P(Q)$ is an injective envelope of $P(X)$ in $\mathscr{C}/\mathscr{A}$.

7.5. Let $\mathscr{C}$ be an Ab 3-category with injective envelopes and let $\mathscr{A}$ be a full subcategory of $\mathscr{C}$. Show that the following are equivalent:

1) $\mathscr{A}$ is a localizing subcategory.

2) There is a class $\{Q_i\}_i$ of injective objects of $\mathscr{C}$ such that $X \in \mathscr{A}$ if and only if $\mathscr{C}(X, Q_i) = 0$ for all i.

7.6. Let A be a commutative ring and Σ a multiplicative set of elements of A. The category of additive fractions of A relative to Σ (which exists by Theorem 7.5) is again a ring, which is denoted by A_Σ, and called the *ring of fractions* of A relative to Σ. Let $u: A \to A_\Sigma$ be the canonical ring morphism, $u_*: \operatorname{Mod} A_\Sigma \to$ $\to \operatorname{Mod} A$ the restriction functor (see Exercise 5.2), and u^* an adjoint of u_*.

Show that u_* is full and faithful and u^* is exact. (Hint: Denote by $\mathscr{A}$ the full subcategory defined as follows. An object X belongs to $\mathscr{A}$ if and only if for each $x \in X$ there is an $s \in \Sigma$ such that $sx = 0$. Then $\mathscr{A}$ is a localizing subcategory and $\operatorname{Mod} A/\mathscr{A}$ is canonically equivalent to $\operatorname{Mod} A_\Sigma$.)

4.8. Left exact functors. The embedding theorem

The aim of this section is to show that abelian categories are "locally" categories of modules. This result, due to Mitchell [1], reduces the study of "atomistic" properties of objects and morphisms in any abelian category to modules. In proving this result we make use of a result from Section 4.7. We show that the category of all left exact functors defined on a small abelian category $\mathscr{C}$ and with values in $\mathscr{Ab}$ is equivalent to a quotient category of $\operatorname{Mod}\mathscr{C}$ relative to a suitable localizing subcategory. This result is essentially due to Gabriel [1], Freyd [1] and Mitchell [2].

References: Cartan—Eilenberg [1], Freyd [1], Gabriel [1], Mitchell [2], Popescu—Sevici [1], N. Popescu [3].

Let $\mathscr{C}$ be a small abelian category and let M be the set of all monomorphisms of $\mathscr{C}$. By Corollary 2.6 we see that M is a left and right permutable multiplicative system. Consider the category $\operatorname{Mod}\mathscr{C}$ and let

$$h^0: \mathscr{C} \to \operatorname{Mod}\mathscr{C}$$

be the canonical cofunctor defined by $h^0(X) = h^X$. By Theorem 1.14, h^0 is full and faithful and commutes with colimits (i.e. carries colimits into limits).

Furthermore, let $\mathscr{L}$ denote the full subcategory of Mod $\mathscr{C}$ which is defined as follows: An object H of Mod $\mathscr{C}$ belongs to $\mathscr{L}$ if and only if, for each $X \in \mathscr{C}$ and for each $x \in H(X)$, there is a monomorphism $f: X \to Y$ such that $H(f)(x) = 0$.

LEMMA 8.1. *$\mathscr{L}$ is a localizing subcategory of* Mod $\mathscr{C}$.

Proof. Since Mod $\mathscr{C}$ is a Grothendieck category, we shall use Example 7.14. We first show that $\mathscr{L}$ is a dense subcategory. For that, let

$$0 \to H' \xrightarrow{u} H \xrightarrow{p} H'' \to 0$$

be an exact sequence of Mod $\mathscr{C}$ such that $H \in \mathscr{L}$. Then it is clear that H' also belongs to $\mathscr{L}$. Furthermore, if $X \in \mathscr{C}$ and $x'' \in H''(X)$, then, by the definition of the category Mod $\mathscr{C}$ (see Example 2.3), we see that there is an $x \in H(X)$ such that $p_X(x) = x''$. Now let $f: X \to Y$ be a monomorphism such that $H(f)(x) = 0$ (this is possible since $H \in \mathscr{L}$). Then by the canonically defined commutative diagram:

$$\begin{array}{ccc} H(X) & \xrightarrow{p_X} & H''(X) \\ {\scriptstyle H(f)}\downarrow & & \downarrow{\scriptstyle H''(f)} \\ H(Y) & \xrightarrow{p_Y} & H''(Y) \end{array}$$

we see that $H''(f)(x'') = 0$. Hence $H'' \in \mathscr{L}$. We leave it for the reader to show that H belongs to $\mathscr{L}$, provided H' and H'' belong to $\mathscr{L}$.

Furthermore, by Example 7.14, to prove that $\mathscr{L}$ is a localizing subcategory, it will suffice to show that $\mathscr{L}$ is closed under coproducts Hence, let $\{H_i\}_i$ be a set of objects of $\mathscr{L}$, X an object of $\mathscr{C}$ and $x = \{x_i\}_i$ an element of $(\coprod_i H_i)(X) = \coprod_i H_i(X)$. By Note 4.1, we have that $x_i = 0$ for all but a finite number of i's, say, $x_{i_1}, \ldots, x_{i_n}$, are not 0. By hypothesis, there are monomorphisms $f_j: X \to X_j$, $j = 1, \ldots, n$, such that $H_{i_j}(f_j)(x_{i_j}) = 0$ for all j. Since H is a right permutable system, there exist morphisms $g_j: X_j \to Y$, $j = 1, \ldots, n$ such that $g_j f_j$ is a monomorphism for every $1 \leqslant j \leqslant n$ and $g_j f_j = g_{j'} f_{j'} = t$ for all $1 \leqslant j, j' \leqslant n$. Hence an easy computation shows that $(\coprod_i H_i)(t)(x) = 0$, i.e. $\coprod_i H_i \in \mathscr{L}$. ✳

LEMMA 8.2. *Let* $0 \to X' \xrightarrow{f} X \xrightarrow{p} X'' \to 0$ *be an exact sequence of* $\mathscr{C}$. *Then we have the exact sequence in* Mod $\mathscr{C}$:

$$0 \to h^{X''} \xrightarrow{h(p)} h^X \xrightarrow{h(f)} h^{X'},$$

in which Coker $h(f) \in \mathscr{L}$.

Proof. Let $Y \in \mathscr{C}$ and $y \in (\operatorname{Coker} h(f))(Y)$; then an element $g \in h^{X'}(Y) = \operatorname{Hom}_{\mathscr{C}}(X', Y)$ can be chosen such that $r_Y(g) = y$, where $r: h^{X'} \to \operatorname{Coker} h(f)$ is the canonical epimorphism. Consider (in $\mathscr{C}$) the cocartesian square:

$$\begin{array}{ccc} X' & \xrightarrow{g} & Y \\ {\scriptstyle f}\downarrow & & \downarrow{\scriptstyle f'} \\ X & \xrightarrow{g'} & Z \end{array} \tag{19}$$

in which f' is a monomorphism (see Corollary 2.6) and the following induced commutative diagram

$$\begin{array}{ccccc} h^X(Y) & \xrightarrow{h(f)_Y} & h^{X'}(Y) & \xrightarrow{r_Y} & (\operatorname{Coker} h(f))\,(Y) \\ {\scriptstyle h^X(f')}\downarrow & & \downarrow{\scriptstyle h^{X'}(f')} & & \downarrow{\scriptstyle (\operatorname{Coker} h(f))\,(f)} \\ h^X(Z) & \xrightarrow{h(f)_Z} & h^{X'}(Z) & \xrightarrow{r_Z} & (\operatorname{Coker} h(f))\,(Z). \end{array}$$

Directly from this commutative diagram and square (19) we get that $(\operatorname{Coker} h(f))\,(f')\,(y) = 0$. ✻

LEMMA 8.3. *An object H of* $\operatorname{Mod}\mathscr{C}$ *is $\mathscr{L}$-closed if and only if it is a left exact functor.*

Proof. We shall use Lemma 7.9. Let us assume that H is $\mathscr{L}$-closed. For every exact sequence

$$0 \to X' \xrightarrow{f} X \xrightarrow{p} X'' \to 0$$

in $\mathscr{C}$ we get the sequence of abelian groups:

$$H(X') \xrightarrow{H(f)} H(X) \xrightarrow{H(p)} H(X'') \tag{20}$$

for which we must show that $H(f)$ is an equalizer of $H(p)$.

First, since H is $\mathscr{L}$-closed, and f is a monomorphism, we see that $H(f)$ is a monomorphism.

Furthermore, let $x \in H(X)$ be such that $H(p)\,(x) = 0$ and let $u: h^X \to H$ be the morphism associated with x (i.e. $u_X(1_X) = x$). From the canonically constructed commutative diagram

$$\begin{array}{ccc} h^X(X) & \xrightarrow{h^X(p)} & h^X(X'') \\ {\scriptstyle u_X}\downarrow & & \downarrow{\scriptstyle u_{X''}} \\ H(X) & \xrightarrow{H(p)} & H(X'') \end{array}$$

we get that $u_{X''}h^X(p)(1_X) = u_{X''}(p) = 0$. But this means that in the diagram

$$\begin{array}{ccccccc} 0 & \longrightarrow & h^{X''} & \xrightarrow{h(p)} & h^X & \xrightarrow{h(f)} & h^{X'} \\ & & & & \downarrow{\scriptstyle u} & & \\ & & & & H & & \end{array}$$

$uh(p) = 0$. Hence there is a unique morphism $v\colon \operatorname{Coker} h(p) \to H$ such that $vq = u$, where $q\colon h^X \to \operatorname{Coker} h(p)$ is canonical. Now if $t\colon \operatorname{Coker} h(p) \to h^{X'}$ is the unique morphism such that $tq = h(f)$, then it follows from Lemmas 8.2 and 7.9 that there is a unique morphism $s\colon h^{X'} \to H$ such that $sh(f) = u$. This means that $H(f)s_{X'}(1_{X'}) = x$, so that sequence (20) is exact, i.e. H is a left exact functor.

Conversely, assume that H is a left exact functor. It is easy to see that H has no nonzero subobjects of $\mathscr{L}$. Furthermore, let $m\colon H \to R$ be a monomorphism such that $\operatorname{Coker} m \in \mathscr{L}$. We define a left inverse of m: Let $X \in \mathscr{C}$ and $x \in R(X)$. Let $p\colon R \to \operatorname{Coker} m$ denote the canonical morphism. By hypothesis, there is a monomorphism $f\colon X \to Y$ such that $(\operatorname{Coker} m)(f)(p_X(x)) = 0$. Consider the canonically defined commutative diagram:

$$\begin{array}{ccccccccc} 0 & \longrightarrow & H(X) & \xrightarrow{m_X} & R(X) & \xrightarrow{p_X} & (\operatorname{Coker} m)(X) & \longrightarrow & 0 \\ & & \downarrow{\scriptstyle H(f)} & & \downarrow{\scriptstyle R(f)} & & \downarrow{\scriptstyle (\operatorname{Coker} m)(f)} & & \\ & & H(Y) & \xrightarrow{m_Y} & R(Y) & \xrightarrow{p_Y} & (\operatorname{Coker} m)(Y) & \longrightarrow & 0 \end{array}$$

By hypothesis $p_Y R(f)(x) = 0$. Hence $R(f)(x) = m_Y(y)$, where $y \in H(Y)$. Now let $q\colon Y \to Z$ be a coequalizer of f. Since H is left exact, and $m_Z\colon H(Z) \to R(Z)$ is a monomorphism, it follows that $H(q)(y) = 0$ and there is a unique element $x' \in H(X)$ such that $H(f)(x') = y$.

Let us define the map $s_X\colon R(X) \to H(X)$ by $s_X(x) = x'$. We leave it for the reader to show that $s_X(x)$ depends only on x (this follows from the fact that M is a right permutable system), and that the maps $\{s_X\}_{X \in \mathscr{C}}$ define a left inverse s of m. ✳

The quotient category $\operatorname{Mod}\mathscr{C}/\mathscr{L}$ is denoted by $\operatorname{Lex}\mathscr{C}$ (left exact functors). Also we let $\bar{h}\colon \mathscr{C} \to \operatorname{Lex}\mathscr{C}$ denote the composition Th, where $T\colon \operatorname{Mod}\mathscr{C} \to \operatorname{Lex}\mathscr{C}'$ is the canonical functor.

We have the following important result.

THEOREM 8.4. *If $\mathscr{C}$ is a small abelian category, then the canonical functor*

$$h\colon \mathscr{C} \to \operatorname{Lex}\mathscr{C}$$

is a full and faithful exact cofunctor.

Proof. The proof follows from Lemmas 8.2 and 8.3 since the functor $h^X\colon \mathscr{C} \to \mathscr{Ab}$ is always left exact. ✳

Lemma 8.5. *Let $\mathscr{D}$ be an abelian category and let U be a projective generator of $\mathscr{D}$. Let A denote* $\mathrm{End}_{\mathscr{C}}(U)$. *Then the Mal'cev functor*

$$S: \mathscr{D} \to \mathrm{Mod}\, A^0$$

is exact and faithful. Moreover, if X is an object of $\mathscr{D}$ such that there is a natural number n and an epimorphism $p: U^n \to X$, then for each object Y of $\mathscr{D}$, the morphism, induced by S,

$$S(X, Y): \mathscr{D}(X, Y) \to \mathrm{Hom}_A(S(X), S(Y))$$

is an isomorphism of abelian groups.

Proof. Since U is a projective generator, S is exact and faithful. Hence, we must show that every morphism $g: S(X) \to S(Y)$ of A^0-modules can be written as $S(g')$, where $g' \in \mathscr{D}(X, Y)$. In other words, we must find a g' such that $g(f) = g'f$ for all $f \in S(X)$.

If $X = U$, then $g(f) = g(1_U f) = g(1_U)f$ since g is a morphism of A^0-modules. Furthermore, let us consider the situation when we have an exact sequence

$$0 \to M \xrightarrow{j} U^n \xrightarrow{p} X \to 0.$$

Let $\{u_i\}_{1 \leqslant i \leqslant n}$ be the structural morphisms of the coproduct, and $\{p_i\}_{1 \leqslant i \leqslant n}$ the associated projections. Let g_i denote the composition

$$S(U) \xrightarrow{S(u_i)} S(U^n) \xrightarrow{S(p)} S(X) \xrightarrow{g} S(Y).$$

Since g_i is a morphism of A^0-modules, for each i we can find a unique morphism $t_i: U \to Y$ such that $g_i(f) = t_i f$ for all $f \in S(U)$. Let

$$t \text{ denote } \Sigma_i t_i p_i: U^n \to Y.$$

Then, for each $f \in S(U^n)$,

$$gS(p)(f) = g(pf) = g(p(\Sigma_i u_i p_i)f) = \Sigma_i g(pu_i p_i f) = \Sigma_i t_i p_i f = tf.$$

We claim that $tj = 0$. Indeed, if $tj \neq 0$ then, since U is a generator, there exists a morphism $s: U \to M$ such that $tjs \neq 0$. But then

$$tjs = gS(p)(js) = g(pjs) = g(0) = 0,$$

which is a contradiction. Hence $tj = 0$, and so there exists a unique morphism $g': X \to Y$ such that $t = g'p$.

Finally, let $f \in S(X)$. Since U is projective, we can write $f = ph$ where $h: U \to U^n$. Then

$$g(f) = g(ph) = gS(p)(h) = th = g'ph = g'f.$$

In other words, g' is the required morphism. ⁎

THEOREM 8.6. (Mitchell [1]). *Let $\mathscr{C}$ be a small abelian category. Then there exist a ring A and a full and faithful exact functor $H: \mathscr{C} \to \text{Mod } A^0$.*

Proof. Let $\mathscr{D}$ denote $(\text{Lex } \mathscr{C})^0$ and let $D: \text{Lex } \mathscr{C} \to \mathscr{D}$ be the duality cofunctor. Denote $D\bar{h}$ by K, where $\bar{h}: \mathscr{C} \to \text{Lex } \mathscr{C}$ is defined as in Theorem 8.4. Since $\bar{h}$ and D are full, faithful, and exact cofunctors, it follows that K is a full faithful and exact functor.

Now, since $\text{Lex } \mathscr{C}$ is a Grothendieck category (see Exercise 5.6), then by Theorem 6.10 it is a category with enough injective objects, so that by Exercise 6.4 it has an injective cogenerator. But it is clear that we can choose an injective cogenerator V such that for every $X \in \mathscr{C}$ there is a monomorphism $h(X) \to V$. Furthermore, it is clear that $U = D(V)$ is a projective generator of $\mathscr{D}$. Finally, let A denote $\text{End}_{\mathscr{D}}(U)$ and let $S: \mathscr{D} \to \text{Mod } A^0$ be the Mal'cev functor. By Lemma 8.5, we see that $H = SH: \mathscr{C} \to \text{Mod } A^0$ is a full, faithful, and exact functor. ✳

PROPOSITION 8.7. *Let $\mathscr{D}_0$ be a small subcategory of an abelian category $\mathscr{D}$. Then there is a small full abelian subcategory $\mathscr{D}'$ of $\mathscr{D}$ such that $\mathscr{D}_0$ is a subcategory of $\mathscr{D}'$.*

Proof. We define inductively a sequence $\{\mathscr{D}_n\}_{n>0}$ of subcategories of $\mathscr{D}$ as follows. $\mathscr{D}_{n+1}$ is the full subcategory of $\mathscr{D}$ consisting of the objects of $\mathscr{D}_n$ together with, for each morphism of $\mathscr{D}_n$, a single representative for the kernel and the cokernel in $\mathscr{D}$, and a single representative for all the finite products in $\mathscr{D}$ of objects of $\mathscr{D}_n$. Since $\mathscr{D}_0$ is small, $\mathscr{D}_n$ is small, for each n, and hence

$$\mathscr{D}' = \bigcup_{n>0} \mathscr{D}_n$$

is small. It is easy to see that $\mathscr{D}'$ is abelian. ✳

COROLLARY 8.8. *Let $\mathscr{D}$ be any abelian category. If a theorem is of the form "P implies Q" where P is a categorical statement about a diagram in $\mathscr{D}$ with a finite number of edges, and Q states that a finite number of additional morphisms exist between the objects over designated vertices in the diagram, so that some categorical statements are true for the extended diagram.*

If the theorem is true Mod A, for each ring A, then the theorem is true for any abelian category $\mathscr{D}$. ✳

Exercises

8.1. Let $\mathscr{C}$ be a small abelian category. Show that any injective object of $\text{Lex } \mathscr{C}$ is an exact functor, and that the converse is true if for each object X of $\mathscr{C}$, every set of its subobjects has a minimal object.

8.2. Prove: If $\mathscr{C}$ be a small abelian category such that every monomorphism of $\mathscr{C}$ is a section, then $\text{Lex } \mathscr{C} = \text{Mod } \mathscr{C}$.

8.3. Let $\mathscr{C}$ be the full subcategory of $\mathscr{Ab}$ generated by all finite groups. Show that $\mathscr{C}$ is a small abelian category and Lex$\mathscr{C}$ is equivalent to the category $\mathscr{Tors}$ (see Exercise 4.1).

8.4. Show that the category $\mathscr{C}$ of all finitely generated abelian groups is a small abelian category and Lex$\mathscr{C}$ is equivalent to $\mathscr{Ab}$.

8.5. Let A be a ring and let $\mathscr{C}$ be the full subcategory of Mod A generated by all objects which are subobjects of finitely generated objects. Show that there is a canonical full and faithful functor

$$P\colon \mathrm{Mod}A \to \mathrm{Lex}\mathscr{C}, \qquad P(X) = \mathrm{Hom}_A(X, ?).$$

This functor is an equivalence of categories if and only if each set of left ideals of A has a minimal element.

8.6. Let $\mathscr{C}$ be an additive category and let $\mathscr{U} = \{U_i\}_i$ be a set of separators of $\mathscr{C}$. Let X be an object of $\mathscr{C}$. A family $\{U_j, f_j\}$ of objects of $\mathscr{C}/X$, where $U_j \in \mathscr{U}$ for all j, is called *epic*, if for each morphism $g\colon X \to Y$ the condition $gf_j = 0$ for all j implies $g = 0$. A cofunctor $F\colon \mathscr{C} \to \mathscr{Ab}$ will be called a P-cofunctor if for each object X of $\mathscr{C}$ and every epic family $\{U_j, f_j\}_j$ over X, the family $\{F(f_j), F(U_j)\}_j$ has the following property: for each abelian group D and each morphism $h\colon D \to F(X)$, the condition $F(f_j)h = 0$ for all j, implies $h = 0$.

a) Let $F, G\colon \mathscr{C} \to \mathscr{Ab}$ be cofunctors, and let $u, v\colon F \to G$ be functorial morphisms. Assume that G is a P-cofunctor and $u_{U_i} = v_{U_i}$ for all $U_i \in \mathscr{U}$. Show that then $u = v$.

b) Let $\tilde{\mathscr{C}}$ be the category all of whose objects are P-cofunctors. Show that $\tilde{\mathscr{C}}$ is an additive category and that there is a canonical full and faithful functor

$$\tilde{h}\colon \mathscr{C} \to \tilde{\mathscr{C}},$$

and, moreover, that $\tilde{\mathscr{C}}$ is a complete category.

c) Let $F\colon \mathscr{C} \to \mathscr{Ab}$ be a cofunctor. We associate with F a P-cofunctor $\tilde{F}$ in the following way. Let $X \in \mathscr{C}$; for $x, y \in F(X)$, write $x \sim y \Leftrightarrow F(f_i)(x) = F(f_i)(y)$, for each $U_i \in U$ and for each $f_i \in \mathscr{C}(U_i, X)$. It is clear that we obtain an equivalence relation $R(X)$. Denote $F(X)/R(X)$ by $\tilde{F}(X)$. Show that the assignment $F(X) \rightsquigarrow \tilde{F}(X)$ gives the required P-cofunctor associated with F.

d) Using c), show that $\tilde{\mathscr{C}}$ is a Grothendieck category.

References

ALMKVIST, G.
[1] Fractional categories, Arkiv. Math. 7 (1968) 449–476; MR **40** # 2723.
ANDREOTTI, A.
[1] Généralités sur les catégories abéliennes, Séminaire A. Grothendieck (1957), Fac. Sciences Paris, 1958.
ANDRÉ, M.
[1] Categories of functors and adjoint functors, Batelle Dept. Genève, 1964.
BAER, R.
[1] Abelian groups that are direct summands of every containing Abelian group, Bull. Amer. Math. Soc. 46 (1940) 800–806; MR **2**, 126.
BĂNICĂ, C., N. POPESCU.
[1] Quelques considérations sur l'exactitude des foncteurs, Bull. Math. Soc. Sci. Math. Phys. R.P.R. 7 (1963) 143–147; MR **33** # 2700.
[2] Categories 'quotient', St. Cerc. Mat. 17 (1965) 951–985 (Romanian).
[3] Sur les catégories préabéliennes, Rev. Roum. Math. Pures et Appl. 10 (1965) 621–633; MR **33**, # 2698.
BARON, S.
[1] Note on epi in T_0, Canad. Math. Bull. 11 (1968) 503–504; MR **38** # 3315.
[2] Reflectors as compositions of epi-reflectors, Trans. Amer. Math. Soc. 136 (1969) 499–508; MR **38** # 4535.
BÉNABOU, J.
[1] Critères de représentabilité des foncteurs, C.R. Acad. Sci. Paris, 260 (1965) 752–755; MR **31** # 222.
BIRKOFF, G.
[1] On the structure of abstract algebras, Proc. Cam. Phil. Soc. 31 (1935) 433–454.
[2] *Lattice theory*, AMS Colloquium Publications, vol. 25, 3rd ed., Providence, 1967; MR **37** # 2638.
BIRKOFF, G., S. MACLANE.
[1] *Algebra*, MacMillan Company, New York, 1967; MR **35** # 5266.
BOGNAR, M.
[1] On ordered categories, Ann. Univ. Sci. Budapest, Eötvös, Sec. Math. 11 (1968) 59–70.
BOURBAKI, N.
[1] *Théorie des ensembles*, 2ᵉ éd. Hermann, Paris, 1963; MR **27** # 4758.
[2] *Algèbre* Ch. 2, 3ᵉ éd. Hermann, Paris, 1962; MR **27** # 5765.
[3] *Topologie générale*, Ch. 1–2, 3ᵉ éd. Hermann, Paris, 1961; MR **25** # 4480.
[4] *Topologie générale*, Ch. 3–4, 3ᵉ éd. Hermann, Paris, 1960; MR **25** # 4021.
[5] *Algèbre commutative*, Ch. 1–2, Hermann, Paris, 1961.
[6] *Algèbre commutative*, Ch. 3–4, Hermann, Paris, 1961; MR **30** # 2027.
BUCHSBAUM, D. A.
[1] Exact categories and duality, Trans. Amer. Math. Soc. 80 (1955) 1 – 34; MR **17** 579.
BUCUR, I., A. DELEANU
[1] *Introduction to the theory of categories and functors*, Wiley, London, 1968; MR **38** # 4534
BUNGE, M.
[1] Categories of set value functors, Dissertation, Univ. Pennsylvania, 1966.
[2] Relative functor categories and categories of algebras, J. Algebra 11 (1969), 64–101; MR **38** # 4536.

CALENKO, M. S.
[1] On the foundations of the theory of categories, Uspehi. Mat. Nauk., XV, 6 (96), (1960) 53–58 (Russian); Russian Math. Surveys (English) **15** (1960), 47–51; MR **26** # 2480.
[2] Completion of a category with products and coproducts, Math. Sb. 60 (1963) 235–256 (Russian); MR **27** # 1484.
[3] Representation of concrete categories in the category of sets, Math. Zametki 6 (1969) 125–127 (Russian); MR **40** # 1444.
CALENKO, M. S., E. G. ŠULGEIFER
[1] *Lectures on the theory of categories*, Univ. of Moscow, 1970 (Russian).
CARTAN, H., S. EILENBERG
[1] *Homological algebra*, Princeton University Press, Princeton, 1956; MR **17** 1040.
CHEVALLEY, C.
[1] *Fundamental Concepts of Algebra*, Academic Press, New York, 1956; MR **18** 553.
COHN, P. M.
[1] *Universal Algebra*, Harper and Row, New York, London, Tokyo, 1965; MR **31** # 224.
[2] *Free Rings and their Relations*, LMS Monographs No. 2, Academic Press, London and New York, 1971.
DWINGER, PH.
[1] On a class of reflective subcategories, Indagat. Math. 30 (1968) 30–45; MR **37** # 1431.
ECKMANN B., P. J. HILTON
[1] Group-Like Structures in General Categories I, Multiplications and Comultiplications, Math. Ann. 145 (1962) 227–255; MR **25** # 108.
[2] Group-Like Structures in General Categories II, Equalizers, Limits, Lengths, Math. Ann. 151 (1963) 150–186; MR **27** # 3681.
[3] Group-Like Structures in General Categories III, Primitive categories, Math. Ann. 150 (1963) 165–187; MR # **27** 3682.
[4] Unions and intersections in homotopy theory, Comment. Math. Helv. 38 (1964) 293–307; MR **29** # 5247.
ECKMANN, B., A. SCHOPF
[1] Über injektive Moduln, Archiv. Math. 4 (1963) 75–78.
EHRESMANN, CH.
[1] *Catégories et structures*, Paris, Dunod, 1965; MR **35** # 4274.
EILENBERG, S., S. MACLANE.
[1] Natural isomorphisms in group theory, Proc. Nat. Acad. Sci. U.S.A. 28 (1942) 537–543; MR **4** 134.
[2] General theory of natural equivalences, Trans. Am. Math. Soc. 58 (1945) 231–294; MR **7** 109.
EILENBERG, S., J. C. MOORE
1[] Adjoint functors and triples, Ill. J. Math. 9 (1965) 381–398; MR **32** # 2455.
EILENBERG, S., N. E. STEENROD
[1] *Foundations of algebraic topology*, Princeton University Press, Princeton, 1952; MR **14** 398.
FREYD, P.
[1] *Abelian categories: An introduction to the theory of functors*, Harper and Row, New York, 1964; MR **29** # 3517.
[2] Algebra-valued functors in general and tensor products in particular, Colloq. Math. 14 (1966) 89–106; MR **33** # 4116.
GABRIEL, P.
[1] Des catégories abéliennes, Bull. Soc. Math. France 90 (1962), 323–448.
GABRIEL, P., U. OBERST
[1] Spektralkategorien und reguläre Ringe in von Neumannischem Sinn, Math. Z. 92 (1962) 389–395.
GABRIEL, P., N. POPESCU
[1] Caractérisation des catégories abéliennes avec générateurs et limites inductives exactes, C.R. Acad. Sci. Paris, 258 (1964) 4188–4191; MR **29** # 3518.
GABRIEL, P., A. RENTSCHLER
[1] Sur la dimension des anneaux et ensembles ordonnés, C.R. Acad. Sci. Paris 265 (1967) 712–715; MR **37** # 243.
GABRIEL P., F. ULMER
[1] Lokal präsentierbare Kategorien, Lecture Notes in Math., 221, Springer, Berlin, 1971; MR **48** # 6205.

GABRIEL, P., M. ZISMAN
[1] *Category of fractions and homotopy theory*, Ergebnisse der Math., Springer, Berlin, 1967; MR **35** # 1019.
GIRAUD, J.
[1] Analysis situs, Séminaire Bourbaki, 15 (1962/63), Exposé 256, 256–266.
GODEMENT, R.
[1] *Topologie algébrique et théorie des faisceaux*, Hermann. Paris, 1958; MR **21** # 1583.
GRAY, J. W.
[1] Sheaves with values in a category, Topology 3 (1965) 1–18.
GRÄTZER, G. A.
[1] *Universal algebra*, D. Van Nostrand and Co. Inc., Princeton, 1968; MR **40** # 1320.
GROTHENDIECK, A.
[1] Sur quelques points d'algèbre homologique. Tôhoku Math. J. 9 (1957) 119–221; MR **21** # 1328.
[2] Technique de descente et théorèmes d'existence en géométrie algébrique I, Séminaire Bourbaki 12 (1959/60), Exp. 190; MR **23** # A 2273.
[3] Techniques de construction et théorèmes d'existence en géométrie algébrique, III. Séminaire Bourbaki 13 (1960/61), Exp. 212; MR **27** # 1339; Benjamin, New York, 1966; MR **33** # 5420 h,
GROTHENDIECK, A., J. DIEUDONNÉ
[1] Éléments de Géométrie Algébrique, III, Publ. Math. I.H.E.S. 1961, No. 11.
GROTHENDIECK, A., J. L. VERDIER
[1] Théorie des Topos et Cohomologie Étale des Schémas. Exposé I, Préfaisceaux et Topos Exposé IV. Lecture Notes in Math., Vol. 269, Springer, Berlin, 1972.
GRUSON, L.
[1] Complétion abélienne, Bull. Soc. Math. 90 (1966) 17–40; MR **34** # 7609.
HASSE, M., L. MICHLER
[1] *Theorie der Kategorien*, VEB Deutscher Verlag der Wissenschaften, Berlin, 1966; MR **35** # 4275.
HERRLICH, H.
[1] Topologische Reflexion und Coreflexionen. Lecture Notes in Math., Vol. 78, Springer, Berlin 1968, MR **41** # 988.
[2] Factorizations of morphisms f: B → FA, Math. Z. 114 (1970) 180–186; MR **42** # 1875.
HILTON, P. J.
[1] Remark on the free product of groups, Trans. Amer. Math. Soc. 96 (1960) 478–488; MR **22** 12136.
[2] Note on free and direct products in general categories, Bull. Soc. Math. Belg. 13 (1961) 38–49; MR **24** # A 3189.
[3] Fundamental group as a functor, Bull. Soc. Math. Belg. 14 (1962) 153–177; MR **25** # 5096.
[4] *Catégories non-abéliennes*, Les presses de l'Université de Montréal, Montréal 1964.
HILTON, P. J., V. STAMMBACH
[1] *A Course in Homological Algebra*, Springer, N. Y.-Heidelberg-Berlin, 1971.
HUBER, P. J.
[1] Homotopy theory in general categories, Math. Ann. 144 (1961) 361–385; MR **27** # 187.
HUQ, S. A.
[1] Semivarieties and subfunctors of the identity functor, Pacific J. Math. 29 (1969) 303–309; MR **40** # 7328.
ISBELL, J. R.
[1] Some remarks concerning categories and subspaces, Canad. J. Math. Vol. 9 (1957) 563–577; MR **20** # 923.
[2] Natural sums and direct decompositions, Duke Math. J. 27 (1960) 507–512; MR **22** # 9525.
[3] Adequate subcategories, Illinois J. Math. 4 (1960) 541–552.
[4] Two set-theoric theorems in categories, Fund. Math. 53 (1963) 43–49; MR **28** # 127.
[5] Subobjects, adequacy completeness and categories of algebras, Rozprawy Mat. 36 (1964); MR **29** # 1238.
[6] Natural sums and abelianizing, Pacific J. Math. 14 (1964) 1265–1281; MR **31** # 3478.
[7] *Uniform Spaces*, Amer. Math. Soc., Providence 1964.
[8] Structure of categories, Bull. Amer. Math. Soc. 72 (1966) 619–655; MR **34** # 5896.

[9] Small subcategories and completeness, Math. Systems Theory 2 (1968) 27–50; MR **37** # 269.
[10] Remarks on decompositions of categories, Proc. Amer. Math. Soc. 19 (1968), 899–904.
KAN, D. M.
[1] Adjoint functors, Trans. Amer. Math. Soc. 87 (1958) 294–329; MR **24** # A 1301.
KELLEY, J. L.
[1] *General Topology*, D. Van Nostrand, Princeton, 1955; MR **16** 1136.
KELLY, G. M.
[1] Tensor products in categories, J. Algebra 2 (1965) 15–37; MR **31** # 1288.
[2] Monomorphisms, epimorphisms and pull-backs, J. Austral. Math. Soc. 9 (1969) 124–142; MR **39** # 1515.
KENNISON, J. F.
[1] Reflective functors in general topology and elsewhere, Trans Amer. Math. Soc. 118 (1965) 303–315; MR **30** # 4812.
[2] A note on reflexion maps, Illinois J. Math. 11 (1967) 404–409; MR **35** # 1649.
[3] Full reflective subcategories and generalized covering spaces, Illinois J. Math. 12 (1968) 353–365; MR **37** # 2832.
[4] On limit-preserving functors, Illinois J. Math. 12 (1968) 616–619; MR **38** # 208.
[5] Normal completions of small categories, Canad. J. Math. 21 (1969) 196–201; MR **39** # 1516.
KLEISLI, H.
[1] Every standard construction is induced by a pair of adjoint functors, Proc. Am. Math. Soc. 16 (1965) 544–546; MR **31** # 1289.
KUROŠ, A. G.
[1] *Theory of Groups*, Moscow 1955 (Russian); English edition, Chelsea Publishing Co., New York, 1960; MR **22** # 727.
[2] Direct decompositions in algebraic categories. Trud. Mosk. Math. Obš. 8 (1959) 391–412 (Russian); AMS Translation; MR **27** # 1480.
KUROŠ, A. G., A. R., LIVŠIČ, E. G. ŠULGEIFER.
[1] Fundaments of the theory of categories, Uspehi Mat. Nauk. XV, 6 (96), (1960), 3–52 (Russian); Russian Math. Surveys (English) 15 (1960) 1–46; MR **23** # A 1688.
LAMBEK, J.
[1] Completions of categories, Lecture Notes No. 24, Springer, 1966; MR **35** # 228.
[2] A fixpoint theorem for complete categories, Math. Z. 103 (1968) 151–161; MR **37** # 270.
LAWVERE, F. W.
[1] Functorial semantics of algebraic theories, Proc. Nat. Acad. Sci. U.S.A. 50 (1965) 869–873; MR **28** # 2143.
[2] An elementary theory of the category of sets, Proc. Nat. Acad. Sci. U.S.A. 52 (1964) 1506 | 1511; MR **30** # 3025.
[3] The theory of categories as a foundation for mathematics, Proc. Conf. Categorical Algebra (La Jolla, 1965) 1–21, Springer, Berlin, 1966.
[4] Some algebraic problems in the context of functorial semantics of algebraic theories, Reports of the Midwest Category Seminar II, Lecture Notes in Math. No. 61, Springer, Berlin, 1968.
LIAPIN, E. S.
[1] *Semigroups*, Moscow, 1960 (Russian); MR **22** # 11054; AMS (English).
LINDERHOLM, C. E.
[1] A group epimorphism is surjective, Amer. Math. Monthly 77 (1960) 176–177.
LINTON, F. E. J.
[1] Autonomous categories and duality of functors, J. Algebra 2 (1965) 315–349; **MR 31** # 4821.
[2] Some aspects of equational categories, Proc. Conf. Categ. Alg. (La Jolla **1965**) 84–94, Springer, Berlin, 1966.
[3] An outline of functorial semantics, 7–25. Applied functorial semantics, 53–74. Coequalizers in categories of algebras, 75–90. Seminar on triples and categorical homology theory, Lecture Notes in Math. No. 80, Springer, 1969; MR, **39** # 5655.
MACDONALD, J. L.
[1] Relative functor representability, Pacific J. Math. 23 (1967) 311–320; MR **36** # 5189
MACLANE, S.
[1] *Homology*, Berlin – Göttingen – Heidelberg, Springer, 1963; MR **28** # 122.
[2] Categorical algebra, Bull. Am. Math. Soc. 71 (1965) 40–106; MR **30** # 2053.

[3] *Categories for the Working Mathematician.* Graduate Texts in Math., Springer, 1971; MR **50** # 7275.
MAL'CEV, A. I.
[1] On the immersion of an associative system into a group, Math. Sb. 6 (1939) 331–336 (Russian); MR **2** 7.
[2] Algebraic systems, Trud. 3-ty Union Math. Symposium 2 (1956) Moscow (Russian).
[3] *Algebraic Systems*, 1970, Moscow (Russian); MR **44**#142 and MR **52**#13380, MR **50**# 1878.
MARANDA, J. M.
[1] Some remarks on limits in categories, Canad. Math. Bull. 5 (1962) 133–146; MR **29**#135.
[2] Formal categories, Canad. J. Math., 17 (1965) 758–801.
MITCHELL, B.
[1] The full embedding theorem, Amer. J. Math. 86 (1964) 619–637; MR **29**#4783.
[2] *Theory of Categories*, Academic Press, New York and London, 1965; MR **34**#2647.
MORITA, K.
[1] Duality for modules and its applications to the theory of rings with minimum condition, Sci. Rep. Tokyo, Kyaiku Daigaku 6 (1958–59) 83–142.
ORE, O.
[1] Linear equations in non-commutative fields, Ann. Math. 32 (1931) 463–477.
PAREIGIS, B.
[1] *Categories and Functors*, Academic Press, London and New York, 1970; *Kategorien und Funktoren*, Teubner, Stuttgart, 1969; MR **42**#337a, b.
POPESCU, A., SEVICI, C.
[1] The embedding of some additive categories (to appear).
POPESCU, D.
[1] Quelques applications de la décomposition triangulaire, Publ. Dept. Math. Lyon 4 (3), (1967) 63–68; MR **39**#5659.
[2] Some remarks on complete cogenerated categories, Rev. Roum. Math. Pures Appl. 15 (1970) 1027–1033.
[3] Les faisceaux d'une théorie, C. R. Acad. Sci. Paris, 259 (1969) 380–382.
[4] Sur les (t, T)-faisceaux, C. R. Acad. Sci. Paris, 269 (1969) 413–415.
[5] Catégories de faisceaux, J. Algebra, 18 (1971) 343–365; MR **44**#6789.
POPESCU, N.
[1] Elements of the theory of sheaves. I, II, III, IV, V, VI. St. Cerc. Mat. Acad. R.S.R. (1966) 267–296; 407–456; 547–583; 647–669; 945–991; 19 (1967) 205–240. (Romanian); I:MR **35**# 4284; II:MR**35**#4285; III: MR**35**#4285; IV: MR**35**#4285; V: MR**37**#1436; VI: MR**40**#1449.
[2] La localisation pour des sites, Rev. Roum. Math. Pures Appl. 10 (1965) 1031–1044.
[3] *Abelian Categories with Applications to Rings and Modules*, LMS Monographs No. 3 Academic
Press, London and New York 1973.
[4] G. C. categories (to appear).
POPESCU, N., A. RADU
[1] *Theory of Categories and Theory of Sheaves*, Bucharest, 1971 (Romanian).
PULTR, A.
[1] Limits of functors and realisations of categories, Comment. Math. Univ. Carolinae, 8 (1967) 663–682; MR **37**#5263.
[2] On full embedding of concrete categories with respect to forgetful functors, Comment. Math Univ. Carolinae, 9 (1968) 281–305; MR **39**#1520.
PUPIER, R.
[1] Sur les catégories complètes, Publ. Dept. Math. Lyon, **2**, *2* (1965) 1–65; MR **33** # 170.
ROSS, J. E.
[1] Introduction à l'étude de la distributivité des foncteurs $\varprojlim$ par rapport aux $\varinjlim$ dans les catégories des faisceaux (topos), C.R. Acad. Sci. Paris 259 (1964) 969–972.
[2] Sur la distributivité des foncteurs $\varprojlim$ par rapport aux $\varinjlim$ dans les catégories des faisceaux (topos), C. R. Adad. Sc. Paris 259 (1964) 1605–1608.

[3] Complément à l'étude de la distributivité des foncteurs $\varprojlim$ par rapport aux $\varinjlim$ dans les catégories des faisceaux (topos), C.R. Acad. Sci. Paris 259 (1964) 1801–1804.
[4] Sur la condition Ab 6 et ses variantes dans les catégories abéliennes, C. R. Acad. Sci. Paris 264 (1967) 991–994.
Roux, A.
[1] Foncteurs d'équivalence dans une catégorie, Publ. Dept. Math.Lyon, 1 (1964), Exp. 16; MR **31**#4828.
[2] Un théorème de plongement des catégories, C.R. Acad. Sci. Paris 258 (1964) 4646–4647.
Samuel, P.
[1] On universal mappings and free topological groups, Bull. Amer. Math. Soc. 54 (1948) 591–598; MR **9**#605.
Schubert, R.
[1] *Categories*, Springer, Berlin and New York, 1972; MR **43** # 311; MR **50** # 2286.
[2] *Topology*, MacDonald Technical and Scientific, London, 1968; MR **40**#1957; MR **49**#11450.
Słomiński, J.
[1] The theory of abstract algebras with infinitary operations, Rozprawy Mat. No. 18, Warszawa, 1959; MR **31**#7173.
Sonner, J.
[1] On formal definition of a category, Math. Z. 80 (1962) 163–176; MR **26**#2483.
[2] Universal and special problem, Math. Z. 82 (1963) 200–211; MR **28**#123.
[3] Canonical categories, Proc. Conf. Categorical Algebra (La Jolla Calif. 1965), 272–294, Springer, New York, 1966; MR **36**#3846.
[4] Lifting inductive and projective limits, Canad. J. of Math. 19 (1967) 1329–1339; MR **37**# 1433.
Stenström, B.
[1] Flatness and localization over monoids, Math. Nach. 48 (1971) 315–334; MR **45**#5252.
Šulgeifer, R. G.
[1] Full embedding of categories. Math. Sb. 61 (103), (1963) 467–503. (Russian).
[2] The lattice of ideals of an object in a category. Math. Sb. 54 (1961) (209–224).(Russian); MR **27**#2452.
Takeuchi, M.
[1] A simple proof of Gabriel and Popescu's theorem, J. Algebra 18 (1971) 112–113; MR **43**# 2048.
Trnková, V.
[1] Completions of small subcategories, Comment. Math. Univ. Carolinae 8 (1967) 581–633; MR **38**#213.
Ulmer, F.
[1] Properties of dense and relative adjoints, J. Algebra 8 (1967) 77–95; MR **36**#5190.
[2] Representable functors with values in arbitrary categories, J. Algebra 8 (1967) 96–129; MR **36**#5191.
[3] Triples in algebraic categories, E.T.H. Zürich, 1969.
[4] A flatness criterion in Grothendieck categories, Inventiones Math. 19 (1973) 331–336; MR **49**#381.
Vincent, Ph.
[1] Construction explicite d'une complétion inductive libre pour une catégorie quelconque, C.R. Acad. Sci. Paris, Ser. A, 265 (1967) 816–819; MR **36**#3847.
Watts, C. E.
[1] Intrinsic characterization of some additive functors, Proc. Amer. Math. Soc. 11 (1960) 5–8; MR **22**#9528.
Wraith, G. C.
[1] Algebraic theories, Lecture Note Series, 22, Aarhus University, 1970; MR **41**#6943.
Yoneda, N.
[1] On the homology theory of modules, J. Fac. Sci. Tokyo, Sec. 1, 7 (1954) 193–227; MR **16** 947.

Index

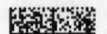